Biosensors and Bioelectronics

BIOSENSORS AND BIOELECTRONICS

D. Dharaneeswara Reddy
Assistant Professor
Department of Biotechnology
Sree Vidyanikethan Engineering College
Tirupati, Andhra Pradesh, INDIA

O.M. Hussain
Professor
Department of Physics
Sri Venkateswara University
Tirupati, Andhra Pradesh, INDIA

D.V.R. Sai Gopal
Professor & Head
Department of Virology
Sri Venkateswara University
Tirupati, Andhra Pradesh, INDIA

D. Muralidhara Rao
Assistant Professor
Department of Biotechnology
Sri Krishnadevaraya University,
Anantapur, Andhra Pradesh, INDIA

K.S. Sastry
Professor & Head
Department of Biotechnology,
Sree Vidyanikethan Engineering College
Tirupati, Andhra Pradesh, INDIA

I.K. International Publishing House Pvt. Ltd.

NEW DELHI • BANGALORE

Published by

I.K. International Publishing House Pvt. Ltd.
S-25, Green Park Extension
Uphaar Cinema Market
New Delhi–110 016 (India)
E-mail: info@ikinternational.com
Website: www.ikbooks.com

ISBN 978-93-82332-19-0

10 9 8 7 6 5 4 3 2 1

Published by Krishan Makhijani for I.K. International Publishing House Pvt Ltd., S-25, Green Park Extension, Uphaar Cinema Market, New Delhi–110 016 and Printed by Rekha Printers Pvt. Ltd., Okhla Industrial Area, Phase II, New Delhi–110 020

Foreword

Our perception of the world around us is a creation of our sensory input. Although human beings possess extremely sophisticated and highly evolved sense organs, our ability to sense, in some instances, is far lower than that of some animals and birds. However, the creative human mind has extended our ability to sense and perceive to a significant extent. It is now possible for us to look at the far reaches of universe and at the same time go deep into single biological cell, cellular architecture, macromolecular assemblies and molecular structures themselves.

Sensors and sensory-based guiding and corrective mechanisms are being used extensively in our day-to-day life. Arrival of the solid state transistor in the late 40s and commercial product in the 60s has revolutionized the electronic industry. Solid state and digital electronics have made dramatic difference to the quality of human life. Now biotechnology is making a major difference to our quality of life including health and nutrition. Combination of biotechnology and electronics is giving rise to molecular electronics, biosensors, biological devices and molecular computations.

This book on "Biosensors and Bioelectronics" provides an exhaustive compilation of almost all known sensors and their applications. Sensors used in healthcare, chemical and pharma industries, water pollution, environmental monitoring are all covered extensively. In addition to describing the sensors and their applications, the book explains basic principles of sensing, discrimination and detection. The materials that go into fabricating sensors including nanomaterials and biomaterials are described extensively. Electronics for signal conditioning and measurement are also described. Thus, the book provides end-to-end solution for biosensors. The reader would be able to understand the principles of sensing, designing, selection of appropriate materials, fabrication, calibration and actual application in different areas. This book should be a valuable resource for students and research scientists interested in biosensors and bioelectronics and their applications in various fields including healthcare, environment, pollution monitoring, food and agriculture.

Dr. Ch. Mohan Rao, FNA, FTWAS
Director
Centre for Cellular & Molecular Biology
Hyderabad
Andhra Pradesh, India

Preface

The past twenty years of Biosensors research has a significant impact on Science and Technology. The integration of electronics and development of packaging technologies makes it possible to manufacture sensors and electronics on a silicon chip no bigger than a pinhead. The sensors contain various micro-sensing elements, measurement electronics and wireless communication circuits. The emerging field of bioelectronics seeks to take advantage of biology in concurrence with electronics in a wider context, for example, biological fuel cells, bionics and biomaterials for information processing, information storage, electronic components and actuators. A key aspect is the interface between biological materials and micro- and nano-electronics which requires a highly interdisciplinary research, involving biologists, chemists, physicists, materials scientists and engineers. This book describes the basic concepts of biosensors, principles, design, transduction technology, important insights into the practical applications of biosensors and commercialization of biosensors and bioelectronics. This book consists of 18 chapters covering different aspects of biosensors and bioelectronics and their applications, as well as fundamental topics of related disciplines. The contents of this book are selected based on UG and PG level syllabi of different Universities.

Chapter 1 describes the overview of sensors and biosensors, their basic principles of operation, different types of sensors and their applications. History of the biosensors along with biosensor industry, its market challenges and innovations are presented. Chapter 2 presents different components of biosensors, i.e., biological materials, membranes, transducers, amplifiers, recorders, their potential applications and performance factors of biosensors. Chapter 3 introduces bio-recognition elements in biosensing process, types of bio-recognition and their advantages and disadvantages. The immobilization techniques along with applications are also discussed in this chapter. Introduction and overview of biosensor technologies such as DNA, Chip based, Wireless and Ambari biosensor technologies and different fabrication methods, principles and applications are discussed in Chapter 4.

The mechanism of biological recognition of biosensors and its potential uses for the developments of various biosensors, viz., Potentiometric, Amperometric, Calorimetric, Piezoelectric, Bioaffinity based, Urea, Ethanol, Glucose, DNA, RNA, Protein, Ammonia, Histamine, Lactose and Methane biosensors are presented in Chapters 5 and 6. The principle of electrochemical transducers and their specifications, especially pH, ion-sensitive and oxygen electrodes, their models, working principles, mechanisms and applications are discussed in Chapter 7. Chapter 8 introduces the fundamentals of transducers, extended with more explanation of their selection factors, their modeling and testing tools. Additionally, the classification of transducers has been significantly enlarged with respect to transduction, materials and technology. Also, this chapter is extended with more explanation on Piezoelectric transducer, Electromagnetic Acoustic Transducers (EMAT), Molecular Electronic Seismometers (MES), Conductometric Transducers, Semiconductors and their types and applications.

Chapter 9 provides an overview of optical biosensors, fibre optic biosensors in combination with different types of spectroscopic techniques such as Absorption, Fluorescence and Phosphorescence. Fundamentals of Surface Plasmon Resonance (SPR) and Automated biosensors, preparation of their probes, design and its operation process are discussed in this chapter. Chapter 10 is further extension of Chapter 9, which deals with optical wave guide biosensors structure and its various architectures, optical biosensors for cell signaling – basic fundamentals, theory and applications. In addition, the basic principle and applications of SU-8 biosensor, Chemiluminescence biosensor and Fluorescence based biosensor have also been included in this chapter.

Microorganisms are suggested to detect a wide range of chemical substances, making them ideal as biological sensing materials and to construct various types of living biosensors. Microorganisms are incorporated with diversity of transducers, viz., Amperometric, Potentiometric, Conductometric, Colorimetric, Luminescence and Fluorescence biosensors. Various types of living biosensors, working principles and applications are included in Chapter 11. The working principles and applications of cell based biosensors and yeast based biosensors are also briefly described in this chapter.

Biochips are to quickly screen large numbers of biological analytes for a variety of purposes, for example, diagnosis of diseases, detection of the bioterrorism agents, etc. Chapter 12 introduces the fundamentals of biochips giving the basic schemes of biochip functions, fabrication process assay formats and principles of biochip technology. In this chapter, biochips development, working mechanisms and their applications in different fields, viz., Potentiometric biochips, Amperometric biochips, Optical fibre biochips, DNA biochips, Brain tumor biochip, Multifunction Microfluidic chip device, RNA biochip, e-NOSE and S4 MS biochips are discussed.

Sensing technology is an efficient screening method for determining analytes or biomarkers. Chapter 13 presents the applications of various types of biosensors in clinical industry, viz., detection of nucleic acids, analysis of blood samples, detection of glucose, lactose and urea. Biosensors are used in healthcare and make them much more effective by reducing the response time and decreasing the granularity of the applications. By using this technology, continuous health monitoring of a human body and real-time data collection would be possible. Various types of biosensors which are used in medicine and healthcare are also discussed in Chapter 13. Chapter 14 introduces various types of biosensors for monitoring the quality of food and agriculture products, for example, Amperometric biosensors for assay of pesticides, Mammalian cell biosensors for detection of pathogens and Carbon biosensors for food industry.

Chapter 15 deals with different types of biosensors that are developed for environmental control and monitoring, to provide fast and specific data of contaminated sites, for example, enzyme based biosensors show the potential for continuous monitoring of compounds such as phenolics in process streams, effluents and groundwater. Commercial biosensors are used to study the cross-reactivity between antibiotics of various groups.

Molecular electronics is an interdisciplinary theme that spans physics, chemistry and materials science. The molecular electronic device consists of organic molecules which are sandwiched between conducting electrodes. The feature of this area is the use of molecular building blocks for the fabrication of electronic components both passive (resistive wires) and active (transistors). Molecular materials for electronics utilize the properties of the molecules to affect the bulk properties of a material and molecular scale electronics focuses on single molecule applications. These are discussed in Chapter 16 along with examples.

Chapter 17 provides the important subject of Photonic computer. Basic fundamentals, principles and applications of Optical or Photonic computer are introduced in this chapter. Finally, the allotropic forms of carbon, viz., Quantum well, Quantum wire and Quantum dot and Graphene, and their applications are briefly described in Chapter 18.

This book is profusely illustrated with simple line drawings, readily accessible by students. Most of the chapters are appended with suitable examples in order to convey the various biosensing techniques. It provides different biosensing systems, various transducers and biological elements that are used in sensors. The contents of this book will be highly useful for all graduate and postgraduate students of biotechnology, engineering, physical, chemical, biological, agriculture and pharmacy sciences.

In the preparation of this book, the authors have utilized relevant materials from various resources. While it is impossible to acknowledge every source individually, mention must, however, we acknowledge all the authors/publishers, who generously afford the permissions. Finally, we want to express our deep appreciation to our family members, who patiently endured many evenings and weekends that were devoted in writing of this book.

Before we take leave, we request the readers, to let us know your annotations, suggestions and productive appreciation to make further improvements to this exciting series.

Authors

Contents

Abbreviations

11-FUT	11-Ferrocenyl-1-Undecane Thiol
1D	One-dimensional
6-APA	6-Amino-Penicillanic Acid
AChE	Acetyl Cholinesterase
ADC	Analog to Digital Converter
ADH	Alcohol Dehydrogenase
ADP	Adenosine 5' Diphosphate
AFM	Atomic Force Microscopy
Al_2O_3	Alumina
ALDH	Aldehyde Dehydrogenase
AMP	Adenosine 5' Monophosphate
ANN	Artificial Neural Networks
AOB	Ammonia Oxidizing Bacteria
AOD	Alcohol Oxidase
APEs	Alkylphenol Ethoxylates
AR	Anomalous Reflection
ARROW	Anti-Resonant Reflecting Optical Waveguide
ATP	Adenosine 5' Triphosphate
ATR	Attenuated Total Reflection
AW	Artificial Wastewater
BAC	Blood Alcohol Concentration
BaP	Benzo[a]Pyrene
$BaTiO_3$	Barium Titanate
BChE	Butyrylcholinesterase
Bio-MEMS	Biological Micro-Electromechanical Systems
BOD	Biochemical Oxygen Demand
BPT	Benzo Pyrene Tetrol
BVDV	Bovine Viral Diarrhea Virus
$CaCO_3$	Calcium Carbonate
CAD	Computer-Aided Design
$CaHPO_4$	Calcium Hydrogen Phosphate

CAM	Computer-Aided Manufacturing
CF	Carbofuran
CFP	Cyan Fluorescent Protein
CFTR	Cystic Fibrosis Transmembrane Conductance Regulator Gene
CFU	Colony-Forming unit
CGM	Continuous Glucose Monitoring
CGMS	Continuous Glucose Monitoring System
CHEMFET	Chemically Sensitized Field Effect Transistor
ChEt	Cholesterol Esterase
ChO	Choline Oxidase
ChOx	Cholesterol Oxidase
CIA	Capillary based Immunoassay
CL	Chemi-Luminescence
CLECs	Cross-Linked Enzyme Crystals
CMOS	Complementary Metal Oxide Semiconductor
CMRR	Common-Mode Rejection Rate
CMS	Cell Monitoring System
CNTs	Carbon Nanotubes
CNW diodes	Crossed Nanowire Diodes
CO	Carbon Monoxide
COD	Chemical Oxygen Demand
COWG	Composite Optical Waveguide
CPG	Controlled Pore Glass
CT	Cholera Toxin
cTnI	Cardiac Troponin I
CuC1TPP	Copper Complexed Porphyrins
CV	Cyclic Voltammetric
DAC	Digital to Analog Converter
DAO	Diamine Oxidase
DDS	Dimethyl Dichloro Silane
DDTC	Diethyl Dithiocarbamate
DH	Dehydrogenase
DMMs	Digital Multi-Meters
DMR	Dynamic Mass Redistribution
DO	Dissolved Oxygen
DOS	Density of States
DPV	Differential Pulse Voltammetry

DSP	Digital Signal Processor
DTA	Direct Toxicity Assessment
DTSP	Dithio-bis Succinimidyl Propionate
DUT	Device Under Test
ECAR	Extracellular Acidification
ECM	Extra Cellular Matrix
ECU	Electronic Control Unit
EDCs	Endocrine-Disrupting Compounds
EDTA	Ethylene Diamine Tetra-acetate
EEG	Electroencephalogram
EGE	Extended Gate Electrodes
egg-PC	egg-Phosphatidyl Choline lipids
EGR	Exhaust Gas Recirculation
ELISA	Enzyme Linked Immunosorbant Assay
EMACs	Extended Metal Atom Chains
EMAT	Electromagnetic Acoustic Transducers
EMF	Electromotive Force
ENFETs	Enzyme Linked Field Effect Transistors
ER	Estrogen Receptor
eT	Electron Transfer
Fc-COOH	Ferrocene Carboxylic Acid
FDA	Food and Drug Administration
FEM	Finite Element Method
FETs	Field Effect Transistors
FFT	Fast Fourier Transform
FITC	Fluorescein Isothiocyanate
fM	Femto Molar
GaN	Gallium Nitride
GDH	Glutamate Dehydrogenase
GFP	Green Fluorescent Protein
GNPs	Gold Nanoparticles
GNRs	Graphene Nanoribbons
GOD	Glucose Oxidase
GPa	Gigapascals
GPCR	G- Protein-Coupled Receptor
GPT	Pyruvate Trans Aminase
GRAS	Generally Recognized as Safe

H_2O_2	Hydrogen Peroxide
HCl	Hydrochloric Acid
HPLC	High Performance Liquid Chromatography
HREC	Hypoglycaemia Related EEG Changes
HRP	Horse Radish Peroxidase
I/O	Input/Output
ICP	Inductively Coupled Plasma
IgG	Immunoglobulin-G
IMEs	Interdigitated Microsensor Electrodes
IP	Isoelectric Point
IR	Infrared Radiation
IS	Ionic Strength
ISE	Ion-selective Electrodes
ISFET	Ion-Sensitive Field-Effect Transistor
ISME	Ion Sensitive Microelectrode
ITO-PANI	Indium Tin Oxide Polyaniline
JFET	Junction Field Effect Transistors
KCl	Potassium Chloride
KO	Knockout
LANs	Local Area Networks
LAPS	Light Addressable Potentiometric Sensor
LASER	Light Amplification by Stimulated Emission of Radiation
LASs	Linear Alkylbenzene Sulfonates
LB	Langmuir-Blodgett
LCD	Liquid-crystal Display
LDH	Layered Double Hydroxide
LDR	Light Dependent Resistor
LED	Light-emitting Diode
LIDAR	Laser Induced Doppler Atmospheric Radar
LPCVD	Low Pressure Chemical Vapor Deposition
LPG	Long Period Grating
LPR	Localized Plasmon Resonance
LVDT	Linear Variable Differential Transformer
M	Metal
MAb	Monoclonal Antibody
MAbs	Monoclonal Antibodies
MAF	Mass Air Flow

MAP	Manifold Absolute Pressure
MAPLE DW	Matrix Assisted Pulsed Laser Evaporation Direct Write
MB	Meldola Blue
MC	Microcontroller
MCLG	Maximum Contaminant Level Goals
mCP	Micro Contact Printing
MDCK	Madin Darby Canine Kidney
MEA	Microelectrode Array
MEG	Multiple Exciton Generation
MEMS	Micro- Electro-mechanical System
MERIE	Magnetically Enhanced using a Reactive Ion Etching technique
merR	mer operon
MET	Molecular Electronic Transducers
MFCs	Microbial Fuel Cells
MIC	Microbiologically Influenced Corrosion
MIMS	Multiple isotope Imaging Mass Spectrometry
MOEMS	Micro-Opto Electromechanical System
MOS	Metal Oxide Semiconductor
MOT	Mechanical and Optoelectronic Transducer
MOVPE	Metallorganic Vapor Phase Epitaxy
MWCNT	Multi-walled Carbon Nanotubes
MWNT	Multi-walled Nanotube
M_xO_y	Metal Oxide
MZI	Mach-Zehnder Interferometers
NaClO	Sodium Hypochlorite
NAD	Nicotinamide Adenine Dinucleotide
NAD^+	Nicotinamide Adenine Dinucleotide
NDE	Non-destructive Evaluation
NDR	Negative Differential Resistance
NHS	N-hydroxysuccinimide
NIAAA	National Institute on Alcohol Abuse and Alcoholism
NICE	National Institute for Health and Clinical Excellence
NMP^+	N-Methylphenazinium cation
NO	Nitric Oxide
NO_2	Nitrogen Dioxide
NSOM	Near Field Scanning Optical Microscopy
NT	Nanotube

NWs	Nanowires
O_2	Oxygen
O_2-MPD	Oxygen Membrane Polarographic Detector
OBD	On-board Diagnostics
OEO	Optical Electrical Optical
OP	Organophosphate
OPH	Organophosphorus Hydrolase
ORMOSIL	Organically Modified Silica
PAbs	Polyclonal Antibodies
PAH	Polycyclic Aromatic Hydrocarbons
PB	Phosphate Buffer
PBT	Poly Butylene Teraphalate
PCA	Principal Component Analysis
PCP	Pentachlorophenol
PCR	Polymerase Chain Reaction
PCS	Polycarbamoyl Sulfonate
PDMS	Poly Dimethyl Siloxane
PEG	Poly Ethylene Glycol
PET	Poly Ethylene Terephthalate
PFV	Protein-Film Voltammetry
pH-FETs	pH-sensitive Field Effect Transistors
pI	isoelectric Point
PIC	Photonic Integrated Circuit
P_m	Permeability
PMMA	Poly (Methyl Methacrylate)
PMT	Photomultiplier Tube
PNA	Peptide Nucleic Acids
POC	Point of Care
poly-Si	Polycrystalline Silicon
PPG	Poly Propylene Glycol
ppm	Parts per million
PSA	Prostate-Specific Antigen
PSAO	Pea Seedling Amine Oxidase
PTFE	Polytetrafluoroethylene
PU-PEO	Polyurethane Polyethylene Oxide
PVA	Poly Vinyl Alcohol
PVC	Poly Vinyl Chloride

PVR	Peak-to-Valley Ratio
PWM	Pulse Width Modulation
QCM	Quartz Crystal Microbalance
R	Resistor
RDE	Rotating Disk Electrode
RTD	Resonant Tunneling Diode
RUs	Resonance Units
RWG	Resonant Wave Guide
SAM	Self Assembled Monolayer
SAM	Scanning Auger Microscopy
Sb	Antimony
Sb_2O_3	Antimony Oxide
SBE	Semiconductor Bloch Equations
SCE	Saturated Calomel Electrode
SDS	Sodium Dodecyl Sulfate
SEEDs	Self Electro-optic Effect Devices
SELEX	Systematic Evolution of Ligands by Exponential Enrichment
SEM	Scanning Electron Microscopy
SGRTD	Strain Gages Resistance Temperature Detector
Si_3N_4	Silicon Nitride
$SiCl_4$	Silicon Chloride
SP	Surface Plasmon
SPE	Screen-Printed Electrodes
SPR	Surface Plasmon Resonance
SPW	Surface Plasmon Waveguides
ssDNA	single stranded DNA
STM	Scanning Tunneling Microscopy
STTs	Short-term Tests
SUV	Small Unilamellar Vesicles
SWNTs	Single Walled Nanotubes
Ta_2O_5	Tantalum Oxide
TAS	Total Analysis Systems
TB	Tuberculosis
TCNQ	7,7,8,8-Tetracyanoquinodimethane
$TCNQ^-$	Tetracyanoquinodimethane Radical Anion
TEM	Transmission Electron Microscopy
TEOS	Tetra Ethyl Ortho Silane

TFD	Thin Film Diode
TIR	Total Internal Reflection
TRITC	Tetramethylrhodamine-5-(and 6)-Isothiocyanate
TSB	Tryptic Soy Broth
USB	Universal Serial Bus
UTIs	Urinary Tract Infections
UV	Ultraviolet
VOC	Volatile Organic Compound
WBAN	Wireless Body Area Network
WDS	Water Deficit Stress
WSNs	Wireless Sensor Networks
WT	Wild-type
WWTW	Wastewater Treatment Works
XD	X-Ray Diffraction
XO	Xanthine Oxidase
ZrO_2	Zirconium Oxide

Authors

D. Dharaneeswara Reddy is Assistant Professor, Department of Biotechnology, Sree Vidyanikethan Engineering College, Tirupati. He was awarded M.Sc., in 2000 with Biotechnology specialization from S.V. University, Tirupati and completed M.Tech in Biotechnology in 2008 from J.N.T.H. University, Hyderabad respectively. He worked as Production Manager in AG Bioteck Laboratories (I) Ltd., Hyderabad during 2000-03. He joined as Assistant Professor, Department of Biotechnology, Alfa College of Engineering & Technology during 2004-06. Later he joined as Scientist and manager in Biotechnology Center, Hyderabad during 2006-08. He has around 5 years of industry and 7 years of teaching and research experience. He worked as a guest faculty to M.Sc., Microbiology course from 2004 – 2006 at Sree Ramakrishna Degree College, Andhra Pradesh. He also served as the subject expert, BOS of the Biotechnology Department at Sri Ramakrishna Degree College during 2004-2006. He guided 7 students for M.Tech in Biotechnology and many students for B.Tech, M.Sc and B.Sc in Biotechnology. He is a life member and fellow in Society for Applied Biotechnology; Biotechnology, Bioinformatics and Bioengineering and membership in Indian Society for Technical Education (ISTE) and the Indian Science Congress Association (ISCA).

O.M. Hussain, is Professor of Physics and Co-ordinator for M.Sc., Nanomaterials and Technologies Course at Sri Venkateswara University, Tirupati. He completed M.Sc. in 1984 and Ph.D. in 1990 from, S.V. University, Tirupati. He worked as Post Doctoral Fellow in the Universite Pierre et Marie Curie, Paris, France during 1991-92. He joined as faculty member in the Department of Physics in 1992 and has 20 years of teaching and research experience. He has published more than 100 research articles in journals of international repute and presented more than 50 papers in conferences. He has successfully guided seven Ph.D. Students and 6 M.Phil students. He has made significant contributions in the growth and characterization of nanocrystalline and polycrystalline metal oxide thin films for their effective utilization in the fields of micro-electronics, opto-electronics and electrochemical micro-devices.

D.V.R. Sai Gopal is Professor and Chairman, BOS in Virology and Microbiology, Head of the Department of Virology, Sri Venkateswara University, Tirupati. He was awarded M.Sc., in 1982 with Virology specialization and obtained his Ph.D., in Virology in the year 1987. He worked as Scientist at Spices Board (Govt. of India) and later as faculty member in the Department of Virology in 1989. He visited IRRI, Phillipines and USA as an invited fellow. He has 22 years of teaching and 27 years of research experience. He has published more than 75 research papers, 6 review articles and 6 books. He has guided 5 students for Ph.D., in Virology. He worked as a guest faculty to M.Sc., Nanomaterial Science course and taught Nanobiotechnology from 2006 – 2010 at SVU Physics department. He is reviewer for Elesiver publication journals, editorial board member to *International Journal of Virology*, reviewer for international biotechnology, microbiology journals. He is a life member in Indian Virological Society; Indian Science Congress Association and Indian Phytopathology Society.

D. Muralidhara Rao is Assistant Professor, Department of Biotechnology, Sri Krishnadevaraya University Anantapur, A.P., India and Co-ordinator for B.Tech Biotechnology programme of SKU College of Engg. & Technology. Dr. Rao is member, Board of Studies in Jawaharlal Nehru Technological University, Anantapur. He obtained M.Sc. from Sri Venkateswara University, Tirupati, (1994), M. Phil. (1999) and Ph.D. (2001) from Sri Krishnadevaraya University, Anantapur and has worked as Senior Research Fellow in Department of Science and Technology, Government of India funded project for a period of two years. Dr. Rao has published one book and fifty research articles both in Indian and Foreign journals. He is also recipient of Young Scientist award at international level. Dr. Rao is also Editorial Board member and reviewer for several national and international journals. Dr. Rao was honored with Visiting Scientist Award by Jawaharlal Nehru Center for Advanced Scientific Research, Bangalore. He is Fellow of Society of Biotechnology and Bioinformatics and Life Member of BRSI. Dr. Rao has guided many M.Tech, B.Tech and M.Sc., biotechnology candidates for their project work. His current areas of interest are Bioprocess technology and DNA finger printing. At present is handling two major research projects funded by UGC and CSIR. He is editorial board member in *Journal of Phytopharmacotheraphy and Natural Products, International Research Journal of Plant Sciences*, and Science Advisory Board, United Kingdom.

K.S. Sastry is Professor and Head of Biotechnology, Sree Vidyanikethan Engineering College, A. Rangampet, Chittor (Dist), Andhra Pradesh, India. He has more than 40 years of experience both in research and teaching. Prof. K.S. Sastry has 80 research publications both in national and international journals and has guided a number of M.Sc and Ph.D students.

Biosensors: Introduction and Overview

Chapter 1

1.0 INTRODUCTION

The living organisms such as humans and animals have primarily five sensors, i.e., nose, tongue, ears, eyes and fingers. One might consider the ears, eyes and fingers to be physical sensors as they detect physical sensations of sound, light, heat, etc., respectively. What we detect with the nose smells are in fact small quantities of chemicals. Tongue is another sensor which can detect the various tastes of the food materials by using tongue buds.

The nose is an extremely sensitive and selective instrument which is very difficult to emulate artificially. It can also distinguish between many different chemical substances qualitatively and can give a general idea of quantity down to very low detection limits. The chemicals to be detected pass through the olfactory membrane to the olfactory bulbs, which contain biological receptors that sense the substrate. The response is an electrical signal which is transmitted to the brain through the olfactory nerves. The brain then transduces this response into the sensation, the smell as we feel. The nostrils collect the smell sample, which is then sensed by the olfactory membrane, i.e., the sensing element. The responses from the olfactory receptors are then converted by the olfactory nerve cell, which is equivalent of the transducer, into electrical signals which pass along the nerve fibre to the brain for interpretation. Thus, the brain acts as a microprocessor, turning the signal into a sensation. The tongue operates in a similar way. (Eggins, B. R. Biosensors: *An Introduction,* 1996, John Wiley & Sons Limited).

One of the best known types of sensor in laboratory is the litmus paper test for acids and alkalis, which gives a qualitative indication by means of a colour reaction. A more precise method of indicating the degree of acidity is the measurement of pH, either by the more extended use of colour reactions in special indicator solutions or even by simple pH papers. However, the best method of measuring acidity is the use of pH meter, which is an electrochemical device by an electrical response which can be read by analog scale or on a digital read out device or input to a microprocessor. In such methods the sensor that responds to the degree of acidity is either a chemical in the dye litmus or a mixture of chemical dyes in pH indicator solutions or the glass membrane electrode in the pH meter. The chemical or electrical response then has to be converted into a signal that can be observed by our eyes. With litmus the above process becomes easy for the measurement of pH. A colour change is observed because of the change in the absorbance of visible light by the chemical itself. In the case of a pH meter, the electrical response (a voltage change) has to be converted into an observable response movement of a meter needle or a digital display. The part of the device which carries out this conversion is called a transducer.

1.1 SENSORS

According to R.W.Catterall: A sensor is a device that detects or measures a physical and chemical property and records, indicates or otherwise responds to it. In general, the sensors are divided into three types: (a) physical sensors (b) chemical sensors and (c) biological sensors.

a. Physical Sensors

Physical sensors for measuring distance, mass, temperature and pressure. The response of a sensor is usually in the form of a physical response.

b. Chemical Sensors

Chemical sensor is a device which responds to a particular analyte in a selective way through chemical reactions and can be used for the qualitative or quantitative determination of the analyte. It can transform chemical information ranging from the concentration of a specific sample component to total composition analysis into an analytically useful signal.

Chemical sensors contain two basic functional units: a receptor and a transducer part. Sensors may include a separator which is a membrane. In the receptor part of sensor, the chemical information is transformed into a form of energy which may be measured by the transducer. The transducer part is a device capable of transforming the energy carrying the chemical information about the sample into a useful analytical signal. In addition to the sensor the analyzer (a component) may contain devices that perform the following functions: sampling, sample transport, signal processing and data processing. An analyzer may be an essential part of an automated system. The analyzer works according to the sampling plan as a function of time and acts as a monitor. The receptor part of sensors is based upon various principles:

1. Physical: where no chemical reaction takes place. Optical examples are those based upon the measurement of absorbance, refractive index, conductivity, temperature and mass change.
2. Chemical: in which a chemical reaction with participation of the analyte gives rise to the analytical signal.
3. Biochemical: in which a biochemical process is the source of the analytical signal. Examples: microbial, potentiometric sensors and immunosensors. They may be regarded as a subgroup of the chemical ones. Such sensors are called 'biosensors'.

c. Biosensors

Biosensors measure the chemical substances by using biological sensing elements. A biosensor can be defined as a device incorporated by a biological sensing element connected to a transducer. An analyte of the sensor is a biological component (target analyte) which detects and measures the chemical components (inorganic/organic). The key difference is that the recognition element is biological in nature.

1.1.1 Structure of Sensors

The sensor devices have to be connected to a transducer of some sort, so that a visibly observable response occurs. Chemical sensors and biosensors are generally concerned with sensing and measuring particular chemicals which may or may not be biological themselves. The general schematic arrangement of a sensor is shown in Fig. 1.1.

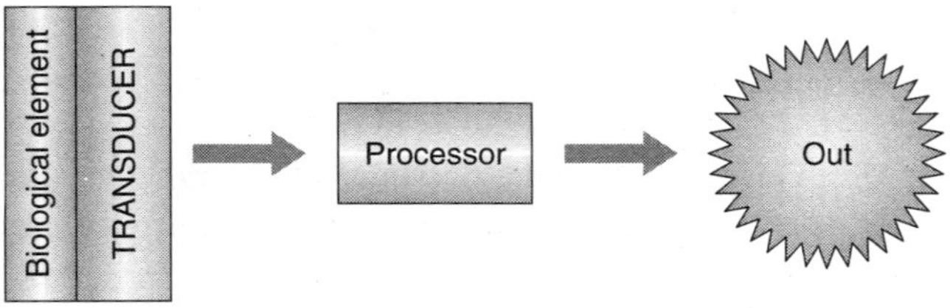

Fig. 1.1 *Schematic layout of a sensor.*

1.1.2 Classification of Sensors

The development of instrumentation, microelectronics and computers makes it possible to design sensors utilizing most of the known chemical, physical and biological principles. The classification of sensors is as follows:

1. **Chemical sensors** may be classified according to the operating principle of the transducer.
2. **Optical devices** transform the changes of optical phenomena, due to the interaction of the analyte with the receptor. This group may be further subdivided according to the type of optical phenomena that take place during the interaction.
   a. **Absorbance** is measured in a transparent medium, caused by the absorptivity of the analytes.
   b. **Reflectance** is measured in non-transparent medium, usually using an immobilized indicator.
   c. **Luminescence** is based on the measurement of the intensity of light emitted by the chemical reaction in the receptor system.
   d. **Fluorescence** is measured as the positive emission effect caused by irradiation. Also selective quenching of fluorescence may be the basis of such devices.
   e. **Refractive index** is measured as the result of a change in solution composition. This may also include a surface plasmon resonance effect.
   f. **Opto-thermal effect** is based on the measurement of the thermal effect caused by light absorption.
   g. **Light scattering** is based on the effects caused by particles of definite size present in the sample.

The application of many of these phenomena in sensors became possible because of the use of optical fibres in various configurations, such devices are called optodes. It should be emphasized that fibre optics now commonly used are the only technical devices applicable in a large group of optical sensors which can be based on various principles.

3. **Electrochemical** devices transform the effect of the electrochemical interaction of an analyte and an electrode into a useful signal. Such effects may be stimulated electrically or may result in a spontaneous interaction at the zero current condition. The following subgroups may be distinguished:
   a. **Voltametric sensors** include amperometric devices in which current is measured in d.c. or a.c. mode. This sensor is based on chemically inert electrodes, chemically active electrodes and modified electrodes. This group includes sensors with and without (galvanic sensors) external current source.
   b. **Potentiometric sensors**, in which the potential of the indicator electrode (ion selective electrode, redox electrode and metal/metal oxide electrode) is measured against a reference electrode.
   c. **Chemically sensitized field effect transistor (CHEMFET),** in which the effect of interaction between the analyte and the active coating is transformed into a change of the source-drain current. The interaction between the analyte and the coating is from the chemical point of view, similar to those found in potentiometric ion selective sensors.
   d. **Potentiometric solid electrolyte gas sensors** work in high temperature and are usually applied for gas sensing measurements.
4. **Electrical devices** are based on measurements, where no electrochemical process takes place but the signal arises from the change of electrical properties caused by the interaction of the analyte.
   a. **Metal oxide semiconductor sensors** are used principally as gas phase detectors based on reversible redox processes of analyte gas components.
   b. **Organic semiconductor sensors** are based on the formation of charge transfer complexes, which modify the charge carrier density.
   c. Electrolytic conductivity sensors are based on the variation of conductivity.
   d. Electric permittivity sensors are based on the variation of dielectric constant of the analyte.
5. **Mass sensitive** devices transform the mass change at a specially modified surface into a change of a property of the support material. The mass change is caused by accumulation of the analyte.
   a. **Piezoelectric devices,** used mainly in gaseous phase but also in solutions, are based on the measurement of the frequency change of the quartz oscillator plate caused by adsorption of a mass of the analyte at the oscillator.
   b. **Surface acoustic wave devices** depend on the modification of the propagation velocity of a generated acoustical wave affected by the deposition of a definite mass of the analyte.
6. **Magnetic devices** are based on the change of paramagnetic properties of the gas being analyzed. These are represented by certain types of oxygen monitors.

7. **Thermometric devices** are based on the measurement of the heat effects of a specific chemical reaction or adsorption which involves the analyte. In this group, the heat effects may be measured in various ways. Example: In catalytic sensors the heat of a combustion reaction or an enzymatic reaction is measured by a thermistor.
8. **High energy irradiation sensor:** The X- or gamma radiation may form the basis for a chemical sensor in case they are used for determination of chemical composition.
9. **Other physical properties:** Sensors can be classified not according to the primary effect but to the method used for measuring the effect, e.g., catalytic devices in which the heat evolved in the primary process is measured by the change in the conductivity of a thermistor. The electrical devices are often put into one category together with the electrochemical devices.

Sensors have also been classified according to the application to detect or determine a given analytes, e.g., pH, for metal ions or for determining oxygen or other gases.

1.1.3 Competitive Sensors

The competitive sensors which are designed and being used in the present-day science and technology are mentioned below. The principles and their applications of certain competitive sensors are briefly outlined.

1.1.3.1 Hydrophone

Hydrophone is a microphone designed to be used for recording or listening to underwater sound. Most hydrophones are based on a piezoelectric transducer that generates electricity when subjected to a pressure change. Such piezoelectric materials or transducers can convert a sound signal into an electrical signal since sound is a pressure wave.

The piezoelectric hydrophone was used late in World War-I by convoy escorts detecting 'U' boats greatly impacting the effectiveness of submarines. The hydrophones are the sole devices to detect the targets while the submarines are submerged.

1.1.3.2 Seismometer

Seismometers are the instruments that measure motions of the ground including those of seismic waves generated by earthquakes, volcanic eruptions and other seismic sources. Records of seismic waves allow seismologists to map the interior of the earth, locate and measure the intensity of the waves from different sources.

Early seismometers are used as optical levers or mechanical linkages to amplify the small motions involved in recording on the soot covered paper or photographic paper. In modern seismometers the signal is digitized using modern electronics and the signal is measured more accurately. Professional seismic observatories usually have instruments measuring three axes: north-south, east-west and the vertical.

1.1.3.3 Curb Feeler

Curb feelers or curb finders are springs or wires installed on a vehicle which act as whiskers to warn drivers that they are too close to the curb or other obstruction. The devices are fitted at the bottom of the body of vehicle close to the wheels. As the vehicle approaches the curb the protruding 'feelers' act as whiskers and scrape against the curb making a noise and alerting the driver in time to avoid damaging the wheels or hubcaps. The feelers are flexible and do not easily break.

1.1.3.4 Defect Detector

A defect detector is a device used on railroads to detect axle and signal problems in passing trains. The detectors are normally integrated into the tracks and often include sensors to detect several different kinds of problems that could occur. Defect detectors which enabled railroads to eliminate the caboose at the rear of the train, as well as various station agents stationed along the route to detect unsafe conditions.

1.1.3.5 Manifold Absolute Pressure (MAP) Sensor

The manifold absolute pressure sensor (MAP sensor) is one of the sensors used as electronic control system in internal combustion engines. Engines that use a MAP sensor are typically fuel injected. The manifold absolute pressure sensor provides instantaneous manifold pressure information to the engine's electronic control unit (ECU). The data is used to calculate air density and determine the engine's air mass flow rate which in turn determines the required fuel metering for optimum combustion. A fuel injected engine may alternately use a MAF (mass air flow) sensor to detect the intake airflow.

MAP sensor data can be converted to air mass data using the speed density method. Engine speed (RPM) and air temperature are also necessary to complete the speed density calculation. The MAP sensor can also be used in OBD II (on-board diagnostics) applications to test the EGR (exhaust gas recirculation) valve for functionality—an application typical in OBD II equipped General Motors engines.

1.1.3.6 Oxygen Sensors

An oxygen sensor or lambda sensor is an electronic device that measures the proportion of oxygen (O_2) in the gas or liquid being analyzed. It was developed by the Robert Bosch GmbH company during the late 1960s under the supervision of Gunter Bauman. The original sensing element is made with a thimble shaped zirconia-ceramic coated on both the exhaust and reference sides with a thin layer of platinum and comes in both heated and unheated forms. The most common application is to measure the exhaust gas concentration of oxygen for internal combustion engines in automobiles and other vehicles.

Scientists use oxygen sensors to measure respiration or production of oxygen and use a different approach. Oxygen sensors are used in oxygen analyzers which find many applications in medical field such as anesthesia monitors, respirators and oxygen concentrators.

1.1.3.7 Parking Sensors

Parking sensor systems use ultrasonic proximity detectors embedded in the front and/or rear bumpers, to measure the distances to nearby objects at low level. The sensors measure the time taken for each sound pulse to be reflected back to the receiver. Depending on the speed of the vehicle and the distance to the obstacle, the system warns the driver by visual or audible means about the risk of collision.

Sensors are usually fitted to the rear of a vehicle but may also be fitted to the front. Rear sensors are activated when reverse gear is selected and then deactivated as soon as any other gear or neutral or park is selected. Front sensors are generally activated by pressing a button and then automatically deactivated when the vehicle exceeds a certain speed, this is to avoid nuisance warnings in slow moving traffic.

1.1.3.8 Radar Gun

A radar speed gun (radar gun and speed gun) is a small doppler radar unit used to measure the speed of moving objects including vehicles, pitched baseballs, runners and other moving objects. Radar speed guns may be hand held, vehicle mounted or static. A radar speed gun measures the speed of objects at which it is pointed by detecting a change in frequency of the returned radar signal due to the Doppler effect. The radar speed gun was invented by Bryce K. Brown of Decatur Electronics in March 1954 and was first used in Chicago, Illinois in April 1954.

Modern LIDAR (Light Detection and Ranging) speed gun devices which use pulsed laser light instead of radar are more commonly used now. Some motorists install radar detectors which can detect signals from these devices to avoid being caught.

1.1.3.9 Water-Fuel Sensor

This sensor detects the presence of water in the fuel. It is installed in the fuel filter and when the water level in the water separator reaches the warning level of the water in fuel sensor sends an electrical signal to the ECU or to dashboard (lamp). The water in fuel sensor is used especially in the common rail engines to avoid the unit injector damage.

The water in fuel sensor uses the difference of electric conductivity through water and diesel fuel by 2 electrodes. First generation water in fuel sensor uses a potting resin to isolate the electronic circuit; last generation of water in fuel sensor (the WS3 sensor in surface-mount technology) is made totally without leakage using an innovative comoulding process. The last generation of water in fuel sensor has a high resistance to vibrations and to thermal excursion cycles.

1.1.3.10 Carbon Dioxide Sensor

This sensor detects and measures the percentage of carbon dioxide. The most common CO_2 sensors are infrared gas sensors (NDIR) and chemical gas sensors. Measuring carbon dioxide is very important in monitoring indoor air quality and many industrial processes.

For air-conditioning applications, these kinds of sensors can be used to monitor the quality of air and the tailored need of fresh air respectively. In applications where direct temperature measurement is not applicable, NDIR sensors can be used. The sensors absorb ambient infrared radiation (IR) given off by a heated surface. CO_2 sensors are also able to determine how many people are in a room and then adjust the air-conditioning accordingly.

1.1.3.11 Carbon Monoxide Detector

A carbon monoxide (CO) detector is a device that detects the presence of the carbon monoxide (CO) gas in order to prevent carbon monoxide poisoning. CO is a colorless and odorless gas produced by incomplete combustion. It is often referred to as the silent killer because it is virtually undetectable without using detection technology. Elevated levels of CO can be dangerous to humans depending on the amount present and length of exposure. Smaller concentrations can be harmful over longer periods of time while increasing concentrations require diminishing exposure times to be harmful.

CO detectors are designed to measure CO levels over time and sound an alarm before dangerous levels of CO accumulate in an environment, giving people adequate warning to safely ventilate the area or evacuate.

1.1.3.12 Hydrogen Sensor

A hydrogen sensor is a gas detector that detects the presence of hydrogen. They contain micro-fabricated point contact hydrogen sensors and are used to locate leaks. They are considered low cost, compact, durable and easy to maintain as compared to conventional gas detecting instruments.

1.1.3.13 H_2S Sensor

The H_2S sensor is a metal oxide semiconductor (MOS) sensor which operates by a reversible change in resistance caused by adsorption and desorption of hydrogen sulfide in a film with hydrogen sulfide sensitive material like tin oxide thick films and gold thin films. Current response time is 25 ppb to 10 ppm in less than one minute. The H_2S sensor is used to detect hydrogen sulfide in the hydrogen feed stream of fuel cells to prevent catalyst poisoning and to measure the quality of guard beds used to remove sulfur from hydrocarbon fuels.

1.1.3.14 Pellistor

A pellistor is a solid state device used to detect gases which are either combustible or which have a significant difference in thermal conductivity to that of air. The word pellistor is a combination of *pellet* and *resistor*.

1.1.3.15 pH Sensor

A pH sensor mainly consists of a glass electrode, which is a type of ion selective electrode made of a doped glass membrane that is sensitive to a specific ion. It is an important part of the instrumentation for chemical analysis and physicochemical studies. In modern practice, widely used membranous ion selective electrodes (ISE, including glasses) are part of a galvanic cell. The electric potential of the electrode system in solution is sensitive to changes in the content of a certain type of ions which is reflected in the dependence of the electromotive force (EMF) of galvanic element concentrations of these ions. Glass electrodes are commonly used for pH measurements.

There are also specialized ion sensitive glass electrodes used for determination of concentration of lithium, sodium, ammonium and other ions. Glass electrodes have been utilized in a wide range of applications from pure research, control of industrial processes, to analyze foods, cosmetics and comparison of indicators of the environment and environmental regulations: microelectrode measurements of membrane electrical potential of a biological cell, analysis of soil acidity, etc.

1.1.3.16 Potentiometric Sensor

A potentiometric sensor is a type of chemical sensor that may be used to determine the analytical concentration of some components of the analyte gas or solution. This sensor measures the electrical potential of an electrode when no current is flowing.

1.1.3.17 Current Sensor

A current sensor is a device that detects electrical current (AC or DC) in a wire and generates a signal proportional to it. The generated signal could be analog voltage or current or even digital output. It can be then utilized to display the measured current in an ammeter or can be stored for further analysis in a data acquisition system or can be utilized for control purpose.

1.1.3.18 Metal Detector

A metal detector is a device which responds to the metal that may not be readily apparent. The simplest form of a metal detector consists of an oscillator producing an alternating current that passes through a coil producing an alternating magnetic field. If a piece of electrically conductive metal is close to the coil, eddy currents will be induced in the metal and this produces an alternating magnetic field of its own. If another coil is used to measure the magnetic field (acting as a magnetometer) the change in the magnetic field due to the metallic object can be detected.

The first industrial metal detectors were developed in the 1960s and were used extensively for mining and other industrial applications. Uses include de-mining (the detection of land mines), the detection of weapons such as knives and guns, especially in airport security, geophysical prospecting, archaeology and treasure hunting. Metal detectors are also used to detect foreign bodies in food and in the construction industry to detect steel reinforcing bars in concrete and pipes and wires buried in walls and floors.

1.1.3.19 Actinometers

Actinometers are instruments used to measure the heating power of radiation. They are used in meteorology to measure solar radiation as pyrheliometers. An actinometer is a chemical system or physical device which determines the number of photons in a beam integrally or per unit time. This name is commonly applied to devices used in the ultraviolet and visible wavelength ranges. For example, solutions of iron (III) oxalate can be used as a chemical actinometer, while bolometers, thermopiles and photodiodes are physical devices giving a reading that can be correlated to the number of photons detected. The actinometer was invented by Sir John Frederick William Herschel to measure the direct heating power of the sun's rays and his work with the instrument is of great importance in the early history of photochemistry.

1.1.3.20 Fish Counters

Automatic fish counters are automatic devices for measuring the number of fish passing along a particular river in a particular period of time. Usually one particular species is of interest. One important species studied by fish counters are *Atlantic salmon*. This species is of interest owing to its ecologically vulnerable status and anadromous life styles.

1.1.3.21 Gas Detector

A gas detector is a device which detects the presence of various gases within an area usually as part of a safety system. This type of equipment is used to detect a gas leak and interface with a control system so that the process can be automatically shut down. A gas detector can also sound an alarm to operators in the area where the leak is occurring, giving them the opportunity to leave the area. This type of device is important because there are many gases, viz., CO_2, CO, NH_3, H_2S, NO_X, CH_4 which are harmful to organic life such as humans, animals and plants.

Gas detectors can be used to detect combustible, flammable and toxic gases, and oxygen depletion. This type of device is used widely in industry and can found in a variety of locations such as on oil rigs to monitor manufacture processes and emerging technologies such as photovoltaic. They may also be used in fire fighting.

1.1.3.22 Leaf Sensor

A leaf sensor is a phytometric device (measurement of plant physiological processes) that measures water loss or the water deficit stress (WDS) in plants by real time monitoring the moisture level in plant leaves. The first leaf sensor was developed by Leaf Sens, in 2001. Leaf Sen has made strides incorporating their leaf sensory technology into citrus orchards in Israel. A solid state smart leaf sensor technology was developed by the University of Colorado at Boulder for NASA in 2007. It was designed to help monitor and control agricultural water demand.

1.1.3.23 Rain Sensor

A rain sensor or rain switch is a switching device actuated by rainfall. There are two main applications for rain sensors. The first is a water conservation device connected to an automatic irrigation system that causes the system to shut down in the event of rainfall. The second is a device used to protect the interior of an automobile from rain and to support the automatic mode of windscreen wipers.

1.1.3.24 Soil Moisture Sensors

Soil moisture sensors measure the water content in soil. A soil moisture probe is made up of multiple soil moisture sensors. One common type of soil moisture sensors in commercial use is a frequency domain sensor such as a capacitance sensor. Another sensor, the neutron moisture gauge, utilizes the moderator properties of water for neutrons. Measuring soil moisture is important in agriculture to help farmers manage their irrigation systems more efficiently. Not only are farmers able to generally use less water to grow a crop they are able to increase yields and the quality of the crop by better management of soil moisture during critical plant growth stages.

Besides agriculture there are many other disciplines using soil moisture sensors. Golf courses are now using sensors to increase the efficiencies of their irrigation systems to prevent over watering and leaching of fertilizers and other chemicals offsite.

1.1.3.25 Tide Gauge

A tide gauge is a device for measuring sea level and detecting tsunamis. Sensors continuously record the height of the water level with respect to a height reference surface close to the geoids. Water enters the device by the bottom pipe (far end of the tube) and electronic sensors measure its height and record it to a tiny computer. Tide gauges are used to measure tides and quantify the size of tsunamis. The measurements make it possible to derive the mean sea level. Using this method, sea level slopes up to several 0.1 m/1000 km and more have been detected. A tsunami can be predicted by using this sensor when the sea level begins to rise.

1.1.3.26 Particle Detector

A particle detector (a radiation detector) is a device used to detect, track, and/or identify high energy particles such as those produced by nuclear decay, cosmic radiation or reactions in a particle accelerator. Modern detectors are also used as calorimeters to measure the energy of the detected radiation. They may also be used to measure other attributes such as momentum, spin, charge, etc., of the particles. Commonly used particle detectors are gaseous ionization detectors, Ionization chamber, Geiger Muller counters, spark chamber, solid-state detectors, calorimeters, micro channel plate detectors and neutron detectors.

1.1.3.27 Scintillation Counter

A scintillation counter measures ionized radiation. The sensor is called a scintillator which consists of a transparent crystal usually phosphor, plastic (usually containing anthracene), or organic liquid that fluoresces when struck by ionized radiation. A sensitive photomultiplier tube (PMT) measures the light from the crystal. The PMT is attached to an electronic amplifier and other electronic equipment to count and possibly quantify the amplitude of the signals produced by the photomultiplier.

The scintillation counter was invented in 1944 by Sir Samuel Curran. Scintillation counters are widely used because they can be made inexpensively yet with good quantum efficiency. The quantum efficiency of a gamma ray detector (per unit volume) depends upon the density of electrons in the detector and certain scintillating materials, such as sodium iodide and bismuth germanate achieve high electron densities as a result of the high atomic numbers of some of the elements of which they are composed. However detectors based on semiconductors notably hyper pure germanium have better intrinsic energy resolution than scintillators and are preferred where feasible for gamma ray spectrometry. In the case of neutron detectors high efficiency is gained through the use of scintillating materials rich in hydrogen that scatter neutrons efficiently. Liquid scintillation counters are an efficient and practical means of quantifying beta radiation.

1.1.3.28 Scintillator

A scintillator is the material which exhibits scintillating, the property of luminescence, when excited by ionizing radiation. Luminescent materials, when struck by an incoming particle, absorb its energy and scintillate, i.e., reemit the absorbed energy in the form of light. Sometimes, the excited state is meta-stable so the relaxation back out of the excited state is delayed: the process then corresponds to either one of two phenomena depending on the type of transition and hence the wavelength of the emitted optical photon, delayed fluorescence or phosphorescence.

A scintillation detector or scintillation counter is obtained when a scintillator is coupled to an electronic light sensor such as a photomultiplier tube (PMT) or a photodiode. PMTs absorb the light emitted by the scintillator and reemit it in the form of electrons via the photoelectric effect. The subsequent multiplication of those electrons results in an electrical pulse which can then be analyzed and yield meaningful information about the particle that originally struck the scintillator. Vacuum photodiodes are similar but do not amplify the signal while silicon photodiodes accomplish the same thing directly in the silicon.

The first device used as scintillator was built in 1903 by Sir William Crookes and used a ZnS screen. The scintillations produced by the screen were visible to the naked eye if viewed by a microscope in a darkened room; the device is known as a Spinthariscope. This technique led to a number of important discoveries but is obviously tedious. Scintillators gained additional attention in 1944, when Curran and Baker replaced the naked eye measurement with the newly developed PMT. This was the birth of the modern scintillation detector.

1.2 BIOSENSOR

A biosensor can be defined as "a device incorporated by a biological sensing element connected to a transducer".

The term biosensor has been variously applied to a number of devices either used to monitor living systems or incorporating biotic elements. Recently, IUPAC committee has been trying to unravel literature, that uses the term to describe a thermometer, a mass spectrometer, daphnia in pond water, electrophysiology equipment, chemical labels for imaging and ion-selective electrodes. The consensus, however, is that the term should be reserved for use in its modern context of a sensor incorporating a biological element such as an enzyme, antibody, nucleic acid, microorganism or cell; this decision is readily endorsed by a cursory examination of a data base such as Chemabs.

A biosensor is defined as a compact analytical device incorporating a biological or biologically derived sensing element either integrated within or intimately associated with a physicochemical transducer. The natural aim of a biosensor is to produce either discrete or continuous digital electronic signals which are proportional to a single analyte or a related group of analytes.

1.2.1 History

1918–2005

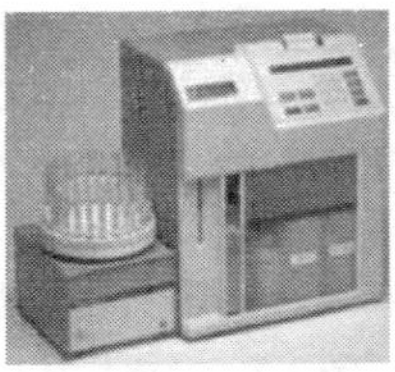

YSI 2300 STAT Plus Glucose & Lactate Analyser.

Prof. Leland C Clark is the father of the biosensor or biosensor concept. In 1956, Clark published his definitive paper on the oxygen electrode. Based on his experience and addressing his desire to expand the range of analytes that could be measured in the body and made a landmark address in 1962 at New York Academy of Sciences symposium in which he described how "to make electrochemical sensors (pH, polarographic, potentiometric or conductometric) more intelligent" by adding "enzyme transducers as membrane enclosed sandwiches". The concept was illustrated by an experiment in which 'glucose oxidase' was entrapped at a Clark oxygen electrode using dialysis membrane (immobilization process). The decrease in measured oxygen concentration was proportional to glucose concentration. Clark and Lyons coined the term enzyme electrode and expanded the experimental details necessary to build functional enzyme electrodes for glucose.

In 1975, YSI (Yellow Spring Instruments Co., Ohio, USA) produced the first of many biosensor based laboratory analysers to be built by companies around the world.

YSI 2300 STAT Plus Glucose & Lactate Analyzer used for measurement of Glucose, Sucrose, Lactose, L-Lactate, Galactose, Cholin, Ethanol, Methanol, Hydrogen Peroxide and Starch. Guilbault and Montalvo (1969) were the first to report the details of a potentiometric enzyme electrode. They described a Urea Sensor based on urease immobilized at an ammonium selective liquid membrane electrode.

The use of thermal transducers for biosensors was proposed in 1974 and the new devices were christened by thermal enzyme probes and enzyme thermistors, respectively.

The biosensor took a further fresh evolutionary route in 1975, when Divis suggested that bacteria could be harnessed as the biological element in microbial electrodes for the measurement of alcohol content. This is the beginning of a major research effort in Japan and elsewhere in biotechnological and environmental applications of biosensors. Lubbers and Opitz coined the term optode in 1975 describing a fibre-optic sensor with immobilized indicator to measure carbon dioxide or oxygen. They extended the concept to make an optical biosensor for alcohol by immobilizing alcohol oxidase on the end of a fibre-optic oxygen sensor. Commercial optodes are now showing excellent performance for *in-vivo* measurements of pH, pCO_2 and pO_2, but enzyme optodes are not yet widely available.

In 1976, Clemens *et al.* incorporated an electrochemical glucose biosensor in a bedside artificial pancreas and this was later marketed by Miles as the Biostator. Although the Biostator is no longer commercially available, a new semi continuous catheter based blood glucose analyser has recently been introduced by Medical Diagnostics.

In 1976, La Roche (Switzerland) introduced the Lactate analyser LA 640 in which the soluble hexa cyanoferrate mediator was used to shuttle electrons from Lactate dehydrogenase to an electrode. Although this was not a commercial success at the time, it turned out in retrospect to be an important forerunner of a new generation of mediated biosensors and of lactate analysers for sports and clinical applications.

Shichiri *et al.* (1982) reported major advances in the *in-vivo* application of glucose biosensors, who described the first needle type enzyme electrode for subcutaneous implantation in 1982. Companies are still pursuing this possibility but no device for general use is available yet. The idea of building direct immunosensors by fixing antibodies to a piezoelectric or potentiometric transducer had been explored since the early 70's. They described the use of surface plasmon resonance to monitor the affinity reactions in real time.

In 1984, the ferrocene and its derivatives as an immobilised mediator for use with Oxidoreductases are employed in the construction of inexpensive enzyme electrodes.

In 1987, a pen sized meter for home blood glucose monitoring formed the basis for the screen printed enzyme electrodes launched by MediSense (Cambridge, USA). The BIAcore (Pharmacia, Sweden) launch in 1990 is based on this technology.

The idea of building direct immunosensors by fixing antibodies to a piezoelectric or potentiometric transducer had been explored since the early seventies, but it was a paper by Liedberg *et al.* (1983) The BIAcore (Pharmacia, Sweden) produced commercial model of this technology in 1990. *(Source: http://www.cranfield.ac.uk/biotech/sensors/history.htm).*

1.2.2 Overview

Biosensors are already thriving in the medical field. External biosensors are used in emergency rooms as point of care diagnostic units such as i-Stat's "lab on a chip," which can reveal almost immediately whether a patient is in cardiac arrest by testing blood chemistry levels. Companies are developing implantable biosensors that track blood glucose levels and delivers insulin. Microchips are used to test a chip implant that offers long term and time controlled drug delivery.

A recent report from market research firm In-Stat revealed that the media spotlight on bio-warfare this application may be premature: Despite the public's anticipation that biosensors with real-time detection will be able to monitor biological and chemical weapons. However, the technology hasn't caught up with expectations. Presently, biosensors in environmental monitoring stations nationwide can detect compounds like anthrax but detection can takes 12 to 24 hours. The best ones on the market take 20 minutes.

Sandia National Laboratories developed the Micro ChemLab, a system that detects biotoxins in 5 minutes. Currently MicroChem Lab is trying to integrate both the gas based and liquid based analyses into a one handheld device. This type of biosensor could be incorporated into military uniforms and eventually into houses as the biowarfare equivalent of a smoke detector.

One of the most important applications is in industry, where biosensors monitor air quality and emissions at chemical refineries and quality control at food processing plants. Currently, testers take random samples off the food line and use biosensors to detect *E.coli* and *Salmonella* species.

1.3 NANOSENSORS

Nanosensor is a device typically comprising nanoscale semiconductor crystals used in combination with biological probes for a variety of applications. Nanosensors are any biological, chemical or surgical sensory points used to convey information about nanoparticles to the macroscopic world. Their use mainly includes various medicinal purposes and as gateway to building other nanoproducts, such as computer chips that work at the nanoscale and nanorobots.

1.3.1 Production Methods

Presently, there are several ways proposed to produce nanosensors including top down lithography, bottom up assembly and molecular self assembly. Top down lithography is the manner in which most integrated circuits are now made. It involves starting out with a larger block of some material and carving out the desired form. These carved out devices, notably put to use in specific micro electromechanical systems used as microsensors, generally reach the micro size but the most recent of these have begun to incorporate nanosized components.

Another way to produce nanosensors is through the bottom-up method, which involves assembling the sensors out of even more very small components, most likely individual atoms or molecules. This would involve moving atoms of a particular substance one by one into particular positions which, though it has been achieved in laboratory tests using tools such as atomic force microscopes. This process would be used mainly for building starter molecules for self-assembling sensors.

The third way to produce faster results involves self-assembly or growing particular nanostructures to be used as sensors. This most often entails one of two types of assembly. The first involves using a piece of some previously created or naturally formed nanostructure and immersing it in free atoms of its own kind. After a given period, the structure, having an irregular surface that would make it prone to attract more molecules as a continuation of its current pattern, would capture some of the free atoms and continue to form more of itself to make larger components of nanosensors.

The second type of self assembly starts with an already complete set of components that would automatically assemble themselves into a finished product. Though this has been so far successful only in assembling computer chips at the micro size, researchers hope to eventually be able to do it at the nanometer size for multiple products, including nanosensors. Accurately being able to reproduce this effect for a desired sensor in a laboratory would imply that scientists could manufacture nanosensors much more quickly and potentially far more cheaply by letting numerous molecules assemble themselves with little or no outside influence rather than having to manually assemble each sensor.

1.3.2 Applications

Medicinal uses of nanosensors mainly revolve around the potential of nanosensors to accurately identify particular cells or places in the body in need. By measuring changes in volume, concentration, displacement and velocity, gravitational, electrical and magnetic forces, pressure or temperature of cells in a body. Nanosensors may be able to distinguish between and recognize certain cells, most notably those of cancer at the molecular level in order to deliver medicine or monitor development to specific places in the body. In addition, they may be able to detect macroscopic variations from outside the body and communicate these changes to other nano products working within the body.

One example of nanosensors involves using the fluorescence properties of cadmium selenide quantum dots as sensors to uncover tumors within the body. By injecting a body with these quantum dots, a doctor could see where a tumor or cancer cell is present by finding the injected quantum dots because of their fluorescence. Developed nanosensor quantum dots would be specifically constructed to find only the particular cell for which the body was at risk. The major disadvantage of cadmium selenide dots is that they are highly toxic to the body.

The potential predicted functions of sensors may also include to detect specific DNA in order to recognize explicit genetic defects especially for individuals at high risk. The implanted sensors can automatically detect glucose levels for diabetic subjects more simply than current detectors.

1.4 ASPECTS OF SENSORS

1.4.1 Recognition Elements

Recognition elements are the key components of any sensor device. They impart the selectivity that enables the sensor to respond selectively to a particular analyte or group of analytes thus avoiding interferences from other substances. Methods of analysis for specific ions have been available for a long time using ion-selective electrodes, which usually contain a selective membrane for the analyte of choice. In biosensors, the most common recognition element is an enzyme. Others include antibodies, nucleic acids and receptors.

1.4.2 Transducers

The transducer devices in chemistry are mainly based on photometric transducers as in spectroscopic and calorimetric methods. However, most sensors have been developed on electrochemical transducers because of simplicity of construction and cost. However, the rapid development of photon driven devices has been done through the use of optical fibre. In addition the use of micro mass controlled devices, based mainly on piezoelectric crystals may become competitive in the near future. Transducers can be subdivided into the following three main types.

1.4.2.1 Optical Transducers

These have taken a new lease of life with the development of fibre optics thus allowing greater flexibility and miniaturization. The techniques used include absorption spectroscopy, fluorescence spectroscopy, luminescence spectroscopy, internal reflection spectroscopy, surface plasmon spectroscopy and light scattering.

1.4.2.2 Piezoelectric Devices

These devices involve the generation of electric currents from a vibrating crystal. The frequency of vibration is affected by the mass of material adsorbed on its surface, which could be related to changes in a reaction. Surface acoustic wave devices are a related system.

1.4.2.3 Thermal Sensors

All chemical and biochemical processes involve the production or absorption of heat. This heat can be measured by sensitive thermistors and hence be related to the amount of substance to be analyzed.

1.5 BIOSENSOR INDUSTRY

Biosensors are one of the most innovative and exciting fields now in the market. Irrespective of the weak economy and doubtful future outlook most industry experts are predicting a 60% annual growth rate for the biosensor industry and most of the drive comes from health care and related industries. There is also an acute shortage of donated livers, pancreas and pancreatic cells for diabetics as well as kidneys for transplants but several companies are racing to provide artificial substitutes which combine human cells with innovative engineering as standby or bridge transplants. Innovative medical biosensor devices which can be used as bridges till organs become available are now the target of several biosensor companies. The potential for more and better sensors and substitute organs is an advantage for patients who are suffering life-threatening diseases.

An ideal biosensor should be tiny, swift, easy to handle, cheap and easily available. It should last for a long time and should be able to be stored in different weather conditions. The possibilities of having own patented innovative design can result in several more biosensors based on the same research model and used for different purposes. The current market is flooded with potentiometric as well as amperometric biosensors which use special calorimetric paper enzyme strips.

The diagnostics represent a very large and well established market, which is continually expanding particularly in the current climate of prevention rather than remedy to the need for detection. Clinical testing is one of the biggest diagnostic markets. Estimates of the market size and future projections tend to be difficult and inaccurate. Several other industries have also turned their interests towards the use of biosensors like the food industry for food quality evaluation as well as environmental agencies for rapid and accurate environmental monitoring.

Putting up a biosensor company however will require extensive research and development into this technology. The research involves several disciplines of physics, biochemistry, human anatomy, laboratory sciences and bioreactor science, physical chemistry, human bodily reactions, electrochemistry, electronics and software engineering. As a result need trained staff of the highest caliber to work on research and development wing to get the best biosensors out in the market for different purposes and also require a precision modern fully automated factory to produce sensitive biosensors for marketing and sales.

Some of the characteristics are required for commercial sensor, viz., relevance of output signal to measurement environment, accuracy and repeatability, sensitivity and resolution, dynamic range, speed of response, insensitivity to temperature/temperature compensation, insensitive to electrical and other environmental interference, amenable to testing and calibration, reliability and self checking capability, physical robustness, service requirements, capital cost, running costs and life, acceptability by user and product safety sample host system must not be contaminated by sensor.

1.5.1 Industry Trends

1. Emerging applications investigate companies to implement comprehensive Research & Development plans.
2. Increased health awareness culminates into development of newer biosensors resulting in the growth of home diagnostic and point of care markets.
3. Integrated collaborative research between academia and industry is on the increase.
4. Focus on security has resulted in the use of advanced biosensor applications for the benefit of public.
5. Growing end-user applications markets ensure growth.
6. Intensive research and development makes Lab on Chip a reality.
7. Glucose meters are the single largest selling biosensor product.

1.5.2 Market

Biosensors are gaining ground in the healthcare market and expanding into the Industrial, Environmental monitoring, Security and Biodefense market segments. During the past five years, biosensors have penetrated diverse markets. Research and Development institutions, universities and industrial sectors have developed through their research efforts which can provide highly accurate, sensitive and pain free diagnostics.

In the recent past R&D investments have focused on miniaturization which has resulted in smaller, more affordable sensitive test monitors specifically blood glucose monitors. Although noninvasive biosensing alternatives are emerging, nanotechnology based biosensors have yet to make a mark in the market.

Biosensor companies have adopted microchip technology (small semiconductor based crystal chips) to reduce the size of their products. A large number of companies are trying to develop a single platform that can be used for multiple applications. Some of them have been successful in achieving this goal, for example by combining diabetic and cholesterol tests. Among biosensors glucose testing biosensors have the largest market presence. One of the factors for their market leadership is the increasing population of diabetic patients and the rising number of point of care applications.

Based on the basic platform a number of different technologies have been developed to construct biosensors. Fig. 1.2 describes biosensor technologies that are currently in use. Electrochemical, piezoelectric and optoelectronic biosensors have experienced technology improvements leading to devices based on surface acoustic wave (SAW) or fibre-optic sensing.

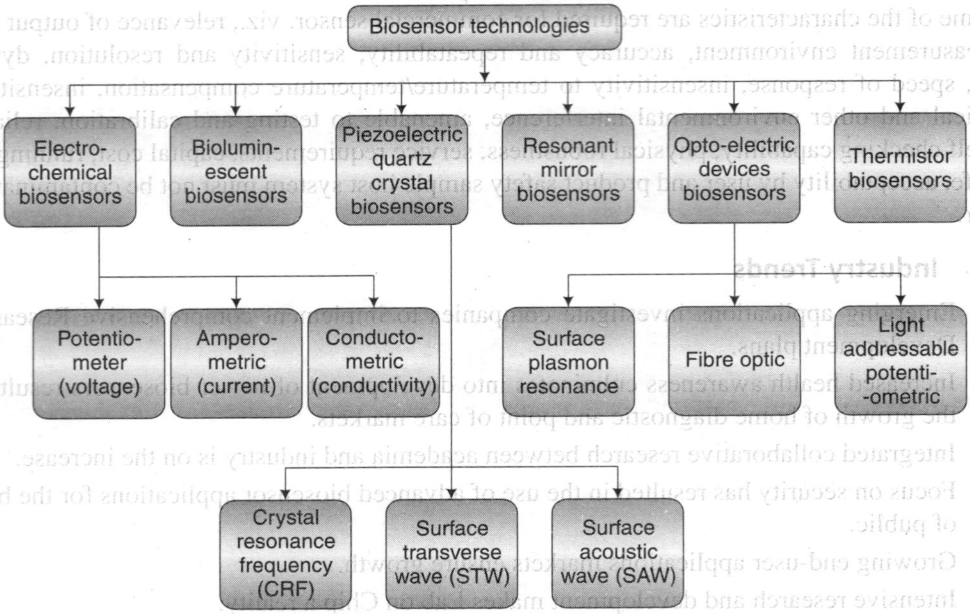

Fig. 1.2 *The different types of biosensor technologies in use. (Source: Frost & Sulvan)*

Figure 1.3 describes that the percent of revenues from markets like environmental, security, bio-defense, home diagnostics etc., have grown from 2006 through 2009 and the forecasts up to 2016 suggest that this growth trend will continue.

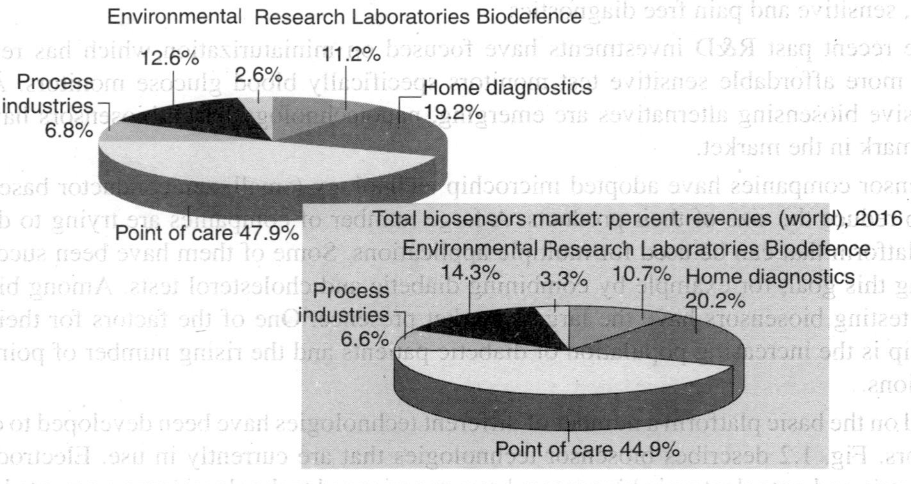

Fig. 1.3 *The total biosensors market, showing the percent of revenues by vertical markets (world) for 2009 and 2016. (Source: Frost & Sullivan)*

Point-of-care continues to be the largest market for biosensors and it is likely to dominate to 2016 and beyond. The development of new biosensor types for new diagnostic tests should contribute to these trends.

1.5.3 Challenges and Innovations

In spite of product and technology innovations and improvements the growth of the biosensors market is still restricted to certain vertical markets. Most biosensors are patented and their market penetration is limited by the resources of the patenting companies. Biosensor awareness and use is limited by a number of factors such as:

1. Sensitivity is the most important performance element in a biosensor. In a biosensor it refers to real time detection and measurement of the reaction of the target analyte and conversion of this measurement into a usable signal.
2. Readout times that vary from one biosensor to another. In some biosensors readout times are very long, in certain cases >20 sec.
3. The life span of a biomolecule is limited.
4. The biosensor requires in certain cases a special pretreatment prior to each use.
5. A number of existing sensors lack long term stability.
6. Miniaturization in sensors poses technical challenges.
7. Some of the biosensors are too expensive for commercial production.
8. Some biosensors are not sufficiently rugged for their intended applications.

Enzyme sensors remain limited to chemistry that allow for a clear signal such as a change in color, pH (hydrogen-ion value), or oxidation state. The applications of these enzymes are not suitable for existing biosensor designs.

The biosensors market is categorized as a growth market with the number of applications increasing as and when each new biosensor is developed. These developments are gradually leading to standardization of equipment, type of biomolecule and test processes. Biosensors are one of the key interdisciplinary research areas targeted for continuous research and development investments. Innovation in biosensors is an ongoing process and this has ensured its penetration into new markets such as security, military bio-defense, environmental monitoring and the process industry. Other potential target markets are automotive and aerospace.

The commercialization of silicon based biosensors has been slow. There has been a gradual increase in the use of micro-fabrication technologies in the development of new biosensors. This trend is expected to lead to the emergence of new markets include clinical analysis, healthcare and environmental monitoring.

The use of an optical fibre as a probe or sensing element in biosensors is also gaining ground due to some of the built in advantages of optical fibres including excellent light delivery, long interaction length, the ability to capture the emitted light from the targets and immunity to electrical and magnetic interference. The key advantage of optical biosensors is their low cost. Their applications include pharmaceutical drug discovery, clinical applications in point-of-care facilities, industrial and military and defense.

The fast detection biosensors are developed in Spain recently. A biosensor can detect bacteria at levels as low as 1 cell per 5 ml of water allowing water to be tested for typhoid fever bacteria in only a few seconds. The researchers used carbon nanotubes containing fragments of DNA (called *aptamers*) that preferentially bonded with *Salmonella typhi*. The aptamers like oligonucleic acid or peptide molecules and fragments of DNA or RNA that bond to particular molecules were secured inside of carbon nanotubes. After being exposed to *Salmonella typhi* bacteria for a few seconds the aptamers latch on to the cells generating a small electrical signal.

Another innovation research work is cell on chip biosensors to detect cell-to-cell interactions. The focus of this research effort was to identify cells in the immune system that suppress tumor growth. The biosensor is capable of identifying interactions between single cells detecting signals from biological activity. This may provide a cost effective tool for the therapeutic monitoring of cancer and is likely to open doors for targeted drug delivery.

In spite of a number of innovative biosensors developed so far, the biosensors market is still far from meeting some of the key end user needs. The key unmet needs include:

1. Development of biosensors capable of multi test detection and monitoring.
2. Development of integrated biosensing platforms.
3. Development of a self configuring biosensor.
4. Complete multiproduct, multivendor interoperability among biosensors.
5. Migration to lab-on-chip biosensors.
6. Availability of wireless options.

1.6 MARKET REVENUES

Figure 1.4 describes that the global revenue for the biosensors market will continue to exhibit strong growth and will exceed the $14 billion mark in the next seven years These revenues are estimated to grow at a CAGR of 11.5% from 2009 to 2016. Annual revenue growth rates are likely to be >12–14% by 2016.

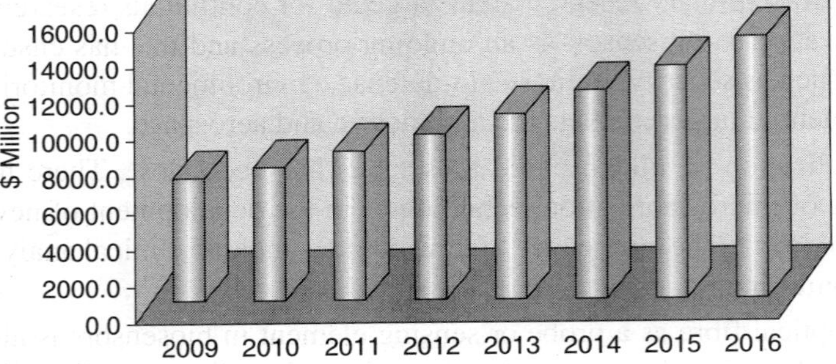

Fig. 1.4 *The total biosensors market showing the world revenue forecast for 2009–2016. (Source: Frost & Sullivan)*

The home diagnostic market continues to be served through retail, online sales, and manufacturer direct sales and is likely to exhibit much higher percent revenues over the forecast period. Home diagnostic market revenues are estimated to exhibit growth at a CAGR of 12.4% during the 2009–2016 time periods. Point–of-care will continue to be the largest market until 2016. Bio-defense and environmental markets are likely to experience the highest CAGRs, estimated to be in the range of 13%–15% for the period from 2009–2016.

Currently, biosensors are used in more than 47 different end-user applications in contrast to the just over 32 end-user applications seven years ago. Undoubtedly, this growth in end-user applications and expansion into other vertical markets is strongly aided by innovations within the biosensor market; more innovations will occur in the years to come. *(Source: Dr. Rajender Thusu, Team Leader,Sensors, Measurement & Instrumentation for Frost & Sullivan, San Antonio, rthusu@frost.com).*

1.6.1 Market Reports

The biosensors in industrial, environmental and especially medical diagnostic applications, world market for biosensors is forecast to reach $6.1 billion by 2012. Growing population, rising incidences of chronic diseases, such as, diabetes and obesity and the growing need for environmental monitoring, are all factors expected to prop up growth in the upcoming years. As stated by the recent report published by Global Industry Analysts Inc, United States and Europe dominate the global market for medical biosensors, collectively capturing 69.73% share estimated in 2008. In Europe, Germany, United Kingdom and France, together, collar 55.3% of the biosensors market. estimated in 2008.The market in Asia-Pacific is projected to reach US$794 million by the year 2012. In medical diagnostics, majority of the biosensors (about 90%) are included in blood gas analyzers, electrolyte analyzers, glucose meters, and metabolite analyzers. Sales in United States the largest market for glucose biosensors is expected to reach US$1.28 billion by 2012. *(Global Industry Analysts, Inc., Email press @ StrategyR.com; Web Site http://www.StrategyR.com/).*

Bioelectronics is the science and technology of integrating biomolecules with electronics to form higher level functional devices such as biosensors, medical prosthetic devices and a range of other devices. According to a updated report (RB-159R Biosensors and Bioelectronics) released by Business Communications Co., the global market for biosensors and other bioelectronics is projected to grow from $6.1 billion in 2004 to $8.2 billion in 2009, at an AAGR (average annual growth rate) of 6.3%. Biomedical and life sciences applications dominated the biosensor and other bioelectronics market in 2003, accounting for 99% of the market. By 2009, biomedical and life sciences applications' share of the market is projected to slip slightly, to 97%. (*Source: http:// www.bccresearch.com*).

The fastest advances will occur in sensor types that are the focus of newer technologies such as optical sensors and biosensors. Almost all products will benefit from improved performance, lower costs and the development of new market opportunities. Growth in the medical/diagnostic market will surpass 8 percent per year, driven by continued expansion of the diagnosed diabetic population.

Blood glucose testing accounts for nearly all of the medical/diagnostic market for chemical sensors, with more than 90 percent of the total market in 2004. It also is the largest single application for chemical sensors overall. Sensors also are used in other diagnostic tests, as well as in a variety of medical equipment, including blood gas analyzers and respiratory monitoring and analysis machines. Optical sensors, biosensors and other emerging technologies such as ultrasound, conductometric and liquid crystal sensors will record the fastest growth. Strong gains in optical and biosensors will come at the expense of more traditional electrochemical and semiconductor sensors. Demand for optical sensors, which include infrared, fibre optic, laser, photoionization and others is being driven by the development of lower-cost, more durable devices capable of competing in cost-conscious and rugged industrial environments. In contrast, demand for reaction-based sensors such as pellistors, and nonspecific sensors such as semiconductors, will see modest gains due to competition from other technologies and maturity in key markets (Table 1.1 and Fig. 1.5).

Table 1.1 U.S. Chemical sensors (in millions). *(Source: The Freedonia Group Inc., © 2005)*

Name of the Item	*1999*	*2004*	*2009*	*04/99*	*09/04*
				Annual growth	
Chemical sensor demand	$2,000	$2,910	$4,165	7.8%	7.4%
Biosensors	$1,138	$1,855	$2,785	10.3%	8.5%
Electrochemical sensors	$702	$832	$1,055	3.5%	4.9%
Optical sensors	$77	$120	$184	9.3%	8.9%
All others	$83	$103	$141	4.4%	6.5%

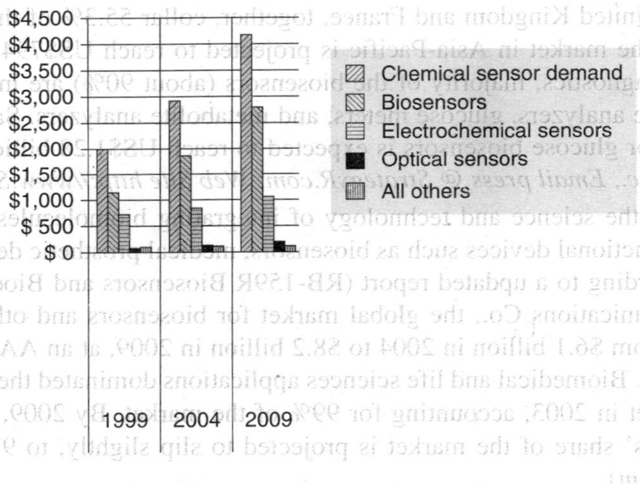

Fig. 1.5 *U.S Chemical sensors (in millions).*

Biosensors Components and Performance Factors

Chapter 2

2.0 BIOSENSOR

Biosensor is a device used for the detection of an analyte, which combines a biological component with a physicochemical detector component (Fig. 2.1). Biosensors are incorporated with a biological material (*Eg.* tissue, microorganisms, organelles, cell receptors, enzymes, antibodies, nucleic acids etc.) or a biological derived material associated with or within a physiochemical transducer or transducer micro-system, which may be optical, electrochemical, thermometric, piezoelectric or magnetic transducer.

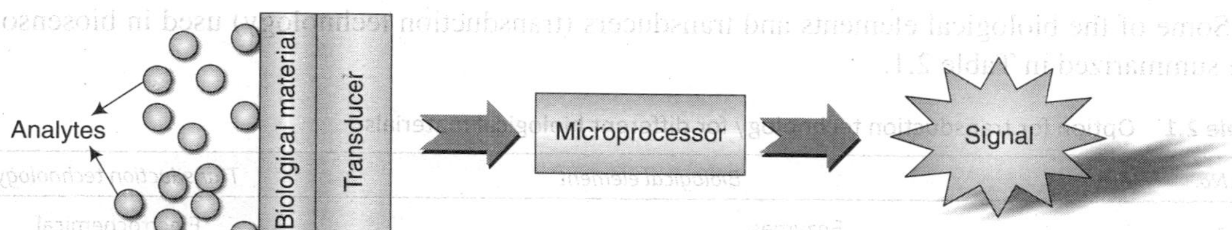

Fig. 2.1 *General layout of a biosensor.*

Biosensor functions by coupling a biological sensing element with a transducer usually yield a digital electronic signal which is proportional to the concentration of a specific analyte or group of analytes.

2.0.1 Principle

Biosensors consist of (i) a biological component for sensing the presence and concentration of a substance and (ii) a transducer device, which works according to the principle shown in Fig. 2.2. The sample is allowed to pass through a membrane so that selection may be exercised and the interfering molecules are retained outside the membrane. The sample then interacts with the biological sensor and forms a product, which may be an electric current/charge, heat, gas or a suitable chemical. The product then passes through another membrane and reaches the transducer, which converts the biochemical signal into an electrical signal that can be further amplified and can be either read on digital panel or recorded. This allows the measurement of the concentration of substance without processing and thus avoiding high consumption of the sample.

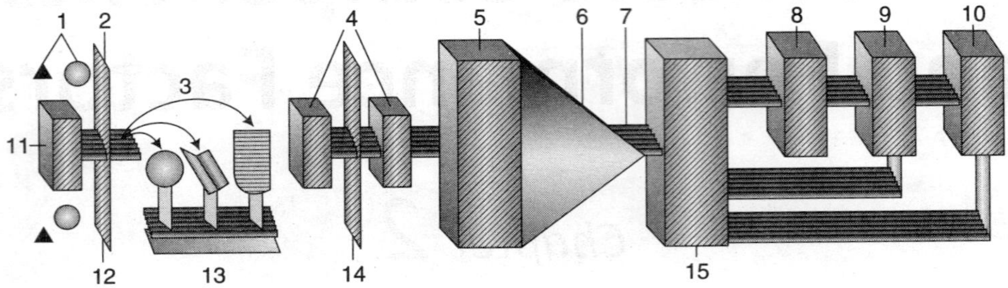

Fig. 2.2 *Schematic outline of a biosensor, showing its different components Interfere and are not allowed passing through membrane.*

1. Analytes
2. Outer Protective membrane
3. Bounded Analytes
4. Product (passes through inner membrane)
5. Transducer
6. Amplifier
7. Signal
8. Data processing
9. Digital Readout
10. Recorder
11. Substrate
12. Protective membrane
13. Sensing part
14. Inner membrane
15. Micro-electronics

Some of the biological elements and transducers (transduction technology) used in biosensors are summarized in Table 2.1.

Table 2.1 Option for transduction technology for different biological materials.

S.No.	*Biological element*	*Transduction technology*
1	Enzymes	Electrochemical
2	Microorganisms	Optical
3	Tissues	Acoustics
4	Antibiotics, Receptors, Nucleic acids etc.,	Calorimetric

Examples:

1. Biosensors for Nutrients : Glucose Sensor
2. Biosensors for Cell Population : Lactate Sensor
3. Biosensors for Product : Alcohol Sensor
4. Biosensors for Environmental Control : Biochemical Oxygen Demand Sensor

The essence of the biosensor is matching the appropriate biological and electronic components to produce a relevant signal during analysis. Isolation of the biological component is necessary to ensure that only the molecule of interest is bound or immobilized on the electronic component or the transducer. The stability of the biological component is critical, since it is being used outside to its normal biological environment.

Attachment of the biological component to the electronic component is very important for the success of biosensors. If the biological component is destroyed in the process of binding or if it binds with the active site unavailable to the analyte, the biosensor will not function. Attachment can be accomplished in a variety of ways, such as covalent binding of the molecule to the detector (usually through a molecular cross-bridge), adsorption onto the surface entrapment in porous material or micro encapsulation. Ultra-thin applications of biological material are usually deposited on transducers by using the Langmuir Blodgett or molecular self assembly techniques.

2.1 COMPONENTS

The biosensor sensing mechanism depends upon the availability of the components. The combination of highly specific biological reactions with an appropriate transducer gives a number of advantages to biosensors, such as the high sensitivity and specificity of the biological element. The characterization of components and assemblies are used in electronic biosensors (Figs. 2.3 & 2.4). The components are designed to detect, record, convert, process and transmit the information regarding a physiological change or process; utilize biological materials to monitor the presence of various chemicals in a substance (*Eg.* an analyte) or combine to an electrical interface (i.e., a transducer) with the biologically sensitive and selective element.

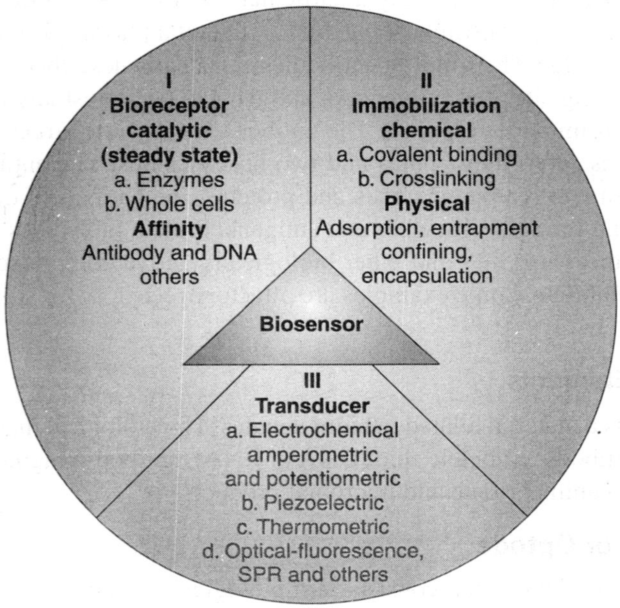

Fig. 2.3 *Biosensor – Components.*
(Source: http://www.lsbu.ac.uk/biology/enztech/biosensors.html)

The required components for a biosensor are bioreceptor molecules, biological material (biocatalyst), transducer, amplifier, processor, recorder and display device (Fig. 2.4).

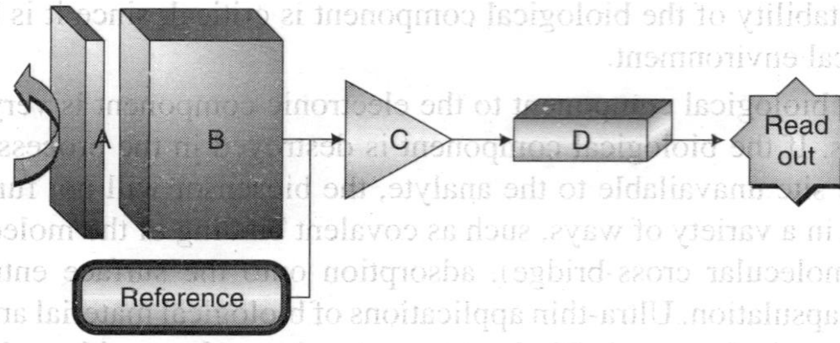

Fig. 2.4 *The main components of a biosensor. (A) biocatalyst (B) transducer (C) amplifier (D) processer.*

2.1.1 Bioreceptor Molecules

A receptor is used to find a target. In biosensors, the target is an analyte and the receptor is some substance which is stuck to the surface and binds to the target to allow it to be detected. Any biomolecule or molecular assembly that has the capability of recognizing a target substrate (i.e. an analyte) can be used as a bio-receptor.

Enzymes have been the most widely used bio-receptor molecules. Recently, antibodies and proteins are also used as bio-receptor molecules in biosensors. The specificity of a biosensor comes from the specificity of the bioreceptor molecule used. An enzyme is a good example. It has a three-dimensional structure that fits only a particular substrate. Antibodies represent one of the major classes of protein; they constitute about 20% of the total plasma protein and are collectively called immunoglobulins (Ig). The simplest antibodies are usually described as Y-shaped molecules with two identical binding sites for antigen. An antigen can be almost any macromolecule that is capable of inducing an immune response. The antibody has a basic structural unit consisting of four polypeptide chains - two light chains and two heavy chains. The antibody binds reversibly with a specific antigen. Unlike the enzymes and proteins, the antibodies do not act as catalysts. Their purpose is to bind foreign substances and antigens. Protein molecules show specific affinity for hormones, antibodies, enzymes and other biologically active compounds. These proteins are mostly bound to a membrane. Some examples are olfactory receptors for smelling, photoreceptors for eyes, etc.

2.1.2 Biological Elements

A biological element is a material related to biomolecules. The biological material may be a protein such as an enzyme, antibody; a nucleic acid (DNA or RNA), antibody fragment, a whole microbial cell, even a plant or an animal tissue and microbial products.

2.1.3 Transducer or Optode

A biosensor is a sensing device that consists of a biological component coupled to a transducer that converts biochemical activity into the most commonly electrical signal. The physico-chemical transducer can be electrochemical (e.g., pH, polarographic, potentiometric or conductometric probes), thermal (*Eg.* thermistors), optical (fibre optic) or piezoelectric crystals.

Transducer is a device that converts energy from one form to another form, for example, telephone companies use transducers to convert sound energy into electrical energy to be carried by long distance through telephone lines and then another transducer at the receiving end to convert the electrical energy back into sound. In biosensors transducers convert the biochemical activity into electrical energy. The transducer converts a biorecognition event into a measurable signal that correlates with the quantity or presence of the chemical or biological target. The key part of a biosensor is the transducer (Fig. 2.5), which makes the use of a physical change accompanying the reaction.

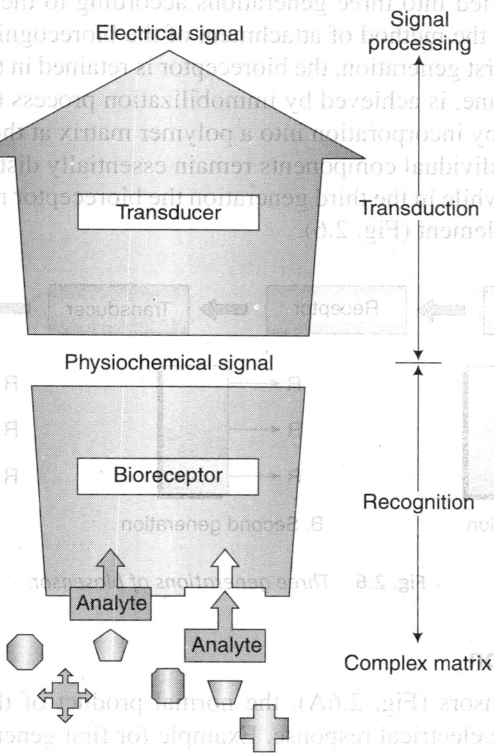

Fig. 2.5 *Transducer signaling process in biosensors.*

Depending on the analyte and bioreceptor, the transducer of a biosensor could utilize one of several mechanisms.

The transducer could rely on amperometric technology, which involves that detects changes in current. They measure currents generated when electrons are exchanged between a biological system and an electrode. Alternately, it may use a potentiometric mechanism which involves reactions that cause a change in voltage (i.e., potential at constant current) between electrodes and this change can be detected or measured.

A transducer that relies on resistive or conductive mechanisms will detect changes in conductivity between two electrodes. It can also utilize a piezoelectric mechanism. In a piezoelectric material, there is a coupling between its mechanical and electrical properties. The coupling can be used to create an electrical oscillator, the frequency of which can be varied and measured by varying a mass applied to its surface. In the case of a biosensor, that mass can change due to the reaction taking place on the surface. Thermal transducers measure changes in temperature. Optical transducers can measure light output during the reaction or a light absorbance difference between the reactants and products.

2.2 EVOLUTION

Biosensors can be classified into three generations according to the degree of integration of the separate components, i.e. the method of attachment of the biorecognition molecule or bioreceptor to the transducer. In the first generation, the bioreceptor is retained in the vicinity of the base sensor behind a dialysis membrane, is achieved by immobilization process (via cross-linking method) at a transducer interface or by incorporation into a polymer matrix at the transduction surface. In the second generation, the individual components remain essentially distinct (*Eg.* control electronics-electrode-biomolecule), while in the third generation the bioreceptor molecule becomes an integral part of the base sensing element (Fig. 2.6).

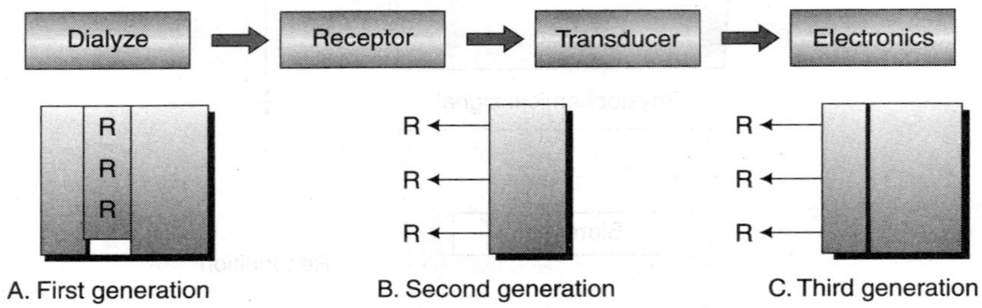

Fig. 2.6 *Three generations of biosensor.*

2.2.1 First Generation

In first generation biosensors (Fig. 2.6A), the normal product of the reaction diffuses into the transducer and causes the electrical response. Example for first generation biosensors are glucose biosensors, which were based on measuring the hydrogen peroxide concentration once it has diffused to an electrode. The electrode acts as the transducer where electron flow would produce a measurable current and was usually made out of either metal or carbon. This approach however had several problems. Firstly, the monitoring of hydrogen peroxide in this way requires a high operating potential. At this potential other unwanted electro-active substances that may be present in biological environments such as uric and other acids can react causing interference. Secondly oxygen has a limited solubility in many biological fluids which in turn limits the enzymatic reaction and thus the response of the biosensor.

Enhanced versions of the first-generation biosensor (Fig. 2.6A) have been developed which counteract these effects. One such design uses carbon electrodes coated in a graphite paste into which metallic particles have been dispersed. Metals such as ruthenium, rhodium, platinum and iridium have been shown to produce improved results over a standard carbon paste and alloys such as ruthenium-platinum enhance the sensitivity of the biosensor even more. Electrodes are constructed in this manner reduce the operating potential required for the reduction of the hydrogen peroxide at the electrode thus eliminating the unwanted reactions of the other electro-active substances present in the environment. The graphite paste that coats the electrodes is a mixture of graphite powder and an organic mixing fluid. Oxygen has a solubility many times greater in certain organic solvents than in water and using fluoro-chemicals in particular as the mixing fluid for the graphite paste has improved the response of the first-generation biosensor to the point where it has been shown to rival many second-generation biosensors. The organic fluid also helps to increase the stability of the biosensor over long periods of time particularly thermo stability. This is due to the fact that enzymes trapped within this organic fluid are much more stable than when they are immersed in water. Dehydrated enzymes have a very rigid configuration and free water molecules are usually required for enzyme inactivation. These improved biosensors are almost very easily prepared as their forerunners making them suitable for use in mass-production and disposable applications.

2.2.2 Second Generation

Second generation biosensors (Fig. 2.6B) involve specific mediators between the reaction and transducer in order to generate improved response. These biosensors avoid the limitations of the first generation devices by replacing oxygen with an external mediating electron acceptor which would shuttle electrons to the transducer from the enzyme. Problems with low oxygen solubility and fluctuating oxygen levels are thus eliminated by this method. Other problems still exist, however the biosensors are still sensitive to interference caused by unwanted substances reacting at the electrode and many mediators are toxic thus limiting the *in-vivo* applications of these biosensors.

2.2.3 Third Generation

In third generation biosensors (Fig. 2.6C), the reaction itself causes the response and no product or mediator diffusion is directly involved. These biosensors have achieved a direct connection between the enzymatic reaction and the transducer without need for diffusion. The enzyme is immobilised at the electrode. Since nothing else needs to be added to the environment, repeated or continuous measurement of blood glucose levels can be much more easily put into effect. This direct connection can be achieved by wiring the enzyme to the electrode with a substance containing electron relays, such as a long-chain polymer.

Only few enzymes or proteins can exhibit direct electron transfer properties and a technique known as Protein Film Voltammetry (PFV) has been shown to be an excellent way of exploiting this. In PFV, the enzyme or protein being utilised is adsorbed as a monolayer film onto the electrode. Each molecule in the film is orientated for optimal electron transfer. Electrons can be driven in and out of the active sites by applying a potential (Fig. 2.7).

Proteins utilized in this technique include cytochrome, myoglobin, haemoglobin and glucose oxidase. Bienzymatic biosensors have also been developed usually, as they are incorporated with oxidase and peroxidase enzymes, the latter catalysing the production of hydrogen peroxide from the reaction. The catalytic reduction of the hydrogen peroxide then provides an electrical signal that can be measured. Glucose oxidase has also been used to create a third generation biosensor by immobilising it directly onto an oxidised boron-doped diamond electrode. Carboxyl groups on the surface of the electrode covalently link with the glucose oxidase.

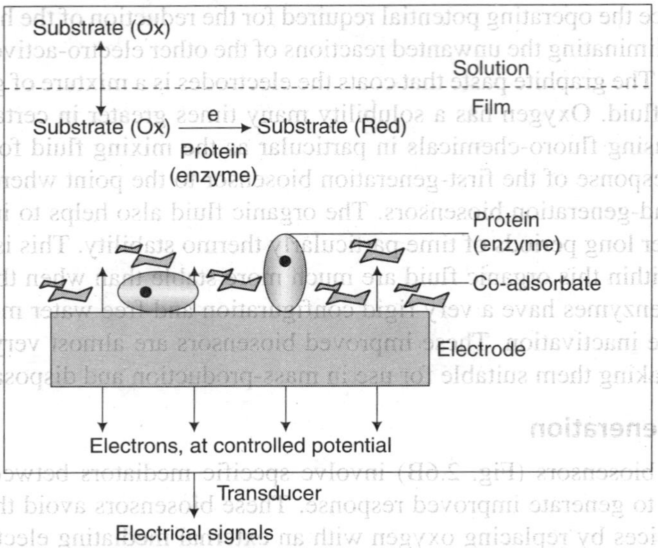

Fig. 2.7 *Biosensor electrode based on Protein-Film Voltammetry (PFV). (Source: http://www.jstage.jst.go.jp/article/analsci/20/4/603/_pdf)*

2.3 LIMITATIONS

1. The first is the instability of the biological sensing component (enzyme, antibody, tissue, etc.), which can lose its activity in hours or days depending on the nature of the molecule and exposure to environmental stresses, such as pH, temperature or ionic strength.
2. The second limitation is on the size of the physico-chemical transducers being used in biosensors.

2.4 ADVANTAGES

1. They can measure nonpolar molecules that do not respond to most measurement devices.
2. They allow rapid continuous control.
3. Speed of response (typically less than a minute) and ease of use is the main advantages offered by biosensors.
4. Typically the smaller the device, the faster and more sensitive is the response.
5. Biosensors can easily detect analytes in the micromolar to nanomolar range.
6. Biosensors can serve exceptionally well in emergency.
7. Situations or for on-site field applications.

2.5 DISADVANTAGES

1. Heat sterilization is not possible as this would denature the biological part of the biosensor.
2. The membrane that separates the reactor media from the immobilized cells of the sensor can become fouled by deposits.
3. The cells in the biosensor can become intoxicated by other molecules that are capable of diffusing through the membrane.
4. Changes in the reactor broth (i.e., pH) can put chemical and mechanical stress on the biosensor that might eventually impair it.

2.6 BENEFICIAL FEATURES

A successful biosensor must possess the following beneficial features:

1. The biocatalyst must be highly specific for the purpose of the analyses be stable under normal storage conditions except in the case of calorimetric enzyme strips and show good stability over a large number of assays (i.e. much greater than 100).
2. The reaction should be as independent of physical parameters as like stirring, pH and temperature. This would allow the analysis of samples with minimal pre-treatment. If the reaction involves cofactors or coenzymes these should preferably be co-immobilized with the enzyme.
3. The response should be accurate, precise, reproducible and linear over the useful analytical range without dilution or concentration. It should also be free from electrical noise.
4. If the biosensor is to be used for invasive monitoring in clinical situations the probe must be tiny and biocompatible having no toxic or antigenic effects. If it is to be used in fermenters it should be sterilisable. This is preferably performed by autoclaving but no biosensor enzymes can presently withstand such drastic wet-heat treatment. In either case the biosensor should not be prone to fouling or proteolysis.
5. The complete biosensor should be cheap, small, portable and capable of being used by semi-skilled operators.
6. The biosensors have been considered to be superior and more sensitive, in comparison to physical instruments due to the following reasons:
   a. In a biosensor the immobilized biological material is present in intimate contact of a suitable transducer so that the biochemical signal is quickly converted into an electrical signal.
   b. The immobilization of biomolecules permits reuse of these molecules (which are expensive) and allows simplification of the entire apparatus.
   c. The biological sensing element is present in a small area and is very sensitive, thus facilitating analysis of substances in small quantities.
   d. Biosensors may be developed according to specific needs and can be highly specific or show broad spectrum.

2.7 DESIRABLE CHARACTERISTICS

Several factors are needed to be considered in the design of a successful biosensor. These include:

1. The biological response element must be highly specific to the substance.
2. Any substances in the biosensor should remain stable when stored under normal conditions.
3. The reaction that makes up the biological response should be as independent of extra physical factors (*Eg.* temperature or pH) as possible. This will cut down on pre-treatment of any samples. Similarly, if the reaction requires extra cofactors or coenzymes, these must be immobilised in the biosensor along with the main biological response element.
4. The response from the biosensor should be as accurate and noise free as possible.
5. Biosensors are used in dangerous monitoring that is non-toxic and non-antigenic.

2.8 PERFORMANCE FACTORS

Electronic components and assembly tests are needed to meet various performance criteria, which include speed, ease of use (i.e., testing, calibration, and maintenance) by non-technical personnel, specificity to the target analyte, sensitivity, accuracy, resolution, and repeatability, dynamic range (i.e., high analyte concentrations will not degrade sensor usability), environmentally robust (i.e., relatively insensitive to temperature, electrical noise, physical shock, vibration, etc.), lifetime, adaptability, safety and integrity.

2.8.1 Selectivity

Selectivity is the most important criteria of biosensors by having the ability to differentiate between the different substances. Such behavior is a function of the selective component although sometimes the operation of the transducer contributes to the selectivity. In biosensors the selectivity improvement may be required at two levels; the interface between the transducer and the biological receptor (reducing interference) and new receptors.

The mediators are used in amperometric biosensors to improve the performance. The pyrroloquinoline quinone is as a natural mediator used with glucose oxidase electrode for the measurement of sugar in drinks. Electrocatalytic detection of the products of enzymatic reactions may be enhanced by the use of chemically modified electrodes such as rhodinised or hexacyanoferrate (modified carbon). Prussian blue, coating on the electrode which may be used for amperometric detection of hydrogen peroxide at both oxidative and reductive potentials in enzyme electrodes for lactate and glucose. Individual ligands or arrays of ligands in conjunction with pattern recognition techniques will be used to design a sensor with improved selectivity.

2.8.2 Sensitivity

The sensitivity is usually sub milli molar, but in special cases it can go down to the 10–15 M range.

2.8.3 Accuracy

Accuracy needs to be better than ±5%. Advanced computational techniques allow modeling both electron transfer reactions and receptor binding interactions with increasing accuracy. This not only enhances understanding of the receptor/transducer interface but also allows designing of new receptors based on biological molecules.

2.8.4 Nature of Solution

Conditions such as pH, temperature and ionic strength must be considered.

2.8.5 Response Time

Biosensors response time is 30 seconds or more.

2.8.6 Recovery Time

Recovery time is the time that elapses before the sensor is ready to analyse the next sample. It must not be more than a few minutes.

2.8.7 Lifetime

The biosensor working lifetime is usually determined by the stability of the selective material. For biological materials this can be as short as a few days although it is often several months or more.

2.9 BIOSENSOR DESIGN

Characterization of electronic sensor components can facilitate the development of sensors. The sensors provide early detection of chemical and biological components. Accurate characterization of electronic sensor components can help to speed up development. Biosensors, transducers and other electronics are being incorporated in a way that allows them to be produced via integrated circuit manufacturing processes. Many types of biosensors, transducers and detection systems are being investigated for various applications and also for more general use. Some of the designs being investigated generate an electrical output which is intended to speed up measurements and meet ease-of-use criteria. Proper testing of the electrical portions of these sensors is needed to qualify the designs for further development.

The components are designed to detect, record, convert, process and transmit information regarding a physiological change or process; utilize biological materials to monitor the presence of various chemicals in a substance (*Eg.* an analyte); or combine an electrical interface (i.e., a transducer) with the biologically sensitive and selective element.

The transducer portion of a biosensor converts a bio-recognition event into a measurable signal that correlates with the quantity or presence of the chemical or biological target of interest (Fig. 2.8).

Immunochemical sensors can accurately recognize protein-based agents and electrochemical sensors can identify DNA. However, most of these sensors require significant sample preparation. Inorganic sensor materials like tin oxide, gold, platinum and semiconductors are used for in-line measurements for toxic airborne gases. These chemical sensors can react quickly to their selected targets and provide a convenient electrical output. However, there is no way to detect the presence of a bio-pathogen with a simple enzyme-based biosensor, or to sense the DNA inside the pathogen. For field use, it is highly desirable to have a biosensor that can provide a high level electrical signal for direct readout on a visual indicator.

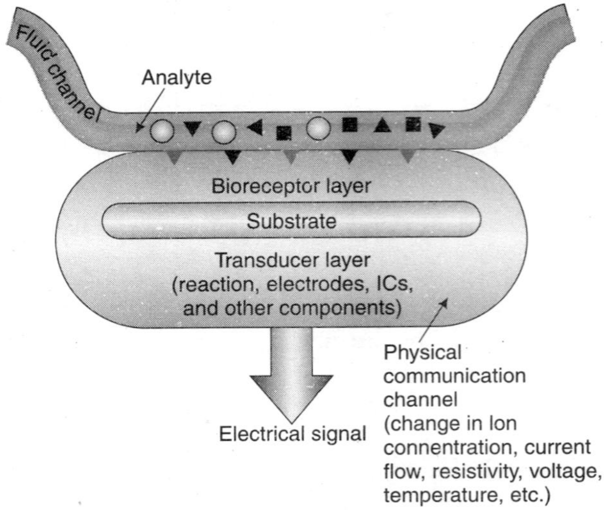

Fig. 2.8 *Representation of a generic biosensor.*

For fast detection with a readable electrical output to develop biosensors that combine various processes, such as pathogen cell lysis, debris removal, DNA cleanup, polymerase chain reaction (PCR) and fluorescence detection. To combine these and similar processes with semiconductor devices to create compact portable units for field must be used. The surface plasmon resonance (SPR) is used to detect biological molecules such as protein and DNA. In addition living cells that can react functionally to the presence of both biological and chemical threat agents are being placed on chips. Liquid-crystal detectors can be used for identification of pathogens.

Depending on the analyte, bioreceptor and transducer portion of a biosensor could utilize one of several mechanisms. The transducer could rely on amperometric technology, which involves devices that detect changes in current. They measure currents generated when electrons are exchanged between a biological system and an electrode (Fig. 2.8). Alternately, it may use a potentiometric mechanism which involves reactions that cause a change in voltage (i.e., potential at constant current) between electrodes and this change can be detected or measured.

A transducer that relies on resistive or conductive mechanisms will detect changes in conductivity or resistivity between two electrodes. A capacitive transducer relies on a biorecognition reaction, which causes a change in the dielectric constant of the medium in the vicinity of the bioreceptor. Such a capacitance measurement method can be used as a transducer.

A transducer can also utilize a piezoelectric mechanism. In a piezoelectric material there is a coupling between its mechanical and electrical properties. The coupling can be used to create an electrical oscillator, the frequency of which can be varied and measured by varying a mass applied to its surface. In the case of a biosensor the mass can change due to the reaction taking place on the surface. Thermal transducers measure changes in temperature.

Biosensors with optical mechanisms correlate changes in concentration, mass, or number of molecules to direct changes in the characteristics of light. One of the reactants or products of the biorecognition reaction has to be linked to calorimetric, fluorescent, or luminescent indicators. An optical fibre might perform this linkage by guiding light signals from the source to the detector.

2.9.1 Device Characterization

Biosensor development programs generally aim to overcome the design limitations in current biosensor systems. For example, one of the challenges involved in biosensor design is achieving a stable, reproducible interface between the biological affinity elements and an inorganic transducer element in the sensor.

The desire to miniaturize biosensors for handheld portability, still achieve adequate sensitivity, imposes significant technical challenges in the coupling of biomolecules to transducer surfaces. Therefore, fast and accurate electrical characterization of biosensor devices and mechanisms in the development lab is essential for qualifying new designs. Because of the complexity in extracting cell and tissue signatures of agent activity and response, it is often desirable to conduct direct current-voltage (I-V) characterization on key components of the biosensor. I-V characterization requires only a small fraction of the time needed for most types of functional testing, but is a powerful predictor of full fledged operation. For example, I-V data can be used to study anomalies, locate maximum or minimum curve slopes, and perform reliable analyses. Depending on design specifications, I-V characterization is often suitable for sensors based on amperometric, potentiometric, conductive, resistive and thermal principles. Usually I-V testing applies a voltage or current to the device under test (DUT) and measures its response to that stimulus (Fig. 2.9). The test procedures may involve probing of integrated circuits to apply the stimulus to certain connection pads and measure the response on others.

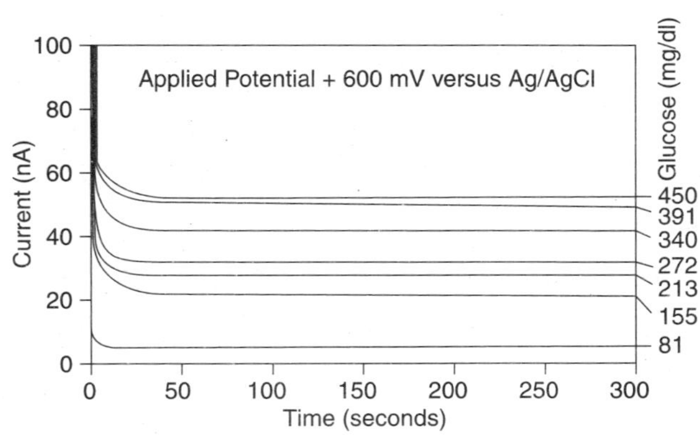

Fig. 2.9 *Electrical characteristics of a prototype electrochemical transdermal sensor based on a glucose oxidase reaction with blood plasma from capillaries.*

Depending on the Device Under Test (DUT) signal the glucose oxidase levels may be estimated. In this case, highly sensitive source, measurement instruments and test techniques that minimize external sources of error are necessary. When an optical mechanism is involved, I-V characterization may also involve simultaneous measurements of the wavelength or intensity of a light output with a photodetector. This is called L-I-V testing.

2.9.2 Device Performance

The biosensors used by medical practitioners, military personnel and public safety forces will be part of a portable system. This places restrictions on the operational power requirements of the sensors and may dictate the level of voltage or current output that can be provided to the measurement circuitry. In battery operated systems sensor output current can range from nanoamps to milliamps and voltage from nanovolts to volts. Different measurement techniques and tools are required for signal levels at the opposite ends of such wide ranges.

2.9.3 Voltage

Characterizing biosensor devices at voltage levels greater than 100 μV should be relatively easy. A sampling data acquisition system based on a PC plug in board may provide adequate resolution as would many programmable digital multimeters (DMMs) and self contained data loggers. For example, most laboratory grade DMMs provide enough range and resolution to make voltage measurements from 1 μV to 1000 V. For the PC based data-acquisition board solution measure over a variety of voltage levels depending on the resolution of the analog-to-digital (A/D) converter and its gain (Table 2.2).

Table 2.2 Voltage measurement resolution and maximum ranges for different A/D converter resolutions. *(Source: Data Acquisition and Control Handbook.)*

Resolution (volts input)	*8[a] 256[b]*	*10[a] 1024[b]*	*12[a] 4096[b]*	*16[a] 65,536[b]*
0-10	39.06 mV	9.765 mV	2.441 mV	152.59 μV
0-5	19.53 mV	4.883 mV	1.221 mV	76.29 μV
±10	78.12 mV	19.530 mV	4.883 mV	305.20 mV
±5	39.06 mV	9.765 mV	2.441 mV	152.59 μV
±2.5	19.53 mV	4.883 mV	1.221 mV	76.29 μV
±1.25	9.76 mV	2.441 mV	610.350 μV	38.15 μV
Maximum input, 1-10[c]	9.960 V	9.990 V	9.9976 V	9.99985 V

[a]= Converter bits (n); [b]=Output states (2^n); [c]= Maximum input=resolution ¥ 2^n-1.

For measurements below 1 μV a nano-voltmeter should be considered instead of a data acquisition board or DMM solution. For example, the model 2182 by Keithley Instruments Inc. (Cleveland) is a low-noise digital nano-voltmeter with an A/D resolution in the 24-bit range. This type of instrument is optimized for accurate low level voltage measurements, even when the signal is approaching the theoretical (lower) limit associated with sensors that have low output impedance. Sensors with low output impedance require a voltmeter with a high input impedance to avoid measurement errors. Although the input impedance of a nano-voltmeter is similar to a DMM, it has much lower voltage noise and drift. This gives it much better voltage sensitivity, and it may be able to read down to 1 nV.

2.9.4 Noise

Noise is often a problem for tools used to measure voltage. Significant errors can be generated by noise that originates in the sensor, the measuring instrument, and sources external to the test circuit. External sources include electromagnetic fields (EMFs), measuring circuit ground loops and thermal EMFs. Johnson noise is another form of thermal noise that occurs in every electrical component. Johnson noise establishes the ultimate limitation of the measurable signal level. For accurate measurements, the noise sources should be minimized as much as possible.

In some test environments, electrical noise is difficult to avoid. When that happens the measurement compensation techniques are needed. These are included in most bench top DMMs and nanovoltmeters to minimize electrical noise from a.c. lines and from random noise sources. These techniques are less likely to be available in PC-based data-acquisition systems but knowledgeable users can program a system for signal averaging to help reduce random noise and use longer A/D integration periods to minimize a.c. line noise.

2.9.5 Current Measurements

Different types of amperometric devices require different characterization approaches. Electrical currents can be measured with data-acquisition systems but the method of acquisition selected will depend on the current level and number of required measurement channels. Otherwise, I-V characterization of a device using a current loop is uncomplicated as long as the current source output voltage is high enough to overcome any test lead resistance. The corollary is that current loops are ideal when there is an appreciable distance between the signal source and the instrumentation (Fig. 2.10).

It may be desirable to qualify the output of a biosensor assembly in terms of engineering units such as concentration of the target analyte in milligrams per deciliter. In the case of an amperometric output the current is an indicator of the phenomenon actually being measured. Therefore it is desirable to have instrumentation that makes it easy to do the conversion. This requires that the instrument be able to perform internal calculation and scaling features that convert current readings to appropriate engineering units. For example, a transducer utilizing a 4–20 mA current loop might be calibrated for a concentration of zero mg/dl at 4 mA and a full scale concentration of 100 mg/dl at 20 mA.

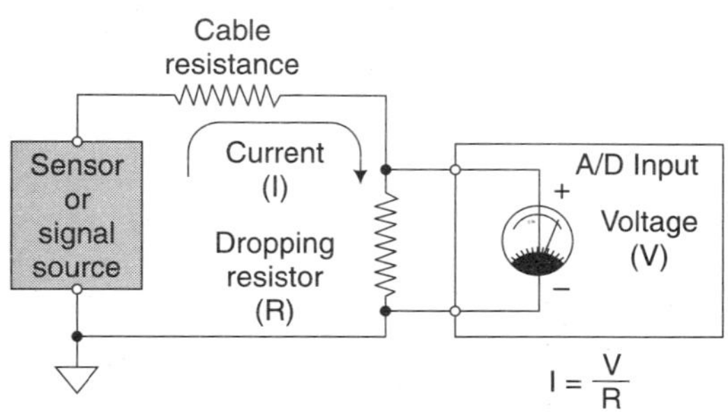

Fig. 2.10 *Current measurement using a dropping resistor and voltmeter.*

In the early stages of product development, it may not be necessary to do this conversion when the significance of the transducers output current level is already understood. In this case, the current reading is the parameter of interest and a voltage stimulus may serve as a proxy for the biological event during I-V characterization. Under these circumstances the current signal can be relatively high but may require care in selecting a dropping resistor for the measurement. The current passes though the dropping resistor at the input stage of a data acquisition system and voltage is measured across the dropping resistor to determine the current level. This type of scaling is a common feature of data-acquisition systems.

However unlike voltage measurements current measurements may be subjected to "voltage burden" errors. Voltage burden is defined as the voltage drop across the input of an ammeter when it is inserted into a circuit. The dropping resistor (R) and A/D voltage input constitute an ammeter and the current flow can be calculated from the voltage drop across the resistor. Noise effects, similar to those for voltage measurements, must also be considered and are exacerbated by voltage burden.

The resistance value will normally be selected to provide a voltage drop corresponding to the full input range of the A/D board when the maximum anticipated current flows. For example, a 20-mA current produces a 10-V drop across a 500-ž resistor. Note that the sensor (i.e., current source) must be capable of developing a minimum output potential of around 10–11 V in order to achieve the full voltage drop across the resistor. This may not be a problem in the lab, but if a portable biosensor circuit is powered by only 6 V, it cannot drive more than 12 mA through the resistor. Furthermore, the resistor must have an adequate power rating (I2R) for the current and resistance values.

Issues such as these can be avoided by performing current measurements with bench top instruments that use either a small dropping resistance or a feedback ammeter circuit. In the former case, highly sensitive voltage measurements allow a small dropping resistance value. A feedback ammeter avoids this problem because its input circuitry consists of an operational amplifier, which has very low input resistance and a voltage burden that typically ranges from about 10 μV to 1 mV. This type of current-measurement circuit is typically used in a complete biosensor instrument.

DMMs typically use dropping-resistor ammeter circuits, whereas pico-ammeters and electrometers use feedback ammeter circuits. In either case, making connections to the signal source is straightforward, and they are suitable for measuring currents up to several amperes. At the other end of the scale, DMMs can measure currents down to about 10 nA, and picoammeters can measure currents as low as 10 fA. Another advantage of these instruments is their built-in signal-conditioning circuitry. Some DMMs can provide voltage, current, resistance, and temperature (thermocouple or thermistor) measurements up to 200 channels.

2.9.6 Source-Measure Instruments

In I-V characterization, the integration of a DC source and measuring instrument can be problematic because of intricate triggering issues. Such issues can often be avoided by using a tightly integrated source-measure unit. These high-precision instruments can act as either a voltage or current source with sweep, pulse, and compliance-limit capabilities, and simultaneously measure I and V parameters. Typical resolutions are in the range of micro volts and picoamperes.

The bipolar voltage and current sources of these instruments are controlled by a microprocessor, which makes I-V characterization much more efficient and simplifies instrumentation setup. Many different test sequences can be stored in its program memory and executed with a simple trigger signal. Test data can be stored in a buffer memory until an I-V sweep is completed and then downloaded to a PC for processing and analysis.

2.9.7 Cabling, Conductors and Capacitance

The connections between the instrumentation and the DUT are important parts of a measurement system. Understanding and managing the limitations of these connections is crucial for accurate measurements. Noise sources, cable length, and cable capacitance can affect the quality of any measurement, but the lower the signal level the more important these issues become. To minimize problems, the measurement circuit, its cables, and its connectors should be matched to the test signals. In addition, cables and test leads should be carefully routed and mounted.

1. Cabling

When evaluating a cable for the measurement application, several issues should be considered. The electrical noise present in the test environment should be quantified. Noise can be defined as any undesirable signal that is impressed upon a signal of interest.

Sources of electromagnetic noise include a.c. power lines, motors and generators, transformers, fluorescent lights, cathode-ray tube displays, computers, radio transmitters, etc. Depending on the nature of the signal and the noise, it may not be possible to separate them once the signal has been acquired at the instrumentation input terminals.

To the extent possible, test leads and cables should be routed to minimize their exposure to noise. The leads should then be mounted rigidly in place so that they cannot move and cause the generation of spurious voltage in the presence of electromagnetic fields.

In addition, the distance between the signal source and measurement system terminals should be measured. Wire has an electrical resistance that is dependent on its composition, length, and diameter. Resistance increases with increasing length and decreasing wire diameter. This resistance is a component of the total cable effects that become part of the analog input of a measurement circuit (Fig. 2.11). High cable resistance in conjunction with low A/D input resistance can result in a significant voltage drop through the interconnect wiring, resulting in measurement errors.

Lastly, it should be determined whether the data-acquisition channel has a single-ended or differential input. Single-ended signals (i.e., those referenced to ground) can be transmitted with two wires or with a shielded cable where the shield is tied to ground. For differential signals, at least two wires are needed to transmit the signal, which consists of a signal high and a signal low, neither of which is referenced to ground. Two individual conductors are sufficient to transmit the signal, but a twisted pair or shielded twisted pair provides greater noise immunity.

2. Conductors

The conductors used in shielded or unshielded cable can be made of solid or stranded wire. Solid wire results in minimum signal attenuation, but stranded conductors provide more flexibility and may be easier to route and mount. Conductors may consist of bare copper that is either plated with silver or tinned with solder. Connector and conductor materials should match to minimize resistance and thermally generated EMFs.

For the highest signal integrity, cables with shielded conductors should be used. Shielding reduces electromagnetic noise picked up by signal leads. It is also helpful in reducing electromagnetic radiation from conductors carrying high frequency signals. Shielding is constructed with different types of wire braid or a combination of wire braid and foil. Multilayer or multibraid shields are more effective than a single layer in attenuating signal pickup and radiation. However, this tends to make cables more stiff and difficult to route and mount.

Several points should be considered when selecting shielded cable. For instance, higher-frequency noise is difficult to attenuate and requires more-elaborate shielding.

Simple spiral wire-wrap foil is the least effective type of shielding. Tight braiding, double braiding, or braiding plus foil offer more effective shielding. Lastly, caustic atmospheres, moisture, etc., can reduce the effectiveness of shielding. In some cases, these contaminants can leach into a cable and degrade the shielding far beneath the outer insulating jacket. Testing in such environments should be avoided, if possible.

3. Capacitance

For many DUTs, the output signal can be modeled as a voltage source in series with a resistance. Similarly, an analog instrument input can be modeled as a meter in parallel with an input resistance. During a measurement, the instrument input absorbs a small bias current that the source must be able to supply. The interconnecting cable is an essential part of this circuit and can introduce resistance, capacitance, and inductive effects that depend on length, gauge, composition, routing, and the physical environment.

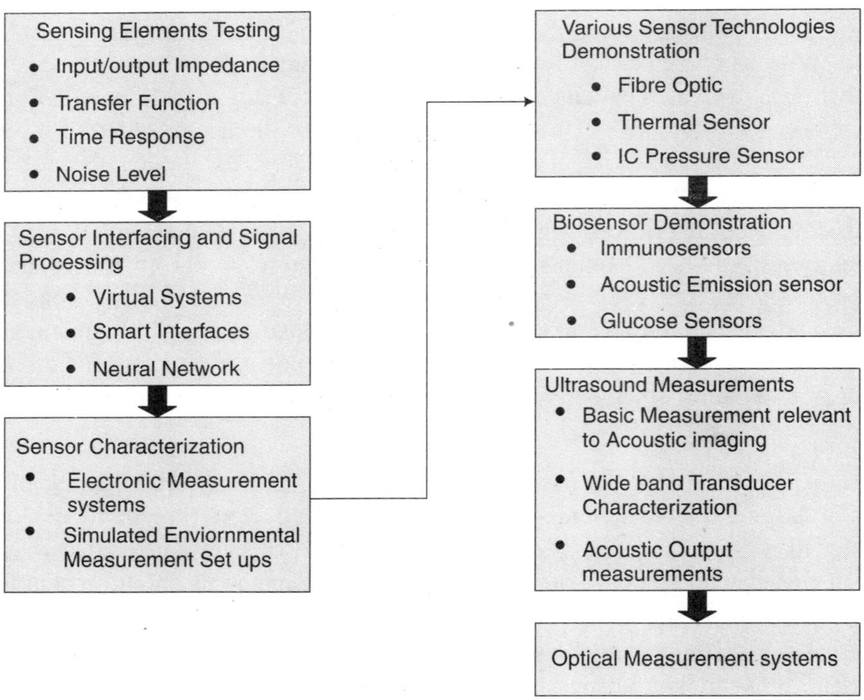

Fig. 2.11 *Testing process of biosensors.*

For high speed DUTs, rapidly changing signals, circuit inductance, and capacitance can be serious obstacles to measurement speed, even if signal source and instrument impedances are properly matched. Often, spurious capacitance is more of a problem than inductance. Signals originating from a high-impedance source take longer to stabilize at the instrument input because the limited current level of the signal requires more time to charge the cable capacitance. In that case, taking a measurement before the signal has settled leads to erroneous readings.

2.10 DESIGN VARIABLES

2.10.1 Immobilization Methods

Four methods are used for immobilizing enzyme for use in a biosensor: (1) adsorption (2) entrapment (3) covalent coupling and (4) cross-linking. These four methods are compared in Table 2.3. Among various methods, the cross-linking method is most frequently used because it has the advantage of the covalent bonding yet the cost is inexpensive.

2.10.1.1 Effects of Immobilization

With immobilized enzymes the measured reaction rate depends not only on the substrate concentration and the kinetic constants K_M and V_{max} but also on so-called immobilization effects. These effects are due to the following alterations of the enzyme by the immobilization process.

2.10.1.1.1 Change in conformation

Conformational changes of the enzyme caused by immobilization usually decrease the affinity to the substrate (increase of K_M). Furthermore, a partial inactivation of all, or the complete inactivation of a part of the enzyme molecules may occur (decrease of V_{max}). These two cases of a conformation-induced drop of V_{max} may be distinguished by measuring the activity of the resolubilized enzyme or by titration of the active center with an irreversible inhibitor.

2.10.1.1.2 Change in microenvironment

Ionic, hydrophobic, or other interactions between the enzyme and the matrix (micro-environmental effects) may also result in changed K_M and V_{max} values. These essentially reversible effects are caused by variations in the dissociation equilibria of charged groups of the active center.

2.10.1.1.3 Non-uniform distribution

A non-uniform distribution of substrate and/or product between the enzyme matrix and the surrounding solution affects the measured (apparent) kinetic constants.

2.10.1.1.4 Reaction and diffusion

In biosensors the biocatalyst and the signal transducer are spatially combined, i.e., the enzyme reaction proceeds in a layer separated from the measuring solution. The substrates reach the membrane system of the biosensor by convective diffusion from the solution. The rate of this external transport process depends essentially on the degree of mixing. In the multilayer system in front of the sensor the substrates and products are transferred by diffusion. Slow mass transfer to and within the enzyme matrix leads to different concentrations of the reaction partners in the measuring solution and in the matrix. Diffusion and the enzyme reaction do not proceed independently of one another; they are coupled in a complex manner.

2.10.2 Outer Membrane

The outer membrane has to be compatible with the medium into which it will be placed. Therefore, the requirement will be different depending on the nature of the measurement medium. For example, the outer membrane for biosensor used in liquid samples should be different from that intended for implantation application. For the latter application, bio-compatibility becomes an issue (the rejection of the sensor by body may occur).

The outer membrane should offer low diffusional resistance to analytes while the resistance should be high for macromolecules.

For long term continuous use applications, the fouling of the membrane must be minimal. The fouling causes an increase in the diffusional resistance of the analyte and thus the signal of the sensor changes as the fouling progresses. If microorganism grows on the surface of the outer membrane, the passage of oxygen to the enzyme layer is hindered which makes the sensor to behave erroneously.

Table 2.3 Comparison of four enzyme immobilization methods.

	Adsorption	*Entrapment*	*Covalent coupling*	*Crosslinking*
Matrix material	ion exchange resins, active - charcoal, silica - gel, clay, aluminum oxide, porous glass.	alginate, carageenan, collagen, polyacrylamide, gelatine, silicon-rubber, polyurethens	agarose, cellulose, PVC, ion-exchange resins, porous glass	Cross-linking agents: glutaraldehyde, bisisocyanate,
Nature of bonding	reversible; changes in pH, ionic strength may detach the enzyme	physical entrapment	chemical bonding	entrapment; functionally inert proteins are often used together (BSA, gelatin)
Enzyme loading	low	low	high	high
Enzyme leakage	some	some	very low	low
Loss of activity	negligible	negligible	significant	small
Cost	inexpensive	inexpensive	expensive	inexpensive

2.10.3 Optimization of Biosensor

Variations of the diffusion resistance of the semipermeable membrane are being used to optimize the sensor performance and semipermeable membrane influences the response time and sensitivity. In contrast, thicker membranes such as polyurethane or charged material, significantly increase the measuring time, but may also lead to an extension of the linear measuring range. Table 2.4 lists some of the commercially available membranes that can be used as the outer membrane.

Table 2.4 Available pre-cast membranes.

	Collagen	*Polycarbonate (Nucleopore)*	*Cellulose acetate*
Characteristics	A hydroxylic natural protein	Uniform pore size	Slightly negative due to $-COO^-$
Derivatizability	easy		Easy
Temperature stability	O.K. at room T unstable at 37°C	Stable	Stable
Selectivity	Exclude protein	Exclude protein	Exclude protein, retard transport of anionic species
Source	Sigma	Nucleopore	Amicon

2.10.4 Inner Membrane

The inner membrane should be permeable and selective to target species (for example H_2O_2 only for the current glucose sensor). Also, it should be as thin as possible and stable for long-term use. Some of the solution castable membranes and their characteristics are compared in Table 2.5.

Table 2.5 Solution castable membranes.

	Cellulose acetate	*Nafion*	*Polyurethane*
Characteristics	–	Perfluorosulphonic acid ionomer, Negatively charged	Widely varying MW dep. on source
Solvent	Acetone, Cyclohexanone	Low MW alcohols	Dissolves in 98% tetrahydrofuran, 2%Dimethylform-amide
Coating method	Dip coating, in-situ formation	Dip coating as thin as 1000A (with 5% solution)	Dip coating
Other characteristics	–	Tend to adsorb proteins & Cations, Not useful as an outer membrane	Biocompatible, retard glucose access to enzyme layer, passes O_2 well

2.10.5 Effect of Enzyme Loading

2.10.5.1 Internal diffusion

Usually in the operation of biosensors, the flow conditions are adjusted to provide a mass transfer rate from the solution to the membrane system faster than that of in the enzyme layer (the internal mass transfer). In the immobilized enzyme layer, reaction and diffusion occur simultaneously. Therefore, rigorous modeling is required to fully characterize the behavior of a biosensor. The key question in designing a biosensor is: (1) how thick should the enzyme layer be? and (2) how much enzyme has to be placed in the layer? Although rigorous modeling is required to fully characterize the behavior of biosensors, the design can be carried out by considering limiting cases.

2.10.5.2 Enzyme loading factor (f_e)

The key variable for determining limiting cases is the enzyme loading factor defined by f_E:

$$fz = \frac{(V_{max}/K_M)}{D_S/d^2} \qquad \textit{... eq.-2.1}$$

where D_s and d are diffusivity of the substrate and the enzyme immobilization layer thickness.

The enzyme loading factor is essentially the ratio of the reaction rate (for the case when $K_M >>$ S; or first order reaction) to the diffusion rate (when $k = D_s/d$ and the mass flux J is expressed as k(dS/dx)).

2.10.5.3 Diffusion control

f_E > 25 When the rate of reaction is much faster than the rate of diffusion, the bottleneck of the transport process is the diffusion. In such a case, we say, the biosensor is operated in a diffusion-controlled regime. The condition is:

$$f_E > 25 \qquad \textit{... eq.-2.2}$$

When this condition is met, the sensor signal depends on the diffusion process. This means the sensor output signal is linear with the analyte concentration, and is independent of the reaction rate of the enzyme layer.

2.10.5.4 Reaction control

When the rate of reaction is much slower than the rate of diffusion, the bottleneck of the transport process is the reaction. In such a case, we say, the biosensor is operated in a reaction-controlled regime. The condition is:

$$f_E > 25 \qquad \textit{... eq.-2.3}$$

When this condition is met, the sensor signal depends on the reaction rate and independent of the diffusion rate. This means that the sensor output signal depends on the reaction rate expression i.e., the Michaelis-Menten equation. Therefore, the sensor output signal will be non-linear with the analyte concentration.

2.10.5.5 Advantages of diffusion control

There are many advantages of diffusion-controlled biosensor. These include:

1. Sensitivity is independent of enzyme content and activity.
2. Sensitivity is independent of inhibitors and pH variations.
3. There is small temperature effect because the activation energy of diffusion is much less than that of reaction.
4. Extended linear range. Extendible beyond $[S] = K_M$. For kinetics controlled case, linear range is achieved only for $[S] < K_M$.
5. Slower response time. The response time is determined by the diffusion time of the reaction product (sensed): d^2/D_p this is the major disadvantage of the diffusion controlled biosensor.

6. Greater functional stability. Due to the excess of enzyme in the membrane, the sensor will have higher functional stability. With diffusion controlled sensor, 2000-10,000 measurement is possible, while kinetically controlled sensors permit only 200-500 measurement.

2.10.5.6 Apparent K_m

When an enzyme is immobilized, the K_M value changes because of the changes in enzyme conformation and the micro-environment. Table-2.6 lists the K_M values for some of the immobilized enzyme systems.

Table 2.6 Apparent enzyme activity and K_M values of adsorbed layers and enzyme membranes.

Enzyme	*Immobilization*	*Apparent enzyme activity, mU/cm²*	*K_M value*
GOD*	gelatine entrapment	110	7.5
GOD	collagen, covalent	60-80	3.0
GOD	cellulose acetate	340	
GOD	cellulose acetate	>1000	
GOD	PVA entrapment	160-700	
GOD	spectral carbon, adsorbed	150-200	
GOD	carbon, covalent	50-170	3.1-19.1
-galactosidase	gelatin entrapment	1000	
Urease	cellulosetriacetate, entrapment	3-30	2.4
Cholesterol oxidase	collagen, crosslinked	3	
Creatinine amidohydrolase	cellulose acetate, covalent	1140	278
Creatine amidohydrolase	cellulose acetate, covalent	110	64.9
Sarcosine oxidase	cellulose acetate, covalent	13	2.4

2.11 APPLICATIONS

In the field of medicine, industry, agriculture, environment monitoring and biotechnology research, routine analyses using physical instruments are conducted for estimation and monitoring the levels of certain analytes (an analyte is a compound or molecule, whose presence and concentration needs to be determined and monitored). Conventional physical methods for this routine analysis do not involve the use of any living organisms or molecules of biological origin. However, for this purpose, biological molecules or living cells have been used to develop sensitive devices that are described as biosensors (Fig. 2.12).

2.11.1 Health Care

Health care is the main area of application of biosensors and chemical sensors (chemisensors). Measurements of blood, gases, ions and metabolites are regularly needed to show a patient's metabolic state.

Biosensors have been determined by samples of urine and blood being taken away to a medical analytical laboratory for classical analysis, which may not be complete for hours or even days. The use of on-the-spot sensors and biosensors enable results to be obtained in minutes at most. The latest ExacTech glucose sensor gives a reading in 12 seconds.

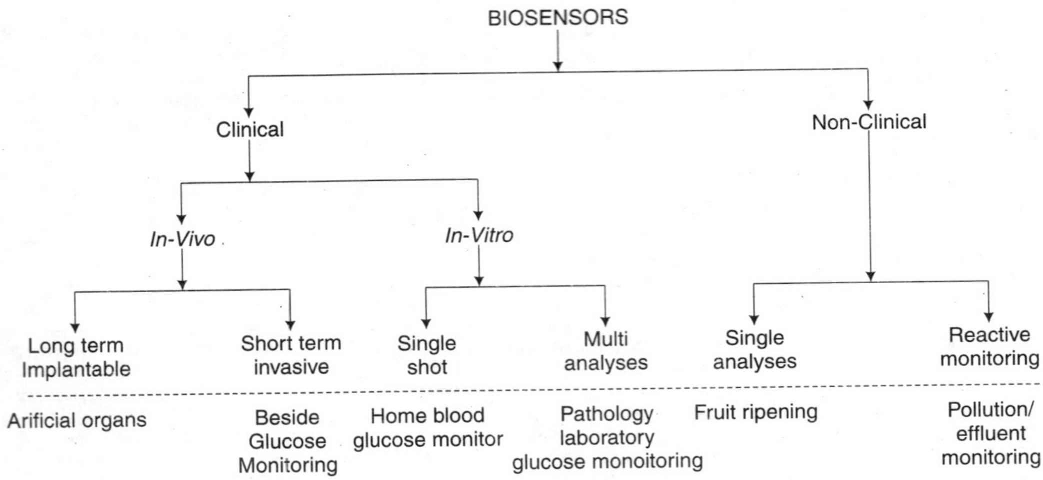

Fig. 2.12 *Potential applications of biosensors.*

Modern smart sensors, based on field-effect transistors (FETs), may combine several measurements in one sensor unit. This particularly applies in the case of ion sensors for sodium, potassium, calcium and pH. Attempts are also being made to make combination biosensors, e.g. for glucose, lactate and urea. Table 2.7 shows a list of common assays that are routinely needed for diagnostic work with patients.

Table 2.7 Common assays that is required in diagnostic medicine.

S.No.	*Analyte*	*Method of Assay*
1	Glucose, Lactate, Cholesterol	Amperometric biosensor
2	Urea, Penicillins	Ptoentiometric biosensor
3	Hepatitis B	Chemiluminescent immunoassay
4	Sodium, pH	Glass ion selective electrode

A potential dream application is to have an implanted sensor for continuous monitoring of a metabolite. This might then be linked via a microprocessor to a controlled drug-delivery system (e.g. an iontophoretic system) through the skin. Such a device would be particularly attractive for chronic conditions such as diabetes.

The blood glucose sensor would be monitored continuously and, as the glucose level reached a certain value, insulin would be released into the patient's blood stream automatically. This type of system is sometimes referred to as an artificial pancreas. The latter would be far more beneficial for the patient than the present system of discrete blood glucose analyses which involve pricking a thumb every time, followed by injection of large doses of insulin every few hours.

2.11.2 Control of Industrial Process

Sensors are used in various aspects of fermentation processes in three different ways, i.e.

1. Off-line in a laboratory

2. Off-line but close to the operation site
3. On-line in real time

At present, the main real-time monitoring is confined to such measurements as temperature and pH, plus carbon dioxide and oxygen measurements. However, biosensors which monitor a range of direct reactants and products are available, such as those for sugars, yeasts, malts, alcohols, phenolic compounds and perhaps, undesirable by-products.

Such monitoring could result in improved product quality, increased product yields, checks on tolerance of variations in quality of raw material, optimized energy efficiency, i.e. improved plant automation, and less reliance on human judgment.

Biosensors can also be used as analysers for specific chemicals for process control in a variety of industries including pharmaceuticals, chemical and petroleum, mining and construction industries. Although, the conditions in some of these industries may be too harsh for biological sensors, there is considerable promise in this area. In fermentation industry, biosensors can be used for monitoring the cultured micro-organisms producing drugs, hormones, vaccines, single cell proteins, etc. Some biosensors have already been produced for fermentation industry (e.g. Threshold TM from Molecular Devices Corporation). Biosensors can also be developed for alerting workers for the presence of hazardous substances in the industrial atmosphere.

2.11.3 Environmental Monitoring

There is an enormous range of potential analytes in air, water, soils and other environmental situations. Such measurements in water include Biochemical Oxygen Demand (BOD), Chemical Oxygen Demand (COD), acidity, salinity, nitrate, phosphate, calcium and fluoride, while pesticides, fertilizers and both industrial and domestic wastes require extensive analyses.

A current concern is for endocrine disruptors that can be active at very low levels of concentration, due to a wide range of estrogens and estrogenic mimics. Continuous real-time monitoring is required for some substances, and occasional random monitoring for others. In addition to the obvious pollution applications, farming, gardening, veterinary science and mining are all areas where sensors are needed for environmental monitoring.

The whole cell biosensors that make use of immobilized cells of *Trichosporon cutaneum* (yeast) can measure Biological Oxygen Demand (BOD) within 30 minutes as against 5 days needed by conventional method. Similarly, whole cell biosensors with immobilized cells of *Salmonella typhimurium* and *Bacillus subtilis* coupled with an oxygen electrode can be used for testing the mutagenicity and carcinogenicity of various chemicals. Portable pesticide monitors for measuring the level of organophosphate and carbamate pesticides in aquatic environment are also available.

2.11.4 Agriculture and Food Industry

Various possible applications of biosensors in agriculture include detection of viral, fungal and bacterial diseases, fertilizers and pesticides, even though many of these biosensors will be too expensive to be viable. The most viable use of biosensors in food industry will include rapid detection of total microbial contaminants and sugar quantification in soft drinks. For instance, in advanced countries biosensors have been developed to determine the freshness of fish and other food items including beef and other foods. In future, biosensors may be developed for on-line use, where analysis will be carried out on food samples, which are removed from the production line if found unsatisfactory. Such an analysis would facilitate real-time control of products during manufacture.

2.11.5 Military and Defense Industry

In military and defense organizations, portable biosensors can be very useful for detection of toxic gases and the agents of chemical warfare, such as mustard and nerve gas.

'Bacteria' are our invisible friends and enemies. Some bacteria aid our digestion, others destroy our poisons. Still other "bugs" make us sick. Natural bacteria are a fact of life, living inside and outside bodies of living organisms. We have learned to live with them and they live with us. Soldiers and sailors in action worry that even more dangerous bacteria are lurking in the environment, these are named as killer bacteria, used as weapons. Such biological warfare agents are viewed by many to be as threatening to human life as nuclear weapons. The key to protecting a military unit or community from dangerous bacteria is to detect them before they reach their intended victims. People can then be warned to leave the area or wear protective gears. Bacteria can be detected using biosensors.

Biorecognition and Immobilization

Chapter 3

3.0 BIORECOGNITION

3.0.1 Introduction

Sensors generate a measurable material property, specifically contain a recognition process that is characteristic of a material sample at the molecular-chemical level and a sensor incorporates a transduction process to create a useful signal. Chemical sensors use chemical processes in the recognition and transduction process. A pure physical sensor generates and transduces a parameter that does not depend on the chemistry, but is a result of the sensor responding as an aggregate of point masses or charges. Biosensors are also chemical sensors, but they use particular classes of biological recognition/transduction processes.

The chemical recognition process focuses on molecular-chemical entity, usually a kind of chemical structure, this structure may be a simple functional group: SiO^- in a glass electrode surface, a chromophore in an indicator dye or a metallic surface structure, such as silver metal that recognizes Ag^+ in solution. The biological recognition processes have been better understood nowadays and the general concept of recognition by receptor or chemoreceptor. Although these are often large molecules bound to cell membranes, they contain specific structures that permit a wide variety of different molecular recognition steps including recognition of large and small species and of charged and uncharged species.

A biosensor can be defined as a compact analytical device or unit incorporating a biological or biologically derived sensitive recognition element integrated or associated with a physiochemical transducer. Biorecognition in biosensors has especially stressed on using receptors and their categories (Fig. 3.1). Principal receptors are enzymes because of their extraordinary selectivity. Other receptors can be more fine species of biochemistry, viz., antibodies, organelles, microbes and tissues. Initially, biosensor recognition elements were isolated from living systems. However, many biosensor recognition elements now available are not naturally occurring but have been synthesized in the laboratory. Now the modern biotechnology techniques are used to construct DNA polynucleotides, thus opening new paths for generation of biosensor recognition elements arising from paths not taken by nature.

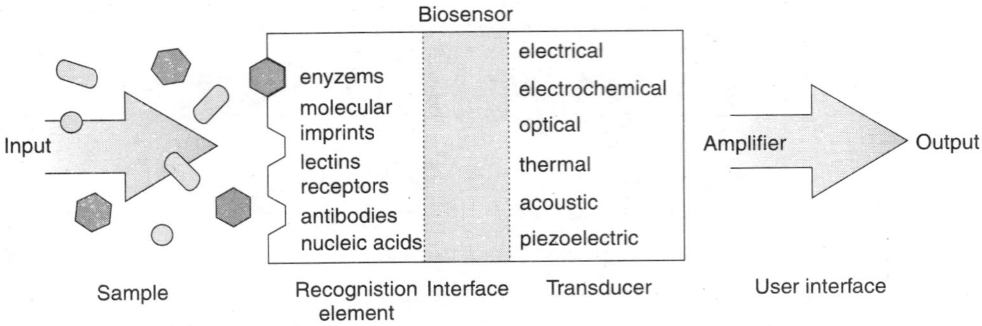

Fig. 3.1 *Bio-recognition process of a biosensor.*

3.0.2 Enzyme Based Recognition

Catalytic enzyme based sensor recognition elements are very attractive for biosensor applications due to a variety of measurable reaction products arising from the catalytic process, which include protons, electrons, light and heat. Enzymes are amazing molecular devices that determine the patterns of chemical transformations and mediate the transformation of one form of energy into another of energy. Enzymes regulate the rate at which chemical reactions proceed because of their catalytic power and specificity, while not being consumed by them. Because of their high efficiency and specificity, as well as the broad range of analytes, the enzymes are the most commonly used biological components of biosensors.

Enzymes do not change basic optical properties when they interact with the analyte therefore to monitor the enzymatic reaction. Fibre optic enzyme based biosensors commonly operate in indirect or direct detection mode. For indirect detection biosensors use an indicator or reactant to indirectly generate the optical signals. Direct detection means sensors detect the optical changes of the products or intermediates or cofactors including some inorganic ions and coenzymes such as Nicotinamide Adenine Dinucleotide (NAD^+).

Oxidases which catalyze the oxidation of compounds using oxygen are usually co-immobilized with oxygen sensitive luminescent materials such as polycyclic aromatic hydrocarbons, porphyrins and transition metal complexes. Among them, the ruthenium di-imine complexes, such as $Ru(phen)_3Cl_2$ and $Ru(dpp)_3Cl_2$ offer definite advantages including high photo stability, long excited state lifetimes, high quantum yields and high quenching rate constants.

In 1996 Rosenweig and Kopelman developed the first micro-sized sensor for glucose by immobilizing glucose oxidase with $Ru(phen)_3Cl_2$ in a polymeric support and then attaching an acrylamide support to a silanized optical fibre tip surface by photo-initiated polymerization. This sensor has a response time at least 25 times faster than larger sensors and it shows as much as 5-6 orders of magnitude improvement in absolute sensitivity. Fibre optic biosensors have also been developed to measure serum concentrations of bilirubin and cholesterol by co-immobilizing the ruthenium di-imine complexes with their specific enzymes.

Monitoring the pH change is another strategy of indirect detection which has been successfully applied in measuring many metabolites such as penicillin, creatinine, heavy metals and chlorinated herbicide atrazine. The enzymes immobilized in this type of biosensors are not restricted to one type any enzyme that can produce acid or base to cause a localized pH change can be co-immobilized with pH-sensitive chemicals. In addition to single enzyme fibre optic biosensors, dual- or tri-enzyme sensors have been developed in the last few years. Multiple enzymatic sequential reactions can significantly improve both the sensitivity and detection limit of a sensor or can be used to detect several related chemicals simultaneously.

The allosteric enzymes afford great potential for use as biosensor recognition elements. The regulatory allosteric enzymes subunit functions, as the recognition element, either in positive or negative fashion through conformational changes, the catalytic site serving as the transducing element (O'Connell and Guilbault, 2001). An engineered enzyme biorecognition element has been used for detection of HIV antibody in serum entailing two overlapping epitopes (P1 and P2 from the gp41 envelope glycoprotein) inserted adjacent to β-galactosidase active site resulting in a hybrid enzyme (Ferrer-Miralles *et al.*, 2001). A number of engineered allosteric catalytic biorecognition elements are now available and include β-galactosidase, alkaline phosphatase, β-lactamase and neural protease (O'Connell and Guilbault, 2001). In similar fashion, green fluorescent protein (GFP) is now used in many allosteric-like sensing element applications. Because the fluorophore is an intrinsic part of the GFP polypeptide chain, no covalent modification of the protein is required. Doi and Yanagawa have made hybrid fusion GFP containing specific molecular recognition sites by inserting protein domains containing the desired molecular recognition, binding site into a GFP surface loop. Random mutation of the ligand specific fusion protein insert gives rise to a combinatorial library from which binding of specific analytes can be selected via changes in fluorescence arising from the binding event (Doi and Yanagawa, 1999).

Of all enzyme recognition element based biosensors, the glucose biosensor is the most widely studied and much-admired sensor success story. Glucose encounters an immobilized enzyme and transduction is achieved amperometrically through sensitive electrode. Most of the glucose biosensors utilize glucose oxidase as their recognition element that catalyzes the oxidation of glucose to gluconolactone (Fig. 3.2):

$$\text{glucose} + O_2 \rightarrow \text{gluconolactone} + \text{Hydrogen peroxide} \qquad \ldots \textit{eq.-3.1}$$

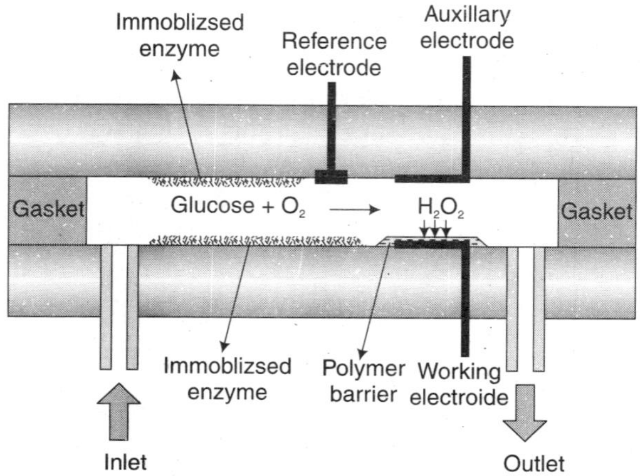

Fig. 3.2 *Diagram of the glucose sensor showing the electrode configuration, the polymer barrier deposited onto the working electrode and the surface where the enzyme (glucose oxidase) is immobilized. (Source: © Horizon Scientific Press. Offprints from www.cimb.org)*

The enzyme urease has been widely used as a sensor biorecognition element due to the need for urea determination/monitoring for both medical and environmental applications (Barhoumi *et al.*, 2006).

3.0.3 Proteins

Many proteins in a cell such as receptors and carrier proteins are bound by a specific ligand. The interactions between them are strong and specific which makes them excellent elements in biosensors. Based on this biorecognition principle Barker *et al.*, (1999) developed the fiber optic nitric oxide selective micro- and nanosensors using Cytochrome-C and Guanylate cyclase as bioreceptors.

3.0.4 Ligands

The simplest example of a receptor is a ligand. One particular ligand is EDTA. EDTA will bind to metal ions. Thus, the presence of metal ions in a solution (such as lead in drinking water) can be detected by mounting EDTA on a sensor chip.

3.0.5 Antibody Based Recognition

The majority of rapid detection systems employ antibodies for recognition, identification and quantitation of target analytes. Antibodies have been used extensively for detection purposes; however, their popularity increased significantly following Kohler and Milstein's seminal work establishing monoclonal antibody (MAb) technology (Jayasena, 1999). Using cell clones that specifically produce MAbs of choice, large quantities of antibody can now be produced. Antibody recognition elements make use of the sensitivity and specificity of bimolecular antibody–antigen interactions. The major advantage of antibody sensor biorecognition elements is that the immunogen, i.e. target, need not be purified prior to detection. A variety of signal transduction (optical and electrochemical) techniques have been developed and the most useful has been enzyme-fluorescence based with catalytic turnover resulting in amplification of signal, thus increasing the sensitivity of the assay.

Like many routinely used diagnostic assays, the majority of current PSA (prostate-specific antigen) assays are variations of enzyme linked immunosorbent assays (ELISA) reporting via the specific formation of PSA immune complex. Wu and coworkers' (2001) observation of nanomechanical motion generated by protein ligand interactions on micro-cantilevers has led to immobilization of PSA antibody recognition element for detection of PSA in serum (Fig. 3.3). Molecules adsorbed on a micro-cantilever cause vibrational frequency changes referred to as curling due to adsorption stress on one side of the cantilever. Surface plasmon resonance based optical transduction by noble metals has also been used as basis of antibody recognition element assays (Hirsch *et al,* 2003). Such technology readily lends itself to well established array micro-fabrication techniques, thereby offering the promising prospect of high throughput, biosensor based analysis of clinical samples.

Commonly used immunosensors are direct assay, competitive assay, binding inhibition assay and sandwich assay (Fig. 3.4). In a direct assay the antigens are incubated with excess amounts of immobilized antibodies. The antigen could be a fluorescent compound in fluorescence-based measurement. A variety of techniques have been applied for unlabeled direct assay which include ellipsometry, optical waveguide light mode spectroscopy, resonant mirror, surface plasmon resonance and interferometry. Unlabeled direct assay is a relatively quick and convenient method since it does not require the time consuming and expensive labeling process. In order to use competitive assay analytes or their analogs must be labeled with detectable chemicals such as Cy_5 and fluorescein (FITC). Upon incubating the analyte and labeled derivative will compete for limited binding sites of the antibodies. The amount of labeled derivative bound is inversely proportional to that of unlabeled analyte present in the sample. Thus the signal decreases as the analyte concentration increases. Binding inhibition assay is an alternative test format of competitive assay with the antigens being immobilized and the antibodies being labeled. In the sandwich assay antibodies to analytes are first immobilized to a sensor.

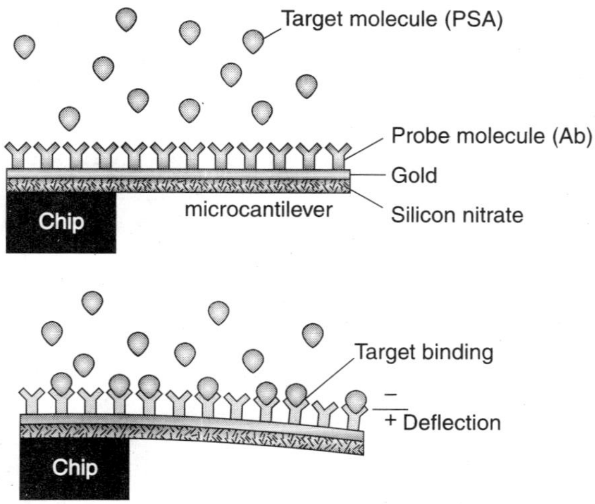

Fig. 3.3 *Diagram of interactions between target and probe molecules on a micro-cantilever beam. (Source: © Horizon Scientific Press. Off prints from www.cimb.org)*

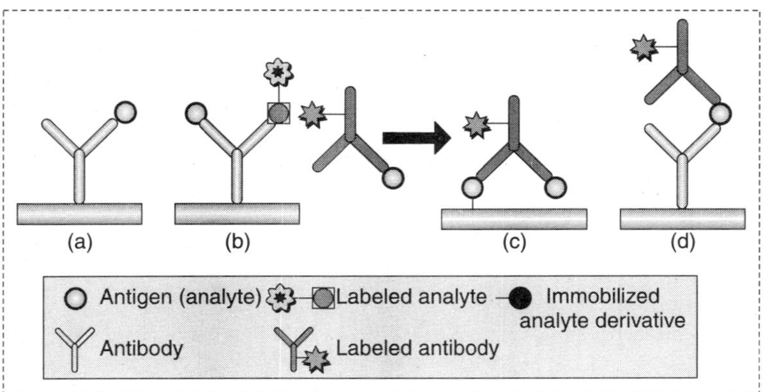

Fig. 3.4 *Assay format in immunosensors. (a) Direct assay (b) Competitive assay (c) Binding inhibition assay (d) Sandwich assay. (Source: Miniature fiber-optic multicavity Fabry-Perot interferometric biosensor, Yan Zhang, December 6, 2005, Blacksburg, Virginia)*

Next the sensor is incubated with the analytes and complexes are formed between the analytes and the immobilized antibodies. Finally a secondary labeled antibody recognizes and binds to another epitope of the analytes. In this case the extent of reaction is directly proportional to the amount of antigen present. The sandwich assay requires two antibodies that bind to epitopes that do not overlap each other on the antigen. This can be accomplished with either two monoclonal antibodies that recognize discrete sites or one batch of affinity purified polyclonal antibodies. The cross reactivity is significantly reduced due to the recognition of different epitopes.

3.0.6 Aptamer Based Recognition

Biosensors using aptamers as biorecognition elements are referred to as aptasensors. Aptamers are nucleic acid ligands (RNA, ssDNA, modified ssDNA, modified RNA) that are isolated from libraries of oligonucleotides by an *in vitro* selection process called SELEX (Systematic Evolution of Ligands by Exponential enrichment) (Ellington and Szostak, 1990, Tuerk and Gold, 1990). These DNA/RNA ligands are thought to recognize their target primarily by shape (i.e. conformation) and not sequence. Since they are short, single stranded oligonucleotides, they are capable of folding into three-dimensional structures due to their self-annealing properties. Aptamers bind with high affinity and specificity to a broad range of target molecules, and have proven suitable for analytic and diagnostic applications.

A library of DNA oligonucleotides (Fig. 3.5) containing a portion of randomized sequence is synthesized. The library is then converted into dsDNA by PCR and into RNA by *in vitro* transcription using the T7 RNA polymerase. After incubation of the target analyte with the nucleic acid pool, the non-specific or low-affinity binding nucleic acid molecules are removed by washing steps and the captured RNA molecules are eluted, recovered and amplified by RT-PCR and subsequently transcribed back to RNA. The whole cycle is repeated until obtaining a specific population of high affinity binding RNA which is isolated and characterized (Fig. 3.6). Aptamers have been produced against a wide range of targets including small molecules, proteins and whole cells, making them very useful for detection purposes. These nucleic acid recognition elements are more flexible than their protein counterparts. Predominantly unstructured in solution, aptamers fold upon associating with their ligands into molecular architectures in which the ligand becomes an intrinsic part of

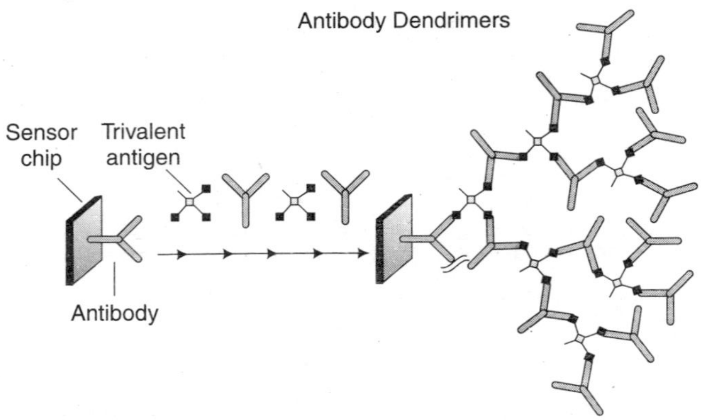

Fig. 3.5 *A proposed structure of an antibody dendrimer complex. (Source: © Horizon Scientific Press. Off prints from www.cimb.org)*

the nucleic acid structure. Due to their high binding affinity, simple synthesis, easy storage, and wide applicability, aptamer sensor recognition elements (Fig. 3.6) are emerging as a new class of molecules that rival commonly used antibody biosensor recognition elements.

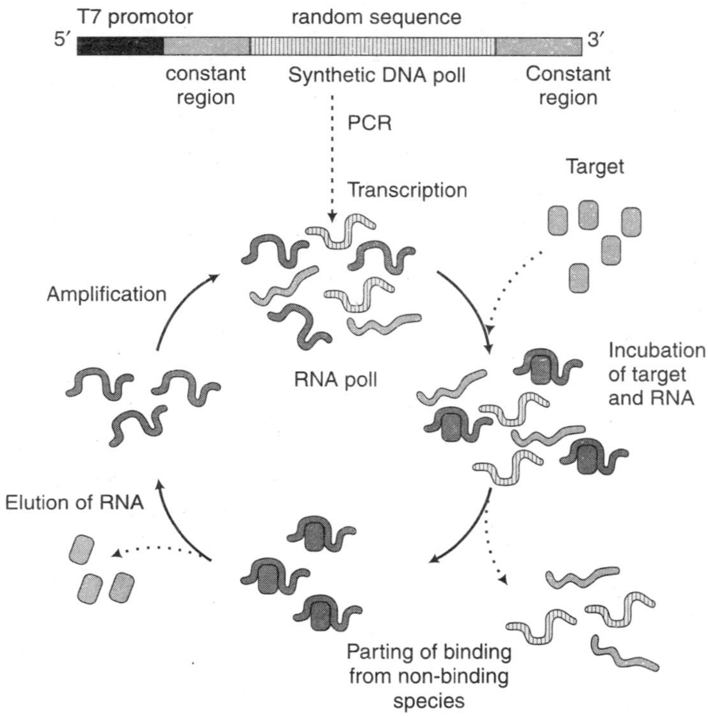

Fig. 3.6 *Scheme of in vitro selection of an RNA aptamer (SELEX). (Source: © Horizon Scientific Press. Off prints from www.cimb.org)*

3.0.7 Peptide Nucleic Acid Based Recognition

Peptide nucleic acids (PNA) are synthetic DNA analogues or mimics with a polyamide backbone instead of a sugar phosphate bone (Egholm *et al.*, 1993). Of significant importance to biosensing, PNAs exhibit superior hybridization characteristics and improved chemical and enzymatic stability compared to nucleic acids. Both double and triple stranded complexes are capable of being formed by PNA in association with nucleotides.

The negatively charged Ribose phosphate backbone of nucleic acids (Fig. 3.7) is replaced by an uncharged N-(2-aminoethyl)-glycine scaffold to which the nucleo-bases are attached through a methylene carbonyl linker. Because the intramolecular distances and configuration of the nucleo-bases are similar to those of natural DNA molecules, specific hybridization occurs between PNAs and cDNA or RNA sequences. The uncharged nature of PNAs is responsible for a better thermal stability of PNA–DNA duplexes compared with DNA–DNA equivalents and as a result, single-base mismatches have a considerably more destabilizing effect. As with DNA, the decrease in duplex stability depends on the position of the mismatch within the sequence. The neutral amide backbone also enables PNA to hybridize to DNA molecules in low salt conditions because no positive ions are necessary for counteracting the interstrand repulsion that hampers duplex formation between two negatively charged nucleic acids. Consequently, the abundance and stability of intramolecular folding structures in the DNA or RNA analytes are significantly reduced, making the molecules more accessible to complementary PNA oligomers. Unlike DNA, which depurinates at acidic conditions, PNAs are stable across a wide range of temperatures and pHs.

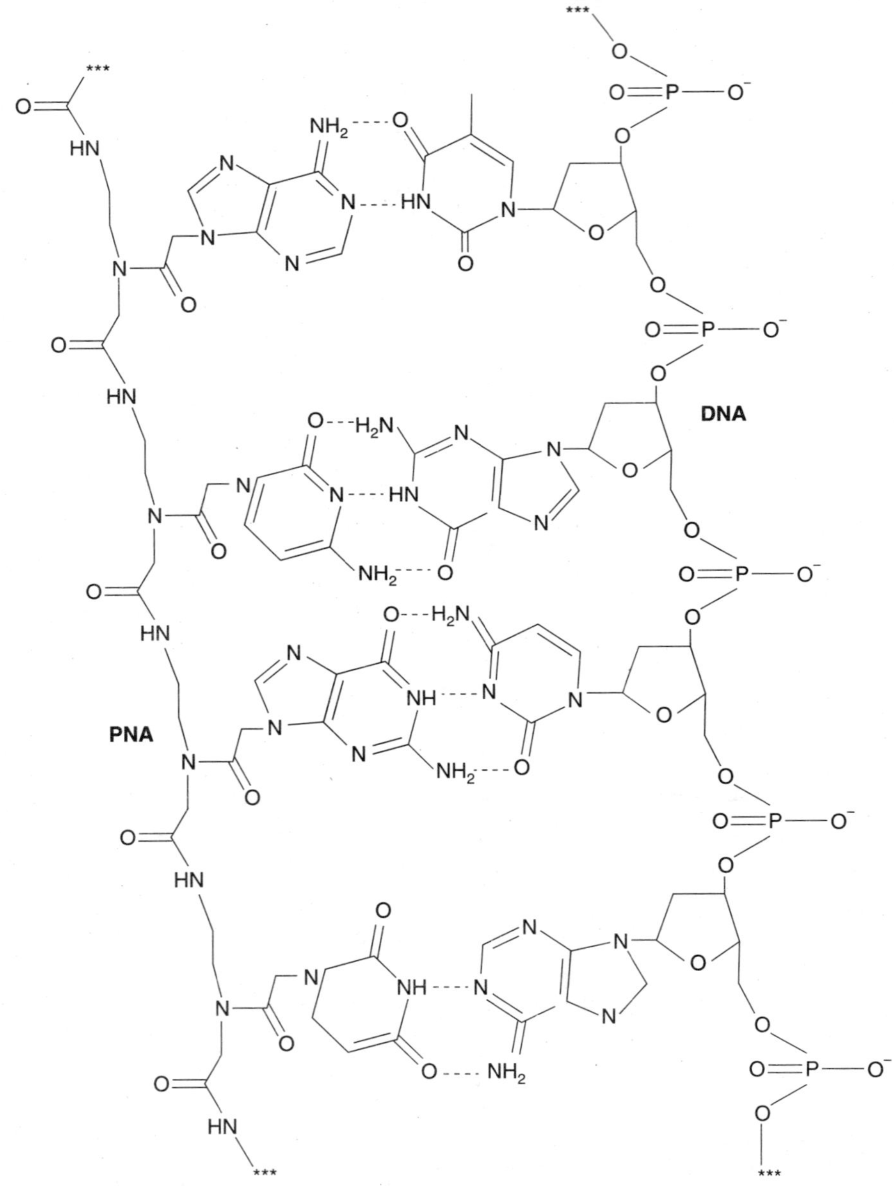

Fig. 3.7 *Comparison of the structures of peptide nucleic acid (PNA) and DNA. (Source: © Horizon Scientific Press. Off prints from www.cimb.org)*

PNAs have now been used to replace DNA to functionalize gold nanoparticles and upon hybridization to complementary DNA strands and formation of nanoparticle aggregates, results in a red-to-blue color transition and high discrimination of DNA single-base mismatches. The change of color results from the shift of surface plasmon band (cf. aptamer based recognition) upon aggregation and this property is now the basis of calorimetric biosensors for selective detection of DNA.

3.0.8 Molecular Imprint Based Recognition

Molecular imprinting is a method for making selective binding sites in synthetic polymers using molecular templates. Molecular imprinted polymers offer great promise for development of very stable solid-state like artificial biosensing elements. The technology of molecular imprinting has proliferated as an inexpensive, accessible and effective strategy for developing sorbent materials exhibiting high specificity for selected substrate materials.

A generalized scheme describes synthesis of a molecular imprint receptor molecule (Fig. 3.8). The possibility of imprinting against a wide range of analytes raises the possibility of generation of robust, artificial biological receptors making possible multiple clinical sample analysis without pretreatment, effectively reagent less chemistries. Molecular imprint quartz crystal microbalance detection is rapidly gaining acceptance for transducing the presence of a variety of analytes, especially glucose. Tai and coworkers (2005) have demonstrated recognition of Dengue viral protein using an epitope-mediated molecular imprinted film. This is significant since rapid diagnosis of this disease with high accuracy is crucial especially in the light of the current diagnostic methodologies that are time and labor intensive and the lack of effective vaccines and drugs for treatment.

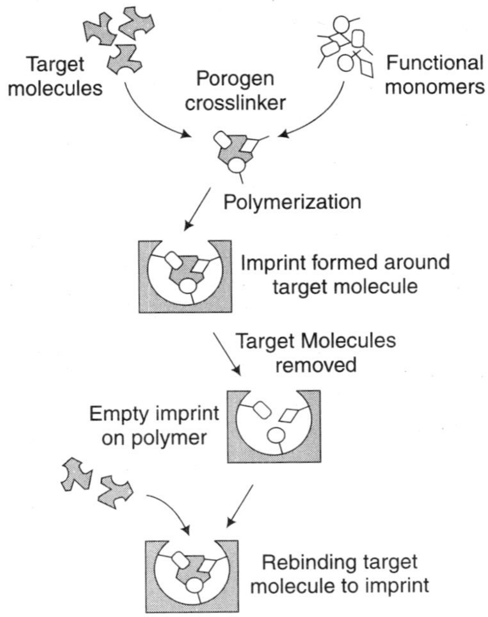

Fig. 3.8 *Schematic representation of the preparation of molecular imprints. (Source: © Horizon Scientific Press. Off prints from www.cimb.org)*

3.0.9 Langmuir-Blodgett Techniques Films

The techniques allowing forming ordered monolayers and multilayers of molecules find wide applications in electronics and bioelectronics. One of the methods for building such layers is based on transferring them from the liquid-gas interface onto a solid substrate. Monolayers of many insoluble compounds on liquid-gas interface can be created by spreading their solutions in a volatile solvent over another liquid. The first systematic study of monolayers of insoluble compounds at the water-air interface was carried out by Langmuir (1917).

Typical representatives of amphiphilic compounds capable to form the above mentioned monolayers are aliphatic carboxylic acids which have two distinct moieties in the molecule: a hydrophilic head group (-COOH) and a hydrophobic long alkyl chain (CnH2n+1, where n≥ 5). If the distance between amphiphilic molecules on the surface of water is large then their interactions are small hence they can be regarded as forming a two-dimensional gas having almost no effect on the surface tension. When a barrier system reduces the area of available surface for the monolayer molecules start to exert a repulsive effect on each other leading to the necessity to introduce the two-dimensional analogue of a pressure termed the surface pressure which is equal to the difference between the surface tension in the absence of a monolayer and that with the monolayer present. The maximum value of the surface pressure for a monolayer of any amphiphilic compound at the water-air interface at room temperature is less than 73 mN/m. On compressing the surface areas beyond the limit of monolayer compressibility, the phenomenon of collapse takes place, which is explained in terms of breaking the monolayers with the formation of disordered multilayers.

A method for transferring multilayers of long chain carboxylic acids on solid substrate was invented by Blodgett (1934) who added to the trough of Langmuir a device which lowers a solid substrate beneath a monolayer at the water-air interface and then withdraws it, therefore the resulting layers, deposited on the substrate, are referred to as the Langmuir-Blodgett (LB) films. An example of deposition of amphiphilic molecules on a hydrophilic substrate is shown in Fig. 3.9.

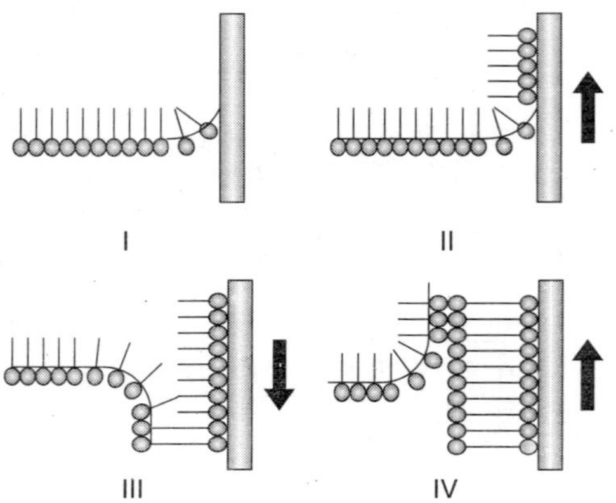

Fig. 3.9 *Deposition of Langmuir-Blodgett films on a solid substrate: (I) first dipping, (II) first withdrawal, (III) second dipping, (IV) second withdrawal.*

During the first dipping (I) water wets the hydrophilic substrate's surface with no deposition at this stage because the meniscus is turned up. Only during the first withdrawal the first monolayer is deposited in which hydrophilic tails are orientated toward the hydrophilic substrate. The second dipping (III) leads to the deposition of the second layer because the resulting surface now is hydrophobic and the meniscus is turned down.

The Langmuir-Blodgett (LB) technique is widely used to immobilize redox amphiphilic compounds on solid electrodes yielding photo-current generating devices, chemical sensors and biosensors. For example the photo-cells were produced by immobilization of LB layers of pyrrolidinofullerene derivatives or ampiphilic stilbazolium dye dimer on a semi-conducting transparent indium-tin-oxide. Amphiphilic molecules also can be employed for the immobilization of enzymes. The pioneering work demonstrating that enzymes retain their activity when entrapped into reversed micelles of surfactants were published in the beginning of the 80's. Next a novel LB method was developed for producing and transferring a uniform protein thin film layer of a protein sandwiched between layers of a lipid.

A glucose sensor was made on the basis of adsorbed layers of glucose oxidase on LB monolayers of N, N-dimethyl acetylacetyl methacryloyl ethyl on platinum. Choline oxidase was inserted in a hydrophilic or hydrophobic environment of behenic acid LB films and the biosensing layer was directly coated on an especially designed amperometric transducer to yield a choline biosensor. An electroactive film was self assembled upon which a solution of proteo-liposomes was fused, the oxidation of reduced Cytochrome-c by the membrane bound Cytochrome-c oxidase being transduced to a gold electrode via a mediator bound to the polymer cushion where the lipid membrane rested. A glucose biosensor was fabricated by LB deposition of a phospholipid analogous and vinyl polymer together with glucose oxidase. LB films containing butyryl cholinesterase were fabricated to produce an enzymatic field effect transistor for the detection of organophosphorus pesticides in water.

3.0.9.1 Advantages of LB films

1. Creation of multilayered structure which allows to deposit all the components of a biosensor by one standard procedure leading to easy automation of the industrial production .
2. Easy inclusion of protein layers in the desired position of the layered structure.
3. Deposition of enzymes without any contact with air which allows to avoid protein denaturation.
4. Minimization of the amount of tansoactive materials consumed for the production of biosensors.
5. High stability in aqueous solutions of the deposited tansoactive bilayers.

3.0.9.2 Disadvantages of LB films

a. Long time of deposition.
b. Limitations of the transport of matter between hydrophobic chains of the deposited lipid bilayers which in some cases can deactivate mediators requiring protons or hydroxyls for their reactions.

3.0.10 Membrane Receptors and Transporters

The glycerophospholipids are the main structural component of biological membranes such as the cellular plasma membrane and the intracellular membranes of organelles. In animal cells the plasma membrane physically separates the intracellular components from the extracellular environment. All eukaryotic cells are compartmentalized into membrane bound organelles which carry out different functions. These glycerophospholipids are amphipathic molecules that contain a glycerol core linked to two fatty acid derived tails by ester or ether linkages and to one head group by a phosphate ester linkage. While glycerophospholipids are the major component of biological membranes, other nonglyceride lipid components such as sphingomyelin and sterols (mainly cholesterol in animal cell membranes) are also found in biological membranes. In plants and algae the galactosyl diacylglycerols and sulfoquinovosyl diacylglycerol which lack a phosphate group are important components of membranes of chloroplasts and related organelles and are the most abundant lipids in photosynthetic tissues, including those of higher plants, algae and certain bacteria.

Self organization of phospholipids: a spherical liposome a micelle and a lipid bilayer. A biological membrane is a form of lipid bilayer as a liposome. The formation of lipid bilayers is an energetically preferred process when the glycerophospholipids described above are in an aqueous environment. In an aqueous system the polar heads of lipids orientate towards the polar, aqueous environment, while the hydrophobic tails minimize their contact with water. The lipophilic tails of lipids (U) tend to cluster together, forming a lipid bilayer or a micelle. Other aggregations are also observed and form part of the polymorphism of amphiphile (lipid) behavior. The polar heads (P) face the aqueous environment, curving away from the water. Phase behaviour is a complicated area within biophysics and is the subject of current academic research. Micelles and bilayers form in the polar medium by a process known as the hydrophobic effect. When dissolving a lipophilic or amphiphilic substance in a polar environment, the polar molecules (i.e. water in an aqueous solution) become more ordered around the dissolved lipophilic substance, since the polar molecules cannot form hydrogen bonds to the lipophilic areas of the amphiphile. So in an aqueous environment the water molecules form an ordered "clathrate" cage around the dissolved lipophilic molecule.

3.0.11 Biomaterials and Biorecognition

The surface force apparatus used for determining the fundamental molecular interactions that govern biomolecular and cell interactions with a variety of materials. These findings are in turn used to guide the fabrication of materials for the tailored growth of cells and tissues on engineered scaffolds. Direct force measurements identify the molecular mechanisms underlying the function of cell adhesion proteins and the relationships between their composition, architecture, and adhesive function. Leckband's work identified novel bioadhesion mechanisms, which may be generally applicable to a variety of different adhesion proteins in several different tissues. The Leckband group is using this molecular level approach, together with protein design and mammalian tissue culture, to identify and exploit biologically active protein fragments for the control or manipulation of tissues.

Biological recognition at surfaces is exploited not only in tissue engineering, but also in sensor design, drug targeting, and cellular communication with the environment. Leckband's group also studied the interactions of biological molecules with surfaces or with surface-immobilized species, because the recognition of immobilized species often occurs in a complex environment, interfacial properties, structural factors and mass transport in addition to specific recognition can alter the rates and strengths of these selective binding reactions. This is of critical importance in such areas as biosensor design and targeted drug delivery. Leckband's group is determining how surface force fields modulate biorecognition at interfaces. This group is also developing mathematical models, in conjunction with experimental studies, to model transport-kinetic processes in biosensors and microfluidic devices. The incorporation of measured intermolecular potentials into these models provides a rigorous basis for engineering surface recognition reactions in such applications as the design of fibre optic biosensors.

Tissue engineering is emerging as a new therapeutic strategy aimed toward understanding the principles of tissue development and translating this fundamental knowledge into clinical restoration, repair, maintenance, and even improvement of tissue and whole organ function. This innovative strategy has been performed by combining new regenerative medicines including protein, genes, cells, and biomaterials.

Kong's group specializes material chemistry to design biomaterials that mimic the function and structure of Extra Cellular Matrix (ECM) and regulate diverse activities of various stem and progenitor cells. They ultimately combine the synthetic ECM with growth factors and genes to promote the regeneration of bone, skeletal muscle and vascular network using both host and transplanted cells.

Harley pursued the fabrication, characterization and testing of scaffold-based systems for *in vivo* and *in vitro* orthopedic and soft tissue engineering applications. They specifically utilize a cellular solids and poroelastic modeling to relate the microstructural features of tissues and materials to cellular activities in the design of tissue engineering scaffold. In addition, they are using a scaffold as a synthetic stem niche for the *in vitro* cell culture of hematopoietic stem cells and mesenchymal stem cells.

The ability to control chemical composition with great precision over very short distances allows microfluidic chips to be used for a wide range of biological studies. The lack of turbulence at the microscale leads to a laminar flow regime in which mixing between merging streams (Fig. 3.10) is controlled by diffusion only. Kenis *et al.,* used microfluidic platforms with integrated electrodes to study the role of the intracellular redox state on cell cycle kinetics (*Eg*. cell differentiation vs. proliferation).

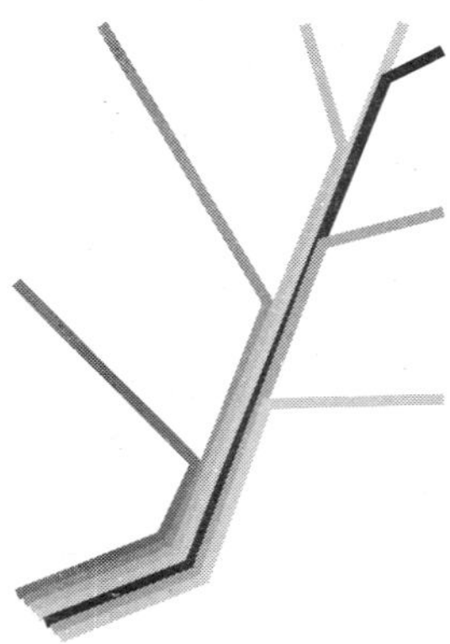

Fig. 3.10 *VLSI - Microfluidic network electro wires system.*

At present, combinatorial screening for leads for pharmaceuticals uses well plates with micro liter sized wells (Fig. 3.11). Kenis and collaborators use VLSI-microfluidic networks to create next generation combinatorial chemistry chips comprised of large arrays of well that are only one nano liter in volume each. These chips have integrated gratings for surface-interaction based optical detection of binding events. How immune cells respond to multiple chemical signals (*Eg.* cytokines) in real time. These data are used to understand how immune cells migrate to sites of infection and also how aberrations in signaling lead to allergy and inflammation.

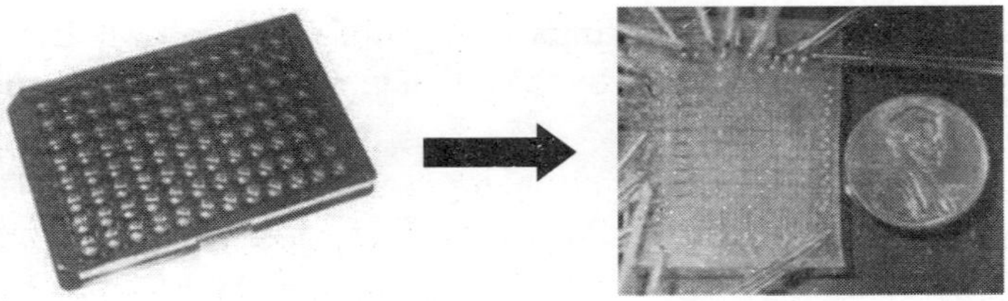

Fig. 3.11 *Plates with microliter sized wells.*

The ability to engineer tissues, prevent viral infection, and develop biomedical devices hinges upon an understanding of the basic molecular interactions that drive cellular processes. Molecular imaging technologies are critical for determining the structure and integrative functions of molecules and macromolecular assemblies within cells.

The lateral organization of the cell membrane and how that organization influences cellular processes, such as viral infection, cell-cell recognition, and signal transduction, is poorly understood. Multiple isotope Imaging Mass Spectrometry (MIMS) is a direct approach to membrane composition analysis. MIMS reveals the isotopic and elemental composition in a sample with high lateral resolution. Kraft and collaborators at Lawrence Livermore National Laboratory are being used MIMS to quantitatively image the distributions of multiple isotopically labeled components in phase separated model membranes (Fig. 3.12). These results are being used to understand the physical and molecular properties that induce lateral heterogeneity within cell membranes.

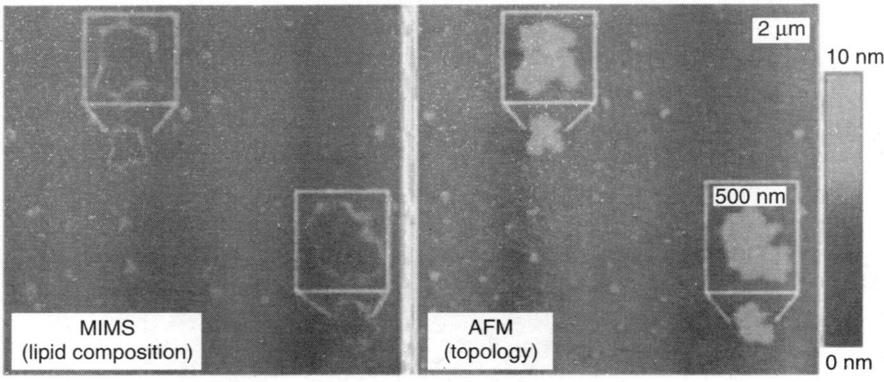

Fig. 3.12 *Multiple isotope imaging mass spectrometry (MIMS) approach.*

3.1 IMMOBILIZATION

3.1.1 Introduction

Immobilization means associating the biocatalysts with an insoluble matrix, so that it can be retained in proper reactor geometry for its economic reuse under stabilized conditions. Immobilization thus allows, by essence, to decouple the enzyme location from the flow of the liquid carrying the reagents and products. Immobilization helps in the development of continuous processes allowing more economic organization of the operations, automation, decrease of labour and investment/ capacity ratio. Immobilized biocatalysts offer several other advantages, notable among them is the availability of the product in greater purity. Purity of the product is very crucial in food processing and pharmaceutical industry since contamination could cause serious toxicological, sensory or immunological problems. The other major advantages include greater control over enzymatic reaction as well as high volumetric productivity with lower residence time, which are of great significance in the food industry, especially in the treatment of perishable commodities as well as in other applications involving labile substrates, intermediates or products.

Immobilization of biocatalysts helps in their economic reuse and in the development of continuous bioprocesses. Biocatalysts can be immobilized either using the isolated enzymes or the whole cells. Immobilization often stabilizes the structure of the enzymes thereby, allowing their applications even under harsh environmental conditions of pH, temperature and organic solvents, and thus enable their uses at high temperatures in non-aqueous enzymology and in the fabrication of biosensor probes.

Biotechnology is currently considered as a useful alternative to conventional process technology in industrial and analytical fields. This is mainly because, unlike the chemical catalysts, the biological systems have the advantages of accomplishing complex chemical conversions under mild environmental conditions with high specificity and efficiency. Biological systems help in ingredient substitution, processing aid substitution, more efficient processing, less undesirable products, increased plant capacity, increased product yields, and improved or unique products. The variety of chemical transformations catalysed by enzymes has made these catalysts a prime target of exploitation by the emerging biotech industries. Despite these advantages, the use of enzymes in industrial applications has been limited by several factors, mainly the high cost of the enzymes, their instability, and availability in small amounts. Also the enzymes are soluble in aqueous media and it is difficult and expensive to recover them from reactor effluents at the end of the catalytic process. This restricts the use of soluble enzymes to batch operations, followed by disposal of the spent enzyme containing solvent. The newer technological developments in the field of immobilized biocatalysts can offer the possibility of a wider and more economical exploitation of biocatalysts in industry, waste treatment, medicine and in the development of bioprocess monitoring devices like the biosensor.

3.1.2 Technique

A large number of techniques and supports are now available for the immobilization of enzymes or cells on a variety of natural and synthetic solid supports. The choice of the support as well as the technique depend on the nature of the enzyme, nature of the substrate and its ultimate application. Therefore, it will not be possible to suggest any universal means of immobilization. It can only be said that the search must continue for matrices which provide facile, secure immobilization with good interaction with substrates and which conform in shape, size, density and so on to the use for which they are intended. Care has to be taken to select the support materials as well as the reagents used for immobilization, which have GRAS status, particularly when their ultimate applications are in the food processing and pharmaceutical industries. Macromolecular, colloidal, viscous, sticky, dense or particulate food constituents or waste streams also limit the choice of reactor and support geometries. Commercial success has been achieved when support materials have been chosen for their flow properties, low cost, nontoxicity, maximum biocatalysts loading while retaining desirable flow characteristics, operational durability, ease of availability, and ease of immobilization.

Techniques for immobilization have been broadly classified into four categories, namely entrapment, covalent binding, cross-linking and adsorption. It must be emphasized that in terms of economy of a process, both the activity and the operational stability of the biocatalysts are important. They determine its productivity, which is the activity integrated over the operational time.

A biocatalyst is termed immobilized, if its mobility has been restricted by chemical or physical means. This limitation of mobility may be achieved by widely different methods. *Eg*. Trapping in the network of a polymer matrix or by membrane confinement. An essential criterion for defining a system as immobilized is that human interference has to be involved. A variety of immobilization methods have been used in the development of successful biosensors. Immobilized enzyme preparations may be more effective since they are recoverable and possibly more stable than free enzymes. Moreover, being in their native environment, operational denaturation of the enzymes can be minimized.

Enzymes are largely used as biocatalysts in chemical, pharmaceutical and food industries, and as specific ligands in clinical and chemical analysis. Since the recovery and the reusability of the free enzyme are limited, immobilization of enzyme has attracted the attention. Immobilized enzymes have the advantages of using in batch and continuous systems, removing easily from the reaction medium and providing the facility of controlled production. However, immobilized enzyme systems also have limitations, such as loss activity due to the immobilization technique and decrease in the apparent activity due to mass transfer resistance for the substrate and product. An optimal support material should provide large surface area per unit volume or mass of the carrier for effective immobilization of the desired amount of enzyme, it should allow substrate and product transports with the least diffusional resistance and also should be easily available and non-toxic. The immobilization technique should also lead to high immobilization and activity yields.

In order to construct a viable biosensor, the biological material has to be properly attached to the transducer. There are various techniques used in biological material immobilization (*Eg*. Enzyme) (Fig. 3.13).

3.1.2.1 Adsorption

This is the simplest of all the techniques and one which does not alter the activity of the bound enzyme. In case of enzymes immobilized through ionic interactions, adsorption and desorption of the enzyme depends on the ability of the ion exchanger. Moreover, a dynamic equilibrium is normally observed between the adsorbed enzyme and the support which is often affected by pH as well as the ionic strength of the surrounding medium. This property of reversibility of binding

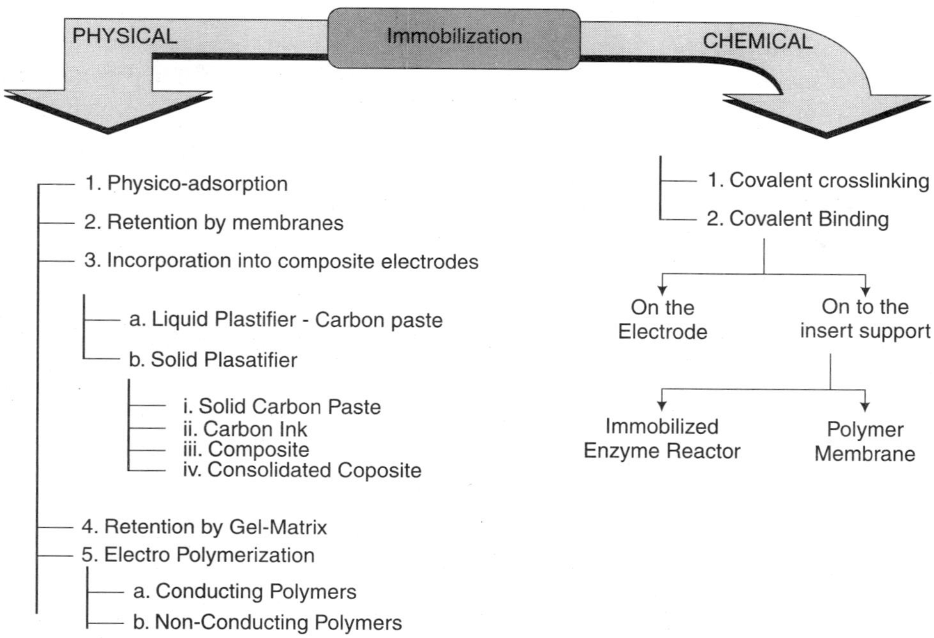

Fig. 3.13 *Immobilization techniques.*

has often been used for the economic recovery of the support. This has been successfully adapted in industry for the resolution of racemic mixtures of amino acids, using amino acid acylase. A variety of commercially available ion exchangers have been investigated for this purpose. One of the techniques, which have gained importance more recently, is the use of polyethylenimine for imparting polycationic characteristics to many of the neutral supports based on cellulose or inorganic materials. Enzymes with low pI, like invertase, urease, glucose oxidase, catalase, and other enzymes have been bound through adsorption followed by cross-linking on polyethylenimine coated supports.

Immobilization of enzymes with hydrophobic interaction is one of the important features of adsorption, which is of great significance. The hydrophobic interactions are usually stabilized by high ionic concentrations, thus enabling the use of high concentrations of substrates as desired in an industrial process. Other types of strong interactive binding techniques have also been reported for the reversible immobilization of enzymes. A typical example is the immobilization of soybean b-amylase on phenyl-boronate agarose, which can be reversed for the recovery of the support using sorbitol. Varieties of biospecific interactions have also been investigated for the reversible immobilization of enzymes by adsorption. Enzymes like acetylcholine esterase, ascorbic acid oxidase, invertase, peroxidase, glucose oxidase, etc. have been immobilized by biospecific-reversible immobilization on lectin bound supports and invertase, using polyclonal anti-invertase antibodies.

Traditional enzyme immobilization procedures involve isolation of the enzyme, followed by use of several steps for the immobilization. Costs of enzyme purification, the immobilization procedure, bioreactor operational stability, and bioreactor regeneration are the major factors that determine the cost of a bioreactor process. Development of techniques for the simultaneous isolation and immobilization of enzymes from crude extracts has obviated these problems. Some typical examples include the immobilization of a Streptavidin-B-galactosidase fusion protein expressed in *Escherichia coli* and bioselectively adsorbed from a crude cell lysate to biotin which is covalently immobilized on controlled pore glass, and the simultaneous purification and immobilization of D-amino acid oxidase from *Trigonopsis variabilis* cell lysate adsorbed on phenylsepharose. The technique can have future potentials especially in the downstream processing and immobilization of enzyme/proteins obtained by recombinant DNA technology.

Techniques for the adhesion of whole cells on polymeric surfaces are also currently gaining considerable importance. The major advantage for the cells immobilized through adhesion is reduction or elimination of the mass transfer problems associated with the commonly used gel entrapment method. The technique of immobilization usually followed is the microbial colonization by recycling of the cell suspension along with nutrients, such that a biofilm is gradually formed. This often results in the immobilization of cells in a viable form for use in heterogeneous fermentations. Useful techniques have been developed also for the immobilization of nonviable cells to be used as an enzyme source for simple chemical conversions. Notable among them include treating the cells or the support with trivalent metal ions like Al^{3+} or Fe^{3+} or charged colloidal particles, and use of polycationic polymers like Chitosan. Novel techniques have been developed to adhere cells strongly on a variety of polymeric surfaces including glass, cotton cloth, cotton threads, and other synthetic and inorganic surfaces using polyethylenimine. This technique has also been recently used for the simultaneous filtration and immobilization of cells from a flowing suspension, thus integrating downstream processing with bioprocessing. This technique may have future potentials for the immobilization of cells for food applications.

3.1.2.1.1 Adsorption of lipid vesicles

Adsorption of lipid vesicles will yield different results on different surfaces. Using egg- Phosphatidyl Choline lipids (egg-PC) treated to form Small Unilamellar Vesicles (SUV). A SUV is a small sphere, radius 25 nm, encased by a lipid bilayer (Fig. 3.14). Egg-PC has two carbon chains of different lengths and some have a double bond somewhere along the chain.

When vesicles in solution come into contact with a surface several scenarios are possible dependent on both surface and lipid properties, has studied three different surfaces: oxidized gold, thiolated (hydrophobic) gold and silicon oxide. Both oxidized gold and silicon oxide are hydrophilic surfaces, but they do not show the same kind of vesicle adsorption. Thus, although hydrophilicity seems to be one of the major parameters that govern adsorption it is not the only one of importance.

Most sensors especially if they can be miniaturized derive their sensing signal at some kind of surface interface. A common problem is that surfaces interact with sensing molecules bonded to the surface. For example, if thiolated single strands of DNA are bonded directly to an Au surface they will interact strongly and lie flat down instead of standing up. This prevents hybridizing with complementary DNA strands from solution, a common sensor situation.

Lipid bilayer provides a very good surface to build a biosensor and modify the bilayer by including 5% biotin-phospholipids. Biotin lipids have a small vitamin called biotin attached to their head groups. Biotin will in turn form a strong, selective bond with a protein called avidin.

The selective bond between biotin and avidin is used to adsorb a layer of protein complexes of avidin named streptavidin. Since the bilayer is liquid crystalline streptavidin moves on the surface and interacts with each other to form a close packed two-dimensional crystal on top of the bilayer at saturated adsorption. Streptavidin has biotin binding sites facing both inward, towards the bilayer, and outwards, towards the solution. By modifying other molecules with biotin make them bind selectively to the streptavidin crystal in a well-defined way protected from interacting with the surface by the inert bilayer and protein crystal.

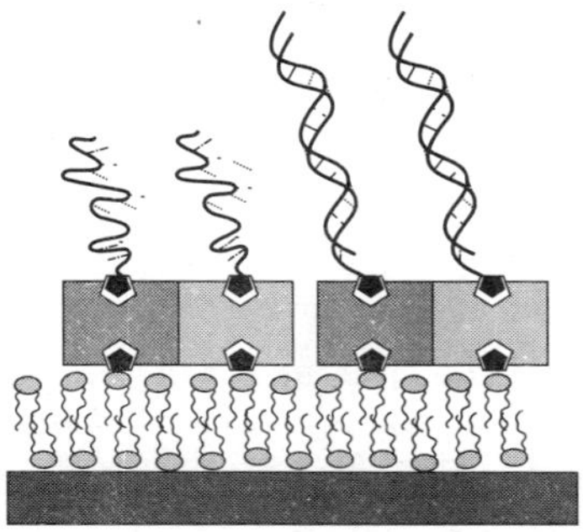

Fig. 3.14 *Schematics of the biosensor in lab. (Source: Lab-PM laboration III in Biomaterials and Biophysics on Surfaces (4p) Skapad av Karin Glasmästar och Erik Reimhult, 2000 Uppdaterad av Erik Reimhult, 2001.)*

To the left the sensor consisting of a bilayer, a streptavidin crystal on top bonded to biotinylated lipids in the bilayer and biotinylated single DNA-strands bonded to the protein crystal. Introducing complementary single DNA-strands are into the solution which, bind to the DNA-strands on the sensor surface. The molecule bonded to the protein crystal gives the recognition capabilities of the sensor, *i.e.,* determines what molecules are able to sense. Single strands of DNA recognize complementary single strands of DNA. Another biological recognition pair is antibody-antigen.

3.1.3 Microencapsulation

This was the method used in the early biosensors. The biomaterial is held in place behind a membrane, giving close contact between the biomaterial and the transducer. It is adaptable. It does not interfere with the reliability of the enzyme. It limits contamination and biodegradation. It is stable towards changes in temperature, pH, ionic strength and chemical composition. It can be permeable to some materials, e.g. small molecules, gas molecules and electrons.

3.1.4 Entrapment

Entrapment has been extensively used for the immobilization of cells, but not for enzymes. The major limitation of this technique for the immobilization of enzymes is the possible slow leakage during continuous use in view of the small molecular size compared to the cells. Biocatalysts have been entrapped in natural polymers like agar, agarose and gelatin through thermo reversal polymerization, but in alginate by ionotropic gelation. A number of synthetic polymers have also been investigated. Notable among them are the photo-cross linkable resins, polyurethane, prepolymers, and acrylic polymers like polyacrylamide. Among these, the most widespread matrix made from monomeric precursors is the polyacrylamide gel. Polyacrylamide may not be a useful support for use in food industry in view of its toxicity, but can have potentials in the treatment of waste and in the fabrication of analytical devices containing biocatalysts.

One of the major limitations of entrapment technique is the diffusional limitation as well as the strict hindrance, especially when the macromolecular substrates like starch and proteins are used. Diffusional problems can be minimized by entrapment in fine fibres of cellulose acetate or other synthetic materials or by using an open pore matrix. Recently, the development of so-called hydrogels and thermoreactive water-soluble polymers, like the albumin-poly (ethylene glycol) hydrogel, has attracted attention in the field of biotechnology. In the area of health care, they offer new avenues for enzyme immobilization. Such gels with a water content of about 96% provide a microenvironment for the immobilized enzyme close to that of the soluble enzyme with minimal diffusional restrictions.

3.1.5 Covalent Binding

Covalent binding is an extensively used technique for the immobilization of enzymes, though it is not a good technique for the immobilization of cells. The functional groups extensively investigated are the amino, carboxyl, and the phenolic group of tyrosine. Enzymes are covalently linked to the support through the functional groups in the enzymes, which are not essential for the catalytic activity. It is often advisable to carry out the immobilization in the presence of its substrate or a competitive inhibitor so as to protect the active site. The covalent binding should also be optimized so as not to alter its conformational flexibility. Some of these problems however, can be obviated by covalent bonding through the carbohydrate moiety when a glycoprotein is concerned.

A number of industrially useful enzymes are glycoproteins wherein the carbohydrate moiety may not be essential for its activity. In general, functional aldehyde group can be introduced in a glycoprotein by oxidizing the carbohydrate moiety by periodate oxidation without significantly affecting the enzyme activity. The enzyme could then be covalently linked to a support containing an alky amine group through Schiffs base reaction. Enzymes like glucose oxidase, peroxidase, invertase, etc. have been immobilized using this technique. Covalent binding has been extensively investigated using inorganic supports. Enzymes covalently bound to inorganic supports have been used in the industry. Enzymes have also been bound to synthetic membranes, thus integrating biconversion and downstream processing.

3.1.6 Cross-Linking

Biocatalysts can be immobilized through chemical cross-linkage by using homo- and hetero-bifunctional agents, among these, glutaraldehyde which interacts with the amino groups through a base reaction has been extensively used, in view of its GRAS (Generally Recognized as Safe) status, low cost, high efficiency and stability. The enzymes or the cells have been normally cross-linked in the presence of an inert protein like gelatin, albumin and collagen. The use of raw hen egg white as an economic, easily available novel proteinic support rich in lysozyme for the immobilization of enzyme or nonviable cells either in a powder, bead or highly porous foam form by using glutaraldehyde as a cross-linker. The unique feature of this support is the large concentration of lysozyme naturally present in hen egg white which gets co-immobilized, thus imparting the bacteriolytic property to the support.

Adsorption followed by cross-linking has also been used for the immobilization of enzymes. The technique of cross-linking in the presence of an inert protein can be applied to either enzymes or cells. The technique can also be used for the immobilization of enzymes by cross-linking the cell homogenates. The osmotic stabilization of cellular organelles or halophilic cells is prior to immobilization using cross-linkers.

3.1.6.1 Cross-linked enzyme aggregate

Cross-linked enzyme aggregate is an immobilized enzyme prepared through cross-linking. They can be used as stereo selective industrial biocatalysts.

Enzymes are proteins that catalyze (*i.e.* accelerate) chemical reactions. They are natural catalysts and are ubiquitous, in plants, animals and microorganisms where they catalyze processes that are vital to living organisms. They are intimately involved in numerous biotechnological processes, such as cheese making, beer brewing and wine. Recent advances in biotechnology, particularly in genetic and protein engineering and genetics have provided the basis for the efficient development of enzymes with improved properties for established applications and novel, tailor-made enzymes for completely new applications where enzymes were not previously used.

Nowadays enzymes are widely applied in many different industries and the number of applications continues to increase. Examples include food (baking, dairy products, starch conversion) and beverage (beer, wine, fruit and vegetable juices) processing, animal feed, textiles, pulp and paper, detergents, biosensors, cosmetics, health care and nutrition, waste water treatment, pharmaceuticals and chemicals and more recently, biofuels such as biodiesel. The main driver for the widespread application of enzymes is their small environmental footprint.

Many traditional chemical conversions used in various industries suffer from inherent drawbacks from both an economic and environmental viewpoint. Non-specific reactions can afford low product yields, copious amounts of waste and impure products. The need for elevated temperatures and pressures leads to high energy consumption and high capital investment costs. Disposal of unwanted by-products may be difficult and/or expensive and hazardous solvents may be required. In stark contrast, enzymatic reactions are performed under mild conditions of temperature and pressure, in water as solvent, and exhibit very high rates and are often highly specific. Moreover, they are produced from renewable raw materials and are biodegradable. In addition, the mild operating conditions of enzymatic processes mean that they can be performed in relatively simple equipment and are easy to control. In short, they reduce the environmental footprint of manufacturing by reducing the consumption of energy and chemicals and concomitant generation of waste. Notwithstanding all these desirable characteristic features of enzymes their widespread industrial application is often hampered by their lack of long term operational stability and shelf-storage life, and their cumbersome recovery and re-use. These drawbacks can generally be overcome by immobilization of the enzyme and a major challenge in industrial biocatalysis is the development of stable, robust and preferably insoluble biocatalysts.

The use of cross-linked enzyme crystals (CLECs) as industrial biocatalysts was pioneered by Altus Biologics in the 1990s. CLECs proved to be significantly more stable to denaturation by heat, organic solvents and proteolysis than the corresponding soluble enzyme or lyophilized (freeze-dried) powder. CLECs are robust, highly active immobilized enzymes of controllable particle size, varying from 1 to 100 micrometer. Their operational stability and ease of recycling, coupled with their high catalyst and volumetric productivities, renders them ideally suited for industrial bio-transformations.

Disadvantages of CLECs: Enzyme crystallization is a laborious procedure requiring enzyme of high purity, which translates to prohibitively high costs. The more recently developed cross-linked enzyme aggregates (CLEAs), on the other hand they are produced by simple precipitation of the enzyme from aqueous solution, as physical aggregates of protein molecules by the addition of salts or water miscible organic solvents or non-ionic polymers. The physical aggregates are held together by non-covalent bonding without perturbation of their tertiary structure, which is without denaturation. Subsequent cross-linking of these physical aggregates renders them permanently insoluble while maintaining their pre-organized superstructure. This discovery led to the development of a new family of immobilized enzymes: cross-linked enzyme aggregates (CLEAs). Since precipitation from an aqueous medium, by addition of ammonium sulfate or polyethylene glycol, is often used to purify enzymes, the CLEA methodology essentially combines purification and immobilization into a single unit operation that does not require a highly pure enzyme. It could be used for the direct isolation of an enzyme in a purified and immobilized form suitable for performing bio-transformations from a crude fermentation broth.

CLEAs are very attractive biocatalysts, owing to their facile, inexpensive and effective production method. They can readily be reused and exhibit improved stability and performance. The methodology is applicable to essentially any enzyme, including cofactor dependent oxidoreductases. Application to penicillin acylase used in antibiotic synthesis showed large improvements over other type of biocatalysts.

3.1.6.2 Applications of CLEAs

1. Synthesis of pharmaceuticals, flavors and fragrances, agrochemicals, pharmaceuticals, fine chemicals, bulk monomers and biofuels.
2. Animal feed, *Eg*. Phytase for utilization of organically bound phosphate by pigs and poultry.
3. Food and beverage processing, *Eg*. Lipases in cheese manufacture and laccase in wine clarification.
4. Cosmetics, *Eg*. in skin care products.
5. Oils and fats processing, *Eg*. in biolubricants, bioemulsifiers, bioemollients.
6. Carbohydrate processing, *Eg*. Laccase in carbohydrate oxidations.
7. Pulp and paper, *Eg*. in pulp bleaching.
8. Detergents, *Eg*. Proteases, amylases and lipases for removal of protein, carbohydrate and fat stains.
9. Waste water treatment, *Eg*. for removal of phenols, dyes and endocrine disrupters.
10. Biosensors/diagnostics, *Eg*. glucose oxidase and cholesterol oxidase biosensors.
11. Delivery of proteins as therapeutic agents or nutritional/digestive supplements *Eg*. beta-galactosidase for digestive hydrolysis of lactose in dairy products to alleviate the symptoms of lactose intolerance.

3.1.7 Poly Vinyl Alcohol (PVA)

Poly vinyl alcohol is manufactured by the hydrolysis of polyvinyl acetate. Even after a prolonged hydrolysis, PVA generally retains around 1 to 2 mole percent of acetate groups. The amount of residual acetate groups affects the physical and chemical properties of PVA, as they are hydrophobic relative to the hydroxyl groups. The residual acetate groups can interfere with inter-molecular and intra-molecular forces such as hydrogen bonding. Highly hydrolyzed PVA has strong hydrogen bonds within and between molecules.

PVA is used as a basic material for a variety of biomedical applications because of their inherent non-toxicity, non-carcinogenicity, good biocompatibility and desirable physical properties, such as elastic nature, good film forming property, high degree of swelling in aqueous solutions, and their water content matches that of biological tissue. PVA gives a hydrophilic tough membrane which can stabilize the activity of various enzymes, including diamine oxidase (DAO) and horseradish peroxidase (HRP), through the inhibition of the formation of non-functional conformations due to the extensive hydrogen bonding between the H atoms of alcohol groups in PVA and the O atoms of the carbohydrate groups in DAO. Moreover, PVA is water-soluble polymer that readily reacts with different cross-linking to form gel. However, owing to strong internal hydrogen bond, it only goes into solution at higher temperature, around 90°C. Aqueous solutions of PVA exhibit non-Newtonian behavior at room temperature.

PVA can be cross-linked chemically or physically to form a hydrogel. The polymer can be cross-linked chemically by any bifunctional agent that can react with organic hydroxyl groups. Some of the various chemicals that can cross-link PVA are glutaraldehyde, formaldehyde, maleic acid and boric acid. Cross-linking can also be achieved physically using ultraviolet (UV) light in the presence of photo-sensitizers, by electron beam or by γ-radiation. Another physical method of cross-linking PVA is through freeze-thaw cycles, where physical bonds are formed.

Cross-linking causes PVA to be insoluble in water. Furthermore, by controlling the cross-link density of this material, variety of transport properties can be obtained. In addition, poly vinyl alcohol is also considered biocompatible.

Protein adsorption onto cross-linked PVA has been shown to be negligible, thus the potential of minimizing fibrotic capsule development around implantable PVA exists. Hence, due to its promised biocompatibility, ease of manipulation and hydrophilicity, PVA has been used extensively in biomedical applications. In the biomedical area, PVA has found use in hernia treatment, in artificial heart valve replacement and as a drug carrier in controlled drug release system among others.

In dense hydrogels, diffusion of solutes is determined by the cross-linking density of the network, or mesh size and the degrees of swelling. The 'pores' are the open spaces between the cross-link points and are not fixed. Different theories have been proposed to describe solute diffusion in hydrogels. Polymer chains are hypothesized to hinder solute diffusion through one or a combination of these methods: by physically obstructing the passage of solute, by increasing the hydrodynamic drag on the solute molecule or by reducing the available free volume for the solute.

3.1.8 Freeze Thawed PVA

Physically cross-linked PVA through freeze-thawing method may be a good choice for immobilizing the enzymes. It can minimize the problems associated with chemically cross-linked PVA. The exposition of aqueous PVA solution to several freeze-thawing cycles leads to reinforced gels owing to a densification of the macromolecular structure, which is a function of the cycling time and temperature. After the freeze-thawing process, fine crystallites are formed due to the slow heat treatment. The chains are physically cross-linked by semi-permanent entanglements, molecular associations or crystalline. Formations of crystallites serve as physical cross-links to render the material insoluble in water.

Some characteristics of the physical cross-linked PVA gels include high degree of swelling in water, a rubbery and elastic nature, and a high mechanical strength because the mechanical load can be distributed along the crystallites of the three-dimensional structure.

The freeze-thawing method as characterized by the absence of chemical cross-linking agents that could compromise is biocompatibility or physical agent, such as gamma radiation that could deactivate the biological substrates, due to the damage caused mostly by the indirect effect of water radiolysis. Generally, physical entrapment of enzyme molecules in polymeric membranes is one of the most advantageous methods because it is rapid, simple and the retained activity is high.

3.1.9 Immobilization for Histamine Biosensor

Several immobilization methods have been used and investigated in order to improve the sensitivity of biosensors. Lange *et al.* (2001) were firstly used 20% of transglutaminase solution for immobilization of diamine oxidase, plasma amine oxidase and tyramine oxidase. The immobilization technique improved the sensitivity of sensor, but unfortunately, when lots of transglutaminase was finished, difficulties occurred with the regular quality (sensor reproducibility was low).

Therefore, they investigated the conventional immobilization method, which was based on glutaraldehyde-albumin cross-linking. In their research, they compared the results obtained (to determine biogenic amines: specifically histamine and tyramine) between enzyme sensor array and high performance liquid chromatography (HPLC). From the results, it can be concluded that by using enzyme sensor array, less time was required to conduct experiments and was not tedious as when handling HPLC, but the reproducibility, data validity, detection limit and so on were still poor.

Glutaraldehyde was always chosen as the cross-linking agent to entrapped diamine oxidase. Tombelli and Mascini (1998) used glutaraldehyde solution on cellulose acetate membrane to immobilize Diamine oxidase on a platinum electrode. Poly vinyl alcohol is a non-toxic water-soluble synthetic material that has good film forming properties, resulting in tough membranes. Glutaraldehyde, a bifunctional agent that can react with organic hydroxyl groups, was used as the cross-linking agent. Glutaraldehyde can also react with the lysine amino acid residues in the enzyme.

The cross-linking process overcomes the loss of enzyme activity due to diffusional loss, which is a prevalent problem for enzyme immobilized in physical entrapment. PVA can stabilize the activity of various enzymes such as horseradish peroxidase. The stabilization effect is achieved through the inhibition of the formation of non-functional conformations due to the extensive hydrogen bonding between the H-atoms of the alcohol groups in PVA and the O-atoms of the carbohydrate groups in Diamine oxidase. These properties make PVA an appropriate matrix for diamine oxidase immobilization.

3.1.10 Immobilization of Antibodies

Immuno-sensors are based on selective bio-affinity interactions between antibodies and antigens. Covalent immobilization of antibodies was developed using thiol-terminal silanes and hetero bi-functional cross-linkers. This process and modified protocols are widely adopted in fibre optic immuno-sensors. A fiber optic immuno-sensor for cocaine detection was developed by covalently immobilized the monoclonal antibody on silanized fibres. In a multi channel fibre-optic immuno-sensor for detection of explosives, antibodies were cross-linked to the silanized fibre with active thiol groups. Brummel *et al.* (1997) developed a fibre optic immuno-sensor for detection of cyclodiene insecticides by covalently binding antibodies on quartz fibres. Before immobilization of the antibody, the fibre was first silanized to introduce active thiol groups, and then reacted with cross-linking reagent. Cullum *et al.* (2000) utilized submicron fibre optic immuno-sensors to measure toxic chemicals in single cells. Fibre tips were coated with silver and then silanized and cross-linked with antibodies.

Antibodies or antigens can be immobilized on the fibre by means of a biotin-streptavidin-biotin bridge. The fibre was first silanized and then incubated in a succinimidyl-derivatized biotin solution. Next, the fibre was transferred to a streptavidin solution. The resulting fibre can be used to immobilize any biotinylated antibody or antigen. In a similar scheme, antibody was immobilized on an optical fibre by means of an avidin-biotin bridge. Avidin was immobilized on the tapered fibres by the hetero bi-functional cross-linker. The captured antigen was then detected by a fluorescently labeled secondary antibody. Polymeric biotinylated thin film on optical fibres can facilitate the links to avidin and further biotinylated antigens or antibodies. Biotin-avidin/streptavidin technique provides a specific and strong non-covalent binding and a passivation layer over the substrate to prevent non-specific adsorptions.

The sol-gel process also provides a suitable host matrix for antibody immobilization. Fluorescein-labeled D-dimer antibodies were immobilized on the optical fibre tip by the sol-gel technique. Sol-gel encapsulation not only retains the activity of the protein by maintaining an aqueous environment around the protein but also enhances the sensitivity and selectivity with high surface areas and optimal pore sizes.

Antibodies can directly adsorb onto the substrate by hydrophobic and electrostatic interactions. Hydrophobic interactions refer to the tendency of hydrocarbons to form intermolecular aggregates in an aqueous medium, which appears as the dehydration of the protein and substrate, and leads to conformational changes of immobilized molecules. Hydrophobic interactions are influenced by the hydrophobicity of the surface. Electrostatic interactions vary with the surface charge, isoelectric point of proteins, pH and ionic strength of the solution.

Immunoglobulin-G (IgG) as one of the major antibodies is commonly applied in immunosensing. Each IgG molecule consists of four polypeptides with two heavy chains and two light chains joined to form a Y shaped molecule (Fig. 3.15). IgG molecules have three protein regions. Two arms of the Y (F(ab')2) are receptor sites that recognize the specific partner. The third region (Fc) forms the base of the Y and it is important in certain aspects of immune response. The adsorption of IgG molecules and their F(ab')2 fragments on hydrophilic surfaces are retarded when repulsive. This effect is stronger for F(ab')2 than for the whole IgG molecule, whereas Fc adsorption is not significantly affected. Therefore the whole IgG molecule tends to adsorb by its Fc fragment in an end-on orientation which is preferred for antigen binding. However, the adsorption of IgG by its

Fc fragment is relatively weak under conditions of low ionic strength. Proteins are more tightly bound to hydrophobic surface than hydrophilic surface. By surface modification techniques such as polyelectrolyte self-assembly, a close-packed IgG layer could be produced which is less time-consuming, cost-effective and simpler than other immobilization techniques.

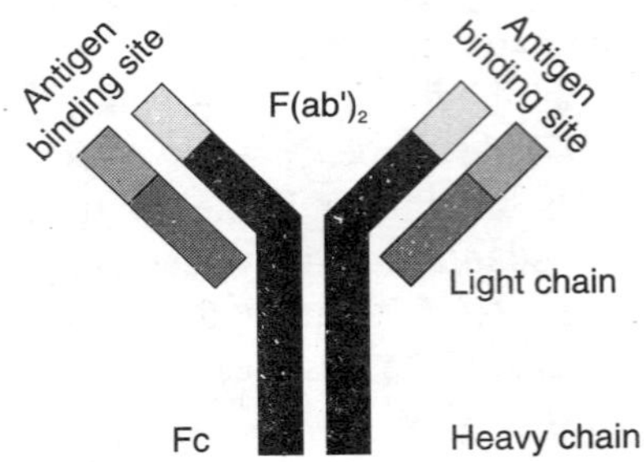

Fig. 3.15 *Structure of IgG molecule.*

In fiber optic biosensors, proteins such as enzymes or antibodies immobilized directly on a solid substrate or encapsulated in a porous matrix are used to detect the presence of a target substrate or antigen. In general, immobilized proteins exhibit lower biological activity compared to those in aqueous media, which may be due to blocking of the receptor's binding sites. In addition, immobilization leads to heterogeneity of the protein activity due to steric effects and conformational changes. Antibodies, enzymes and other receptors exhibit uniform kinetic rate constants and binding affinities. But non-uniform kinetic and thermodynamic properties appear after immobilization. Vijayendran *et al.* (2000) compared the binding heterogeneity of the analyte with antibodies immobilized by different strategies.

Optical fibres were used as the substrate for both random and structured immobilization schemes. Binding affinities of immobilized antibodies were at least an order of magnitude lower than the solution value. Random immobilized antibodies displayed heterogeneous binding characteristics, while more homogeneous protein preparations exhibited more uniform behavior. The decreased biological activity and heterogeneity limit the sensitivity and reproducibility of biosensors. Therefore, considerable interest is focused on increasing the biological activity and eliminating the heterogeneity associated with immobilized proteins. Surface modification techniques are essential for efficient adsorption of biological molecules onto optical fibres. Oriented and site specific immobilization of biological compounds are still in progress.

3.1.11 Immobilization of Enzymes for the Fabrication of Biosensors

Most of the techniques described above have been used for the immobilization of biocatalyst for biosensor applications. The choice of the support and the technique for the preparation of membranes has been dictated by the low diffusional resistance of the membrane coupled with its ability to incorporate optimal amount of enzyme per unit area. In this respect, stable membranes have been prepared by binding glucose oxidase to cheese cloth in the fabrication of a glucose biosensor. Enzymes entrapped inside the reversed micelle have also shown promise in the fabrication of biosensors. Cross-linked enzyme crystals (CLCs) described above provides their own supports and so achieves enzyme concentration close to the theoretical packing limit in excess of even highly concentrated enzyme solutions.

Sensors based on small transducer or thinner enzyme immobilized membranes (miniature biosensors) are also emerging. The development of molecular devices incorporating a sophisticated and highly organized biological information processing function is a long-term goal of bioelectronics. For this purpose, it is necessary in the future to develop suitable methods for micro-immobilizing the proteins/enzymes into an organized array/pattern, as well as designing molecular structures capable of performing the required function. A typical example is the micro-immobilization of proteins into organized patterns on a silicone wafer based on a specific binding reaction between streptavidin and biotin combined with photolithography techniques. Immobilized enzymes have also been used for various other analytical purposes.

3.1.11.1 Immobilized multienzymes and enzyme-cell co-immobilizates

A common feature of metabolic pathways in intermediary metabolism is that the product of one enzyme in sequence is the substrate for the next and so forth. One of the advantages of such an arrangement is that a favorable high local concentration of intermediates in the microenvironment of an enzyme system can be created. Such an approach perhaps will be useful in the development of enzymatic processes requiring multiple enzymes. The enzymes in question can be simultaneously bound to the same support or could be simultaneously immobilized through entrapment.

Often a cell may not contain all the necessary enzyme compliments for carrying out a required process. Thus *Saccharomyces cerevisiae*, a potential ethanol fermentor, cannot utilize lactose or starch since it lacks the enzyme b-galactosidase and amylase, respectively. There is considerable interest in genetically modifying such cells for the expression of the deficient enzymes. The success will depend on the plasmid stability, and quite often on the ability of these cells to secrete these enzymes for their action on the raw materials present in the growth medium. Recently, genetically engineered cells have been obtained which cannot only produce the deficient enzyme but can also transport it to the periplasmic place and naturally immobilize it. Another more practical approach is to immobilize the deficient enzyme directly on the cell wall. This approach which is termed enzyme-cell co-immobilizate permits the tailoring of a whole cell for specific complex chemical conversion by combining the biochemical potential of both the cell and the additional enzyme from another source. Enzymes can also be immobilized on the microbial cell surfaces coated with polyethylenimine through adsorption followed by cross-linking. The different techniques for the preparation of enzyme cell co-immobilizate, and their applications have been reviewed. (Hartmier, W., *Immobilized Biocatalysts; An Introduction*, Springer Verlag, Berlin, 1988).

3.1.11.2 Immobilized enzymes for the regeneration of cofactors

Enzyme technology is mainly with simple reactions which do not require coenzymes. One of the major challenges for enzyme technologists is in the application of cofactor-dependent enzymes. Over 25% of the known enzymes require cofactors like NAD/NADH, ATP/ADP, etc. which, unlike the enzymes, undergo stoichiometric changes during a biochemical reaction. The living cells have circumvented this problem either by introducing certain cofactor-recycling enzyme systems in their metabolic pathways or through their recycling via the electron transport system in the case of aerobic metabolism. *In vitro* applications of such enzymes will largely depend on developing efficient *in vitro* cofactor recycling systems. These can be obtained by co-immobilizing another enzyme so that it can recycle the cofactors, e.g. alcohol dehydrogenase in the presence of ethanol can be used for the recycling of NAD to NADH in a bioprocess catalysed by a NADH-dependent enzyme. Choice of the coupling enzyme for such application has been made based on its ability to use an economical substrate like ethanol or formate (formate dehydrogenase) as well as one which results preferably in a volatile byproduct like CO_2 or acetaldehyde so as to minimize the downstream processing problems. In addition to the recycling of the cofactor other important criterion for development of cofactor dependent biochemical process is the retention of cofactors.

Methods are available for the covalent binding of the cofactors along with the enzymes on polymeric supports or directly on enzymes or on soluble high molecular weight polymers like dextran or polyethylene glycol. Membrane reactors containing continuous cofactor regenerating systems have been used in the large-scale production of sugar derivatives, and optically active a hydroxyl acids and amino acids. Cofactor regenerating system can be obtained by immobilization of permeabilized yeast cells as a source of alcohol dehydrogenase. Multi-enzymes and cofactors immobilized within microcapsules have been used for the multi-step conversion of lipophilic and lipophobic substrates.

3.1.11.3 Applications of immobilized enzymes

The first industrial use of an immobilized enzyme is amino acid acylase, for the resolution of recemic mixtures of chemically synthesized amino acids. Amino acid acylase catalyses the deacetylations of the L form of the *N*-acetyl amino acids leaving unaltered the *N*-acetyl-d amino acid, that can be easily separated, racemized and recycled. Some of the immobilized preparations used for this purpose include enzyme immobilized by ionic binding to DEAE-sephadex and the enzyme entrapped as micro droplets of its aqueous solution into fibres of cellulose triacetate by means of fiber wet spinning developed by Snam Progetti. Rohm GmbH has immobilized this enzyme on macro-porous beads made of flexi-glass like material (Table 3.1).

Table 3.1 Commercial immobilized glucose isomerase preparations.

Enzyme source	*Immobilization procedure*	*Industry*
Steptomyce ribigenosas	Enzyme adsorbed on ion-exchanger	Corn Processing
Bacillus coagulaas	Enzyme mixed with inorganic diluent and formed into solid spheres or cell lysate cross-linked with glutaraldehyde	Novo Industry
Streptomyces ulivaceas	Glutaraldehyde cross-linked whole cells	Miles Laboratory
Actinoplanes missouriensis	Cells entrapped in gelatine and cross-linked by Glutaraldehyde.	Gist Brocades
Arthrobacter sp.	Flocculated cells extruded and dried as cylindrical pellets.	ICI Americas Inc.
Streptomyces olivochromogenes	Adsorption on alumina or ion exchange resin.	CPC Int. Inc.
Streptomyces olivochromogenes	Enzyme adsorbed on controlled pore alumina	Corning Glass Works

The most important application of immobilized enzymes in industry is for the conversion of glucose syrups to high fructose syrups by the enzyme glucose isomerase. Some of the commercial preparations have been listed in Table 3.2. It is evident that most of the commercial preparations use either the adsorption or the cross-linking technique. Application of glucose isomerase technology has gained considerable importance, especially in non-tropical countries that have abundant starch raw material. Unlike these countries, in tropical countries like India, where sugarcane cultivation is abundant, the high fructose syrups can be obtained by a simpler process of hydrolysis of sucrose using invertase. Compared to sucrose, invert sugar has a higher humectancy, higher solubility and osmotic pressure. Invertase is perhaps the first reported enzyme in an immobilized form. The possible use of whole cells of yeast as a source of invertase was demonstrated by D'Souza and Nadkarni, 1978.

Table 3.2 Commercial immobilized lactase preparations.

Enzyme source	*Immobilization procedure*	*Industry*
Saccharomyces lactis	Enzyme entrapped in cellulose triacetate fibres	Gist Brocades
Aspergillus niger	Enzyme covalently bound to silica beads	Corning Glass Works
Aspergillus niger	Enzyme adsorbed to phenol formaldehyde resin	Valio Laboratory
Aspergillus niger	Enzyme covalently bound to silanized Mn-Zn ferrite	Sturge
Aspergillus oryzae	Enzyme covalently bound to macroporous amphoteric ion exchange resin.	Sumitomo

L-aspartic acid is widely used in medicines and as a food additive. The enzyme aspartase catalyses a one-step stereo specific addition of ammonia to the double bond of Fumaric acid. The enzymes have been immobilized using the whole cells of *Escherichia coli.* This is considered the first industrial application of an immobilized microbial cell. The initial process made use of polyacrylamide entrapment which was later substituted with the Carragenan treated with Glutaraldehyde and Hexamethylenediamine. A phenol formaldehyde resin is used for adsorbing aspartase in their continuous process. These processes make use of immobilized nonviable cells of *Brevibacterium ammoniagenes* or *B. flavus* as a source of Fumarase.

One of the major applications of immobilized biocatalysts in dairy industry is in the preparation of lactose hydrolysed milk and whey, using β-galactosidase. A large population of lactose intolerants can consume lactose-hydrolysed milk. This is of great significance in a country like India where lactose intolerance is quite prevalent. Lactose hydrolysis also enhances the sweetness and solubility of the sugars, and can find future potentials in preparation of a variety of dairy products. Lactose-hydrolysed whey may be used as a component of whey-based beverages, leavening agents, feed stuffs, or may be fermented to produce ethanol and yeast, thus converting an inexpensive byproduct into a highly nutritious, good quality food ingredient.

The process makes use of a neutral lactase from yeast entrapped in synthetic fibres. Unlike the milk, the acidic β-galactosidase of fungal origin has been used for this purpose. An immobilized preparation obtained by cross-linking β-galactosidase in hen egg white (lyophilized powder) has been used in laboratory for the hydrolysis of lactose. A major problem in the large-scale continuous processing of milk using immobilized enzyme is the microbial contamination which has necessitated the introduction of intermittent sanitation steps. A co-immobilizate obtained by binding of glucose oxidase on the microbial cell wall using Con A has been used to minimize the bacterial contamination during the continuous hydrolysis of lactose by the initiation of the natural lacto-peroxidase system in milk.

One of the major applications of immobilized enzymes in pharmaceutical industry is the production of 6-aminopenicillanic acid (6-APA) by the deacylation of the side chain in either penicillin G or V, using penicillin acylase (penicillin amidase). More than 50% of 6-APA produced today is enzymatically using the immobilized route. One of the major reasons for its success is in obtaining a pure product, thereby minimizing the purification costs. The first setting up of industrial process for the production of 6-APA was in 1970s simultaneously by Squibb. A number of immobilized systems have been commercially produced for penicillin acylase which make use of a variety of techniques either using the isolated enzyme or the whole cells. This is also one of the major applications of the immobilized enzyme technology in India.

Immobilized oxido-reductases are gaining considerable importance in biotechnology to carry out synthetic transformations. Of particular significance in this regard are oxido-reductase mediated asymmetric synthesis of amino acids, steroids and other pharmaceuticals and a host of specialist chemicals. They play a major role in clinical diagnosis and other analytical applications like the biosensors. Future applications for oxido-reductases can be in areas as diverse as polymer synthesis, pollution control, and oxygenation of hydrocarbons. Keto-acids can be obtained using either L- or D-amino acid oxidases.

The use of D-amino acid oxidase has the advantage of simultaneous separation of natural L-isomer from DL-recemates along with the conversion of D-isomer to the corresponding keto acid which can then be transaminated in the body to give the L-amino acid. Of the several microorganisms screened, the Triangular yeast (*Trigonopsis variabilis)* was found to be the most potent source of D-amino acid oxidase with the ability to deaminate most of the D-amino acids. The permeabilized cells entrapped either in radiation polymerized acrylamide Ca-alginate or gelatins have shown promise in the preparation of alpha keto acids. Another interesting enzyme that can be used profitably in immobilized form is catalase for the destruction of hydrogen peroxide employed in the cold sterilization of milk. A few reports are available on its immobilization using yeast cells (Immobilization of yeast cells in several conducting polymer matrices, selmiye Alkan, 2002).

Lipase catalyse a series of different reactions. Although they were designed by nature to cleave the ester bonds of triacylglycerols (hydrolysis), lipase are also able to catalyse the reverse reaction under micro aqueous conditions, viz. formation of ester bonds between alcohol and carboxylic acid moieties. These two basic processes can be combined in a sequential fashion to give rise to a set of reactions generally termed as inter esterification. Immobilized lipases have been investigated for both these processes. Lipases possess a variety of industrial potentials starting from use in detergents; leather treatment; controlled hydrolysis of milk; fat for acceleration of cheese ripening; hydrolysis, glycerolysis and alcoholysis of bulk fats and oils; production of optically pure compounds, flavors, etc. Lipases are spontaneously soluble in aqueous phase but their natural substrates (lipids) are insoluble.

A variety of biologically active peptides are gaining importance in various fields including in pharmaceutical industries and in food industries as sweeteners, flavorings, antioxidants and nutritional supplements. Proteases have emerged over the last two decades as powerful catalysts for the synthesis and modification of peptides. The field of immobilized proteases may have a future role in this area. One of the important large scale applications will be in the synthesis of peptide sweetener using immobilized enzymes like Thermolysin. Proteolytic enzymes, such as Subtilisin, alpha-Chymotrypsin, Papain, Ficin etc., which have been immobilized by covalent binding, adsorption or cross-linking to polymeric supports are used (Bayer AG) to resolve *N*-acyl-DL-phenylglycine ester racemate, yielding *N*-acyl-D-esters or *N*-acyl-D-amides and *N*-acyl-L-acids. Immobilized Amino peptidases have been used to separate DL-Phenylgycinamide racemates.

Biosensor Technology and Fabrication

Chapter 4

4.0 BIOSENSOR TECHNOLOGY

4.0.1 Introduction

To meet the demands in indigenous development of sensors for food industry, drink industry and environmental analysis, the research activities have been made possible through the establishment of a multidisciplinary lineup in Biochemistry, Microbiology, Chemistry, Biotechnology, Electrical Engineering, Instrumentation and Signal Processing.

An amperometric sucrose biosensor was constructed for food and drink industries. This prototype is composed of a microprocessor analysis system, a potentiostat and an enzyme electrode. The biocatalytic components are glucose oxidase and invertase, co-immobilized on carbon paper. The linear range is 0-8 milli molar and the response time was 1-2 minutes. The enzyme electrode could be stored longer than 12 months. Sucrose biosensors are advantageous in that they provide simple and rapid automated analyses, with low running cost. There are many food industries in which real benefits will be obtained from their introduction such as the sugar industries, sweetened condensed milk industry, canned juice industry and distillery industry.

Biosensor technology uses antibodies to detect biomarkers (molecules) in the human body which are often a marker for disease much faster than current testing methods. The technology could be used in doctors' surgeries for more accurate recommendation to consultants and in hospitals for rapid diagnosis. Tests have shown that the biosensors can detect a wide range of analytes (substances being measured), including biomarkers present in prostate and ovarian cancer, stroke, multiple sclerosis, heart disease and fungal infections, also believes that the biosensors are versatile enough to test for diseases such as tuberculosis and HIV.

Currently blood and urine are tested for disease markers using a method called ELISA (Enzyme Linked Immunosorbant Assay), developed in the 1970s, the process takes an average of two hours to complete, is costly and can only be performed by highly trained staff. Now new techniques are developed for attaching antibodies to innovative surfaces and novel electronic measurement methods that need no reagents or labels. New technology which provides results in 15 minutes or less could be developed in a small mobile sized device into which different sensor chips could be inserted, depending on the disease being tested for.

4.0.2 DNA Biosensor Technology

DNA biosensors are being used in identifying genetic diseases, cancer research, identification of dangerous microorganisms and forensic science. A significant advantage of DNA based sensing devices is that no external monitoring is needed. DNA biosensors are complicated mini machines consisting of sensing elements (probes that match up in sequence with the DNA to be detected), micro lasers and a signal generator. The probe can be free in a solution but in all cases the principle of specific hybridization, between single stranded DNA (ssDNA) known as the probe and target sequence which must be detected, plays a key role in DNA based biosensors. The converted signal is a response to hybridization of the probe and target sequence can be optical, electrochemical or pizoelectrical DNA biosensor function is the fact that two strands of DNA stick to each other by virtue of chemical attractive forces. On such a sensor, only an exact fit that is, two strands that match up at every nucleotide position gives rise to a fluorescent signal (a glow) that is then transmitted to a signal generator. Ideally, the sensor would be a little square of a chip that could be immersed in a test fluid blood, for instance to pick up traces of disease causing bacteria or viruses.

More DNA fragments can be immobilised in smaller geometric areas, typically a few milli meters square. When the chip is exposed to a sample of unknown DNA the complementary strands join up revealing the sequence of the unknown DNA. This technology is not only applicable to DNA chips but might allow the production of biosensors using a wide range of biomolecules which may be used as miniature implantable sensors for monitoring conditions within the body, *e.g.* the development of an artificial pancreas which could both measure glucose and control insulin delivery would be of major benefit to diabetics.

Innovative methodologies for bacterial detection are being developed to improve sensitivity and speed in detection. The polymerase chain reaction (PCR) methodologies include multiplex and real time PCR analysis which require 5-7 days for microbial detection. Similarly, the solid phase microarray format emerged as the preferred method for high throughput, high parallel hybridization testing for DNA and RNA samples. Currently available microarray technologies suffer from certain limitations that prohibit the exploitation of the full range of life science applications. The outcomes are not promising; the majority of the methods developed require enrichments in culture media before the analysis can begin. Problems associated with the established fluorescence based optical detection technique include the high equipment costs and the need to use sophisticated numerical algorithms to interpret the data. DNA biosensor is one of the proposed solutions for the detection of food-borne pathogens. Over the past decade research efforts have advanced the development of DNA biosensors but improvements are still needed before DNA biosensors become a real and reliable choice.

4.0.3 Wireless Biosensor Technology

Biosensors will provide significant advantages over more conventional methods. The biggest problem with conventional measuring systems is poor skin sensor contact. In subcutaneous measuring systems the sensor to body contact is more stable. Furthermore, external electrical interference of the measurement signal is reduced, which improves the measurement result. Health care costs are saved, when monitoring is not time and place dependent, patients will no longer have to make an appointment with the physician for a consultation or tests. Patients under remote supervision can continue living their normal lives for a longer period of time.

The integration of electronics and development of packaging technologies make it possible to manufacture sensors and electronics on a silicon chip no bigger than a fingernail. Due to its light weight and small size several of these types of chips can be implanted in a sizable area. The sensors might also contain various micro sensors, measurement electronics and wireless communication circuits. In addition to the small size, the packaging is also challenged by the environment into which it is placed. The wireless developed sensors and technologies which pose no risk to the patient's health. The work is extremely challenging because the electronics have to work reliably for long periods of time under the skin, in a moist, corrosive environment and they must not pose any health hazards, even if the protective coating were to be damaged for some reason.

Advances in biomaterial technologies allow the biocompatible coatings of sensors to be customized for each application. It is even possible to incorporate functional elements by enhancing the implant coating with a layer that releases antibiotics. Sensors should also be flexible, so that they can follow the patient's movements. This requires that the sensor circuit boards are flexible and its components are thin enough to bend with the circuit board. A silicon chip reduced to a thickness of less than 0.1 millimetre will be flexible. When this flexible package is coated with a thin, protective and biocompatible material, the entire unit will effectively flex with and withstand the patient's movements while implanted.

4.0.4 Chip-Based Biosensor Technology

The rapid and accurate identification of harmful biological agents in food, water and the environment has never been so important. As well as adventitious diseases, which claim over 20 million lives every year, the terrorist threat of biological warfare now looms over us. In the event of an attack involving organisms such as Anthrax (*Bacillus anthracis*) or Plague (*Yersinia pestis*) only rapid detection and containment will allow effective control measures to be implemented. Similarly, it is important to identify plant diseases very quickly so that control strategies can be put into place and crop losses can be minimized.

Conventional methods for the detection and identification of microorganisms (and the products derived from them) are based on culture enrichment followed by biochemical and microbiological tests which in some cases take more than a week to perform. This means that positive identification often comes too late to prevent a major disease outbreak. In humans this also means that medical staffs are unable to make informed choices concerning the drugs administered to critically ill patients.

Over the last decade, several new technologies have emerged that facilitate the rapid and accurate detection and identification of pathogens. This has culminated in the development of a range of highly sensitive, miniaturized devices that are easy to use and allow biological agents to be detected and identified in non-enriched field samples within minutes or hours. These biosensors are mechanistically diverse but each employs a sensitive and specific molecular interaction assay to detect the target pathogen and converts the interaction into a signal that alerts the operator to the presence of that pathogen in real time. Therefore, the successful cooperation of biotechnology, micro systems technology, microelectronics and information processing plays a key role within the current scientific research. The interdisciplinary team of biologists, chemists, computer specialists, engineers and physicists is connecting the excellence research into theoretical relationships and development of processes for analysis, diagnosis as well as screening on a biomolecular and cellular level.

4.0.5 Urinary Tract Biosensor Technology

According to the American Urological Association more than 7 million urinary tract strips are using in medicine and health care, urinary tract infections (UTIs) afflict an estimated 40% of women and 12% of men at least once during their lifetime. Though the infections are easily cured with antibiotics, diagnosis requires a lengthy bacterial identification process that can compromise patient care. However, the development of DNA biosensors may enable specific bacteria identification in as little as 45 minutes.

A micro fabricated electrochemical sensor array developed by Gene Fluidics correctly identified the gram negative bacteria in 98% of UTI tainted urine samples. The results of the study, conducted by the David Geffen School of Medicine at UCLA and the Veterans Affairs Greater Los Angeles Healthcare System, mark the first species specific detection of bacteria in human fluid samples by this type of instrument. Culture specimens until bacteria are large enough to visually identify a process that can take 48 hours. The sensor array used in the UCLA study produced results in just 45 minutes. These expedited results could enable physicians to prescribe targeted antibiotics. In order to alleviate patient discomfort, physicians frequently prescribe antibiotics to patients exhibiting symptoms of UTIs without knowing the specific bacteria causing the infection. If this sensor technology is adopted by the healthcare system, doctors will be able to identify the most effective antibiotic for the identified bacteria.

Fig. 4.1 *A sensor chip coated with species specific genetic probes is placed in a multi channel reader instrument that measures electrical currents coming from sensors on the chip that detect bacteria.*

Gene Fluidics 16-sensor chips coating with UCLA designed species specific genetic probes. Each sensor featured three single layer gold electrodes and each electrode contained one representative capture probe for a bacterial urinary pathogen. Urine samples were then applied directly to the sensor chips. Gene Fluidics multi channel reader instrument measured electrochemical signals which enabled researchers to identify UTI pathogens by observing elevated signals on the chip.

4.0.6 Ambri Technology

The biosensor converts a biological binding event into a digital electrical signal, this enables the biosensor to employ computer technology to analyze and define this biological event. By eliminating the visual reading or human interpretation of the result of a biochemical binding event, the biosensor is capable of producing a response which is objective, more accurate and reproducible than alternative technologies.

AMBRI has built an Ion Channel Switch, a membrane which can detect the presence of specific molecules and signal their presence by triggering an electrical current (Fig. 4.2). Biosensor is a two molecular layer self assembled membrane based on the ion channel gramicidin. The AMBRI Biosensor required various components which are the molecule to be detected, the antibody fragments, the linker protein Streptavidin and biotin linkers, the membrane layers with included gramicidin molecules.

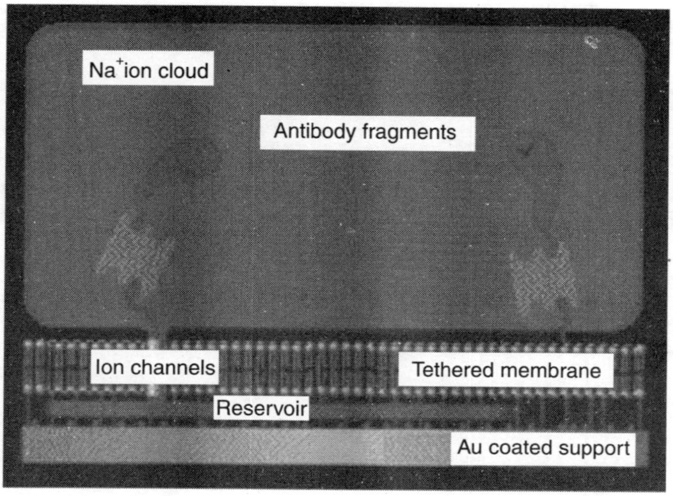

Fig. 4.2 *The various components of the AMBRI biosensor. (Source: AMBRI® biosensor, Gene Ostrovsky, 2005)*

The AMBRI biosensor operates as a synthetic mimic of a nerve cell membrane. The required elements for this artificial membrane which include: membrane forming molecules chemically tethered to a surface, simple ion channels within the membrane which facilitate the transport of ions like sodium, an ionically conducting reservoir space between the electronically conductive gold surface and the membrane to store ions when they have crossed the membrane and receptors such as antibodies attached to the membrane to recognize target molecules.

The detection mechanism operates by the binding of the target molecule to the antibody fragment, altering the population of conduction ion channels pairs within the tethered membrane. This result is a change in the membrane conduction. The single crystal structures of component proteins are useful in explaining the mechanism of AMBRI Biosensor (Fig. 4.3).

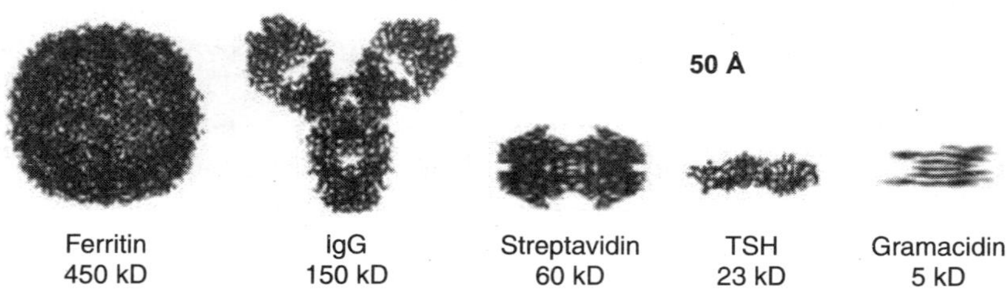

Fig. 4.3 *Single crystal structures of the component proteins. (Source: AMBRI® biosensor, Gene Ostrovsky, 2005)*

The binding of the analyte (green molecule) to the antibody fragments (red molecules) causes them to cross-link the mobile gramicidin. The ion channels to tethered sites on the membrane preventing the formation of conducting dimers. This causes a loss of conduction of ions across the membrane between the outer solution and the inner reservoir space (Figs. 4.4 to 4.6). The measurement of current conduction of DNA, *E.coli* and TSH and compared with the control, and decreasing current measurement shown in the Figs. 4.4 to 4.6.

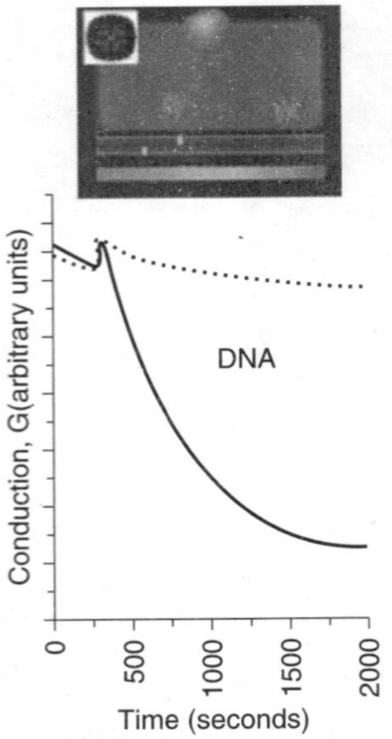

Fig. 4.4 *Reaching the binding anaylte to antibody fragments.*

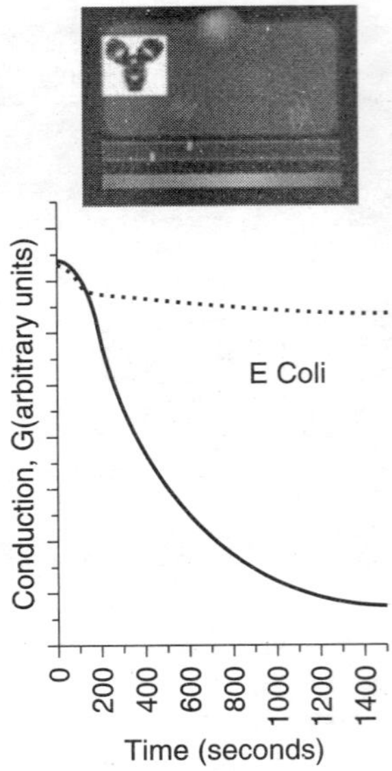

Fig. 4.5 *Interaction of analyte and antibody fragments causes signal product.*

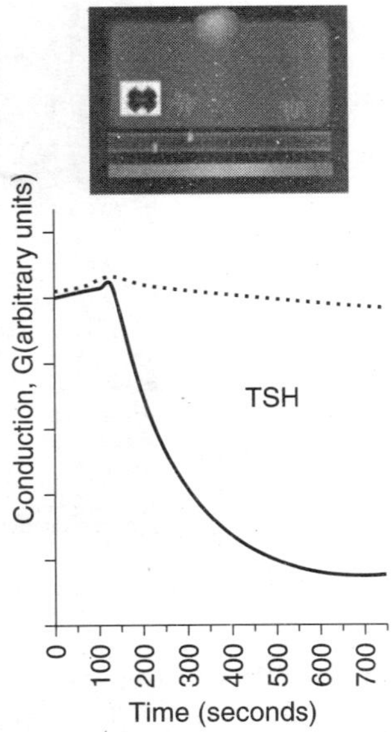

Fig. 4.6 *The single product to cross-link the mobile gramiciding A ion channels.*

4.1 FABRICATION

The method of fabricating a biosensor composed of a biological material and transducer. In direct immobilization process, the biological material on the signal transducer by the use of hydrophilic polyurethane, thereby fabricating a biosensor which is superior in specificity, selectivity and stability. The most important factors for the analysis of biosensors include the specificity and sensitivity for the targets to be analyzed. While the specificity for targets is dependent on the biochemical properties of the bioactive reagents and the sensitivity is determined by the recognition efficiency of sensing elements. Biosensors have a characteristic advantage different from those which chemical sensors have of showing excellent specificity for targets as well as maintaining high sensitivity by virtue of the molecular recognition of the bioactive reagents.

The biosensors also suffer from a disadvantage, in that it is very difficult to obtain highly reliable and accurate analysis results because the targets exist at trace amounts with great difference in biochemical properties and molecular weights and the signals are generated through complex reaction pathways. In order to maximize the sensitivity of biosensors with efficiency in the detection of analytes, the bioactive reagents are preferably immobilized in a layer adjacent to the sensing surface of the basic electrochemical transducer. More preferably, the bioactive reagents are immobilized in such a way as to react with their targets at maximal magnitudes. Another requirement is that the bioactive reagents should maintain their entities without reduction or extinction of their biochemical activity. Therefore, the immobilization of the bioactive reagents such as enzymes or functional proteins on an appropriate support occupies an important position in the research on biosensors. In addition, active research has been directed to the introduction of the biosensors into solid-state electrodes because they can offer the advantage of miniaturization and mass fabrication for cost reduction.

Sensor structures of the invention have higher sensitivity than previous structures due to the use of nanoporous low refractive index material instead of a polymer sub-wavelength period grating structure. When the refractive index of the sensor structure (directly beneath the sub-wavelength grating structure) is reduced below the refractive index of any liquid used in a sample, the electromagnetic field of the photonic crystal interacts more strongly with the test sample, yielding a structure whose reflected wavelength is more strongly tuned by a given amount of adsorbed biological material and can be detecting, *Eg.,* a single cell attached to its surface.

A sol-gel derived low index nanoporous silica thin film was incorporated into a sensor in the place of the UV cured epoxy. Since the low index material cures by heat rather than UV exposure, it was necessary to develop a new fabrication process. It was desirable to retain a low cost imprinting method, though it was obvious that a plastic substrate could not sustain the requisite high temperatures for porous glass annealing. One possible approach to sol-gel glass imprinting was to use a Poly Dimethyl Siloxane (PDMS) mold and a glass substrate.

The sub-wavelength grating structure of the low-k biosensor was fabricated using a combination of lithography, molding, and imprinting processes. Sylgard 184 PDMS (Dow Corning) daughter molds are first replicated from a silicon master wafer patterned with a positive image of the surface structure desired in the sensor. To facilitate release of the PDMS mold from the silicon wafer, the wafer was surface treated with Dimethyl Dichloro Silane (DDS), to improve stamps for 10 nm level wafer scale nano-imprint lithography. The PDMS replicas are then used to imprint a thin film of uncured nano glass, a low-index sol-gel glass, spun-on to a glass substrate. Once the low-index dielectric becomes rigid, the flexible PDMS mold is removed and the sol-gel glass is fully cured by further baking. The sensor structure is completed by evaporating 175nm of TiO_2 onto the patterned surface. A subsequent surface treatment with Dimethyl Dichloro Silane encourages bio-adsorption and promotes sensor stability.

A plastic calorimetric resonant optical biosensor used for multi parallel detection of label free biochemical interactions. Sensors and actuators structures used a 550nm and 170 nm imprint depth though the polyester or polymer and low index porous glass devices use 120nm and 165nm TiO_2 coatings, respectively. The two devices will be referred to as the polymer and porous glass sensors. The polymer devices were provided as an array of sensors aligned and attached to bottom less 96 well standard micro titer plates. The porous glass devices are fabricated on 75 mm × 25 mm × 1 mm glass microscope slides. Adhesive rubber wells are attached to the glass surface to provide liquid containment for 5-6 sensors on each slide. Deionized water and Isopropyl Alcohol were used to determine the bulk shift coefficient of each sensor. First, DI H_2O was pipetted onto the surface of the sensor and the PWV was measured. A broad wavelength light source was coupled to an optical fiber that illuminates a ~2 mm diameter region of the photonic crystal surface from below the substrate at normal incidence. Reflected light was collected by a second optical fiber that is bundled next to the illuminating fibre and measured by a spectrometer.

4.1.0 Photolithography

Fabrication started by depositing 2 μm and 4 μm of oxide on the top and backside of the packaging wafer respectively. Standard photolithographic protocols, proximity photolithography and etching in an inductively coupled plasma (ICP) tool were used to pattern the oxide film on the top of the packaging wafer (Fig. 4.7: step 3), once confirmed that the oxide was imprinted down to the silicon using optical microscopy, patterned the backside oxide film using a double side mask aligner (Fig. 4.7 step 4).

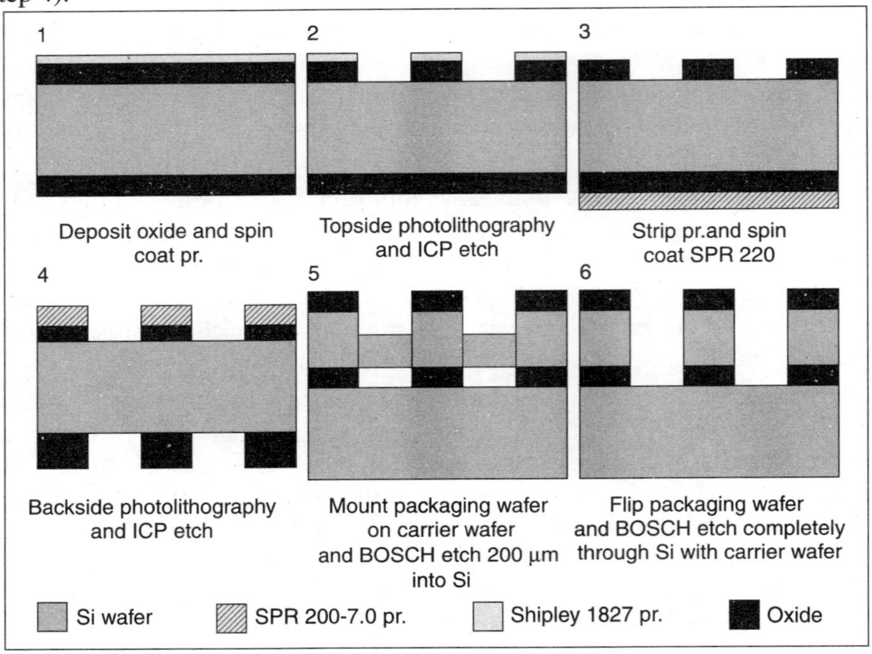

Fig. 4.7 *Diagram of fabrication process. (Source: Glass-Frit Bonding for Selective Functionalization and Packaging of Biosensors; bauerstatus@gmail.com)*

Using the Bosch process then imprinted 200 μm into the silicon wafer from the topside and then etched completely through the wafer from the backside. This process allowed wedging the resonators in the windowed areas on the packaging wafer. Continuing the fabrication, screen printed the glass-frit material onto the packaging wafer (Fig. 4.8). In order to eliminate the organic binder in the glass-frit material, then went through an organic burnout process. This process step fixed the glass frit material to the substrate, an initial step towards completing the bonding process. The fabrication was then completed by bonding the packaging wafer to a test wafer.

The packaging wafer was successfully fabricated and the steps were carefully characterized. For the silicon etching with the Bosch process, acquired an etch rate of 800Å per cycle and an etch rate selectivity to SiO_2 of 178 to 1. In 1827 Shipley, photo- resist was used for the topside photolithography while Mega posit SPR 220-7.0 photo-resist was used for the backside photolithography due the difference in oxide thickness. The thickness of the screen printed glass frit ranged from 30-50 μm. In order to confirm whether this thickness will be sufficient the packaging wafer has to be bonded to the device wafer.

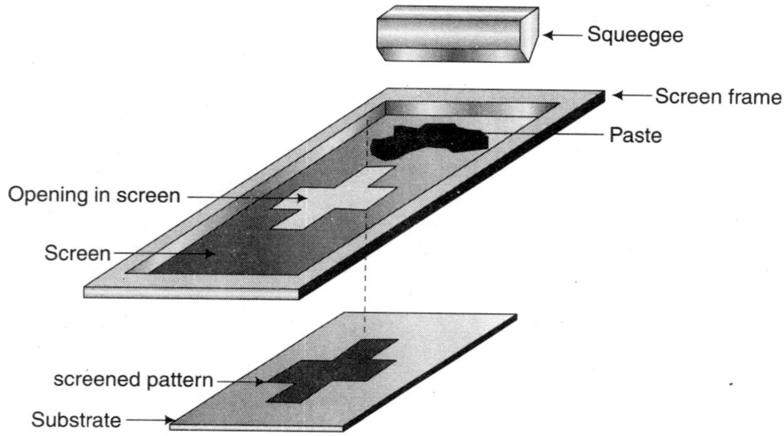

Fig. 4.8 *Diagram of screen printer. (Source: Glass-Frit Bonding for Selective Functionalization and Packaging of Biosensors; bauerstatus@gmail.com)*

4.1.1 Complementary Metal Oxide Semiconductor Technology (CMOS)

Optical biosensors represent a recent trend in biosensor development outstanding to their high sensitivity, ease of operation, high accuracy and wide detection capacity. The need for an expensive setup involving a photomultiplier tube and associated equipment, limits the application of the fiber-optic biosensors to the development of simple portable devices for fast screening and in-field/on-site monitoring.

The fabrication of a small optical biosensor was made by replacing the photomultiplier tube with a photodiode to detect luminol luminescence directly (Preuschoff *et al.,* 1994). Although the achievable detection limit of hydrogen peroxide was low, the overall setup of this optical biosensor was complex. Powerful and miniature photodiode at low cost can be fabricated by employing the mature complementary metal oxide semiconductor (CMOS) technology (Fig. 4.9). The miniature optical biosensor was developed by integrating the CMOS photo sensor, the sol–gel technique and the Chemi-Luminescence (CL) detection method. This miniature CMOS based optical biosensor not only exhibits high detection capability, short detection time and high reproducibility, but also requires a small sample, with these characteristics the developed CMOS based optical biosensor has great potential in the future development of a simple portable biosensing system for rapid screening and in-field/on-site monitoring.

In fabrication process of CMOS the required chemicals are D-Glucose, Horseradish Peroxidase, Luminol, Tetra Ethyl Ortho Silane (TEOS), Silanol-terminated Poly Dimethyl Siloxane (PDMS, molecular weight 1700), Tris Base, H_2O_2 (30%), Glucose oxidase from *Aspergillus niger* and Chlorobutadiene.

HRP stock solution (15.5 mg/ml) was prepared in 100mM Tris–HCl buffer, pH 8.6; Glucose oxidase stock solution (144 mg/ml) was prepared in 100mM phosphate buffer, pH 7.0. The stock enzyme solutions were stored at 20ºC. The HRP and GOx solutions were melted on ice just before use and were diluted with pH 7.4 phosphate buffer saline (1× PBS). One unit of HRP corresponds to the formation of 1.0 mg of purpurogallin from pyrogallol in 20 seconds at pH 6.0 and 20º C.

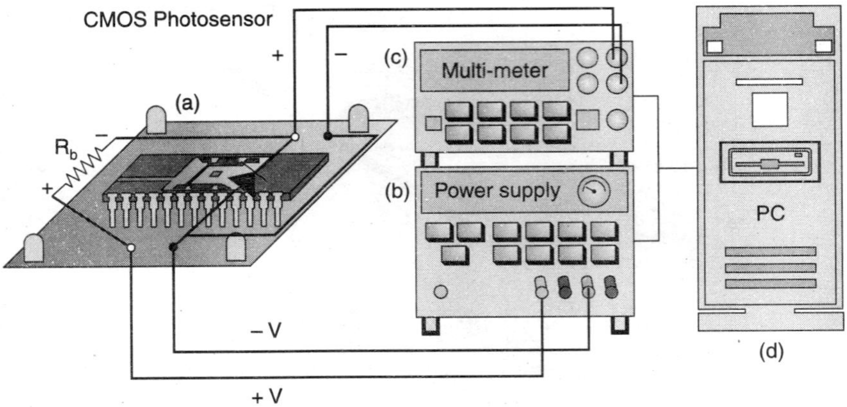

Fig. 4.9 *Experimental setup for fabricating a miniature CMOS-based optical biosensor. (Fabrication of a miniature CMOS-based optical biosensor, Chiun-Jye Yuan et al., 2007)*

(a) The CMOS photodiode was fixed on the PCB board, which was connected to E3646 and 34401A. (b) An E3646A power supply (Agilent) was utilized to apply a revise bias to the CMOS photodiodes. (c) A 34401A multimeter (Agilent) was utilized to measure the luminescence generated current. (d) A PC connected to E3646 and 34401A via a GPIB constituted the automatic detection system.

4.1.1.1 Fabrication process

Silica sol–gel has many advantages in the immobilization of enzymes but there is a limitation in industrial application. A Tetra Ethyl Ortho Silane (TEOS) or Poly Dimethyl Siloxane (PDMS) hybrid material was developed and shown to have great potential to immobilize enzymes. This organic–inorganic hybrid material was demonstrated to overcome the drawbacks of conventional silica sol–gel by improving the stability of entrapped enzymes. Therefore, the TEOS/PDMS ORMOSIL (Organically Modified Silica) was adopted to immobilize Horse Radish Peroxidase (HRP) on the surface of the CMOS photodiode to form a miniature optical H_2O_2 biosensor (Fig. 4.10A). The HRP catalyzed, H_2O_2 dependent luminol luminescence was detected in real time on the developed miniature optical H_2O_2 biosensor (Fig. 4.10B). The luminescence rapidly reached its peak in about 5 seconds before rapidly declining which result is consistent with the fact that HRP catalyzed a flash type luminol luminescent reaction, for this only a small sample volume (10μl) was required. This reveals that the fabricated optical H_2O_2 biosensor is useful for monitoring the real time responses to HRP catalyzed luminol luminescence in a small volume.

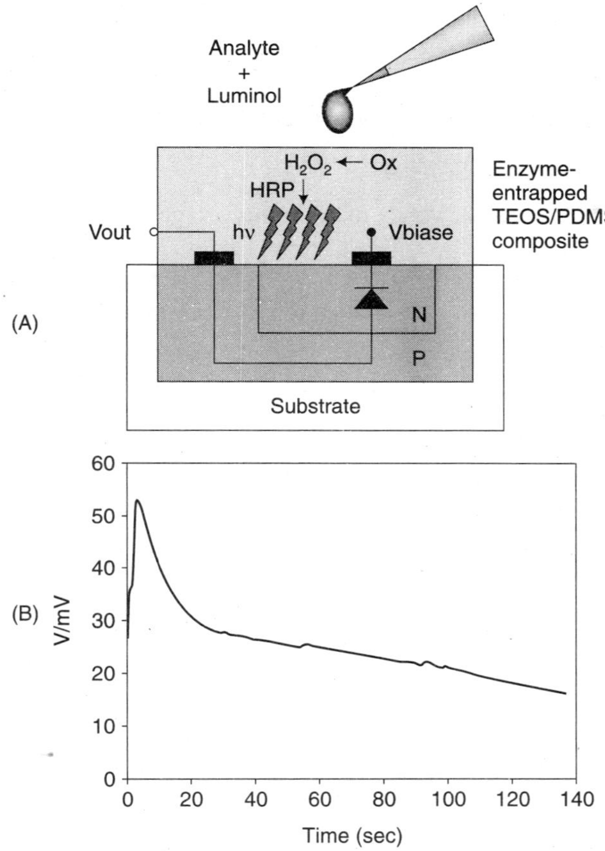

Fig. 4.10 *Real-time measurement of HRP-catalyzed luminol luminescence on a miniature optical H_2O_2 biosensor. (Fabrication of a miniature CMOS-based optical biosensor, Chiun-Jye Yuan et al., 2007)*

CMOS based optical biosensor (Fig. 4.10.A), the P–N junction diode was implemented with N^+ diffusion and a P-well and was manufactured using a standard 0.35μm CMOS process. The enzyme ORMOSIL composite was coated directly onto the photodiode. A mixture of analyte and reagent (10μl) was added to the enzyme ORMOSIL composite to start the reaction. Luminescence generated during the enzyme reaction was detected by the bottom CMOS photodiode, which transforms luminescence to photo current.

Real-time detection curve of HRP catalyzed luminescence on CMOS-based H_2O_2 biosensor (Fig. 4.10.B), the reaction was performed at 25°C in 100mM TrisHCl buffer (pH 8.5) in the presence of 1mM H_2O_2 and 10mM luminol. The total reaction volume was 10μl.

Directly coating enzyme TEOS/PDMS Ormosil composite onto the CMOS photodiode reduced the distance between the photodiode and the emitting source of luminol luminescence. The Inverse Square Law states that the illuminance (E) or light intensity per unit area of a target surface varies in inversely proportional to the square of the distance (d) ($E = I/d^2$). In the developed luminescence sensing systems, the distance between the illuminance source and photodiode is around 2 mm.

Direct immobilization of enzyme on the surface of CMOS photodiode eliminate the light scattering effect and allows a photo current to occur over the threshold of the CMOS photodiode even with a volume of around 10 μl. Hence, the miniature optical H_2O_2 biosensor has great potential for the future development of a portable luminescence sensing system.

4.1.2 Fabrication of Glucose Sensor

Glucose biosensors were fabricated which took an advantage of the catalytic action of glucose oxidase with proton sensing wire type solid state electrodes. A small amount of poly lysine was added when directly immobilizing glucose oxidase in hydrophilic polyurethane membrane with the aim of enhancing the permeability of the resulting polyurethane membrane.

The glucose oxidase catalyzed the oxidation of glucose to gluconic acid while the signal transducers detected the resulting pH change. A response curve to glucose was obtained a slow response behavior was observed with the resulting glucose sensor which may be attributed to a slow diffusion of glucose through the polyurethane membrane and to a certain amount of time required for the pH change of the buffer within the layer adjacent to the pH sensing surface. The method using hydrophilic polyurethane which allows the simplification of the fabrication of biosensors without conducting complicated chemical reactions and washing steps such as cross-linking. The enzyme entrapped within the hydrophilic polyurethane film can retain its high activity for an extended period of time and the intrinsic potentiometric response of the underlying ion selective polymeric membrane is not affected by the enzyme immobilized polyurethane film coated on its sensing surface. Therefore the biosensors fabricated according to the present invention are superior in specificity, selectivity and stability. Further the adoption of solid state type electrodes by virtue of the direct immobilization of the bioactive reagents allows the biosensors to be miniaturized offering the advantages of the detection of small quantities of samples and mass production for low cost.

GOx (from *Aspergillus niger*), Potassium ferricyanide and β-D(+) glucose. Glucose solutions were prepared before use and stored at 4°C. The 0.1 M, pH 7.4 phosphate buffer solution (PBS) was prepared by stock standard solutions of NaH_2PO_4 and adjusting the pH with 0.5 M NaOH.

The electrode surface was examined by using an Olympus STM6 microscope. The electrochemical measurement and cyclic voltammetric experiments were carried out at a CHI1221 electrochemical analyzer. A two electrode configuration was used in which one Au-BPE served as the working electrode and the other Au-BPE as the reference/counter electrode. The Au-BPE with a diameter of 1.145 mm was dried at room temperature (25 ± 2°C) to 50% relative humidity before use. All applied potentials were versus the reference/counter electrode and carried out at room temperature. Reference values measured by using the YSI 2300 STAT Blood Glucose Analyzer were used to estimate the accuracy of capillary blood glucose.

Fabricating of glucose sensors by screen printing technology, the fabrication steps are largely reduced. Barrel plating of gold was performed by immersing the electrodes into a 2 g/L KAu $(CN)_2$ solution in a hexagonal plating barrel and applying a potential in the range of 6–8 V for 50 min. The final diameter of the Au-BPE was ~1.145 mm. The electrodes are plugged into the hollow holes physically without the use of adhesion to connect the electrodes with the base. The key is that the aperture of the holes should be smaller than the diameter of the Au-BPE. Large number of electrodes, i.e., 0.8 million electrodes/per lot can be processed in barrel plating manufacture process. Then chemical reagents containing 48.5% GOx, 8.5% potassium ferricyanide and 43% non-reactive ingredients are immobilized on the surface of the Au-BPEs. The final step is to place a hydrophilic film (3M) around the rectangle area of integrated reagent/blood separation film to finish fabricating the electrode strips. Glucose can be detected as the blood is drawn gently into the strip by capillary action.

The amperometric current response of glucose was measured at a fixed potential of +0.2 V versus the reference/counter electrode and collected with a 2 seconds delay time. The laboratory results were analyzed by linearity test and Clarke error grid analysis. The use of ferricyanide as a redox mediator to lower the applied potential to discriminate from interfering species is first evaluated at the proposed sensor. The typical cyclic volta metric responses are of ferricyanide at the Au-BPE. As can be seen, well defined and reversible one electron transfer voltammograms up to 20 mM ferricyanide with a good linear calibration plot for the measured oxidation current with respect to the concentration of ferricyanide (slope = –0.484 μA/mM, r^2 = 0.997) are observed. Since ferricyanide exhibits an excellent electrochemical activity as an electron transfer mediator at the Au-BPE, glucose oxidized by GOx and the electrons involved in the redox reaction can be relayed through the mediator to the electrode. The resulting current responses proportional to the level of glucose in sample solutions can thus be used for quantification.

The strip structure was designed for mass production to ensure sample introduction by capillary action for simple and universal usage. To verify this, 10 sensor strips were used for the detection of each glucose concentration by measuring the amperometric responses with varying concentration of glucose at various times of 2, 5 and 10 seconds. The observed representative dynamic responses of the sensors toward each concentration are shown good correlation of current signal versus the concentration of glucose in pH 7.4 PBS. Most importantly, in the proposed strip, a 2 second delay time is enough for the sample to reach the working electrode for good correlation between the response current and glucose concentration (slope = –0.0397 μA/mM, r^2 = 0.999). Compared to the most known glucose sensor this is a fast-response sensor with good linearity in the measurement of glucose concentration particularly for clinical use in diabetic patients. As to the electrode-to-electrode reproducibility, the CV of <2% (n = 10) is better than most commercially marketed glucose sensors. Overall the sensor-to-sensor reproducibility, sensitivity and linearity obtained with the proposed sensor system appear to be adequate for glucose measurements in physiological samples.

4.1.3 Field Effect Transistors (FET) Fabrication

GaN made nanowires such as Field Effect Transistors (FETs) because that the properties of GaN has a wide direct band gap of 3.4eV at room temperature, can be doped to be either a p- or n-type semiconductor and electronic devices like diodes. Using standard lithography processes small dots of Ni (0.5-1 μm squares) were patterned over a wafer. NWs were grown from the Ni pattern and 15nmTi/ 150nmAl/50nmTi/50nmAu source/drains were patterned on top of the NWs. The distance between the source and the drain was varied from 1-3 μm. The natural oxide layer was etched off the back of the samples before current/voltage measurements were taken.

All of the FET devices that looked at with the SEM had more than 2 NWs connecting the source to the drain (Fig. 4.11). The device with the fewest connecting NWs is shown in Fig. 4.11. In the case where there was only 1 μm between the source and the drain the NWs are so crossed and tangled that it was not possible to count how many NWs were connecting the source and drain. For a 1 NW connection, only one Ni dot is necessary for each FET and that the distance between the source and drain (for geometrical configuration) should probably be between 4-6 µm. Current/voltage curves showed little modulation as the back gate voltage was changed. This may be because we had not put a metal contact on the back of the devices. Some of the devices were asymmetrical about the y-axis, which may be caused by poor contact with the GaN NW. Despite these imperfections, almost every single device measured demonstrated FET electrical behavior: a linear relationship between source and drain current and voltage until high voltages were reached and current was close to constant.

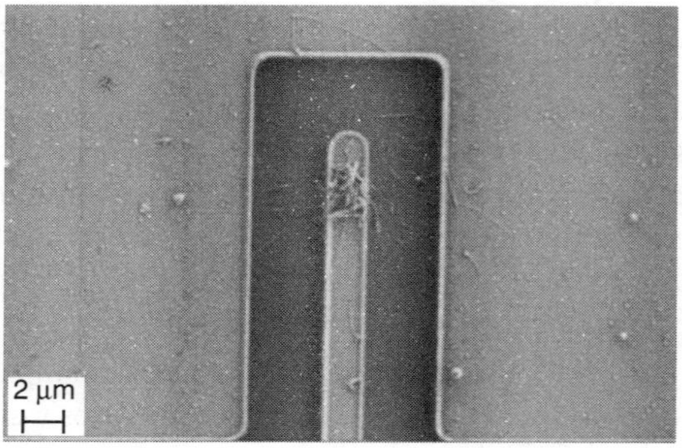

Fig. 4.11 *Connection NWs in Field Effect Transistor (FET) device. (Source: Gallium Nitride Nanowire Growth, and FET Device and Biosensor Fabrication Kaylee Sill McElroy, Physics, Brigham Young University, 2005 NNIN REU Research Accomplishments)*

Ion-sensitive field-effect transistor (ISFET) based biosensors are energetically studied because of their attractive features such as small size, low cost and multiple integration of various kinds of biologically active materials *Eg.*, enzymes, antibodies, DNA and cells. The latest investigation concerns that the applications of the ISFET based biosensor extended to immune-sensing and DNA hybridization sensing. The ISFET was introduced by Bergveld (1970) it has been applied to biosensors in the fields of clinical medicine and biotechnological and environmental applications (Schöning and Poghossian, 2002). The first ISFET was applied by Bergveld (1972) to a biosensor for measuring ion concentration around nerve tissues. The concept of an enzyme sensor based on ISFET was proposed by Janata and Moss (1976). The first result of an ISFET based biosensor for measuring penicillin was reported by Caras and Janata (1980).

The ISFET based enzyme sensor is constructed by immobilizing an enzyme on the gate insulator of the ISFET. When a target substrate is present in a sample solution the immobilized enzyme catalyzes the substrate which results in the production of proton and then a local pH change around the gate insulator. The ISFET based enzyme sensors are pH-sensitive the concentration of the target substrate can therefore be determined by monitoring the local pH change. In addition, as ISFET based enzyme sensors can be combined with any kind of enzyme in principle, many kinds of ISFET based enzyme sensors have been developed for detecting various kinds of analytes such as glucose and urea. However, the obtained results were not satisfactory in early phases of ISFET based enzyme sensors; they had a slope of less than 16 mV/decade and of about 10 mV/decade

(Miyahara and Moriizumi, 1985) even though the interfacial potential of a pH sensitive membrane theoretically has a sensitivity of 59 mV/pH. Since an ISFET based enzyme sensor works on the principle of detecting the local pH change caused by an enzyme catalyzed reaction, the ISFET-based enzyme sensor would not work well because of interference from buffer conditions such as pH or buffer capacity. Several approaches to overcome the above-mentioned problem have been developed, for example, applying additional charged polymeric membranes (Volotovsky *et al.*, 1996) and using an optimized low capacity poly mix buffer solution. However few ISFET-based enzyme sensors have been able to be applied commercially for biomedical analysis because of their lower sensitivity and instability.

The extended gate FET sensors with a gold sensing electrode were developed for DNA detection (Kim *et al.*, 2004; Ishige *et al.*, 2006). DNA probes were immobilized on the gold electrode of the sensor using gold thiol binding chemistry. This extended gate FET sensor was also applied as a novel immune-sensor based on potentiometrically detecting the adsorption of enzyme catalyzed products on the gold electrode. The extended gate FET sensor detects a chemical reaction on its gold electrode as a change in its interfacial potential. Therefore applying an extended gate FET sensor as an enzyme sensor would make it possible to detect an enzyme catalyzed reaction through chemical reactions other than pH change.

An enzyme sensor based on an extended gate FET sensor with a ferrocenyl-alkane thiol immobilized gold electrode in order to detect an enzyme reaction through a redox reaction instead of a change in pH. The FET sensor sensitivity theoretically examined with an 11-ferrocenyl-1-undecanethiol (11-FUT) immobilized gold electrode to hexacyanoferrate (III) ions and hexacyanoferrate (II) ions which are standard redox reagents of an enzymatic assay in blood tests. After the theoretical examination conducted a detection experiment in which the FET sensor experimentally detected a redox reaction of both the reduction of hexacyanoferrate (III) ions and the oxidation of hexacyanoferrate (II) ions as a change in the interfacial potential of the gold electrode in accordance with the Nernstian response at a slope of 59 mV/decade at 25° C. This FET sensor used for measuring cholesterol levels. The FET based enzyme sensor successfully detected cholesterol in concentrations from 33 to 233 mg/dL using a serum sample of 2.5 L and showed a Nernstian slope of 57 mV/decade.

Fabrication of extended gate FET sensors and gold electrode chips (Fig. 4.12) (5mm × 7.5mm), which has an extended gate FET sensor and an Ag/AgCl electrode as a quasi-reference electrode. The extended gate FET sensor consists of two parts: the gold electrode and the FET structure that transduces the change of the interfacial potential on the gold electrode into a change of the drain current. The method of fabricating the sensor chip is described as follows. The gold electrode (400 μm ×400 μm) was fabricated by sputter deposition of an adhesion layer of nickel–tungsten alloy and then a gold layer (100 nm, 99.99% purity) on aluminium wire that was connected with the gate area of the FET. The FET had an n-channel depletion type structure (formed by As^+ ion implantation at 120 keV with a dose of 6.0×10^{11} molecules/cm^2) with p-type source and drain areas. The width and length of the channel region (W/L) were 2400 and 5 μm respectively and the gate insulator of the FET consisted of a 17.5 nm SiO_2 layer formed by a wet oxidation method under a 1.3 μm Si_3N_4 layer formed by plasma enhanced chemical vapor deposition. An Ag/AgCl electrode was fabricated by deposition of Ag/AgCl paste on an underlying gold electrode which was also fabricated in the same semiconductor process.

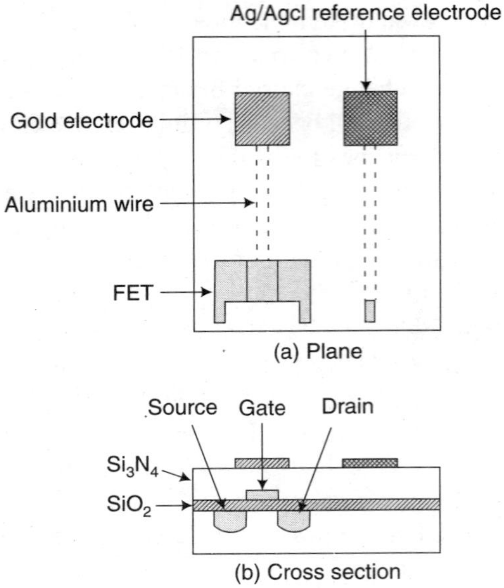

Fig. 4.12 *Diagram of sensor chip (5 mm × 7.5 mm), which has an extended-gate FET sensor and an Ag/AgCl reference electrode. (Source: Yu Ishige et al., (2009). Extended-gate FET-based enzyme sensor with ferrocenyl-alkanethiol modified gold sensing electrode. Biosensors and Bioelectronics, Elsevier. 1096-1102)*

Gold electrode chips 4 mm × 4 mm in sizes were fabricated using the same semiconductor process as for the extended gate FET sensor. A gold layer was sputtered on a silicon wafer to a thickness of 100 nm. The alkanethiol modification procedure of the gold electrode was as follows. The gold electrode chips were kept in an alkanethiol solution for 1 hour. The concentration of ferrocenyl-alkanethiol solution was 5mM in ethanol. After alkanethiol modification, the gold electrode chips were rinsed with ethanol and deionized water twice to remove unreacted alkanethiol molecules. The chips were kept in a 0.1M sodium sulfate solution at room temperature until use. The alkanethiol modification for the gold electrodes of extended gate FET sensor chips was performed in the same way as for the gold electrode chips.

The FET based enzyme sensor consists of an enzyme reaction part and a detection part with both an extended-gate FET sensor and an Ag/AgCl reference electrode (Fig. 4.13). Substrates in the sample solution were catalyzed by the oxidoreductase in the enzyme reaction part and the redox compound in the oxidation state was converted to redox compound in the reduction state, which results in changing the ratio of redox compound in the oxidation state to that in the reduction state. The change in the ratio of redox compound in both states is detected by the extended gate FET sensor as the change in the interfacial potential of the ferrocenyl-alkanethiol modified gold electrode.

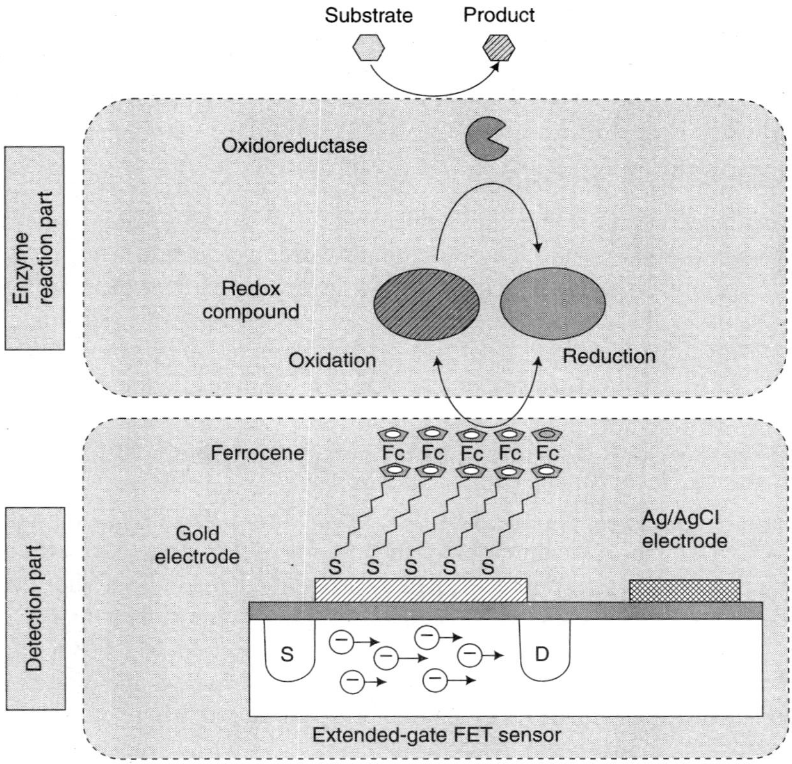

Fig. 4.13 *Structure of extended-gate FET-based enzyme sensor with Ferrocenyl alkanethiol modified gold electrode. (Source: Yu Ishige et al., (2009). Extended-gate FET based enzyme sensor with ferrocenyl-alkanethiol modified gold sensing electrode. Biosensors and Bioelectionics, Elsevier. 1096-1102)*

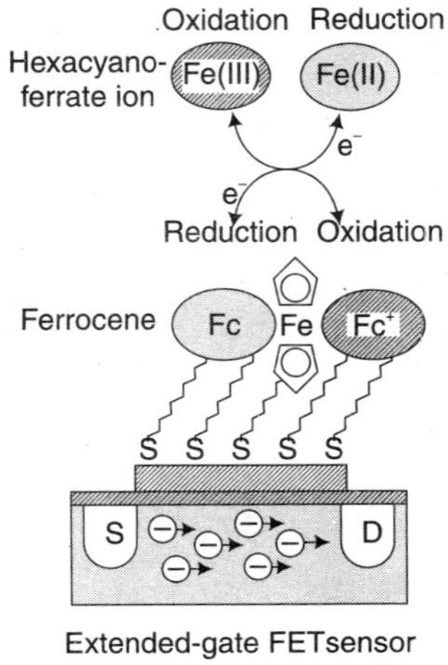

Fig. 4.14 *Reaction system between hexacyanoferrate ions in solution and immobilized ferrocene molecules on the gold electrode. (Source: Yu Ishige et al., (2009). Extended-gate FET based enzyme sensor with ferrocenyl-alkanethiol modified gold sensing electrode. Biosensor and Bioelectronics, Elsevier. 1096-1102)*

Both hexacyanoferrate ions and ferrocene molecules are in equilibrium state of one electron transfer, the change in redox state of hexacyanoferrate ions can be detected as the interfacial potential change of ferrocene immobilized gold electrode based on the Nernst equation regarding hexacyanoferrate ions (Fig. 4.14).

4.1.4 Self Assembled Fabrication

Micro-cantilevers are used in commercial atomic force microscopy (AFM), have been developed as sensitive stress sensors. Enzyme linked immunosorbent assays (ELISAs) have been used for the detection of specific biomolecular interactions, however this approach required fluorescent tag labeling of samples, necessitated high analytical skills, expensive and time consuming. To improve these disadvantages, several label free methods were developed such as surface plasmon resonance (SPR), Quartz Crystal Microbalance (QCM), Scanning Tunneling Microscopy (STM) and Atomic Force Microscopy (AFM).

Self assembly has been one of the most important properties of biomolecules, it has been utilized with numerous DNA hybridizations or antigen antibody reactions. In recent years, self assembly principle has largely been applied to memory devices and biomedical fields, these interactions can collectively produce the mechanical motion of the micro-cantilever and have also been adapted to elucidate the micro-cantilever deflection created by protein ligand binding. Micro-cantilever based biosensor for the analysis of biochemical reactions between a protein and a ligand. The mechanical response of the micro-cantilever can be monitored as a deflection of the micro-cantilever with respect to a change in the surface stress on the protein-ligand binding side of the micro-cantilever relative to the other side. The quantities of bending were measured using both optical and electrical detection methods. The optical beam deflection method has been the most widely used technique to measure the deflection of cantilevers. It has sub-angstrom resolution and can be easily operated. The electrical detection method included integrated piezo-resistive readout which greatly facilitated operation in solutions since the refractive indices of the liquids did not influence detection. Thus, measurements could be performed even in non-transparent liquids.

The micro-cantilevers were fabricated through a surface micromachining technique that had attractive advantages in terms of cost efficiency, simplicity and ability of fabricating in an array. A bottom-up method, a surface micromachining technique was widely used in the micro- electro-mechanical system (MEMS) in comparison with the bulk micromachining of a top-down method. The proposed micro-cantilever based biosensor utilized silicon nitride as the main substrate material. For detection of biomolecules, the deflections of the micro-cantilever were measured by both an optical method and an electrical method. The micro-cantilever detection system required a small quantity of the solution and thus provided fast reaction time. In addition, it was advantageous in terms of miniaturization, portability and the array design of the devices.

4.1.5 Fabrication of Piezoresistive Microcantilever

Amorphous silicon with the thickness of 2 μm was deposited on a thermally oxidized layer as a sacrificial layer with Low Pressure Chemical Vapor Deposition (LPCVD). A 300 nm thick polycrystalline silicon (poly-Si) layer was deposited on a 300 nm thick LPCVD silicon nitride as a piezoresistive material using LPCVD. The ion implantation was performed on a poly-Si layer. A doped poly-Si layer was processed by lithography and ICP dry etching.

A silicon nitride film was deposited under a low stress condition to eliminate any curling distortion. Since the thermal processing considerably affected the mechanical properties of the cantilever, silicon nitride was further processed by lithography and Magnetically Enhanced using a Reactive Ion Etching technique (MERIE). The metallization was performed with piezoresistive material to pad the micro-cantilever for the recognition of the change in resistance of the piezoresistive material integrated in the micro-cantilever. A contact hole was filled with aluminum by a sputtering technique. The Cr/Au (3 nm/30 nm) (Fig. 4.15) layer was deposited on the upside surface of the micro-cantilever by sputtering then lifted it off. Finally, the sacrificial layer was removed by XeF_2 dry etching to release the micro-cantilever.

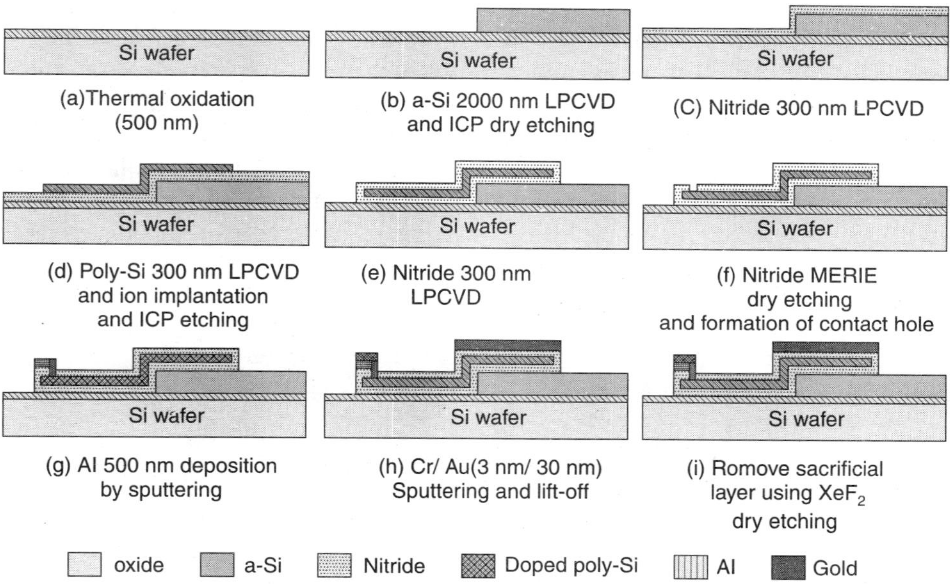

Fig. 4.15 *The schematic fabrication process flow of the piezoresistive micro-cantilever for optical and electrical measurements. (Source: Fabrication and characteristics of microcantilever-based biosensor for detection of the protein-ligand binding Kyung-Ah Yoo1, Jun-Hyuk Kim1, B. H. Nahm2, C. J. Kang1,3, Yong-Sang Kim1,4, IOP Publishing Journal of Physics: Conference Series 61 (2007) 1308–1311)*

The protein ligand binding on the gold coated micro-cantilever was situated in a fluid cell system. Cystamine dihydrochloride, Glutaraldehyde and Streptavidin solutions were sequentially allowed to flow over the micro-cantilever. The principle in Gold/Cystamine dihydrochloride/Glutaraldehyde/Streptavidin binding formation is the self assembly concept. Initially, Cystamine dihydrochloride was injected into the system to facilitate its interaction with a gold forming Self Assembled Monolayer (SAM). The interfacial contact of gold with Cystamine dihydrochloride was the alkanethiol group (-SH) located at the terminal amine. The Glutaraldehyde solution which contained terminal aldehyde groups was injected next over the micro-cantilever. One aldehyde group formed a covalent bond with the amine group of Cystamine dihydrochloride immobilized on the gold surface while the other aldehyde groups bonded with Streptavidin. According to this

SAMs of gold Cystamine dihydrochloride-Glutaraldehyde-Streptavidin were formed. The micro-cantilever deflection was caused by the protein-ligand binding process on the binding site.

4.1.6 Fabrication of Nanobiosensors

Carbon nanomaterials have been extensively used in electro analysis and the most common forms are spherical fullerenes, cylindrical nanotubes and carbon fibres and blacks. Individual carbon nanotubes (CNTs) can be used as nanoscale transistors. These transistors have outstanding potential for electronic detection of biomolecules in solution possibly down to single-molecule sensitivity. To detect biologically derived electronic signals, the CNTs are often functionalized with linkers such as proteins and peptides to interface with soluble biologically relevant targets.

Recently, the graphite nanoplatelets (thickness of 10 nm would contain approximately 30 graphene sheets, considering an interlayer spacing of 0.335 nm) have been used in biosensor applications. This type of biosensor was among the best both in sensing performance and production cost and an inexpensive alternative to carbon nanotubes. Nano-sized functional particles represent an area of great research and commercial interest. In the field of biosensors and biocatalysts, nano-sized materials offer the potential for extremely high surface area to volume ratios. This allows the immobilization of large amounts of biomolecules per unit projected area and can result in high efficiency biosensors or biocatalysts. Unfortunately, the cost of nanomaterials is a big practical bottleneck. In contrast, the graphene based nanomaterials are inexpensive but with high performance.

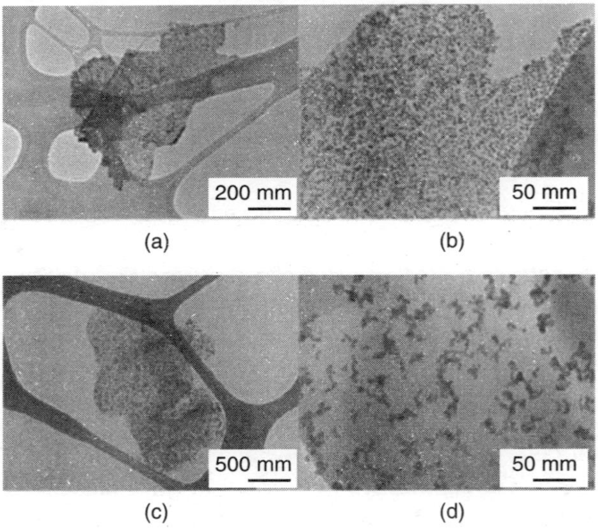

Fig. 4.16 *TEM images of 20 wt% platinum-xGnP at (a) low magnification, (b) high magnification, and 20 wt% palladium-GnP at (c) low magnification and (d) high magnification. (Reprinted with permission from American Chemical Society).*

The graphene has unique properties such as the highest thermal oxidation resistance and the highest degree of graphitization compared to carbon nanotube and carbon black that are widely used. Graphene also provides excellent support for nanometal catalysts. Use of these properties by decorating exfoliated graphite nanoplatelets with platinum and palladium nanoparticles as a transducer for a glucose biosensor. The Graphene nanoparticles (xGnP) incorporated into the biosensor interface increased the effective electrode surface area and served as an excellent support for metal nanoparticles. The platinum and palladium nanoparticles that were immobilized on the xGnP provided a large electro active surface area that effectively catalyzed redox reactions involving hydrogen peroxide (Fig. 4.16). The glucose biosensors produced using this approach exhibited sensitivity up to 61.5 ± 0.6 μA/ (mM-cm^2) and a response time less than 2 seconds. These performance properties are among the best reported for glucose biosensors. Both the metal loading and the nano-particle size (2 nm) had a significant effect on biosensor performance. The combination of high performance and inexpensive nanomaterials components much less expensive than carbon nanotubes suggests that this approach may be very suitable for commercial applications. Of course, the incorporation of metal decorated xGnP in biosensor applications would not be limited to glucose biosensing but could be applied to many other biosensors and biocatalytic systems as well.

4.1.6.1 Fabrication process

Heterogeneous functionalization of solid surfaces by micro and nanopatterning provides powerful strategies to generate active bio-interfaces. Different technologies can be applied: layer by layer, molecular imprinting, Langmuir Blodgett technique, plasma spray, micro contact printing, laser-particle guidance, liquid or droplet micro dispensing approaches. Most of the listed immobilization techniques even if they give reproducible results, they lack in proving that the native conformation of the deposited biomaterial and hence its functionality is preserved. Also, many times they require stepwise processes, high quality starting materials, modified surface coatings, expensive organic precursors, solvents and surfactants.

Micro Contact Printing (mCP) referred to as soft lithography has emerged as a fast stamping technique with capability to process large surfaces in one run. In mCP an elastomer polydimethylsiloxane (PDMS) is used to form a stamp to be coated with an ink containing the biological molecules to be transferred. The main function of this technology is to establish the direct contact with the substrate in a precise manner and a given printing time by reducing as much as possible any mechanical forces that might affect the viability of the transferred material. With mCP different biological materials have been printed: alkanthiols, proteins, DNA or cells. However, it was shown that mCP printing could lead to ill-defined patterns, to aggregations and loss of activity of the printed material. Therefore, alternative methods have to be exploited. Ideally, any alternative methods should bring a defined quantity of biomolecules in contact with a surface and as such induce adsorption with high spatial control and minimum loss of functionality.

The direct writing of biological material by laser transfer technology matrix assisted pulsed laser evaporation direct write (MAPLE DW). Generally, MAPLE DW uses a laser pulse to transfer material from a ''donor'' to an acceptor surface. The material is transferred in micron sized voxel patterns with high precision, high resolution and in a computer-aided design/computer aided manufacturing (CAD/CAM) manner. In order to preserve the functionality and to prevent high evaporation rates or possible cross contamination a buffer is used as embedding medium. The mCP and MAPLE DW to manufacture patterns of streptavidin on glass surfaces.

4.1.6.1.1 Micro contact printing

Polydimethylsiloxane (PDMS) stamps were fabricated by casting and curing Sylgard 184 elastomeric polymer against photo resist micro patterned silicone masters. Different stamp configurations were used for these experiments stripes or squares geometries. The PDMS stamps were first cured for 4 hours at 65°C and then exposed to inking Streptavidin protein solution (fluorescein labeled and unlabelled; 100–200 mg/ml) for various amounts of time (1, 3 and 5 minutes respectively). Excess solution was removed with a stream of nitrogen gas. Subsequently, the inked stamp was brought in direct contact with the cleaned glass by applying a low mechanical force. Once the stamp was removed, the glass was analyzed in epifluorescence or used to form a perfusion chamber.

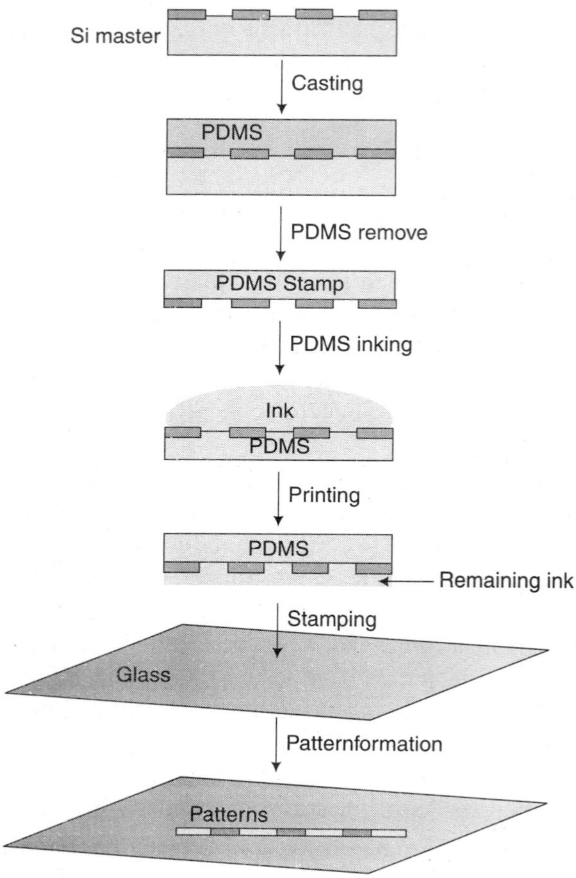

Fig. 4.17 *Schematic diagram of micro contact printing technique. (Source: Printing technologies for fabrication of bioactive and regular microarrays of streptavidin, C.Z. Dinu a, V. Dinca b,*, J. Howard a, D.B. Chrisey c, Applied Surface Science 253 (2007) 8119–8124)*

In order to determine whether the deposited patterns are functional performed a fluorescence assay in a perfusion chamber. The perfusion chamber was built up from a cleaned glass and the patterned cover glass separated by a double side tape spacer. The chamber was precoated with a casein containing solution (0.8 mg/ml,) for 5 min (to avoid any subsequent unspecific binding) and subsequently with a solution containing biotin Rhodamine labeled (1 mg/ml) for 5 minutes, in dark at room temperature. After the incubation, any unbound biotin was washed away and the sample was analyzed with the epifluorescence microscope.

Using mCP patterns of proteins were printed on a glass substrate. For this, a PDMS stamp was used to deliver labeled Streptavidin (Fig. 4.18). In some cases, unlabelled Streptavidin was printed and served as template for further ligand recognition reaction (Fig. 4.17). Fluorescein labeled biotin was incubated with the printed Streptavidin in a perfusion chamber. During the incubation process the Streptavidin target specifically bound biotin capture molecules from the aqueous solution. Any unbound molecules were washed away. The fluorescence was quantified with an inverted epifluorescence microscope.

Micro-contact printing was an excellent way of creating active Streptavidin patterns on a solid glass substrate. The mechanism of transfer was due to the fact that binding between the cleaned glass surface and the proteins was stronger than the adhesion force between the proteins and the PDMS stamp. However the transfer was not only localized to the stamp patterns: there were also molecules transferred between the patterns. This was presumably due to the fact that not only the inking molecules were transferred but also parts of the PDMS stamp. With the reduction in pattern size, defects in the transfer appeared: the patterns were neither well defined nor homogenous. More important, they were not anymore continuous (i.e. no free-standing structures). This was mainly due to the instability of the stamp features to mechanical stresses applied at the transfer. Site specific attachment of biotin to the Streptavidin bio-scaffold gave the patterns the property of changing their optical characteristics as they switch between two different states-free/bound. The observed fluorescence signal allowed us to conclude that the deposited patterns were viable (Fig. 4.18).

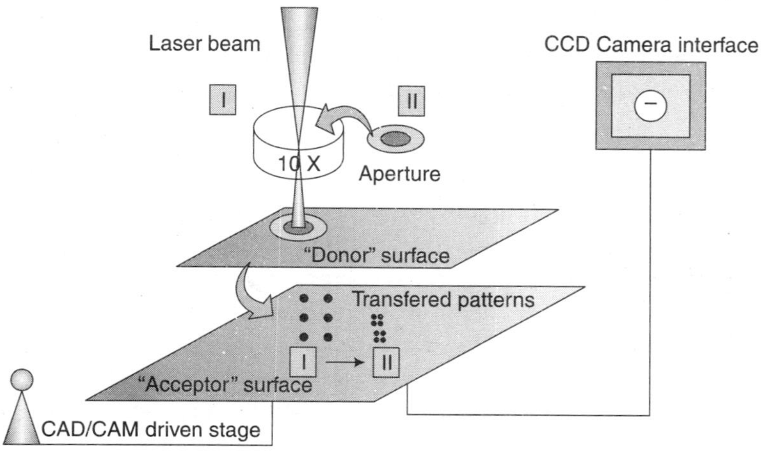

Fig. 4.18 *The MAPLE DW set-up. (Source: Printing technologies for fabrication of bioactive and regular microarrays of streptavidin, C.Z. Dinu a, V. Dinca b,*, J. Howard a, D.B. Chrisey c, Applied Surface Science 253 (2007) 8119–8124)*

With the laser irradiation and through an evaporation process, proteins were transferred from the donor surface to the acceptor surface. An aperture was used to control the diameter of the deposited patterns.

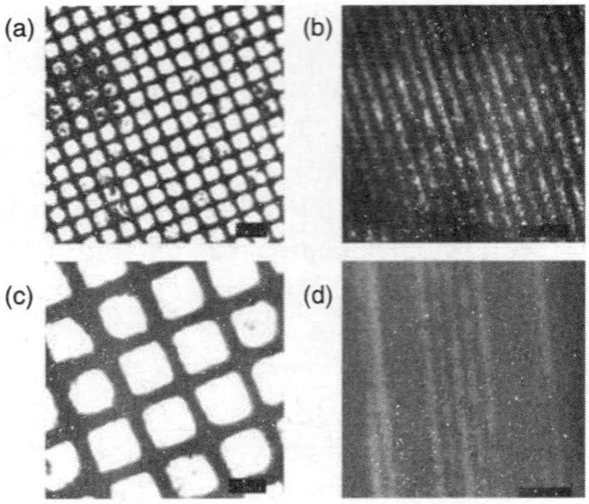

Fig. 4.19 *Deposition of Streptavidin molecules by soft lithography. Herein scale bar was 10 mm. (Source: Printing technologies for fabrication of bioactive and regular microarrays of streptavidin, C.Z. Dinu a, V. Dinca b,*, J. Howard a, D.B. Chrisey c, Applied Surface Science 253 (2007) 8119–8124)*

a. Fluorescence micrographs of fluorescene labeled Streptavidin printed on a glass surface in a square like geometry. The pattern size was about 5 mm.
b. Fluorescence micrographs of fluorescene labeled Streptavidin on a glass surface in a square like geometry. The pattern size was about 10 mm.
c. Fluorescence micrographs of fluorescene labeled Streptavidin on a glass surface in stripes-like geometry. The transferred stripes were about 2 mm.
d. Fluorescence micrographs of Rhodamine labeled biotin bound specifically to Streptavidin previously printed on the substrate. The Streptavidin was transferred in a stripe-like geometry and served as template for further ligand: labeled biotin.

4.1.6.1.2 MAPLE DW

Using Matrix Assisted Pulsed Laser Evaporation Direct Write (MAPLE DW) patterns of Streptavidin were deposited on a glass substrate. The fraction of the molecules expelled from the ribbon was directly correlated to the laser fluence (Fig. 4.21). To determine, if any reduction in pattern size was possible, an aperture with fixed opening was used (Fig. 4.20) to observe how the rate of energy deposition affect desorption of biomolecules from the ribbon. When less energy was used, a smaller degree of overheating was reached at the end of the laser pulse and thus a less violent evaporation occurred. For high fluences (F1 and F2, respectively), the patterns were not homogenous being formed from splashed particles. With decreasing the laser fluence, regular mesoscopic patterns were formed and no observable splashing occurred.

The mechanical or adhesive properties and thus the adhesion force in vapor is always less than in vacuum due to the capillary condensation of the water around the surface contact sites. As such, one specific goal was to control the adhesion of the Streptavidin patterns to the glass at room temperature conditions. It is known that adhesion between protein and glass surfaces varies as a function of Ionic Strength (IS), pH, loading force, Isoelectric Point (IP) and the residence time. As such, the adhesion force between the protein and the surface increases with residence time and decreases with Ionic Strength IS and pH. Changes in pH can alter the conformation of the protein, therefore preserving the IP is mostly desired when preserving the protein activity. This was achieved by using an embedding media that preserved the IP by acting as an energy absorber and causing a limitation of the diffusion and an apparent stabilization of the protein.

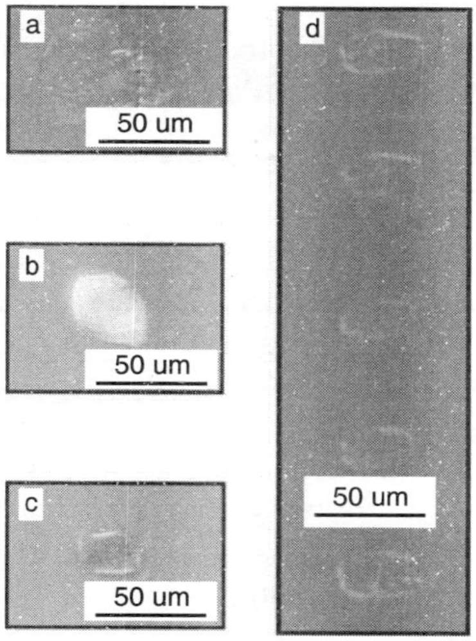

Fig. 4.20 *Optical micrographs of Streptavidin protein patterns deposited on a glass substrate by MAPLE DW. (Source: Printing technologies for fabrication of bioactive and regular microarrays of streptavidin, C.Z. Dinu a, V. Dinca b,*, J. Howard a, D.B. Chrisey c, Applied Surface Science 253 (2007) 8119–8124)*

During the transfer process, the optical fluence was varied from 35 to 25 mJ/cm^2 with F1 (in a) > F2 (in b) > F3 (in c) > F4 (in d). The changes with the fluence were revealed.

Restricting the laser beam by using variable apertures proved that if aimed, reduction in the pattern size was possible. By using a circular aperture one can obtain patterns with the same morphological integrity, well defined, not ill-masked as in the case of mCP and with a reduced size (up to 10 mm). This is presumably due to improving the beam quality of the excimer laser. The preceding spatial filtering and the aperture removed as much of the non-coherent beam as it could. This was critical for focusing and for beam uniformity and combined with the final aperture led to small transferred patterns.

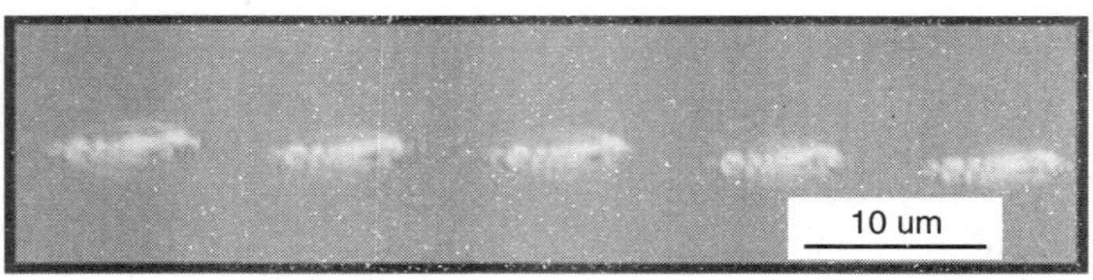

Fig. 4.21 *5. Micrographs of Streptavidin patterns obtained using MAPLE DW and an aperture that limited the laser beam. (Source: Printing technologies for fabrication of bioactive and regular microarrays of streptavidin, C.Z. Dinu a, V. Dinca b,*, J. Howard a, D.B. Chrisey c, Applied Surface Science 253 (2007) 8119–8124)*

When compared with mCP, in MAPLE clearly show a reduction in size complementary to well defined and with the same morphological characteristics patterns. Also, shows that there is no transfer outside the pattern given by the fact that the media presumably favorized the high absorption to the substrate and cancelled the external splashing. In comparison with mCP, MAPLE DW (Fig. 4.21) has also an improved resolution and moreover the system is envisaged as an automated CAD/CAM device, here micro patterns with desired location can be produced. Advantages of the system are the automated interface, the rapid process time, the ease of operation and the ability to dynamically interrupt the process any time after achieving the requested patterns.

4.1.7 Fabrication of ITO-PANI Biosensor

The indium tin oxide polyaniline (ITO-PANI) biosensor consisted of two components, immunosensor and an amperometric measuring device. The immunosensor was constructed from an ITO glass and layered with PANI and antibodies. The amperometric measuring device was a three electrode electrochemical cell that was connected to a potentiostat. The ITO glass is a common substrate used in an amperometric biosensor due to its structural flexibility to bind directly with biological elements, stable electrical property with high charge carrier density and inexpensive production cost. An ITO glass substrate with a dimension of $1.27 \times 2.54 \times 0.1$ cm^3 and a resistance value ranging from 15 to 25 ohms were used for biosensor fabrication.

To fabricate the biosensor, the ITO substrate was first cleaned and treated with a strong oxidizing agent (ammonium hydroxide) to enable the adherence of PANI to the ITO substrate. Figure 4.22 shows the increase in thickness with NH_4OH. The self doped PANI was layered onto the ITO substrate using a spin coating method. Spin coating method was chosen because of its ease of use, rapid processing time, reproducibility and is inexpensive compared to other types of polymer coating mechanism, such as Langmuir-Bloggert technique or layer by layer molecular deposition. With a speed of 500 rpm for 6 seconds, a thickness of 324.62, 234.2 and 91.05 nm were observed on the ITO substrates coated with 65, 50 and 20 KDa PANI, respectively (Fig.-4.23). After the coating procedure, the substrate was then functionalized into a biosensor by immobilizing antibodies onto the surface.

A three-electrode electrochemical cell was used to charge the substrate by applying a small (0.5V) negative potential. This step was essential to promote electrostatic bonding between the negatively charged substrate and the NH^+ site of the antibodies. The use of a divalent cross-linker, Glutaraldehyde also helped facilitate the antibody binding mechanism. The successful fabrication of the biosensor was evaluated by using an atomic force microscope (AFM). AFM images of an ITO-PANI biosensor prepared with 20, 50 and 65 KDa PANI, functionalized with antibodies and incubated with 104 and 106 cell culture infective dose per ml (CCID/ ml) of bovine viral diarrhea virus (BVDV). An increase in height in the z direction (thickness) between plain ITO (Fig. 4.24) and ITO-PANI substrates (Fig. 4.24) indicates successful polymer coating. Fig. 4.24 also shows that the higher molecular weight PANI formed a thicker layer on the ITO substrate. Similarly, an increase in thickness of each of the ITO-PANI substrate was immobilized with antibodies suggesting successful antibody immobilization.

More importantly, Fig. 4.25 show the AFM images of the biosensor surface after incubating them in various concentrations of BVDV cultures. A thicker substrate was observed when incubating the biosensor with the higher concentration of BVDV, supporting the logical phenomenon that the higher the antigen concentration, the more antibody-antigen complex occurs. When the biosensors were tested with the same level of BVDV concentration (e.g., 106 CCID/ml), biosensors coated with the higher molecular weight PANI (e.g., 65 KDa) were observed to be thicker than that with a lower molecular weight (*Eg*., 20 KDa) (Fig. 4.25). This finding shows that the higher the molecular weight of PANI, the more antibody binding sites are available.

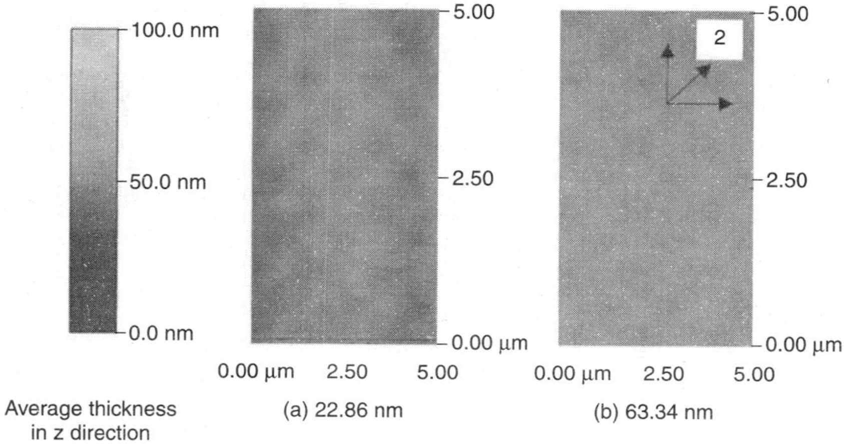

Fig. 4.22 *Atomic force microscopy images of a) untreated ITO glass and b) ITO glass treated with ammonium hydroxide (NH_4OH). (Source: Indium Tin Oxide-Polyaniline Biosensor: Fabrication and Characterization Zarini Muhammad Tahir 1,*, Evangelyn C. Alocilja 1,* and Daniel L. Grooms, Sensors 2007, 7, 1123-1140, ISSN 1424-8220, © 2007 by MDPI,www.mdpi.org/sensors)*

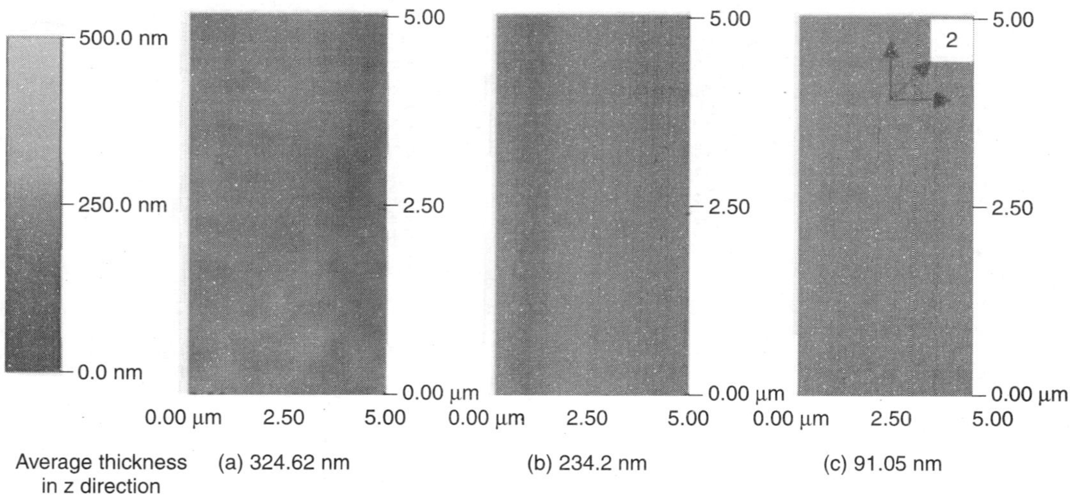

Fig. 4.23 *Atomic force microscopy images of a) ITO glass treated with (NH_4OH) and spin coated with 65KDa, b) 50KDa, c) 20KDa Pani. (Source: Indium Tin Oxide-Polyaniline Biosensor: Fabrication and Characterization Zarini Muhammad Tahir 1,*, Evangelyn C. Alocilja 1,* and Daniel L. Grooms, Sensors 2007, 7, 1123-1140, ISSN 1424-8220, © 2007 by MDPI,www.mdpi.org/sensors)*

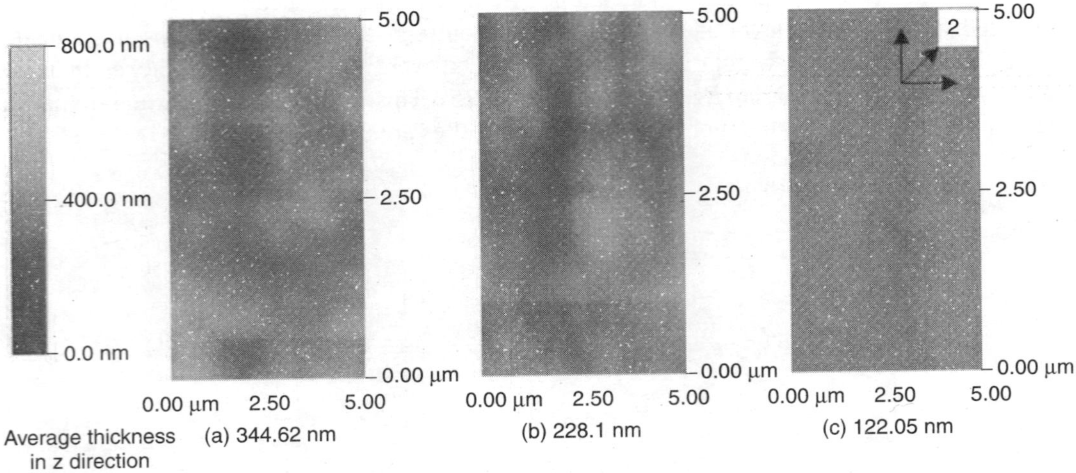

Fig. 4.24 *Atomic force microscopy images of a) ITO glass spin coated with Pani (65KDa) + antibody, b) Pani (50 KDa) + antibody, and c) Pani (20 KDa) + antibody. (Source: Indium Tin Oxide-Polyaniline Biosensor: Fabrication and Characterization Zarini Muhammad Tahir 1,*, Evangelyn C. Alocilja 1,* and Daniel L. Grooms, Sensors 2007, 7, 1123-1140, ISSN 1424-8220, © 2007 by MDPI,www.mdpi.org/sensors)*

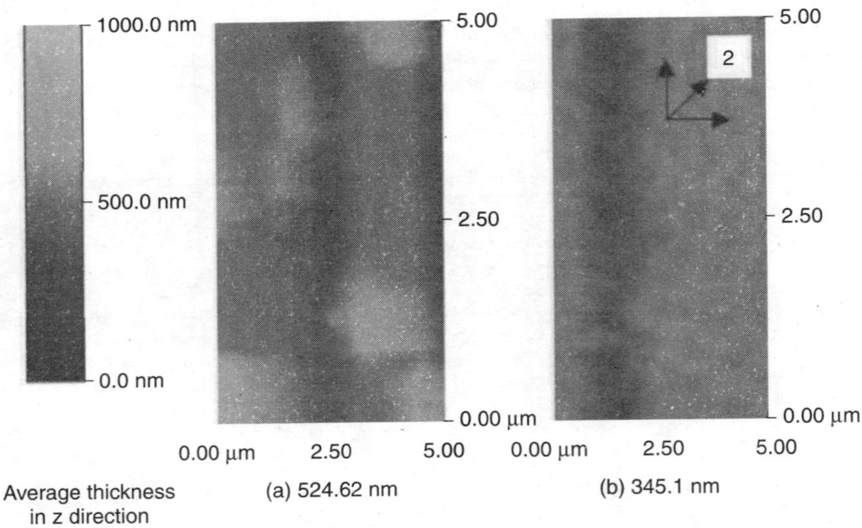

Fig. 4.25 *Atomic force microscopy images of a) ITO-Pani biosensor (65KDa) tested with a) 106 CCID/ml and b) 104 CCID/ml of bovine viral diarrhea virus (BVDV). (Source: Indium Tin Oxide-Polyaniline Biosensor: Fabrication and Characterization Zarini Muhammad Tahir 1,*, Evangelyn C. Alocilja 1,* and Daniel L. Grooms, Sensors 2007, 7, 1123-1140, ISSN 1424-8220, © 2007 by MDPI,www.mdpi.org/sensors)*

To elucidate the use of PANI as a mediator or glue for the binding of antibody on the biosensor platform, ITO glasses undergoing the same fabrication process without the PANI spin-coating step were tested for their amperometric responses (Fig.-4.26). Depending on the fabrication process, the substrate is referred to as substrate A or substrate B thereafter. Fig. 4.26A shows a significant difference between (Io) and (Is) for substrate A. The (Io) and (Is) responses for substrate B (Fig. 4.26B), on the other hand, are statistically insignificant between each other. The latter finding is proposed to be caused by the absence of PANI on the ITO substrate which contributes to the lack of antibody binding and subsequently to the insignificant responses between (Io) and (Is) (Fig. 4.26).

The presence of antibodies on both substrates was confirmed by repeating the above experiment with fluorescent-tagged antibodies. Higher fluorescence emission level (250 out of 256-bit color mode) was observed from substrate A than that from substrate B (20 out of 256-bit color mode), this finding implies that more antibodies are present from substrate A than substrate B, it can be concluded that PANI is not only required as the biosensor transducing system but also as a mediator for the antibody binding. Kim *et al.* (2000) demonstrated the use of PANI as a mediator between the antibodies and gold electrodes in their PANI based biosensor.

An increase in current response (amperometric responses Fig. 4.26) is observed after the ITO substrate is coated with PANI. Then a decrease in current response is observed after antibody immobilization, suggesting the reduction in the electron flow that could be due to the insulating property of antibodies. A much higher drop in current response is observed after the antibody-antigen binding.

The amperometric measurement was conducted by first immersing the ITO glass in a 45-ml of PB. Using the three electrode electrochemical cell setup, a fixed potential of 0.5 volts (V) was applied over 25 minutes. The theoretical value of the current flow between the working electrode and the electrolyte was obtained by performing a linear regression analysis on the charge-versus-time experimental data.

4.1.7.1 PANI preparation

Self doped PANI was prepared by doping emaraldine base with camphorsulfonic acid. Each of the commercially available emaraldine base compounds was doped with camphorsulfonic acid in the ratio of 1.0 p-Phenyleneimine (C_6H_4N) unit per 0.5 mol doping acid. Emaraldine and camphorsulfonic acid were mixed using a mortar and pestle in a nitrogen environment. The mixture was dissolved in chloroform to give a final concentration of 0.5 % by weight. To improve process ability of PANI, phenol in a ratio of 0.5 mol per p-Phenyleneimine was added. The mixture was stirred for 15 minutes before being filtered through a 0.45 μm filter paper. The solution was then stirred for 72 hours before further used.

A mixture of each self doped PANI compound dissolved in chloroform was air dried overnight. The dried PANI was compressed into pellets and its conductivity was measured using a four point probe method. A Transmission Electron Microscope (TEM) and an Atomic Force Microscopy (AFM) were used to inspect the physical morphology of PANI and the biosensor. No pretreatment was performed prior to the analysis.

Indium tin oxide glass with a dimension of 1.27 cm × 2.54 cm was used as the biosensor platform. The ITO glass was first washed with methanol/chloroform (1:1) before being soaked in ammonium hydroxide and air dried. A 100 μl of PANI solution prepared was used to spin coat a thin layer of the polymer on the ITO platform at the speed of 500 rpm for 6 seconds. The substrate was left to air dry and referred to hereafter as ITO-PANI substrate.

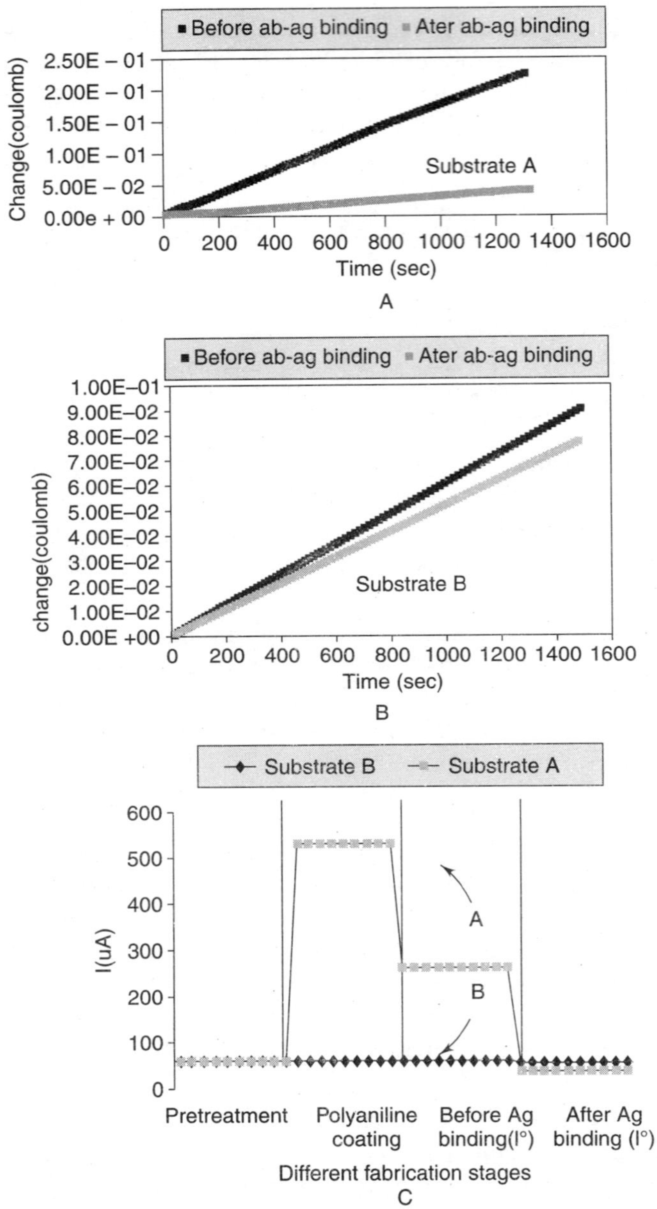

Fig. 4.26 *Current-versus-time response of A) substrate A (ITO biosensor spin-coated with self-doped Pani and B) substrate B (ITO biosensor without PANI coating). Amperometric responses of both substrates at every fabrication stage (C). (Source: Indium Tin Oxide-Polyaniline Biosensor: Fabrication and Characterization Zarini Muhammad Tahir 1,*, Evangelyn C. Alocilja 1,* and Daniel L. Grooms, Sensors 2007, 7, 1123-1140, ISSN 1424-8220, © 2007 by MDPI,www.mdpi.org/sensors)*

The ITO-PANI substrate was functionalized into a biosensor by using target-specific antibodies at 0.5 mg/ml of concentration. All antibodies were prepared in 0.1 M phosphate buffer pH 7.4 (PB). The antibodies were immobilized onto the ITO-PANI substrate using the three electrode cell, the potentiostat and a potential of -0.5 V was applied to the ITO-PANI substrate for 1500 seconds. 45 ml of phosphate buffer was used as the electrolyte. After applying the potential for 1500 seconds, the substrate was removed from the cell and then immersed in 1ml of solution mixture containing the antibody, 1% Glutaraldehyde and phosphate buffer in a volume ratio of 2.5:0.5:1. The substrate was left in the antibody solution at room temperature for 30 minutes. The above immobilization procedure was repeated using fluorescein isothiocyanate (FITC) labeled antibodies to confirm the antibody immobilization. A confocal fluorescence microscope was used to visualize and measure the intensity of the emitted fluorescence and to confirm the immobilization.

4.1.8 Fabrication of Algal Biosensor

Algal biosensor, in which photosynthetic oxygen evolution is monitored by the Clark type oxygen electrode, has been studied for the determination of toxicity in environmental water. However the oxygen electrode is relatively expensive and not very compact. Shitanda *et al.* (2005) were developed a compact and disposable device by using ITO electrode. They entrapped an *Alga* in an alginate gel and immobilized the *Alga* directly on the ITO electrode surface. However the *Alga* was immobilized by casting method which was relatively difficult to thin the membrane thickness. For fabricated a screen printing *Algal* biosensor in which an algal ink was used.

A unicellular *Alga Chlorella vulgaris* strain C-27 was screen printed on carbon electrode and used as a working electrode. *Algal* ink was prepared by mixing the *Algal* cells and 4 wt% sodium alginate solutions. The algal ink was printed on the carbon electrode surface by screen printing. The electrode printed was immersed in 200 mM $CaCl_2$ solution. Amperometric measurements were performed in a $CaCl_2$ solution with an Ag/AgCl and a coiled platinum wire as a reference and counter electrodes respectively. The potential of the *Algal* electrode was set at -0.6 vs. Ag/AgCl. Responses of *Algal* biosensor were evaluated as inhibition ratio of photosynthetic oxygen evolution under periodic visible light illumination.

An algal biosensor (Fig. 4.27) immobilized with *Algal* cells on the carbon electrode surface by using the screen printing technique. *Algal* cells were printed as thin film with high uniformity. Typical current responses of the *Algal* biosensor, the oxygen reduction current increased when the *Algal* biosensor was illuminated. This reflects that oxygen concentration increased at the electrode surface as result of the *Algal* photosynthesis. These results indicated that the printed *Algae* retained photosynthetic activity.

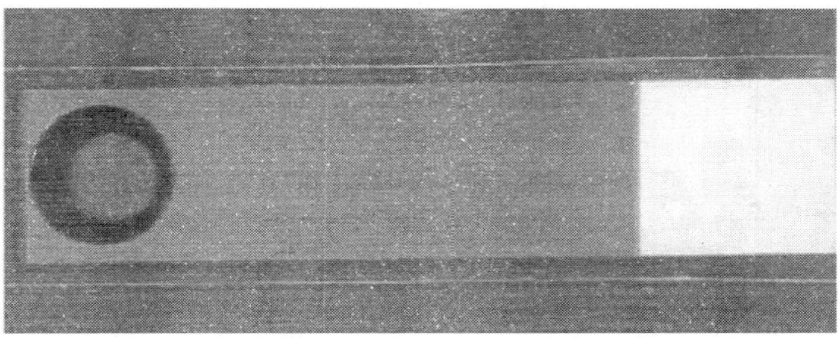

Fig. 4.27 *Algal biosensor fabricated by screen printing.*
(Source: I Shitanda, Fabrication of Algal Biosensor by Screen Printing, Dept. of Pure and Applied Chemistry, Tokyo University of Science, Japan).

4.1.9 Fabrication for Urea Biosensor

To fabricate a biosensor for urea, a urease which catalyzes the hydrolysis of urea to produce carbon dioxide and ammonia is recruited as a sensing material. An enzyme layer is obtained through the direct immobilization, in which the urease is mixed with hydrophilic polyurethanes synthesized with a variety of mole ratios between Poly Ethylene Glycol (PEG) and Poly Propylene Glycol (PPG) and coated over an electrode membrane. The response and detection limit toward ammonium ions of the asymmetric electrode membranes in which the enzyme layers prepared from mixtures of hydrophilic polyurethanes and urease are introduced. They are found to be similar to those of the electrode membrane lacking the enzyme layer indicating that the presence of an additional hydrophilic polyurethane layer does not significantly affect the electrochemical properties of the underlying signal detecting polymeric membrane.

Compared with the asymmetric electrode membranes in which urease are directly immobilized, on the other hand, a relatively slower response behavior is observed in an asymmetric electrode membrane in which a cross linked enzyme film is formed over a hydrophilic polyurethane layer by the glutaraldehyde method owing to the shielding effect of the constituting three layers: hydrophilic polyurethane layer, polylysine layer and enzyme film. For a similar reason, the electrode membrane comprising a cross-linked urease film is slower in the response toward urea than are the membranes employing directly immobilized urease. From the above observation, it should be noted that the electrode membranes which are used to introduce enzymes into the biosensors, exhibit similar response performance regardless of the type of the hydrophilic polyurethane used, indicating that they have no influence on the detection capability of the biosensors.

In order to investigate the influence of the shape of electrodes on the detection capability of the asymmetric urea biosensor in which an enzyme film is immobilized in hydrophilic polyurethane by the direct immobilization with an ammonium ion selective membrane serving as a signal transducer with wire type and planar type electrodes are prepared. Responses toward urea for the wire type and planar type solid state biosensors are observed. Regardless of the type of the base solid state transducer employed a similar response tendency is exhibited. In addition the solid-state urea sensor fabricated by the direct immobilization retained 80% or greater of its initial urea response for two months, indicating the life span of the solid-state biosensor is similar to those of the urea sensors fabricated by the conventional cross-linking method.

The direct immobilization is also applied for the biosensor membrane employing Adenosine deaminase as a bio-active reagent which catalyzes the hydrolysis of adenosine to produce inosine and ammonium ions. A film of Adenosine deaminase mixed with hydrophilic polyurethane is coated over ammonium sensing wire type solid state electrodes. An examination is made of the effect of the film thickness on the biosensor performance, the thinner of enzyme film is the faster response is detected. From this observation it is recognized that the mass transfer of species between the external solution and the ion-selective surface is greatly affected by the thickness of the enzyme film.

Three biosensors for urea, which took advantage of the catalytic action of urease to hydrolyze urea into carbon dioxide and ammonia, were fabricated by mixing the enzyme with three polyurethane species different in the mole ratio of PPE to PPG and coating the mixtures over the electrode membranes. The compositions of the polyurethanes used are given in Table 4.1. Water uptake experiments showed, a water uptake of 42 wt % for the hydrophilic polyurethane A, 100 wt % for the hydrophilic polyurethane B and 206 wt % for the hydrophilic polyurethane C.

Table 4.1 Compositions of hydrophilic polyurethane.

Compositions	*HPU-A*	*HPU- B*	*HPU-C*
PEG	0.005 mol	0.01 mol	0.015 mol
Desmodur	0.52 mol	0.052 mol	-
Ethylene Glycol	0.03 mol.	0.03 mol.	0.03 mol
PEG: PPG	1:3	1:1	3:1

Dynamic responses observed toward ammonium and urea for Urea sensors fabricated by use of conventional type electrodes. The response capability and detection limit toward ammonium ion of the asymmetric membranes in which there were the hydrophilic polyurethane films directly immobilized with urease, were similar to those of the enzyme film lacking membrane. These data demonstrate that the presence of an additional hydrophilic polyurethane film does not significantly alter the intrinsic potentiometric response property of the underlying ammonium sensing polymeric membrane because of having an enzyme film which consists of three layers of hydrophilic polyurethane, polylysine and enzyme, the asymmetric electrode membrane fabricated by the cross linking method using glutaraldehyde is relatively slower in the response time to ammonium ions than is the electrode membrane which has a polyurethane film in which the enzyme is directly immobilized. The response toward urea was slower with the cross-linked urease membrane than with the membrane employing urease mixed with the three kinds of hydrophilic polyurethane. It is apparent that all of the three membranes exhibit virtually the same response performance irrespective of the type of the hydrophilic polyurethane employed, indicating that their physical properties, i.e., water uptake properties do not affect the detection capabilities of the biosensors.

Urea sensors employing two different types of solid state electrodes, i.e., wire type and planar type were fabricated according to the above mentioned process and an examination was made of the response change according to the electrode types. The responses are observed toward urea for the wire type and planar type solid state biosensors in which they are introduced the asymmetric hydrophilic polyurethane films directly immobilized with urease with ammonium ion selective membranes serving as signal transducers. A similar response behavior toward urea was observed for the biosensors with a slope of 52 mV/decade for the wire type and 55 mV/decade for the planar type in the range of 10^{-5} to 10^{-2} M of urea. These values were essentially the same as those of the conventional type biosensors. In addition, the life span of the solid-state biosensor was similar to those of the conventional type biosensors. It was measured that the solid-state biosensor retained 80% or greater of its initial urea response for a two month period of time.

4.1.10 Fabrication for Cholesterol Biosensor

A Dithio bis Succinimidyl Propionate (DTSP) self-assembled monolayer (SAM) prepared on a gold (Au) surface has been utilized for covalent immobilization of Cholesterol oxidase (ChOx) and Cholesterol esterase (ChEt). These ChOx–ChEt electrodes have been characterized by using electrochemical impedance and Cyclic Voltammetric (CV) techniques, respectively.

Differential Pulse Voltammetry (DPV) has been used for enzymatic assay of immobilized ChOx and ChEt onto the DTSP modified gold surface as a function of Cholesterol oleate concentration (fig.-4.28). The response measurement conducted on ChOx–ChEt/DTSP/Au bio-electrode reveal the value of Michaelis Menten constant (K_m) as 0.95 mM suggesting enhanced affinity of enzymes

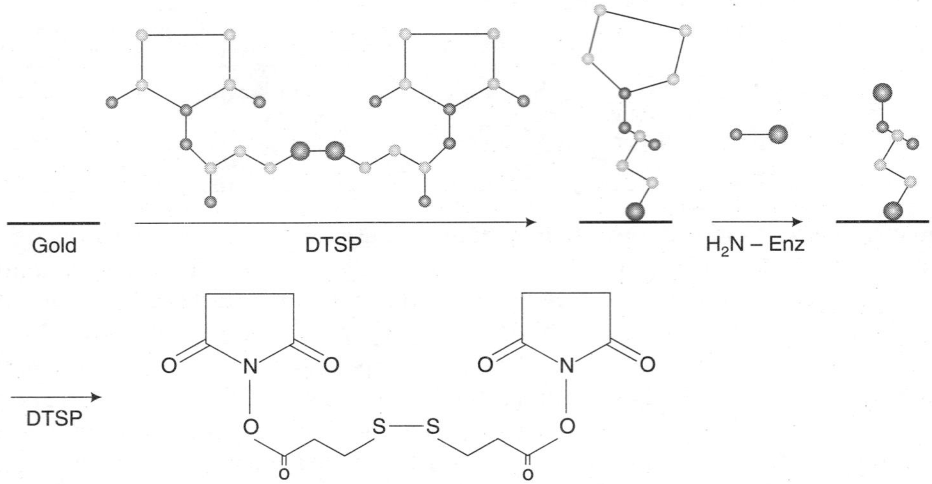

Fig. 4.28 *A Dithio bis Succinimidyl Propionate (DTSP) Self Assembled Monolayer (SAM) prepared onto a Gold (Au) surface has been utilized for covalent immobilization of Cholesterol oxidase (ChOx) and Cholesterol esterase (ChEt).*

(ChOx and ChEt). The ChOx–ChEt/DTSP/Au bio-electrodes show linearity in range of 50 to 400 mg dl^{-1} of Cholesterol oleate and the shelf life of more than 50 days when stored at 4°C. This biosensing electrode shows correlation coefficient of 0.9973 and standard deviation of regression as 0.859 μA.

Biosensors - I

Chapter 5

5.0 INTRODUCTION

Investigation on biosensors is the subject of research for the last five decades. There are many proposed and commercialized biosensors based on their principle including those for detection of pathogens and toxins. The most typical part of electrochemical biosensors is the presence of a suitable biological material, viz., enzyme, antibodies and oligonucleotides, etc. The biorecognition layer consists electro active substances for detection by the physico-chemical transducer providing the measurable signal. Based on their operating principle, the electrochemical biosensors can act as potentiometric, amperometric and impedimetric transducers converting the chemical information into a measurable signal.

The first biosensor described by Clark and Lyons (1962) for the determination of glucose called enzyme electrode. Glucose is of special importance because of its involvement in human metabolic processes. In particular sufferers from *Diabetes mellitus* do not produce sufficient insulin in their pancreas to control adequately, the level of glucose in their blood. Doses of insulin have to be administered and it is very important that the diabetic regularly monitors the level of glucose in the blood. Previously, substantial blood sample had to be taken and analyzed in a medical laboratory with consequent time delays and consequent uncertainty about the diabetic's condition. With the availability of glucose biosensors, the patient himself or herself can extract one small drop of blood and obtain a direct digital readout of the glucose concentration within 60 seconds. Glucose biosensors are based on the fact that the enzyme 'Glucose oxidase' catalyzes the oxidation of glucose to gluconic acid. In the early biosensors, oxygen was used as the oxidizing agent and the consumption of oxygen was followed by electrochemical reduction at a platinum electrode, as in the Clark oxygen electrode (1953) (Fig. 5.1).

The glucose oxidation reaction catalyzed by Glucose Oxidase (GOD) is

$$\text{Glucose} + O_2 + H_2O \xrightarrow{\textit{Glucose oxidase}} \text{Gluconic acid} + H_2O_2 \qquad \textbf{... eq.-5.1}$$

At the electrode:

$$O_2 + 2\,e^- + 2H^+ \longrightarrow H_2O_2 \qquad \textbf{... eq.-5.2}$$

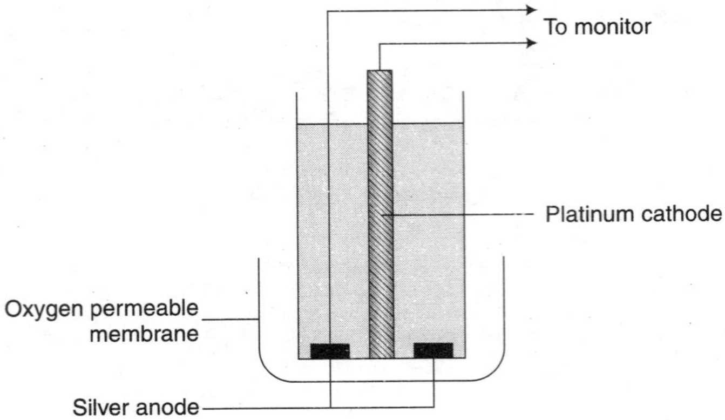

Fig. 5.1 *A Clark oxygen electrode.*

A voltage of –0.7 V is applied between the cathode (platinum) and the anode (silver) electrodes, the measured current is proportional to the oxygen concentration. The concentration of glucose is proportional to the decrease in current (oxygen concentration). The oxygen electrode has an oxygen permeable membrane such as Polytetrafluoroethylene (PTFE), polythene or cellophane covering the electrodes. A layer of enzyme (Glucose oxidase) is placed over this and held in place of second membrane such as Cellulose acetate.

5.1 POTENTIOMETRIC BIOSENSORS

Potentiometric biosensors made the use of ion-selective electrodes in order to transduce the biological reaction into an electrical signal. These biosensors consist of an immobilised enzyme and the probe which is surrounded by membrane from the pH-meter, where the catalysed reaction generates or absorbs hydrogen ions (Table 5.1). The reaction occurred next to the thin sensing glass membrane caused a change in pH which may be read directly from the display. The use of such electrodes is that the electrical potential was determined at very high impedance allowing effectively zero current flow and causing no interference with the reaction. A semi-permeable membrane (a) surrounds the biocatalyst (b) entrapped next to the active glass membrane (c) of a pH probe (d). The electrical potential (e) is generated between the internal Ag/AgCl electrode (f) bathed in dilute HCl (g) and an external reference electrode (h) (Fig. 5.2).

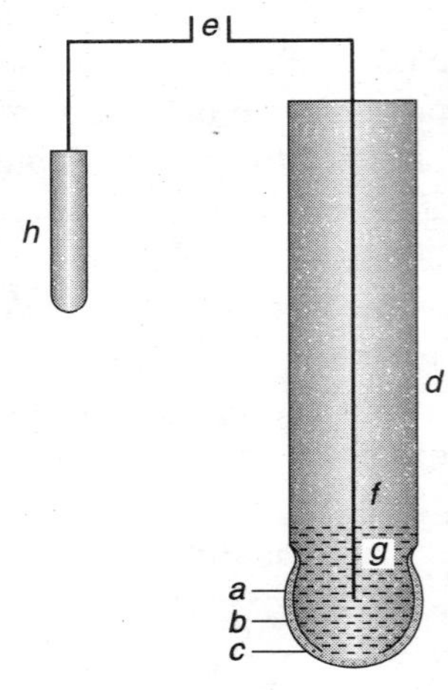

Fig. 5.2 *Schematic diagram of potentiometric biosensor.*

The following types of ion-selective electrodes are used in Potentiometric biosensors.

1. Glass electrodes for cations (e.g. normal pH electrodes) in which the sensing element is a very thin hydrated glass membrane which generates a transverse electrical potential due to the concentration dependent competition between

the cations for specific binding sites. The selectivity of this membrane is determined by the composition of the glass.

2. Glass pH electrodes coated with a gas permeable membrane selective for CO_2, NH_3 and H_2S. The diffusion of the gas through this membrane causes a change in pH of a sensing solution between the membrane and the electrode which is then determined.
3. In solid-state electrodes the glass membrane is replaced by a thin membrane of a specific ion conductor made from a mixture of silver sulphide and a silver halide. The iodide electrode is useful for the determination of Iodide ions (I^-) in the Peroxidase reaction (Table 5.1 c) and also responds to cyanide ions.

Table 5.1 Release or absorption of ions in reactions and these are utilised by potentiometric biosensors.

a. H^+ cation:

1. $\text{D-glucose} + O_2 \xrightarrow{\text{Glucose Oxidase}} \text{D-glucono-1,5-lactone} + H_2O_2 \xrightarrow{H_2O} \text{D-gluconate} + H^+$ ... *eq.-5.3*

2. $\text{Penicillin} \xrightarrow{\text{Penicillinase}} \text{Penicilloic acid} + H^+$ ... *eq.-5.4*

3. $H_2NCONH_2 + H_2O + 2H^+ \xrightarrow{\text{Urease (pH 6.0)}} 2NH_4^+ + CO_2$ ... *eq.-5.5*

4. $H_2NCONH_2 + 2H_2O \xrightarrow{\text{Urease (pH 9.5)}} 2NH_3 + HCO_3^- + H^+$ ... *eq.-5.6*

5. $\text{Neutral lipids} + H_2O \xrightarrow{\text{Lipase}} \text{glycerol} + \text{fatty acids} + H^+$ ... *eq.-5.7*

b. NH_4^+ cation:

1. $\text{L-amino acid} + O_2 + H_2O \xrightarrow{\text{L-amino acid oxidase}} \text{keto acid} + NH_4^+ + H_2O_2$ ... *eq.-5.8*

2. $\text{L-asparagine} + H_2O \xrightarrow{\text{Asparaginase}} \text{L-aspartate} + NH_4^+$ ... *eq.-5.9*

3. $H_2NCONH_2 + 2H_2O + H^+ \xrightarrow{\text{Urease (pH 7.5)}} 2NH_4^+ + HCO_3^-$ ... *eq.-5.10*

c. I^- anion:

$H_2O_2 + 2H^+ + 2I^- \xrightarrow{\text{Peroxidase}} I_2 + 2H_2O$ ... *eq.- 5.11*

d. CN^- anion:

$\text{Amygdalin} + 2H_2O \xrightarrow{\beta\text{-glucosidase}} \text{2glucose} + \text{benzaldehyde} + H^+ + CN^-$ ... *eq.-5.12*

The response of an ion-selective electrode is given by

$$E = E_0 + \frac{RT}{zF} \ln([i]) \qquad \text{... eq.-5.13}$$

where E is the measured potential (in volts), E_0 is a characteristic constant for the ion-selective/ external electrode system, R is the gas constant, T is the absolute temperature (K), z is the signed ionic charge, F is the Faraday constant and [i] is the concentration of the free uncomplexed ionic species. This means, that there is an increase in the electrical potential of 59 mv for every 10-fold increase in the concentration of H^+ at 25°C. The logarithmic dependence of the potential on the ionic concentration is responsible both for the wide analytical range and the low accuracy and precision of these sensors. Their normal range of detection is $10^{-4} - 10^{-2}$ M, typical response time is between 60 and 300 seconds.

Biosensors which involve H^+ release or utilization necessitate the use of very weakly buffered solutions (i.e. < 5 mM) if a significant change in potential is to be determined. The relationship between pH change and substrate concentration is complex, including other non-linear effects as pH activity variation and protein buffering. However, conditions can often be found where there is a linear relationship between the apparent change in pH and the substrate concentration. A recent development from ion-selective electrodes is the production of ion-selective field effect transistors (ISFETs) and their biosensor use as enzyme linked field effect transistors (ENFETs, Fig. 5.3). Enzyme membranes are coated on the ion-selective gates of these electronic devices. The biosensor responding to the electrical potential change via the current output, thus, these are potentiometric devices although they directly produce changes in the electric current. The main advantage of such devices is their extremely small size (<< 0.1 mm^2) which allows cheap mass produced fabrication using integrated circuit technology.

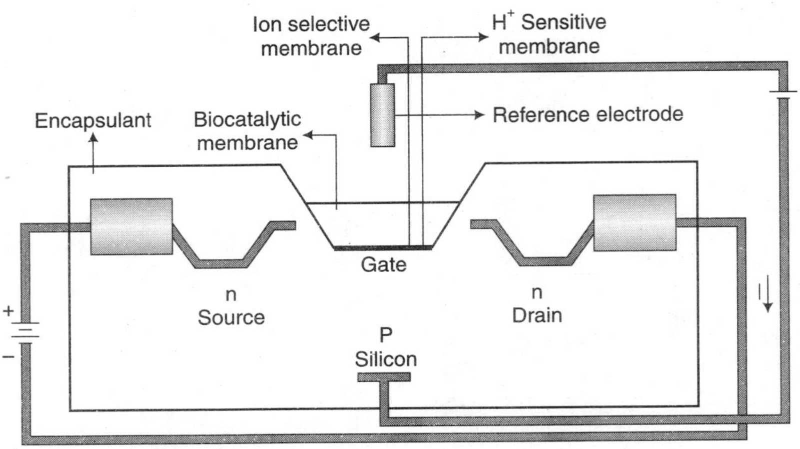

Fig. 5.3 *Schematic diagram of the section across the width of an ENFET.*

The actual dimensions of the active area are about 500 μm long by 50 μm wide by 300 μm thick (Fig. 5.4). The main body of the biosensor is a p-type silicon chip with two n-type silicon areas; the negative source and the positive drain. The chip is insulated by a thin layer (0.1 μm thick) of silica (SiO_2) which forms the gate of the FET. Above this gate is an equally thin layer of H^+ sensitive material (e.g. tantalum oxide), a protective ion selective membrane, the biocatalyst and the analyte solution, which is separated from sensitive parts of the FET by an inert encapsulating polyimide photopolymer. When a potential is applied between the electrodes, a current flows through the FET dependent upon the positive potential detected at the ion-selective gate and its consequent attraction of electrons into the depletion layer. This current (I) is compared with that from a similar, but non-catalytic ISFET immersed in the same solution.

Ion-selective electrodes (ISE) and ion-sensitive field effect transistors (ISFET) based Potentiometric biosensors producing signal is due to ions accumulated at the ion-selective membrane interface. Current flowing through the electrode is equal to or near zero. The electrode follows the presence of the monitored ion resulting from the enzyme reaction. *Eg. G*lucose Oxidase can be immobilized on a surface of the pH electrode. Glucose has only minimal influence on pH in the working medium, however the enzymatically formed gluconate causes acidification. A biorecognition element is immobilized on the outer surface or captured inside the membrane. In the past, the pH glass electrode was used as a physico-chemical transducer, at present these are replaced with semiconductor based physico-chemical transducers. ISFETs and LAPS (light addressable potentiometric sensor) based systems especially are convenient for biosensor construction. The ISFET principle is based on a local potential generated by surface ions from a solution. This potential modulates the current flow across a Silicon semiconductor. The transistor gate surface in ISFET is covered by a selective membrane; for pH detection, this could be made from compounds such as silicon nitride (Si_3N_4), alumina (Al_2O_3), zirconium oxide (ZrO_2) and tantalum oxide (Ta_2O_5).

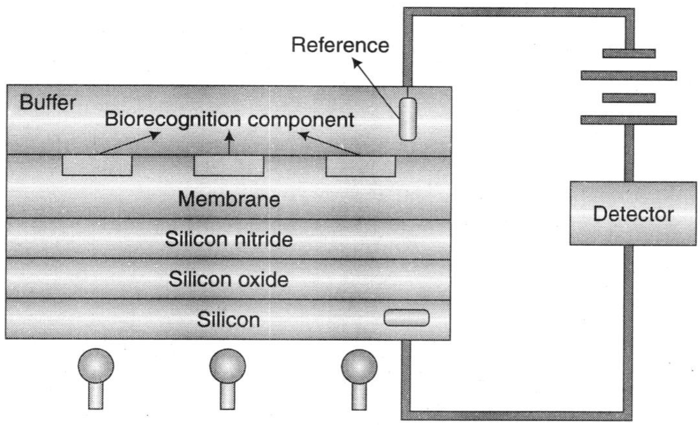

Fig. 5.4 *Block diagram of the light addressable Potentiometric sensor with biorecognition component bound into membrane and with buffered reaction cell.*

The LAPS principle is based on semiconductor activation by a light-emitting diode (LED). The sensor is made from n-type silicon typically coated with 30 nm of silicon oxide, 100 nm of silicon nitride, and indium-tin oxide. The LAPS measures a voltage change as a function of medium pH in the LED activated zone. This opens the way for multi position sensing and construction of an array of biorecognition zones.

A potentiometric biosensor with a molecularly imprinted polymer constructed for the herbicide Atrazine assay allows detecting from 3×10^{-5} to 1×10^{-3} M molecularly imprinted polymer was also used for tracking the level of neurotransmitter Serotonin. Another Potentiometric biosensor with co-immobilized Urease and Creatinase on the poly vinyl chloride ammonium membrane was used for Creatine analysis. ISFET with immobilized butyryl Cholinesterase was used for the detection of glycoalkaloids. A simple pH electrode modified with Acetyl Cholinesterase (AChE) was used for the detection of organophosphate pesticides.

5.2 AMPEROMETRIC BIOSENSORS

Amperometric biosensors function by the production of a current when a potential is applied between two electrodes. These biosensors are generally have response time, dynamic ranges and sensitivities similar to the Potentiometric biosensors. The simplest amperometric biosensors commonly involve the Clark oxygen electrode (Fig. 5.5). This consists of a platinum cathode at which oxygen is reduced and a silver/silver chloride reference electrode. When a potential of -0.6 V, relative to the Ag/AgCl electrode is applied to the platinum cathode, a current proportional to the oxygen concentration is produced. Normally both electrodes are bathed in a solution of saturated potassium chloride (KCl) and separated from the bulk solution by an oxygen permeable plastic membrane (Teflon or Polytetrafluoroethylene). The analyte solution is separated from the biocatalyst (Glucose oxidase) by another membrane, permeable to the substrate(s) and product(s). This biosensor is normally about 1 cm in diameter but has been scaled down to 0.25 mm diameter using a Pt wire cathode within a silver plated steel needle anode and membranes. The following reactions occur at Platinum cathode and Silver anode.

$$\textbf{Ag anode } 4Ag^0 + 4Cl^- \rightarrow 4AgCl + 4e^- \quad \text{... eq.-5.14}$$

$$\textbf{Pt cathode } O_2 + 4H^+ + 4e^- \rightarrow 2H_2O \quad \text{... eq.-5.15}$$

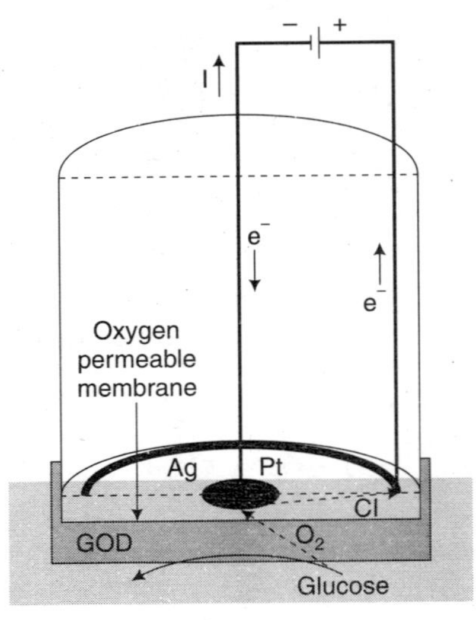

Fig. 5.5 *Schematic diagram of a simple Amperometric biosensor.*

The efficient reduction of oxygen at the surface of the cathode depends on the rate of diffusion of the oxygen from the bulk solution which is dependent on the concentration gradient. The significant proportion of the oxygen present in the bulk solution is consumed at the surface of the oxygen electrode which measures the rate of a process which is far from equilibrium, whereas ion-selective electrodes are used close to equilibrium conditions. This causes the oxygen electrode to be much more sensitive to changes in the temperature than Potentiometric sensors.

5.2.1 APPLICATIONS

A typical application of this biosensor is the determination of glucose concentrations by an immobilised Glucose oxidase membrane; it reduces the oxygen concentration as it diffuses through the biocatalytic membrane to the cathode, this being detected by a reduction in the current between two electrodes (Fig. 5.6). In similar way other Oxidases are also used for the analysis of their substrates. *Eg.* Alcohol oxidase, D-and L-amino acid oxidases, Cholesterol oxidase, Galactose oxidase and Urate oxidase.

Between analyses, the biosensor is placed in oxygenated buffer devoid of glucose. The steady rates of oxygen depletion may be used to generate standard response curves and determine unknown samples. The time required for an assay can be considerably reduced if only the initial transient (curved) part of the response need be used, via a suitable model and software. The wash-out time, which roughly equals the time the electrode spends in the sample solution, is also reduced significantly by this process (Fig. 5.6).

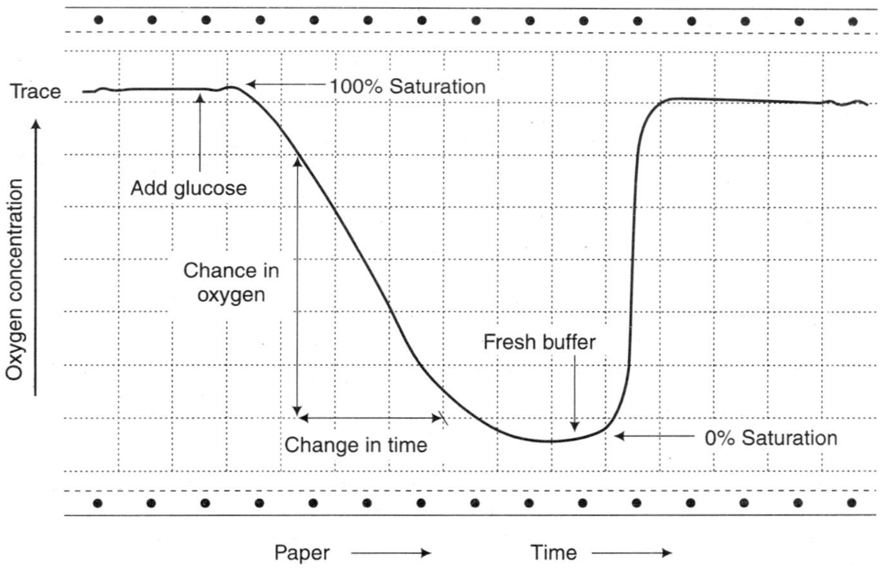

Fig. 5.6 *The response of an Amperometric biosensor utilizing glucose oxidase in the presence of glucose solutions.*

5.2.2 Generations

The amperometric biosensors are categorized into three generations depending on their mode of the measurement of hydrogen peroxide, dependence on the dissolved oxygen concentration and removal of electrons directly from the enzymes without mediators.

First generation: In this generation, the determination of rate of reaction is to measure the production of hydrogen peroxide directly by applying a potential of +0.68 V to the platinum electrode, relative to the Ag/AgCl electrode and causing the following reactions at two electrodes.

$$\text{Pt Anode } H_2O_2 \xrightarrow{\text{Catalase}} O_2 + 2H^+ + 2e^- \qquad \text{... eq.-5.16}$$

$$\text{Ag Cathode } 2AgCl + 2e^- \rightarrow 2Ag^0 + 2Cl^- \qquad \text{... eq.-5.17}$$

Second generation: The major problem with Ampherometric biosensors is their dependence on the dissolved oxygen concentration. This may be overcome by the use of 'mediators' which transfer the electrons directly to the electrode bypassing the reduction of the oxygen co-substrate. These mediators possess a number of useful properties.

1. They must react rapidly with the reduced form of the enzyme.
2. They must be sufficiently soluble in both the oxidised and reduced forms, to be able to rapidly diffuse between the active site of the enzyme and the electrode surface. This solubility should not be so great as to cause significant loss of the mediator from the biosensor's microenvironment to the bulk of the solution. However, the soluble mediator should generally be non-toxic.
3. The over potential for the regeneration of the oxidised mediator at the electrode should be low and independent of pH.
4. The reduced form of the mediator should not readily react with oxygen.

Most important used mediators in Amperometric biosensors are 'ferrocenes' (Fig. 5.7). Their reactions may be represented as follows:

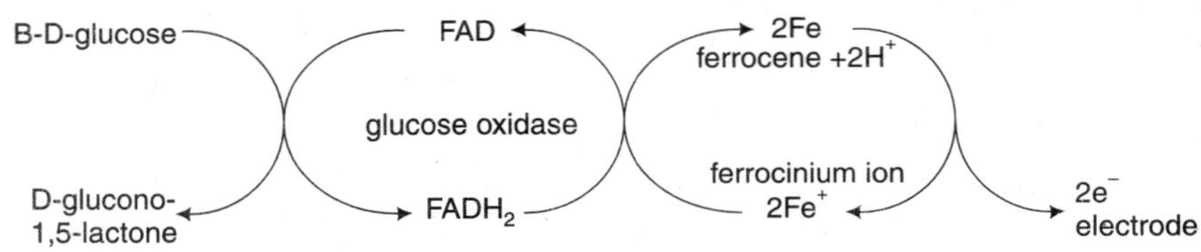

Fig. 5.7 *Transfer of electrons from the reaction to electrode through 'ferrocene' mediators.*

Third generation: At present, electrodes have been developed which can remove the electrons directly from the reduced enzymes without the necessity for such mediators. They utilise a coating of electrically conducting organic salts such as N-methylphenazinium cation (NMP^+, Fig. 5.8) with Tetracyanoquinodimethane radical anion ($TCNQ^-$ Fig. 5.8). Many flavor enzymes are strongly adsorbed by organic conductors due to the formation of salt links, utilising an alternate positive and negative charges within their hydrophobic environment. Such enzyme electrodes can be prepared by simply dipping the electrode into a solution of the enzyme and they may remain stable for several months. These electrodes can also be used for reactions involving NAD $(P)^+$ dependent dehydrogenases as they also allow the electrochemical oxidation of the reduced forms of these coenzymes.

The three types of Amperometric biosensors utilizing, product or mediator or organic conductors represent the three generations in biosensor development (Fig. 5.9). The reduction in oxidation potential, found when mediators are used, greatly reduces the problem of interference by extraneous material.

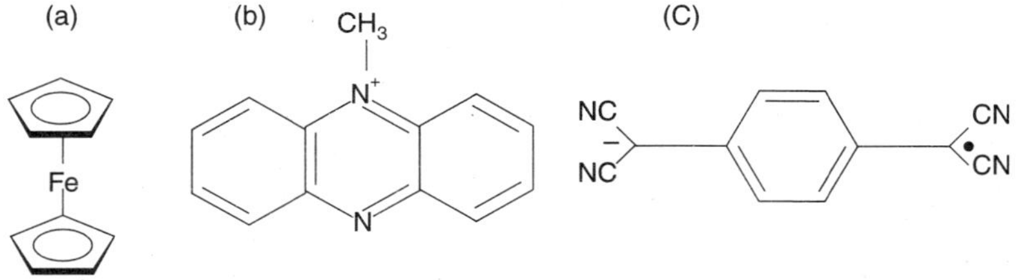

Fig. 5.8 *(a) Ferrocene (5-bis-cyclopentadienyl iron), the parent compound of a number of mediators. (b) TMP^+, the cationic part of conducting organic crystals. (c) $TCNQ^{\cdot-}$, the anionic part of conducting organic crystals. It is a resonance-stabilized radical formed by the one-electron oxidation of $TCNQH_2$.*

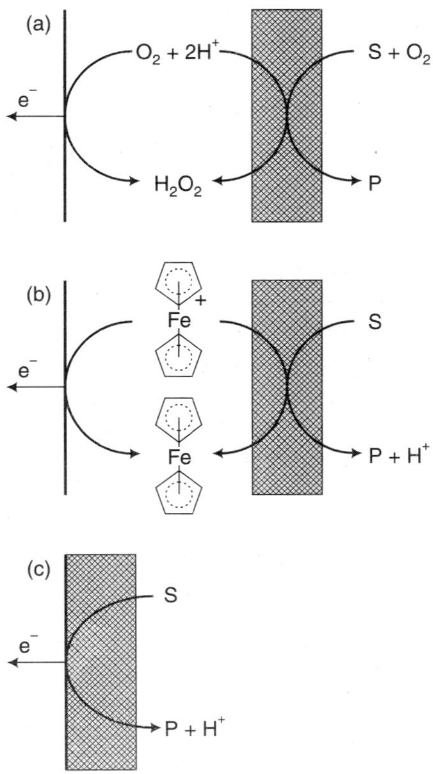

Fig. 5.9 *Generations of Amperometric biosensors.*

a. First generation electrode utilising the H_2O_2 produced by the reaction. (E_0 = +0.68 V).
b. Second generation electrode utilising a mediator (ferrocene) to transfer the electrons, produced by the reaction, to the electrode. (E_0 = +0.19 V).
c. Third generation electrode directly utilising the electrons produced by the reaction. (E_0 = +0.10 V). All electrode potentials (E_0) are relative to the $Cl^-/AgCl, Ag^0$ electrode.

The following reaction occurs at the enzyme in all three biosensors:

Substrate (2H) + FAD-oxidase $\rightarrow$ Product + $FADH_2$-oxidase ... *eq.-5.17(a)*

This is followed by the processes:

(a) **biocatalyst:** $FADH_2$-oxidase + $O_2 \rightarrow$ FAD-oxidase + H_2O_2 ... *eq.-5.18*

Electrode: $H_2O_2 \rightarrow O_2 + 2H^+ + 2e^-$ ... *eq.-5.19*

(b) **biocatalyst:**

$FADH_2$-oxidase + 2 $Ferricinium^+ \rightarrow$ FAD-oxidase + 2 Ferrocene + $2H^+$ ... *eq.-5.20*

Electrode: 2 Ferrocene $\rightarrow$ 2 $Ferricinium^+ + 2e^-$ ... *eq.-5.21*

(c) **biocatalyst/electrode:**

$FADH_2$-oxidase $\rightarrow$ FAD-oxidase + $2H^+ + 2e^-$ ... *eq.-5.22*

The current (i) produced by such amperometric biosensors is related to the rate of reaction (v_A) by the expression:

$$i = nFAv_A \quad \text{... eq.-5.23}$$

Where n represents the number of electrons transferred, A is the electrode area, and F is the Faraday constant. Usually the rate of reaction is made diffusionally controlled by use of external membranes. Under these circumstances the electric current produced is proportional to the analyte concentration and independent both of the enzyme and electrochemical kinetics.

5.3 BIOCATALYSIS BASED BIOSENSORS

At present the most extensive commercial biomedical sensor is Glucose biosensor, is being frequently used to measure the glucose concentration in the blood samples. There are quite a few enzymes which could be used as glucose detectors, many of them require cofactors. *Eg.* The glucose enzyme reaction with Glucose oxidase doesn't require cofactors and is in addition one of the most stable enzymes for the detection of glucose.

$$\text{Glucose} + O_2 \xrightarrow{\text{Glucose oxidase (GOD)}} \text{Gluconic acid} + H_2O_2 \quad \text{... eq.-5.24}$$

$$H_2O_2 \xrightarrow{\text{Catalase}} \tfrac{1}{2}\, O_2 + H_2O \quad \text{... eq.-5.25}$$

In the above shown first reaction, there are two possibilities to measure the glucose concentration in glucose biosensor, first one measures the oxygen consumption and second one measures the hydrogen peroxide production. As a third option it would be also possible to measure the temperature increase due to heat released by the above first reaction. In the second, the presence of catalase enzyme although not really essential for the assay but it enhances the first reaction and this happens to the decomposition of the hydrogen peroxide.

Biocatalysis based biosensors are used for environmental applications, depend on the use of enzymes. These biosensors primarily rely on two operational mechanisms (1) The first mechanism involves the catalytic transformation of a pollutant. (2) The second mechanism involves the detection of pollutants that inhibit or mediate the enzyme's activity.

5.3.1 First Mechanism

The first mechanism of biocatalysis based biosensors in environmental monitoring involve atalytic transformation of the pollutant (from a non-detectable form to a detectable form) and these are simple in design and easy to operate. *Eg*. Tyrosinase and Organophosphate hydrolase enzymes are used for the detection of phenols and organophosphorus pesticides respectively.

5.3.1.1 Limitations

The nature of the enzyme itself and include the limited number of environmental pollutants which happen to be substrates for the enzyme and the relatively high detection limits (compared to those required by many environmental monitoring applications) for environmental pollutants. The detection limits for these sensors are determined by the enzyme's catalytic properties and are defined by K_m and V_{max} values. Biosensor formats have been devised which substantially reduce these inherent limitations.

Eg.: (1) In the case of the Tyrosinase enzyme electrode, catalytic cycling of the enzyme intermediate between the quinone and catachol oxidation states can significantly amplify the biosensor response. This results in lower detection limits for phenols than expected from K_m and V_{max} values. Mechanisms for the Tyrosinase biosensors involve the detection of phenols either through the electrochemical reduction of quinone intermediates or through oxygen consumption (O_2 is a co-substrate) using a Clark electrode. 2) A biosensor was designed to increase both sensitivity and the range of substrates typically measured using the Tyrosinase enzyme electrode. With this biosensor, a wide range of chlorinated phenols are detected as oxidation reduction mediators in a Glucose oxidase electrode system. This is accomplished by chemically oxidizing the chlorophenols which then cycle between the quinone and hydroquinone oxidation states. In this case, the quinone acts as an electron shuttle for the Glucose oxidase catalyzed oxidation of glucose. This configuration results in low nM detection limits for a number of chlorinated phenols.

5.3.2 Second Mechanism

The other primary mechanism used in biocatalytic biosensors for environmental applications is inhibition of the enzyme by the pollutants. Inherent advantages for these biosensors involve the larger number of environmental pollutants, usually of a particular chemical class, that inhibits the enzyme and the low concentrations needed to affect the enzyme activity. Detection limits for these biosensors based on irreversible inhibitors are usually within the range required for a variety of environmental applications. *Eg*. Detection limits for Cholinesterase biosensors are reported to be in the μg/l to ng/l range for compounds such as Aldicarb, Carbaryl and Carbofuran.

5.3.2.1 Limitations

1. In addition to the analyte of interest, these assay formats require the use of substrates, cofactors and mediators.
2. For a number of these assays, pollutants must be chemically oxidized to metabolic intermediates to show maximum sensitivities.
3. The irreversible nature of many analyte-enzyme interactions that result in increased sensitivity also renders the biosensor inactive after a single measurement. This may not be a problem, however, for those systems that can be reactivated or that employ disposable sensing elements.

4. Another potential limitation to this type of biosensor is diverse classes of pollutants that inhibit a specific enzyme.
5. The co-contamination of environmental samples with organics and heavy metals, such interferences may lead to unexpected results.

The interface of these devices to the contaminated matrix is critical to the successful application of biocatalytic based biosensors, particularly for *in situ* applications. These matrices may range from pristine drinking water to highly contaminated industrial sludges. Although a considerable amount of work has been done with respect to use of biosensors in biological matrices (such as blood and fermentation media), little work has focused on the development of membrane barriers for the direct sampling of groundwater or bore water in sludges and saturated sediments.

5.3.3 Advantages

1. More specific than cell based sensors.
2. Faster response due to shorter diffusion paths (no cell walls).

5.3.4 Disadvantages

1. More expensive to produce due to the additional problem of isolating the enzyme.
2. Enzymes are often unstable when those are isolated.
3. Many enzymes need cofactors for the detection of substances.

5.4 CALORIMETRIC BIOSENSORS

Many enzymes catalysed reactions are exothermic, generating heat (Table 5.2) which may be used as basis for measuring the rate of reaction and the analyte concentration. The temperature changes are usually determined by means of thermistors at the entrance and exit of small packed bed columns containing immobilised enzymes at constant temperature environment (Fig. 5.10). Under controlled conditions up to 80% of the heat generated in the reaction may be registered as a temperature change in the sample stream, this may be calculated from the enthalpy change and the amount reacted. If 1 mM reactant is completely converted to the product in a reaction, it generates 100 kJ mole^{-1} of heat (each ml of solution generates 0.1 J of heat). At 80% efficiency this will cause a change in temperature of the solution amounting to approximately 0.02°C, for biosensors useful necessitates temperature resolution is 0.0001°C.

Table 5.2 Heat output (molar enthalpies) of enzyme catalysed reactions.

Enzyme	*Reactant*	*Heat output (kJ/mole)*
Cholesterol oxidase	Cholesterol	53
Chymotrypsin	Esters	4–16
Glucose oxidase	Glucose	80
Catalase	Hydrogen peroxide	100
Penicillinase	Penicillin G	67
Amylase	Starch	8
Invertase	Sucrose	20
Urease	Urea	61

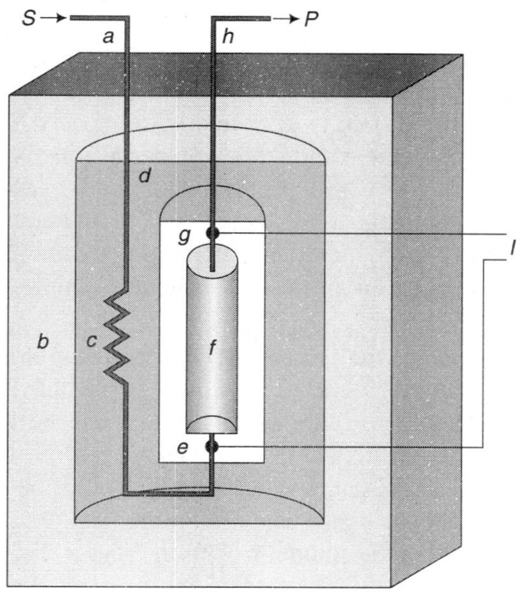

Fig. 5.10 *Schematic diagram of a Calorimetric biosensor.*

The sample stream (a) passes through the outer insulated box (b) to the heat exchanger (c) within an aluminium block (d). From there, it flows past into the reference thermistor (e) and into the packed bed bioreactor (f, 1ml volume), containing the biocatalyst, where the reaction occurs. The change in temperature is determined by the thermistor (g) and the solution passed to waste (h). External electronics (l) determines the difference in the resistance and temperature between the thermistors. The thermistors are used to detect the temperature change function by changing their electrical resistance with the temperature.

$$\ln\left(\frac{R_1}{R_2}\right) = B\left(\frac{1}{T_1} - \frac{1}{T_2}\right) \quad \text{... eq.-5.26}$$

Therefore:

$$\frac{R_1}{R_2} = e^{B\left(\frac{1}{T_1} - \frac{1}{T_2}\right)} \quad \text{... eq.-5.27}$$

$$R_1 = R_2\left\{1 + B\left(\frac{T_2 - T_1}{T_1 T_2}\right)\right\} \quad \text{... eq.-5.28}$$

Where R_1 and R_2 are the resistances of the thermistors at absolute temperatures T_1 and T_2 respectively and B is a characteristic temperature constant for the thermistor. When the temperature change is very small, as in the present case, $B\left(\frac{1}{T_1} - \frac{1}{T_2}\right)$ is very much smaller than one and this relationship may be substantially simplified using the approximation when $x<<1$, As $T_1 \approx T_2$, they both may be replaced in the denominator by T_1.

$$\frac{\Delta R}{R} = -\left(\frac{B}{T_1^2}\right)\Delta T \quad \text{... eq.-5.29}$$

The relative decrease in the electrical resistance ($\Delta R/R$) of the thermistor is proportional to the increase in temperature (ΔT). The changed resistance is converted to voltage using balanced Wheatstone bridge incorporating precision wire-wound resistors, before the amplification. A major problem with this biosensor is the difficulty encountered in closely matching the characteristic temperature constants of the measurement and reference thermistors. An equal movement of only 1°C in the background temperature of both thermistors commonly causes an apparent change in the relative resistance of the thermistors equivalent to 0.01°C and equal to the full scale change due to the reaction. It is clearly of great importance that such environmental temperature changes are avoided which accounts for inclusion of the well-insulated aluminium block in the biosensor design.

The sensitivity (10^{-4} M) and range (10^{-4} to 10^{-2} M) of thermistor biosensors are both quite low for the majority of applications although greater sensitivity is possible using the more exothermic reactions (e.g. Catalase). The low sensitivity of the system can be increased substantially by increasing the heat output by the reaction. In the simplest case this can be achieved by linking together several reactions in a reaction pathway, all of which contribute to the heat output. Thus the sensitivity of the glucose analysis using Glucose oxidase can be more than doubled by the co-immobilization of Catalase within the column reactor in order to disproportionate the hydrogen peroxide produced (Fig. 5.11).

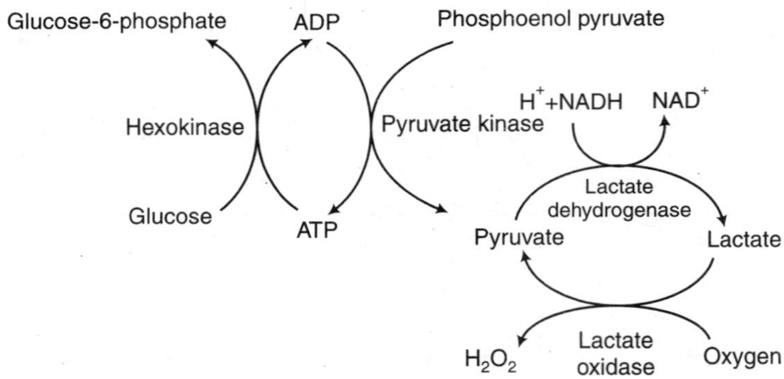

Fig. 5.11 *Recycle scheme for the detection of ADP.*

ADP is the added analyte and excess glucose, phosphoenol pyruvate, NADH and oxygen are present to ensure maximum reaction. Four enzymes (Hexokinase, Pyruvate kinase, Lactate dehydrogenase and Lactate oxidase) are co-immobilised within the packed bed reactor. In spite of the positive enthalpy of the Pyruvate kinase reaction, the overall process results in a 1000-fold increase in sensitivity, primarily due to the recycling between Pyruvate and Lactate. Reaction limitation due to low oxygen solubility may be overcome by replacing it with benzoquinone, which is reduced to hydroquinone by flavo-enzymes. Such reaction systems have the serious disadvantage in that they increase the probability of the occurrence of interference in the determination of the analyte of interest. Reactions involving the generation of hydrogen ions can be made more sensitive by the inclusion of a base having a high heat of protonation. *Eg*. The heat output by the Penicillinase reaction may be almost doubled by the use of Tris (tris-(hydroxymethyl) amino methane) as the buffer. The main advantages of the thermistor biosensor are its general applicability, the possibility for its use on turbid or strongly coloured solutions and are the difficulty in ensuring that the temperature of the sample stream remains constant (0.01°C).

5.5 BIOAFFINITY BASED BIOSENSORS

Bioaffinity based biosensors depend on the use of antibodies which include Monoclonal antibodies (MAbs) and Polyclonal antibodies (PAbs). These MAbs and PAbs are expressed toward a wide range of environmental pollutants, for a specific compound or closely related groups of compounds. Small molecular weight organic pollutants in the environment have distinguishing optical or electrochemical characteristics. The detection of stoichiometric binding of these compounds to antibodies is typically accomplished using competitive binding assay formats. Competitive immunosensor depend on the use of an antigen-tracer which competes with the analyte for a limited number of antibody binding sites. Affinity-based biosensors are accomplished in three formats, which are mentioned below.

1. The first format used antigen-tracer competes with analyte for immobilized antibody binding sites and this format is often used in fluorescence based systems.
2. In this second format the antigen is immobilized to the signal transducer while free binding sites on the antibody, which has been exposed to the analyte bind to the surface immobilized antigen. The antibody is binding to the surface bound antigen which can be detected by signal transduction methods such as Surface Plasmon Resonance, Acoustic Systems and Optical Systems that measure changes in refractive index.
3. The third format required an indirect competitive assay and relies on the use of an enzyme labeled antigen tracer. In this format, the assay is completed in two steps. (a) In first step the enzyme tracer competes with analyte for immobilized antibody binding sites (b) Then in the second step, after removal of the unbound tracer (by means of a washing step), a non-detectable substrate is catalytically converted to an electrochemically or optically detectable product. This assay format is used almost universally with electrochemical signal transduction.

Bioaffinity based biosensors (Immunosensors) are mostly used for environmental applications rather than the fundamental understanding, however, immunosensors for the most part represent technological advances for existing bioanalytical assays. It is important to improve the characteristics of a particular assay with respect to known or anticipated applications, because of the wide variety of scientifically established and commercially available immunoassays (test kits). There are a variety of ways to group immunosensor based on signal transducers, one functionally useful discriminator involves classification based on reusable/regenerable or disposable format configurations because, immunosensors (particularly those using disposable formats) are most closely related to immunoassay test kit technology. These issues which become important for comparison are more practical in nature and involve the potential for multi-analyte capability, format versatility, assay time, assay sensitivity, system cost, assay cost, shelf life, reproducibility, ruggedness, etc. In contrast to the disposable formats, the multi use immunosensors, which can be recharged or regenerated with certain advantages, particularly for use as detectors for chromatographic and flow injection analysis systems.

The fibre optic Immunosensors are used for detection of explosive residues in groundwater. Nucleic acid based affinity biosensors are developed for environmental applications, these biosensors include the detection of chemically induced DNA damage and the detection of microorganisms through the hybridization of species-specific sequences of DNA.

5.6 PIEZOELECTRIC BIOSENSORS

5.6.1 Introduction

The biorecognition processes include bioaffinity recognition and bio-metabolic recognition, in both processes involve the binding of a chemical species with another, which has a complementary structure this is referred to as shape-specific binding. In bioaffinity recognition the binding is very strong and the transducer detects the presence of the bound receptor-analyte pair. The most common types of processes are receptor-ligand and antibody-antigen binding. In bio-metabolic recognition, the analyte and other co-reactants are chemically altered to form the product molecules, these can be recognized by the biorecognition elements which are diverse as the different reactions that occur in biological systems.

The concept of shape-specific recognition is commonly used to explain the high sensitivity and selectivity of biological molecules, especially antigen-antibody systems. The analyte molecule has a complementary structure to the antibody and the bound pair is in a lower energy state than the two separate molecules and this binding is very difficult to dissociation.

The interaction of antibodies with their corresponding antigens is an attractive reason for attempting to develop antibody-based chemical biosensors, (Immunosensors). Theoretically, if an antibody can be raised against a particular analyte, an immunosensor could be developed to recognize it. Despite the high specificity and affinity of antibodies towards complementary ligand molecules, most antibody-antigen interactions do not cause an electronically measurable change. However, the remarkable selectivity of antibodies has fueled much research to overcome this intrinsic problem. The piezoelectric effect in various crystalline substances is a useful property that leads to the detection of analyte (Fig. 5.12).

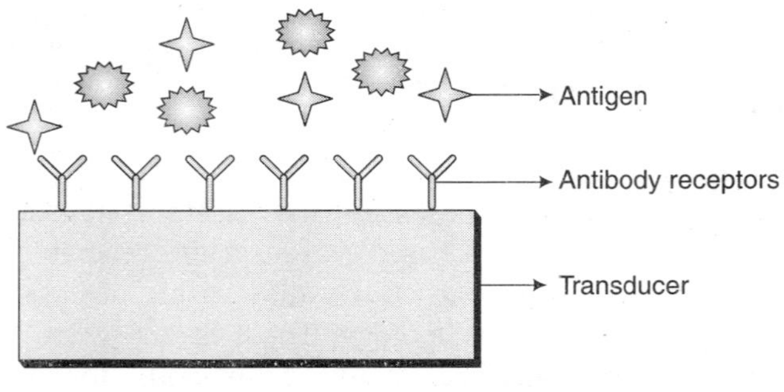

Fig. 5.12 *Schematic diagram of an immunosensor device.*

The piezoelectric immunosensors are one of the most sensitive analytical instruments, which are being capable of detecting antigens in the picogram range. Moreover, this type of device is believed that to have the potential to detect antigens in the aerosol phase as well as in the liquid phase.

The diagnosis of Tuberculosis (TB) at present methods has certain drawbacks in the detection and they tend to be either non-specific or more time consuming. In most cases of pulmonary and extra-pulmonary TB diagnosis depends upon culturing the *Mycobacterial* organism which requires 4-8 weeks. Significant attention has been devoted to developing more rapid diagnostic methods for TB, but some of them do not have the high specificity and sensitivity required for proper diagnosis. A piezoelectric sensor that could consistently detect the *Mycobacterial* antigen in biological fluids would be of enormous use. For instance, detection of the antigen in saliva could constitute a non-invasive method of screening in high-risk groups. The Piezoelectric crystal sensor gives results within a couple of hours after exposing the electrode to a liquid containing antigen. The apparatus would be quite portable, the tests could be performed virtually anywhere and the results could be obtained very quickly. The feasibility of using Piezoelectric immunosensors to diagnose TB based upon the detection of *Mycobacterial* antigens in liquid and on the degree of sensitivity and specificity. The feasibility of aerosol or vapor phase detection of antigen depends upon these same factors plus any difficulties that may be unique to aerosol-phase antigen capture by antibodies.

5.6.2 Principle

The basic equations describing the relationship between the resonant frequency of an oscillating piezoelectric crystal and the mass deposited on the crystal surface have been derived by Sauerbrey and co-workers. Each followed a different path but their final equations are similar, the Sauerbrey equation being the most widely accepted. In 1959, Sauerbrey developed an empirical equation for AT-cut quartz crystals vibrating in the thickness shear mode that describes the relationship between the mass of thin metal films deposited on Quartz crystals and the corresponding change in resonant frequency of the crystal:

$$\Delta F = -2.3 \times 10^6 F^2 \frac{\Delta M}{A} \qquad \text{... eq.-5.30}$$

where, ΔF is the frequency change in oscillating crystal in Hz, F is the frequency of piezoelectric quartz crystal in MHz, ΔM is the mass of deposited film in grams and A is a area of electrode surface in cm^2.

These relationships not only apply to film deposition but also to particulate deposition. When vibrating in the thickness-shear mode, the oscillating frequency of an AT-cut quartz crystal is given by:

$$F = \frac{N}{a} \qquad \text{... eq.-5.31}$$

where F is the frequency of the crystal, N is the material constant (N = 1.66 MHz-mm for AT-cut quartz crystal) and a is the thickness of the crystal plate. Thus,

$$dF = d\left(\frac{N}{a}\right) = -N\left(\frac{da}{a^2}\right) \qquad \text{... eq.-5.32}$$

For finite amount of change, may write:

$$\Delta F = \frac{-N\,\Delta a}{a^2} \quad \text{... eq.-5.33}$$

Dividing Equation 5.33 by Equation 5.31 gives:

$$\frac{\Delta F}{F} = \frac{-N\frac{\Delta a}{a^2}}{\frac{N}{a}} = \frac{-\Delta a}{a} \quad \text{... eq.-5.34}$$

If the thickness is defined as:

$$a = \frac{M}{A.\rho} \quad \text{... eq.-5.35}$$

where M is the mass of the electrically driven portion of the crystal, A is the area of the electrically driven portion of the crystal and ρ is the density of the crystal.
Assuming the constant density, may write:

$$dA = \frac{dM}{A.\rho} \quad \text{... eq.-5.36}$$

If the changes are finite,

$$\Delta a = \frac{\Delta M}{A\rho} \quad \text{... eq.-5.37}$$

Using equation 5.34 yields,

$$\frac{\Delta F}{F} = -\left(\frac{\Delta M}{M}\right) \quad \text{... eq.-5.38}$$

Thus,

$$\Delta F = F\left(\frac{\Delta M}{M}\right) \quad \text{... eq.-5.39}$$

Since F and M are constants, we may say:

$$\Delta F = -k\,\Delta M \quad \text{... eq.-5.40}$$

The oscillating frequency of the crystal changes linearly with the change in mass on the crystal and the mass change occurs due to deposition of materials on the surface of the crystal. This relationship is only valid for small mass changes. For larger mass changes, it would be invalid since the density would change.

Equation 5.39 may also be stated as:

$$\frac{\Delta F}{F} = -\frac{F}{M} \quad \text{... eq.-5.41}$$

This shows that the terms ΔF and ΔM would increase, if the base oscillating frequency of the crystal is increased. The term itself is the sensitivity of the crystal sensor and equ. 5.12 shows that sensitivity is directly proportional to the base frequency of the crystal.

If an aerosol stream is sent over the surface of the piezoelectric crystal and if it contains an analyte with concentration C, then

$$C = \frac{\Delta M}{V} \qquad \textbf{... eq.-5.42}$$

where C is the concentration of the analyte in the aerosol stream, ΔM is the mass of analyte in the aerosol stream and V is the volume of the aerosol in the stream. The volume of the aerosol stream is related to the sampling time by the following relationship:

$$V = q \cdot t \qquad \textbf{... eq.-5.43}$$

where q is the flow rate of the aerosol stream and t is the sampling time. Equation 5.13 can be rewritten as:

$$C = \frac{E \cdot \Delta M}{q \cdot t} \qquad \textbf{... eq.-5.44}$$

where E is the collection efficiency of the coating material. If E is assumed to be equal to one and substituting Equation 5.39 into Equation 5.44, we may say

$$C = \frac{-(\Delta F)(M)}{(F)(q)(t)} \qquad \textbf{... eq.-5.45}$$

or

$$\Delta F = \frac{C \cdot F \cdot q \cdot t}{M} \qquad \textbf{... eq.-5.46}$$

Thus, the change in the oscillating frequency of the crystal is related to both the sampling time and the concentration of the particles in the carrier aerosol. If we keep the sampling time, career aerosol flow rate and the mass of the analyte in the aerosol stream constant, may restate equation-5.46 as:

$$\Delta F = K\,C \qquad \textbf{... eq.-5.47}$$

Equation 5.47 shows that the change in frequency of the crystal is directly proportional to the concentration of the analyte in the aerosol stream flowing over it.

The Piezoelectric crystal detector can be a very powerful analytical tool because of the relationship shown for the change in frequency to the analyte concentration with high sensitivity. Conversely, the above explanation shows that the crystal detector indiscriminately changes the frequency due to the deposition of mass of any material on its surface.

5.6.3 Electrode Fabrication

In Piezoelectric biosensors the most frequently used detector crystal is alpha quartz. These crystals are most suitable for piezoelectric application because they are insoluble in water and resistant to high temperatures. Alpha quartz crystals can be resistant to temperature up to 579°C with no loss of piezoelectric properties. The resonant frequency of quartz crystal depends on the physical dimensions of the quartz plate and the thickness of the electrode deposited. AT and BT cut crystals are most useful as piezoelectric detectors. These cuts refer to the orientation of the plate with respect to the crystal structure. The AT cut crystal is the most stable with a temperature coefficient of 1 ppm per degree centigrade over a temperature range of 10°C to 50°C. The crystals usually take the form of discs, squares and rectangles.

All AT cut quartz crystals with an electrode coating deposited on each side using sputtering method. The crystal was mounted on a holder with stainless steel with leads (Fig. 5.13). Silver composite was used to connect the electrode to wire. The crystals were 14 mm in diameter and the electrodes on both sides of the crystal were 8 mm in diameter. The crystals were mounted on size HC6/u holders.

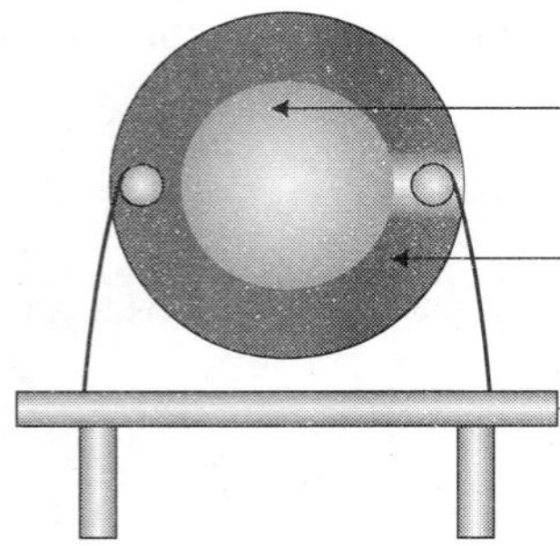

Fig. 5.13 *Quartz crystal and holder.*

The piezoelectric quartz crystal was driven by a low frequency transistor oscillator, powered by a 1-30 V DC regulator power supply and set at 9 V DC (Fig. 5.14). The frequency of the vibrating crystal was monitored by a Protek multifunction frequency counter. The crystal mounted on its holder was connected to the oscillator circuit and the frequency counter was connected to the oscillator device. After each step in the coating process, first with the various metal depositions and then with the biomolecular analytes, the frequency reading was recorded.

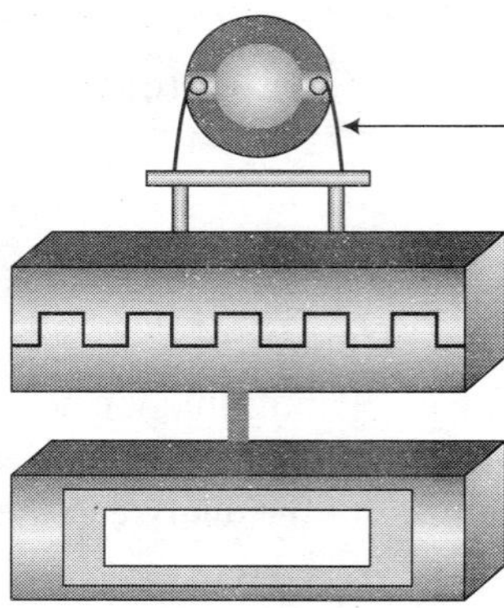

Fig. 5.14 *Experimental apparatus for a piezoelectric sensor.*

5.6.4 Immunosensing

The crystal electrodes were first modified with 5 ml coating of Protein-A for better adhesion of the antibodies to the surface of the transducer. Protein-A is a polypeptide isolated from *Staphylococcus aureus* that binds specifically to the immunoglobulin molecules, especially IgG antibodies without interacting at the antigen site. This property permits the formation of tertiary complexes consisting of Protein- A, antibody and antigen. Prior to modification, the electrodes were anodically oxidized at constant current in 0.5 M NaOH. They were cleaned in 0.5 M HCl and 0.5 M $HCrO_2$, then dried in an incubator for one hour and the antibody (IgG) coating was applied to the Protein-A coating. 10 ml of antibody was applied on both sides of the crystal and kept for drying for 60 minutes. 10 ml of antigens were coated onto the crystal and dried. After each step, the frequency of the crystal was recorded and washed as a precaution against non-specific binding.

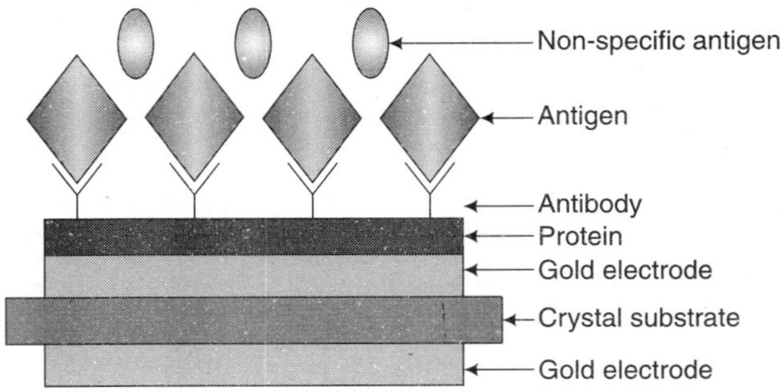

Fig. 5.15 *Schematic diagram of the antibody-antigen binding.*

Control crystals and experimental crystals were coated with antibodies (Fig. 5.15). The former were coated with an irrelevant antibody (HBV-honey bee antibody), one specific for an antigen not present in the solution containing the analyte. The experimental sample was coated with antibody (*M. tuberculosis*) specific for binding to the antigen and both were exposed to the solution containing the analyte. Then the difference in frequency change between the control and experimental crystals were compared, reflecting the immunologically specific binding of analyte (Fig. 5.15). First the crystal was coated with gold substrate and later with magnetic materials. A magnetic field was induced during the investigation. The protein helps the antibody to bind to the electrode and the antigen in aerosol state binds to specific antibody.

The X-ray diffraction patterns of the gold films deposited on quartz and are polycrystalline in nature. The theoretical values at varying gold thicknesses, compared with experimental values at different thicknesses, which clearly shows that the experimental and theoretical values of change in frequency are almost equal for different thicknesses of electrode coating. The deposited coating changes the frequency, approximately in agreement with the projections made by Sauerbrey.

Although platinum is a non-reactive material compared to the gold, the adherence of the platinum films was very poor on quartz substrates and reacted with different buffer solutions, (*Eg.* HCl and NaOH) during the specimen preparation for antigen-antibody binding. Gold electrodes are preferred due to superior adhesion and non-reactive properties and the optimum thickness of the gold electrode layer was estimated at 1,000 A^0. Although the binding between antigen and antibody did show a change in frequency, the results were not always reproducible. The antibody binding to the protein layer was critical to achieve desirable results. The use of magnetic materials underneath the gold coating helped make the antigen detectable.

5.7 IMPEDIMETRIC BIOSENSORS

The impedimetric devices (Fig. 5.16) follow either impedance (Z) or its components resistance (R) and capacitance (C); inductance has a minimal influence in a typical electrochemical setup. The expression of impedance is the inverse value of resistance called conductance, such systems are named conductometric. These biosensors are also including two electrodes with applied alternating voltage, from a few to 100 mV and are commonly a functional part of the Wheatstone bridge. These systems are used for the assay of urea when Urease is used as a biorecognition component. The following reaction takes place in the medium:

$$\underset{\text{Urea}}{(NH_2)_2CO} + \underset{\text{Water}}{3H_2O} \xrightarrow{\text{Urease}} 2NH_4 + HCO_3^- + OH \quad \text{... eq.-5.48}$$

The above equation gives that the urea and water molecules on the left side of the equation exhibit only minimal influence on the measured impedance. The enzymatically produced ions are able to provide a significant increase of impedance. These biosensors have been successfully used for microorganism growth monitoring due to the production of conductive metabolites (Silley and Forsythe, 1996). False positive results due to electrolytes from the samples are the main disadvantage of impedance biosensors.

Impedimetric biosensors are less frequent compared to Potentiometric and Amperometric biosensors. Hybridization of DNA fragments previously amplified by a polymerase chain reaction has been monitored by an impedance assay. An impedance immunosensor containing electrodeposited Polypyrrole film with captured Avidin connected through Biotin to anti-human IgG was able to detect antibodies 10 pg/ml present in a sample. The ethanol level in some alcoholic beverages was evaluated by an Impedance biosensor with immobilized yeast. The impedance based commercial device Malthus, 2000 was used for an assay of the pathogenic fungus *Ichthyophonus hofery* and the *Erwinia carotovora* rot.

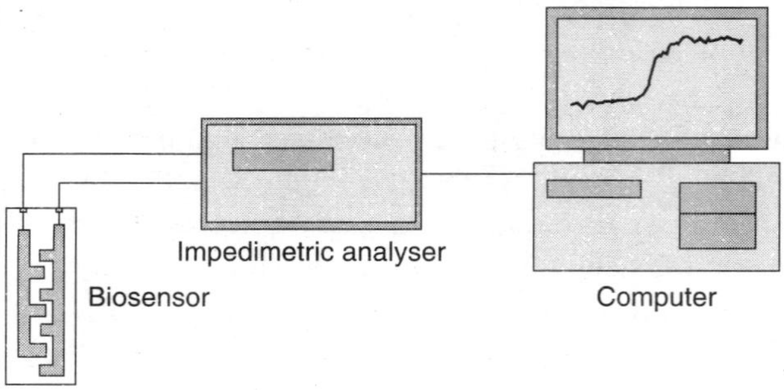

Fig. 5.16 *Simplified scheme of analytical device based on Impedimetric biosensor. (Source: Electrochemical biosensors – principles and applications, Miroslav Pohanka, Petr Sklādal, J. Appl. Biomed.,6: 57–64, 2008, ISSN 1214-0287)*

5.8 CALORIMETRIC SENSOR

The calorimetric sensor belongs to the class of solid state gas sensors. They can owe their name to the fact that they measure produced energy on oxidation of a flammable gas. The combustible gases in the presence of oxygen create hazardous mixtures, hence there are two main reasons why these gases need to be measured (1) Safety requires the monitoring of these hazardous gases and (2) The energy industry combusts such mixtures under controlled conditions. In order to achieve a higher efficiency of combustion these gases need to be burned in a certain ratio to each other.

5.8.1 Principle

The most famous and widely used calorimetric gas sensor is the Pellistor. It consists of a small diameter platinum coil wire supported in a refractory bead (Fig. 5.17). The coil is used for two purposes, firstly, to heat the bead electrically to its operating temperature of approximately 500° C (depending on the catalyst used) and secondly it detects temperature changes which are being caused by exothermic reactions of the flammable gases. The reaction of CO with O_2 liberates approximately 280kJ/mol, taking the flow rate of the gas through the sensor with an efficiency of 10% into consideration. The spiral platinum wire is covered with a porous alumina which contains precious metals as catalysts and the mostly used catalyst in Pellistor is palladium or platinum. The use of the catalysts results in a larger surface area and therefore permits a much lower operating temperature which increases the lifetime of the sensor. A higher temperature leads to possible hot spots in the platinum coil and thus degradation and eventually to a short in the circuit (usually the lifetime is correlated to the temperature of the device by a logarithmic relationship). The catalyst is able to influence the number of active sites and thus the sensitivity and selectivity of the gas sensor. The overall combustible gas sensor consists of a matched pair of resistor elements, the active detector bead carrying the catalyst and a compensating element.

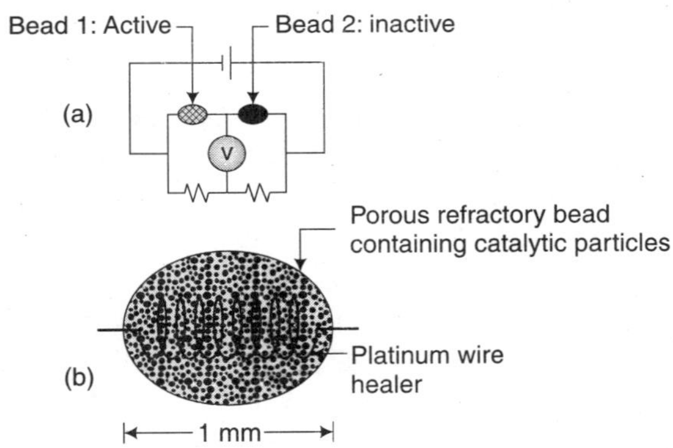

Fig. 5.17 *Typical Pellistor structure.*

The matched pair of resistors forms a Wheatstone bridge circuit and a voltage is supplied to heat the sensor. The output of the Wheatstone bridge is zeroed with a potentiometer when clean air is in the environment. If there is a flammable gas in the environment a reaction takes place at gas leakage pipes thereby temperature increases around the area, this temperature can be detected and monitored by calorimetric sensor. The heat production changes the resistance of the sensing arm of the Wheatstone bridge and thus the bridge is out of balance. The magnitude of the signal change is proportional to the concentration of the flammable gas. The performance parameters, selectivity and sensitivity of the sensor are influenced by the choice of the catalyst and influence the selectivity, sensitivity, response time and stability of the sensor. By varying the catalyst parameters, the number of active sites and their distribution are changed.

The drawback for all existing catalysts is that they are strongly influenced by poisoning gases which reduce their activity dramatically. Choosing the right catalyst is of major importance for this type of sensor, since its operating principle is sensitive to almost every flammable gas. An alternative route for the manufacture of combustible sensors is, thick film print of the above mentioned in the Wheatstone bridge and the principle and operation process is the same (Fig. 5.18).

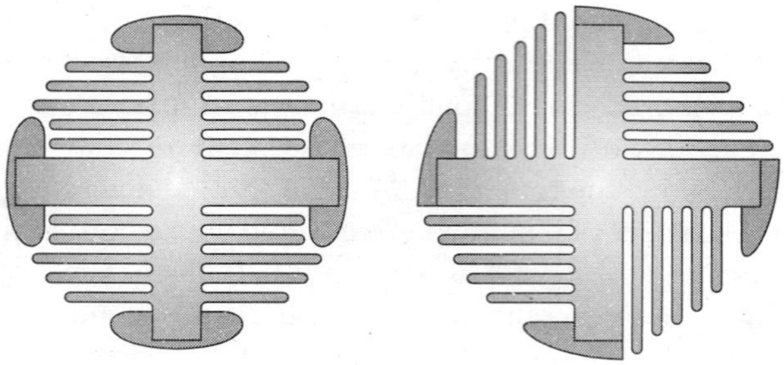

Fig. 5.18 *The outline of the resistor tracks of a Servomex thick film device.*

Each of the four quadrants represents one resistor of Wheatstone bridge circuit (Fig. 5.18). Two of the resistors (quadrants) are covered with a catalyst, this is where the actual sensing of the gas takes place. The exothermal energy drives the bridge out of balance and shows indirectly the concentration of the flammable gas.

5.8.2 Specifications

Using a calorimetric sensor sensitivities of tenths of parts per million can be achieved. The sensitivity of the sensors is mainly dependent on the catalyst used and number of distributed active sites. These sensors show response time less than 10 seconds. The selectivity of the sensor can be influenced by the type of the catalyst however, due to the nature of a combustible sensor, (all combustible gases are interference gases) it is difficult to make sensor selectivity.

The main drawback for combustible sensors is the risk of hazardous gaseous agents released through the gas oxidation process in the environment which effect the sensor catalyst and inhibit or block its action, resulting in a decreased selectivity and sensitivity of the sensor.

The use of calorimetric sensors is not possible when performing a dynamic measurement of the single breath test. The main drawback of these sensors in respect to this application is their slow speed of response to a step change in gas concentration, thus not sufficient information on the shape of the carbon monoxide decay during smell can be collected. The humidity in the gas sample would not necessarily introduce a problem because these sensors are usually operated at elevated temperatures, so condensation effects do not occur.

Biosensors - II

Chapter 6

6.0 ALCOHOL BIOSENSOR

6.0.1 Introduction

The investigation of alcoholism, the physiological effects of alcohol in humans or animals or the efficacy of treatment regimens, have only two ways to collect field measurements of Blood Alcohol Concentration (BAC) from the samples. One is the Breathalyzer, which produces an estimate of BAC by measuring the concentration of alcohol in exhaled breath and then multiplying it by the partition ratio of 2100 to 1, the ratio of breath alcohol to blood alcohol. However, the use of the Breathalyzer has a number of significant drawbacks, it requires voluntary compliance, active participation, inconvenient and obtrusive. More significantly, as a result of factors such as incorrect use, the presence of residual mouth alcohol when the reading is taken and variation in the value of the partition ratio due to differences in body temperature and blood chemistry from individual to individual, it can produce inaccurate measurements.

In an effort to address these obstructions to get accurate BAC measurements in the field, the National Institute on Alcohol Abuse and Alcoholism (NIAAA) has developed an alcohol biosensor, which is portable, non-invasive, relatively unobtrusive, reliably and accurately determines blood alcohol concentrations over minutes to weeks time periods in the field.

The transdermal alcohol is used to measure BAC developed by Giner in 1930s, when it was determined that ingested ethanol can be measured in concentrations that are approximately equal to its concentration in the blood. This sensor is non-invasive, relatively simple and has been proven to work. This sensor consists of membrane bound Alcohol dehydrogenase (ADH), Aldehyde dehydrogenase (ALDH) and an electron transfer system.

6.0.2 Ethanol Sensors

Ethanol sensors use a metal coil inserted into a tubular substrate as the heater, thus keeping the sensor film on the substrate surface at an elevated temperature for detection sensitivity. This device structure has deadly shortcomings of large thermal mass and thermal loss, preventing it from many practical applications due to high power consumption and slow response. Therefore, it is very difficult to design a battery operated breath alcohol.

Recently, Micro Electro Mechanical System (MEMS) technology has developed metal oxide type Alcohol sensor, in which the sensing film and micro heater are formed on low stress membrane of a micro-machined chip. The structure has the advantage of extremely low thermal mass and heat loss. Only 70mW or lower is required for heating to maintain the sensor film at the elevated working temperature, which allows OGS-100 series ethanol gas sensors being suitable for all kinds of portable design.

Portable alcoholic sensor has compact design with a digital display to show the alcohol content and is easy to carry. The Alco Hawk features are, a folding mouthpiece design and two digits Light Emitting Diode (LED) display along with a highly accurate semiconductor oxide sensor. It also contains an advanced electric breathe out pressure monitor to ensure that an accurate breath sample is obtained.

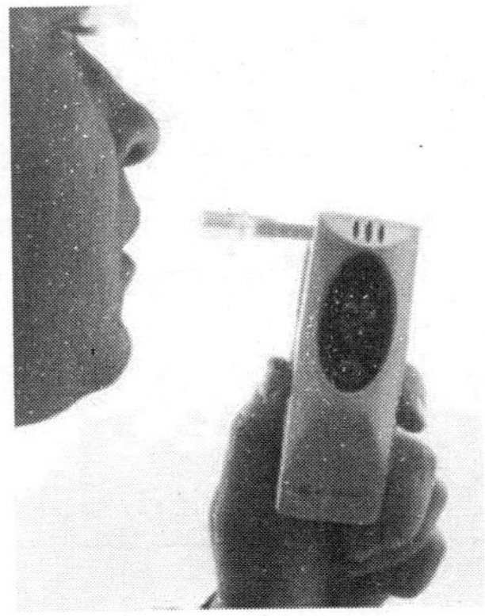

Fig. 6.1 *Alcohol sensor (Portable).*

6.0.3 Features

1. High sensitivity to ethanol
2. High response speed
3. Low power consumption
4. High reliability and stability

6.0.4 Applications

1. Breath alcohol detectors
2. Measurement of bottle wine concentration

Continual measurements of ethyl alcohol concentration in culture broth are required for fermentation industries. Furthermore, in the cultivation of yeast using sugar as a carbon source, it is well known that ethyl alcohol as a by-product decreases the sugar basis yield of whole cells. In the cultivation of microorganisms by using methyl alcohol as a carbon source, the concentration must be maintained at the optimum level to avoid substrate inhibition. Assimilation of alcohols by microorganism can be determined from the respiration activity of microorganisms and it is directly measured with an oxygen electrode. Therefore, it is possible to construct a microbial sensor for alcohols using immobilized microorganisms and oxygen electrode. A microbial electrode consisting of immobilized yeast or bacteria, a gas permeable Teflon membrane and an oxygen electrode can be prepared for the determination of methyl and ethyl alcohol. Furthermore, the microbial sensor can be applied to continuous determination of alcohols in a fermentation broth. To test the reliability of the alcohol sensor, an experiment was conducted by using microbial sensor and gas chromatography for the determination of ethyl alcohol concentration.

The determination of alcoholic compounds, particularly ethanol, is relevant to the food industry, especially in alcoholic beverages such as beer, wines and spirits. Enzyme based electrochemical devices are successfully developed for the detection of alcohols. Two enzymes are commercially exploited - Alcohol oxidase and Alcohol dehydrogenase. The Alcohol dehydrogenase is a NADH depending enzyme, which is required to co-immobilize the co-enzyme. In addition, this co-enzyme requires a potential of 1.0 V for oxidation process and at this potential, a number of other substances present in food samples are also oxidized and can interfere in the alcohol measurement. Different chemical mediators could be used to decrease the over potential as well as to prevent the electrode pasivation.

The Meldola Blue (MB) is a Phenoxazine dye, which facilitates NADH oxidation and it allows achieving high sensitivity for the amperometric determination of NADH. This mediator allows the detection limit as low as 2×10^{-6} mol.L^{-1} with good selectivity since the measurements could be made in an "ideal" potential window (from 0.00 V to –0.200 V vs. Saturated Calomel Electrode (SCE)), where electrochemical interferences are minimal. The formal potential of Meldola Blue assimilated with the arithmetic mean of cathodic and anodic peak potential ranges from –0.051V to –0.175 V vs. Saturated Calomel Electrode (SCE) depending on the sensor material can be used for mediator entrapment. In recent years, number of reports concerned to biosensor based screen printed electrodes is continuously increasing and Meldola Blue is often a preferred modifier when designing NADH detectors.

The Meldola Blue (MB) having a fast rate of electron transfer with NADH and this approach allows not only to shift the equilibrium of the enzymatic reaction to the product side but also to reduce the over potential for the oxidation of NADH. The modification of a screen printed carbon electrode by including Meldola's Blue in the graphite. The sensor was characterized and further developed to produce a low cost biosensor for ethanol based with the association of screen printing technology and chronoamperometry. The problems related to NAD were overcome by the simple design of the sensors: disposable biosensors for ethanol were obtained by direct deposition of small amounts of Alcohol dehydrogenase and NAD on the planar surface of a mediator-incorporating carbon electrode.

The enzyme Alcohol dehydrogenase (ADH) from Baker's Yeast 264 U/mg solid and nicotinamide adenine dinucleotide (NAD) oxidized form is NAD^+ and reduced form is NADH. In amperometry and crono-amperometry determination, the buffer contained 0.1mol L^{-1} KCl in order to insure proper functioning of the screen printing pseudo Ag/AgCl reference electrode. The disposable biosensors for ethanol were prepared by simply depositing on the working area of Melodola Blue-modified electrodes 2.5 μL of a mixture containing 8 IU ADH and 1.3×10^{-3} mol L^{-1} NAD^+ (optimum amounts). The biosensors were left to dry overnight at 4°C.

The electrochemical behaviour of Meldola Blue incorporated in the graphite ink of the screen-printed electrodes reported and the modified electrodes were used in cyclic voltammetry experiments. The potentials scanned between –0.4 V and +0.4 V vs. Ag/AgCl at 0.1 V/s in 0.1 mol L^{-1} phosphate buffer pH 7 and these studies revealed the catalytic properties of MB towards NADH oxidation. From Fig. 6.2, in the presence of 5×10^{-3} mol L^{-1} NADH the magnitude of the oxidation peak for MB increase, whereas that of the reduction peak decreases. This is a typical feature of catalytic processes which follow an E-C mechanism:

$$\text{MBH} \xrightarrow{E>E^\circ} \text{MB}^+ + \text{H}^+ + 2e^- \qquad \textit{... eq.-6.1}$$

$$\text{NADH} + \text{MB}^+ \rightarrow \text{NAD}^+ + \text{MBH} \qquad \textit{... eq.-6.2}$$

$$\text{Alcohol} + \text{NAD}^+ \xrightarrow{\text{ADH}} \text{Acetaldehyde} + \text{NADH} + \text{H}^+ \qquad \textit{... eq.-6.3}$$

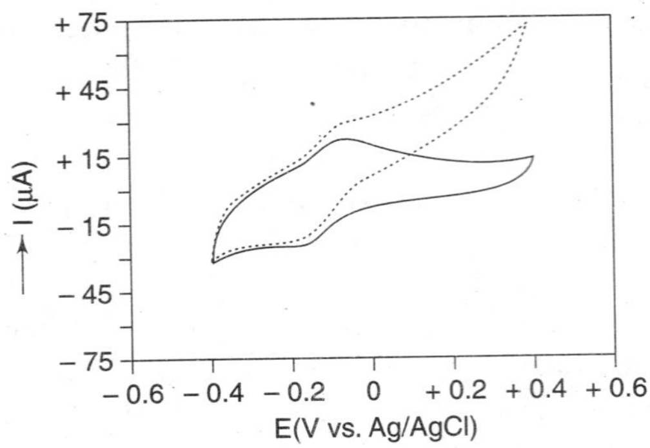

Fig. 6.2 *Voltammograms recorded in 0.1 mol/L phosphate buffer pH 7 with MB-incorporating screen-printing electrodes in the absence ( ) and the presence (----) of 5×10^{-3} mol L^{-1} NADH. (Source: Disposable alcohol biosensor based on alcohol dehydrogenase and screen-printed electrodes, Camelia Bala, L. Rotariu and V. Magearu, Analele Universiti din Bucureti – Chimie, Anul XII (serie noua), vol. I-II, pag. 55–60)*

Therefore, the changes in anodic produced current are directly proportional to the concentration changes in the sample. The sensitive detection of the NADH performed at +50 mV vs. Ag/AgCl and the potential range of the electrochemical interferences are reduced to a minimum. The sensitivity of the sensors towards NADH detection, the noise level and the value of residual current was found an optimum of 2% MB (w/w graphite). The analytical performances of 2% MB modified screen printed electrodes towards NADH are presented in Table 6.1.

Table 6.1 Analytical performance of 2% MB modified screen printed electrodes.

Limit of detection (mol/L x 10^{-6})	*Linear range (mmol/L)*	*Equation*
0.01	0.02 – 0.2	I (nA) = 75 + 2786 C_{NADH}

By using the above method, single use biosensors can be developed on the basis of enzyme deposited on the surface of the electrode. It is an ideal method for the analysis of small amount of liquid (30 μL-50 μL) by chronoamperometry. Taking into account the difficulty of immobilizing the enzymatic cofactor NAD^+, this configuration was found appropriate for fabrication of disposable biosensors based on the NAD dependent dehydrogenase. The response of ethanol biosensor was recorded by chronoamperometry at 30, 60 and 120 seconds after applying the potential +50 mV vs. Ag/AgCl, for different concentration of substrate in the range 0.1 – 20 m mol L^{-1} ethanol. The optimum time for determination is 60 seconds and did not observe significant differences compared with the biosensor response beyond 120 seconds. The coefficient of variation was 8.9% for n = 5 electrodes, which is acceptable for screen-printed electrodes and considering that the deposition of biocatalytic layer was performed manually. The performance of the disposable biosensors was tested to determine the ethanol concentration of different beverages: whisky, gin and white wine. In order to fit the analyse concentration within the linear range of the calibration curve, the sample dilution is required with phosphate buffer 1:20000 for the wine and 1:50000 for the other beverages. The results obtained with the disposable biosensor were compared with Official Method of the European Community (Table 6.2).

Table 6.2 Ethanol determination in beverages.

Beverage	*Biosensor Determination (%) n = 5*	*Official Methods (%) n = 5*	*Label indicated alcohol content(%)*
Whisky	38 ± 0.5	39.5 ± 0.2	40
Gin	37.3 ± 0.4	38.9 ± 0.2	40
Wine	11.7 ± 0.3	10.8 ± 0.1	11

6.1 GLUCOSE SENSOR

The glucose sensor is fabricated by immobilizing Glucose oxidase (GOD) on the membrane of the oxygen electrode. The glucose sensor responds as soon as the glucose solution is injected into the buffer solution and reaches a steady state in 5 to 10 minutes. The sensor responds almost linearly for glucose concentration between 0.2 to 2.0 mM, which is comparable to conventional glucose sensors.

The microbial CO_2 sensor was constructed by Suzuki *et al.* (1987) using this oxygen electrode, the autotrophic bacterium S17, which can grow with only carbonate as the carbon source. Bacterial cells are immobilized on a micro-oxygen electrode. The sensitive area of the oxygen electrode is immersed in 0.2% sodium alginate solution containing S17 bacterial cells, then removed and immediately immersed in 5% $CaCl_2$ solution to form bacteria immobilized calcium alginate gel. The response time is 2 to 3 minutes. Carbon dioxide was supplied by acidification of $NaHCO_3$, the concentration of which can be related to CO_2 concentration. Linear relationship is obtained between the current decrease and $NaHCO_3$ concentration in the range 0.5 to 3.5mM. The lowest detection limit is 0.5mM $NaHCO_3$ within the margin of the noise amplitude.

6.1.1 Blood Glucose Monitoring

Blood glucose monitoring is a way of testing the concentration of glucose in the blood (Glycemia) and it is important in the care of *Diabetes mellitus*. The blood glucose test is performed by pricking the skin on the finger tip to draw the blood. Then placing the blood on a chemically active disposable strip which indicates the result either by changing colour or changing an electrical characteristic, the latter being measured by an electronic meter.

Fig. 6.3 *Blood glucose testing.*

Most of the people with Type 2 diabetes test at least once per day (usually before breakfast) to assess the effectiveness of their diet and exercise. Many people with Type 2 use an oral medication to combat their insulin resistance, and test their blood glucose before and after breakfast to assess the effectiveness of their dosage. Diabetics who use insulin (all Type 1 diabetes and many Type 2) usually test their blood sugar more often (3 to 10 times per day), both to assess the effectiveness of their prior insulin dose and to help determine their next insulin dose. Improved technology for measuring blood glucose is rapidly changing the standards of care for all diabetic people. Currently many methods are available for blood glucose testing.

6.1.2 Visual Test Strips

Visual test strips are medium cost method for monitoring blood glucose. A drop of blood (≈ 20 μL), usually taken from the fingertip, is placed on a test zone which is located at one end of a strip of plastic. The test zone contains chemicals which react with the sugar (glucose) in the blood, changing color according the concentration of glucose, which can then be read by comparing the color with a chart on the side of the test strip container. These strips are suitable for people who occasionally monitor their blood glucose level (prediabetic or type 2) and do not use insulin. They are also useful for people who can't afford test strips that are used in electronic meters and for people who dislike electronic meters or who need a back up for their meter.

6.1.3 Blood Glucose Meters

A blood glucose meter is an electronic device for measuring the blood glucose level. A relatively small drop of blood is placed on a disposable test strip which interfaces with a digital meter. Within several seconds, the level of blood glucose will be shown on the digital display. Sample sizes vary from 30 to 0.3 μL. Test times vary from 5 seconds to 2 minutes (modern meters are typically below 15 seconds).

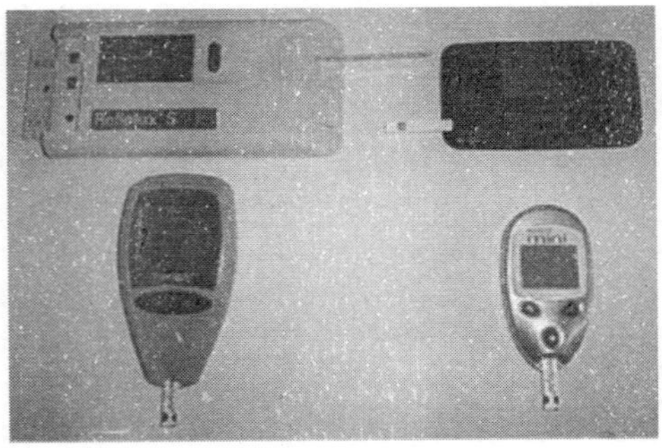

Fig. 6.4 *Four generations of blood glucose meter.*

While more expensive, blood glucose meters seem a breakthrough in diabetes self care. As the drops of blood needed for the meter become smaller, the pain associated with testing is reduced and the compliance of diabetic people to their testing regimens is improved. Although the cost of using blood glucose meters seems high, it is believed to be a cost benefit relative to reduce the medical costs of the complications of diabetes.

A recent and welcome advance is the use of small blood drops for blood glucose testing from other places than the finger tips. This alternate site testing uses the same test strips and meter, is practically pain free. The disadvantage of this technique is that there is usually less blood flow to alternate sites, which prevents the reading from being accurate when the blood sugar level is changing.

6.1.4 Continuous Glucose Monitoring

The first capillary blood test strip Glukotest was developed in 1956. The relationship between metabolic control and complications were recognized in 1977. In 2003, accurate, small and fast devices using tiny (< 1-4 mcL) amounts of blood allowed the generation of statistics and the use of telemedicine. In the early 1980s, a needle type new glucose sensor was developed and it is a wearable artificial pancreas used in a closed loop glycemic control system with a computer. This sensor used to study the mixture rates of insulin, glucagon or both, producing perfect control of blood glucose for up to 7 days and it is tested in pancreatectomized dogs.

The Continuous Glucose Monitoring System (CGMS) named as GlucoWatch and other devices such as Glycemic Holter which helps the diabetologist and the ability to provide real-time continuous glucose monitoring. CGMS device has four separate components: the pager-sized glucose monitor, sterile disposable subcutaneous glucose sensor, cable connector and communication system, enabling data stored in the monitor to be uploaded to a computer.

The glucose sensor has a platinum electrode, a Glucose oxidase layer, and a membrane allowing conversion of glucose in the presence of oxygen to glucuronic acid and hydrogen peroxide (H_2O_2) and generates the electric current. Glucose concentrations are measured every 10 seconds and stored every 5 minutes, for 288 measurements daily, with the capacity to store 14 days of data, although each sensor can only be used for a maximum of 72 hours. At least 4 finger stick glucose measurements are calibrated and must be recorded each day. The presenters also noted that interstitial fluid glucose is not always an accurate reflection of plasma glucose, so it is important not to calibrate when there will be large variation, such as the postprandial period or for approximately 1 hour following hypo- or hyperglycemia. The software is designed to ignore data points that appear to give wrong calibration. However, this assumes that there is no drift in sensor sensitivity, which may occur and lead to incorrect calibration. Furthermore, it may be desirable to use a laboratory plasma glucose measurement for calibration to avoid the potential errors seen with most glucose meters. An important note is that the points of calibration should not be used in calculation of the regression coefficient to assess the validity of results, as this would only represent an internal description of the quality of the calibration procedure. The use of the CGMS device has been extensively studied in pediatric practice and its use is also being explored in gestational diabetes.

6.1.4.1 Glucose monitoring

Glucose monitoring helps people with diabetes manage the disease and avoid its associated problems. A person can use the results of glucose monitoring to make decisions about food, physical activity and medications. The most common way to check glucose levels involves pricking a fingertip with an automatic lancing device to obtain a blood sample and then using a glucose meter to measure the blood sample's glucose level.

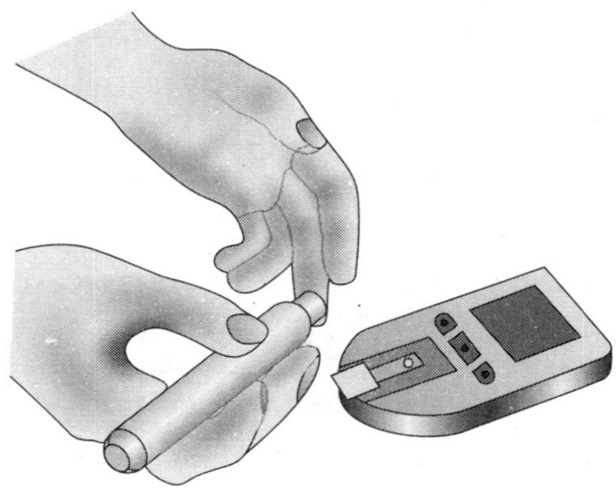

Fig. 6.5 *People with diabetes typically use a lancing device to obtain a blood sample and a glucose meter to measure the glucose level in the sample.*

Many types of glucose meters are available, all are accurate and reliable if used properly. Some meters use a blood sample from a less sensitive area than the fingertip, such as the upper arm, forearm or thigh.

6.1.4.2 Continuous Glucose Monitoring (CGM)

Continuous glucose monitoring (CGM) systems are use a tiny sensor inserted under the skin to check glucose levels in tissue fluid. The sensor stays in that place for several days to a week and then must be replaced. A transmitter sends information about glucose levels via radio waves from the sensor to a pager like wireless monitor (Figs. 6.6 & 6.7). The user must check blood samples with a glucose meter to program the devices, because currently approved CGM devices are not as accurate and reliable as standard blood glucose meters, users should confirm glucose levels with a meter before making a change in treatment.

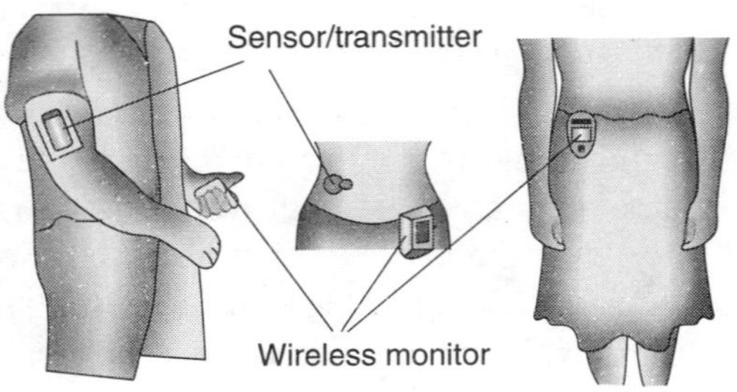

Fig. 6.6 *CGM systems provide glucose measurements as often as once per minute. The measurements are transmitted to a wireless monitor.*

Fig. 6.7 *People who use CGM systems can download data to a computer to see patterns and trends in their glucose levels.*

CGM systems are more expensive than conventional glucose monitoring, but they may enable better glucose control. CGM devices produced by Abbott, DexCom and Medtronic have been approved by the U.S. Food and Drug Administration (FDA) and are available by prescription. These devices provide real time measurements of glucose levels, with glucose levels displayed at 5 minute or 1 minute intervals. Users can set alarms to alert them when glucose levels are too low or too high. Special software is available to download data from the devices to a computer for tracking and analysis of patterns and trends, and the systems can display trend graphs on the monitor screen. Additional CGM devices are being developed and tested.

6.1.4.3 Prospects for an artificial pancreas

To overcome the limitations of current insulin therapy, researchers have long sought to link glucose monitoring and insulin delivery by developing an artificial pancreas. An artificial pancreas is a system that will mimic, as closely as possible, the way a healthy pancreas detects changes in blood glucose levels and responds automatically to secrete appropriate amounts of insulin. Although not a cure, an artificial pancreas has the potential to significantly improve diabetes care and management and to reduce the burden of monitoring and managing blood glucose. An artificial pancreas based on mechanical devices requires at least three components:

1. A CGM system
2. An insulin delivery system
3. A computer program that "closes the loop" by adjusting insulin delivery based on changes in glucose levels.

With recent technological advances, steps have been taken towards closing the loop. The first pairing of a CGM system with an insulin pump, the MiniMed Paradigm Real Time System, is not an artificial pancreas but it does represent the first step in joining glucose monitoring and insulin delivery systems.

6.1.5 Glucose Sensing Bio-Implants

Long term solutions to continuous monitoring, not yet available but under development, use a long-acting bio-implant. These systems promise to ease the burden of blood glucose monitoring for their users but at the trade off of a minor surgical implantation of the sensor that lasts from one year to more than five years depending on the product selected.

6.1.6 Non-Invasive Technologies

Some new technologies to monitor blood glucose levels will not require access the blood to read the glucose level. Non-invasive technologies include near IR detection, ultrasound and dielectric spectroscopy. These will free the person with diabetes from finger sticks to supply the drop of blood for blood glucose analysis. Most of the non-invasive methods under development are continuous glucose monitoring methods and offer the advantage of providing additional information to the subject between the conventional finger stick, blood glucose measurements and over time periods where no finger stick measurements are available (i.e. while the subject is sleeping).

6.1.7 Effectiveness

Patients with diabetes mellitus type-2, the importance of glucose monitoring and the optimal frequency of monitoring are not clear. Self-monitoring of blood glucose did not improve the Hba1c among "reasonably well controlled non-insulin treated patients with type-2 diabetes". A recent study found that a treatment strategy of intensively lowering blood sugar levels (below 6%) in patients with additional cardiovascular disease risk factors poses more harm than benefit. For those who are not on insulin, exercise and diet are the best tools to aid those with type-2, thus blood glucose monitoring is a tool to evaluate the success of diet and exercise. But those with type-2 who use insulin need to monitor their blood sugar frequently just as those with type 1.

The National Institute for Health and Clinical Excellence (NICE), UK released updated diabetes recommendations on the 30th May 2008, which recommend that self monitoring of plasma glucose levels for people with newly diagnosed type-2 diabetes must be integrated into a structured self management education process.

Side effects of continuous glucose monitoring system are piercing the body, particularly on the fingertips. There haven't been any studies investigating the long term damage caused by continual body piercing.

6.2 DNA SENSORS

Currently, DNA assays are used for the identification of large number of genetic and infectious diseases in clinical laboratories. Over the past two decades intensive investigations on electrochemical DNA sensors lead to the development of home used devices. The basic requirements to develop an electrochemical DNA sensor are immobilization of an oligonucleotide probe onto an electrode, hybridization of a complementary target sequence and transduction of hybridization event.

Millan and Mikkelsen (1993) reported the first electrochemical DNA sensor. The transduction of DNA sensor was based on an electro active indicator of Tris (2, 2'-bipyridyl) Cobalt(III) $[Co(bpy)_3{}^{3+}]$. This redox indicator binds more strongly to double strand DNA (dsDNA) than to single stranded DNA (ssDNA) in an intercalative mode, thus the voltammetric current signal increases if the sample contains the complementary target sequence. In addition to $[Co(bpy)_3{}^{3+}]$, other metal complexes *Eg.* Tri (1, 10-phenanthroline) cobalt(III), $[Co(Phen)_3{}^{3+}]$ and organic molecules *Eg.* Methylene blue, and 2, 6-disulfonic acid anthroquinone have been employed as hybridization indicators.

Barton *et al.* (1999) studied the charge transport between the redox intercalator especially Daunomycin and Methylene blue and a gold electrode through DNA. The experiments involved the immobilization of a short duplex (15 bp) via gold thiol linkage (the 5' end of one strand was labeled with a thiol group), forming a densely packed DNA film on the gold surface. Subsequent to a washing step, the DNA modified electrode was immersed in an intercalator solution and studied by cyclic voltammetric measurement. The compact structure resulted in the binding of the intercalator close to the top of the film (farthest from the electrode surface), so charge transport was achieved through DNA. Remarkably, the presence of a single-base mismatch in the duplex caused a huge decrease in the current signal. It should be pointed out that these redox intercalators do not necessarily bind preferentially to dsDNA. *Eg.* Methylene blue has been shown to interact with exposed Guanine base specifically; therefore, the current signal of the probe target electrode was lower than that of the probe only electrode. The assay sensitivity of these redox intercalators depends strongly on their binding properties *Eg.* binding constant and dissociation rate constant. Synthetic threading intercalator which is Naphthalene diimide derivative with better superior binding properties has been functionalized with Ferrocene redox marker, giving rise to a highly sensitive electrochemical hybridization indicator.

Non-DNA binding soluble redox species are also used for hybridization. Thorp *et al.* (1998) demonstrated that tris (2, 2'-bipyridyl) ruthenium (III), $[Ru(bpy)_3^{3+}]$, catalyzed the oxidation of guanine. For construction of the DNA sensor, guanine bases in the probe were replaced by inosine bases, which could base-pair with cytosine but much less reactive toward $[Ru(bpy)_3^{3+}]$. This was significant in that the probe only surface had a low background signal and thus the hybridization of the guanine containing complementary sequence could lead to a larger change in signal (i.e., higher signal to background ratio), thus the sensitivity increases.

The ion channel sensor technique, which took advantage of the electrostatic interactions between a soluble redox marker and DNA modified electrode surface. Gold electrode was modified with a mixed monolayer of Peptide Nucleic Acid (PNA) and 6-mercapto-1-hexanol. PNA contains a neutral *N*-(2-aminoethyl) glycine backbone in contrast to DNA's negatively charged sugar phosphate backbone. Due to the lack of electrostatic repulsion, PNA has a higher binding affinity with its complementary DNA sequence than its DNA counterpart. A negatively charged redox marker, $Fe(CN)_6^{3-/4-}$ could access the electrode surface in the absence of the target sequence, while the hybridization resulted in electrostatic repulsion and the voltammetric current was significantly reduced.

Willner *et al.* (2000) has developed a novel amplification strategy using liposomes to enhance the sensitivity of the repulsion based DNA sensors. The oligonucleotide capture probe was immobilized onto a gold electrode and the capture probe was designed to hybridize with one part of the complementary target sequence. After the target hybridization, the other part of the target was allowed to hybridize in a sandwich format with a detection probe linked to a negatively charged liposome. The interfacial electron transfer resistance for this negatively charged interface with the redox marker of $Fe(CN)_6^{3-/4-}$ was measured by Faradaic impedance spectroscopy. Apart from liposomes, the same group labeled the detection probe with Horseradish peroxidase via biotin–avidin linkage that catalyzed the oxidation of a soluble compound (4-chloro-1-naphthol) to form an insoluble product covering the electrode surface. This created a very high barrier for interfacial electron transfer and was probed by Faradaic impedance spectroscopy with $Fe(CN)_6^{3-/4-}$ redox marker. The electrical wiring of redox oxidase within a redox hydrogel has been extended to DNA sensing, as in glucose sensor by Heller's *et al.* (2002) In this case, a capture probe was immobilized within the redox hydrogel. Upon the hybridization of the target and Horseradish peroxidase-labeled detection probe, the electrical contact between the Horseradish peroxidase and redox hydrogel was established. The amount of target sequence present in the sample was proportional to the hydrogen peroxide electro reduction current.

Mirkin *et al.* (2002) reported a novel electrical DNA detection scheme using oligonucleotide modified gold nanoparticles. The assay involved a pair of microelectrodes (patterned on a SiO_2 coated silicon wafer by photolithography) with a 20 μm gap in between. A capture probe was immobilized onto the gap, followed by the hybridization of the target and oligonucleotide modified gold nanoparticle in a sandwich format. At this point, the gold nanoparticles were brought close to, but separated from one another. A silver enhancement step was then carried out to deposit silver metal onto the gold nanoparticle preferentially, thereby bridging the gap and the resistance across the pair of electrodes dropped dramatically.

Electrochemical detection of specific DNA sequences with gold nanoparticle probes on conventional electrodes has also been achieved, which involved the direct oxidation of the gold nanoparticles or silver metal after an enhancement step either by chemical or electrocatalytic deposition. One drawback of these gold nanoparticle probes is that they can only be used for single target detection, unless an array of electrodes are employed. On the other hand, semiconductor nanocrystals allow simultaneous detection of multiple targets based on their well defined and diverse redox potentials.

Wang *et al.* (2003) demonstrated the detection of three targets in a single tube and voltammetric run by using zinc sulfide, cadmium sulfide and lead sulfide nanoparticles. Furthermore, using four nucleoside monophosphate semiconductor nanoparticle conjugates, they distinguished all the eight possible single base mismatches in a single run. These two studies made use of magnetic beads as the biorecognition support. After stringent washing to remove excess nanoparticle labels, the magnetic bead bound labels were dissolved by an acidic treatment and finally quantified by anodic stripping voltammetry using a glassy carbon disc working electrode.

All above mentioned indicator based approaches require careful washing steps to remove unbound non-complementary sequences and indicators. One wash free indicator based approach was developed by Motorola's Clinical Micro Sensors Division. A gold electrode was modified with a self assembled monolayer (SAM) of capture probe, oligo-phenylethynyl (as molecular wire), and polyethylene glycol (as insulator). The sample was mixed with a detection (or signaling) probe containing ferrocene modified nucleotides. The presence of the target sequence held the signaling probe in close contact with the molecular wire, thereby facilitating electron transfer between the ferrocene redox markers and the gold electrode. The unhybridized probes were effectively blocked by the polyethylene glycol insulator.

Plaxco *et al.* (2003) were developed another wash free format that eliminate the need of a signaling probe and molecular wire, which took advantage of the distance dependent charge transfer property resulting from the conformational change of a molecular beacon type capture probe labeled with a ferrocene marker. The molecular beacon probe had a thiol at its 3' end for immobilization onto a gold electrode and a ferrocene redox marker at its 5' end. The probe had an initial stem loop structure that kept the ferrocene marker close to the electrode surface for efficient electron transfer (Fig. 6.8). When a complementary sequence hybridized with the probe, the stem-loop structure was straightened out and thus the electron transfer was greatly reduced (Fig. 6.8). Modifications of the capture probe design switched the architecture from signal-off to signal on mode.

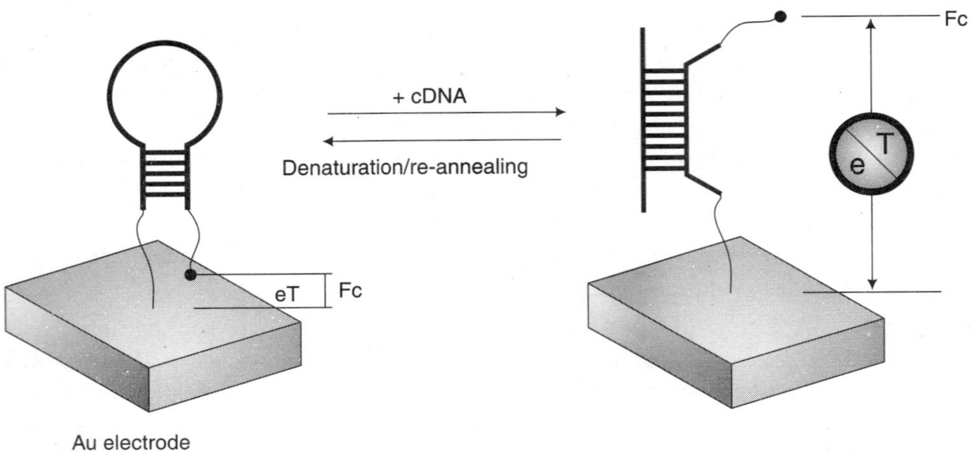

Fig. 6.8 *Schematic diagram illustrating the mechanism of the molecular beacon type electrochemical DNA sensor. (Source: From Fan et al., Proc. Natl, Acad. Sci. USA, 2003, 100, 9134-9137)*

In fact, the hybridization event can be detected without the use of an indicator. The indicator free approach relies on the intrinsic electro activity of DNA bases. The most redox active nitrogenous base in DNA is guanine, which has an oxidation peak at ~+1.0 V with carbon paste electrode versus Ag/AgCl reference electrode and at +0.73 V with gold electrode. In these cases, the guanine bases in the capture probes were replaced by inosine. Another approach takes advantage of the changes in interfacial electrical properties *Eg.* Capacitance and non-Faradaic impedance depend upon the hybridization of negatively charged target DNA on electrode surfaces. One facile method that does not involve laborious and time consuming chemical modification of the electrode surface is based on the doping of capture probe within electro polymerized polypyrrole. This is particularly useful for multiplexed detection with an array of closely-spaced micro-fabricated electrodes. With advanced microelectronics fabrication techniques, silicon field effect transistors have been utilized for electrical transduction of the hybridization event.

Lieber *et al.,* (2004) developed the miniaturization of the field effect DNA sensors with ultra-high sensitivity using silicon nanowires. This silicon nanowire sensor fabrication process is involved on nanowire synthesis by chemical vapor deposition and electrical wiring by electron-beam lithography. These processes do not favor mass production that is essential for successful commercialization. In view of this, recently, Gao *et al.* (2007) reported the fabrication of silicon nanowire arrays using complementary metal-oxide semiconductor compatible technology. The mass production capability is conducive to the production of low-cost devices for routine diagnostics. An array of silicon nanowire sensors was constructed, which could allow multiple analytes to be detected simultaneously (Fig. 6.9).

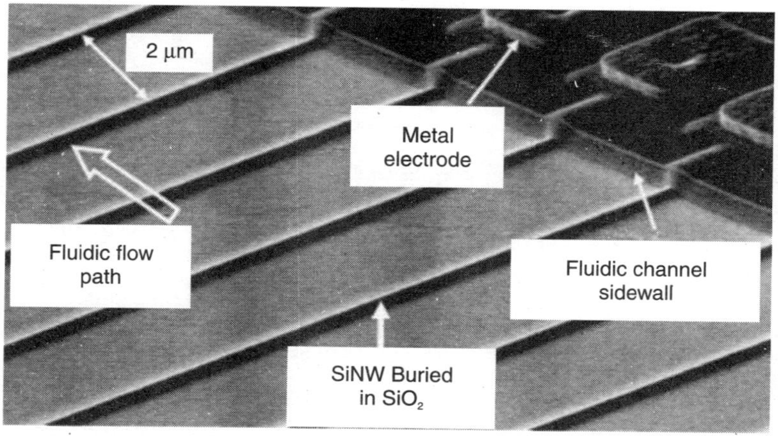

Fig. 6.9 *Scanning electron microscope photograph of a silicon nanowire array fabricated by CMOS compatible technology. (Source: From Gao et al., Anal. Chem. 2007, 79, 3291-3297).*

It must be emphasized that most of the reported electrochemical/electrical DNA sensors use synthetic short oligonucleotide as model targets. Problems could arise when dealing with real samples as a result of huge steric hindrance encountered by very large targets and few studies addressed this issue using polymerase chain reaction (PCR) amplicons as targets. Moreover, coupling electrochemical DNA sensors to PCR could increase the assay sensitivity tremendously. A complete DNA analysis usually involves multiple steps of DNA extraction, isolation and target DNA amplification by PCR and finally detection of PCR product.

In 2004, Motorola's Microfluidics Laboratory developed a fully integrated plastic biochip that contained microfluidic mixers (piezoelectric disk made of lead zirconate titanate, PZT disk), valves, pumps, channels, chambers, heaters, and DNA sensors (microarray chamber) for performing sample preparation, PCR, and electrochemical sequence-specific PCR product detection (Fig. 6.10).

The wash free indicator based eSensor™ electrochemical DNA sensor was also developed by Motorola. The entire assay took about three hours (30 min of sample preparation, 90 min of amplification, and 60 min of detection). In an attempt to shorten the assay time, a special technique termed electrochemical real-time PCR was developed by Hsing *et al.* (2008) which featured simultaneous PCR amplification and electrochemical detection in a silicon glass micro device. The silicon chip contained a PCR micro chamber as well as integrated thin film platinum heater and temperature sensor for precise and fast thermal control. The glass chip, which sealed the micro chamber, had four indium tin oxide working electrodes together with platinum pseudo reference and counter electrodes patterned on its surface for electrochemical measurements. The working electrodes were functionalized with a capture probe (at the annealing step of PCR) hybridized with the PCR amplicon. The conventional PCR recipe was slightly modified by substituting ferrocene-labeled deoxyuridine triphosphate for deoxythymine triphosphate. At the extension step, the capture probe was elongated by the polymerase with the incorporation of ferrocene labels. A progressive accumulation of the redox marker onto the electrode surface in response to the target amplicon generation was observed.

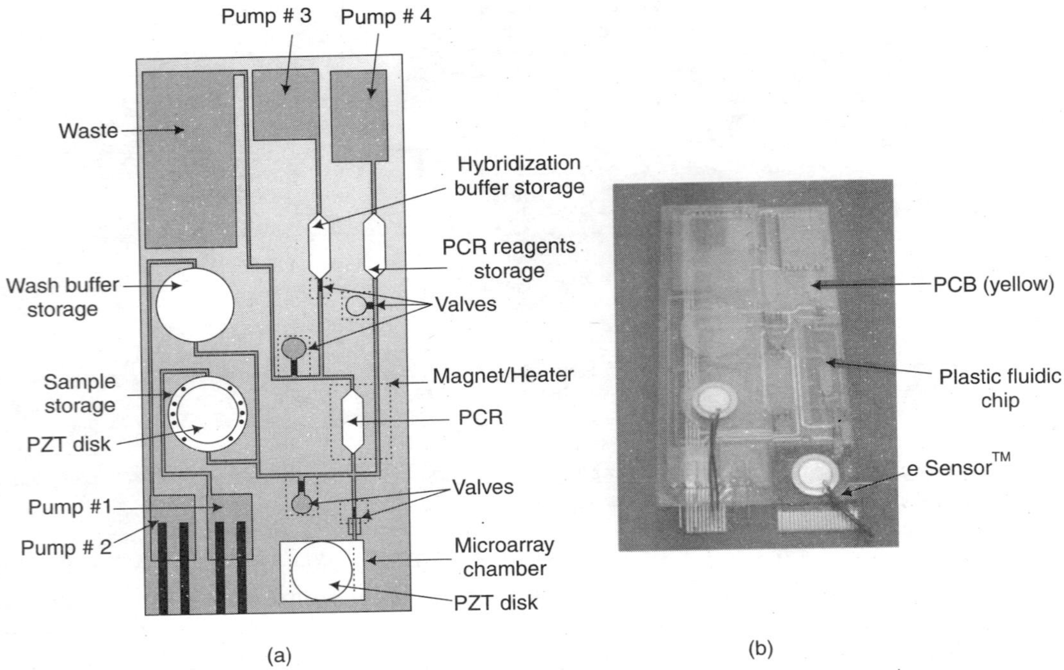

Fig. 6.10 *(a) Schematic and (b) photograph of the self-contained, fully integrated DNA biochip developed by Motorola. (Source: From Liu et al., Anal. Chem.* ***2004****, 76, 1824-1831.Reprinted with permission from ACS.*

6.3 PROTEIN SENSORS

Proteins play an important role in many biological processes such as metabolism (enzymes), cell signaling (receptors) and immune response (antibodies). The presence or absence or the amount of certain proteins within our body have proved to be useful biomarkers of our health status. Some well has known examples include human chorionic gonadotropin (hCG) for pregnancy, prostate specific antigen (PSA) for prostate cancer, cardiac troponin I (cTnI) for myocardial infarction and many others. In clinical laboratories, the most common protein detection method is enzyme linked immunosorbent assay (ELISA), which makes use of one or more antibodies to recognize a target antigen specifically.

The enzyme labels convert the recognition event into colorimetric, chemiluminescence or electrochemical signal. Depending upon the required sensitivity and specificity as well as the antigen itself, several immunoassay formats can be employed including homogeneous/heterogeneous, direct/indirect and sandwich/competitive. Heterogeneous, direct and sandwich ELISA would result in the highest sensitivity and specificity. Typically, a monoclonal antibody (capture antibody) is immobilized onto a 96-well micro titer plate, followed by a blocking step to minimize subsequent nonspecific binding. Then, a sample is introduced and an antibody–antigen complex is formed when the sample contains the target antigen. After that, an enzyme conjugated secondary antibody that binds to a different epitope than the capture antibody is applied. Finally, an enzyme substrate is added and the reaction product is measured. A thorough washing is needed between each of the above steps in order to remove the excess reagent and nonspecifically bound substances. The microtiter plate format is suitable for high throughput analysis, but not for home testing. It should be pointed out that ELISA on microtiter plate is considered as bioassay but not biosensor because the recognition element is not in close contact with the transducer.

Janata in 1975, developed an electrochemical immunosensors, with an antibody or antigen in close contact with an electrode. Different transduction strategies have been developed including potentiometry, amperometry, capacitance, impedance and field-effect transistor (FET). The amperometric immunosensors detect the signal amplification offered by the enzyme labels and hence can detect trace amounts of target analytes (sub ng/mL levels). Analogous to the microtiter plate format, the main disadvantage is the cumbersome washing steps.

To simplify the amperometric assay procedures, various separation free strategies have been developed. A novel method was reported by Duan and Meyerhoff (1994) that enabled preferential measurement of surface bound enzyme labeled antibody than the unbound ones in the bulk solution. A micro porous nylon membrane with a thin film of gold sputtered on one side was mounted between two chambers of a diffusion cell. Anti hCG monoclonal capture antibody was immobilized covalently onto the gold layer via a self assembled monolayer of thioctic acid. The sample together with an alkaline phosphatase-anti-hCG antibody conjugate was added to the gold side of the diffusion cell. After 30-min incubation, the enzyme's substrate (4-aminophenyl phosphate) was added to the other side of the diffusion cell. In the presence of hCG, a sandwich was formed and the alkaline phosphatase was brought close to the gold surface, which served as the working electrode. The substrate diffused through the membrane and reacted with the enzyme bound to the gold surface first. The reaction product (aminophenol) was detected immediately by oxidation at the gold electrode. Another separation free strategy was based on enzyme channeling. It had a dual enzyme design, with one enzyme covalently attached to the electrode and the other one conjugated to a secondary antibody in a sandwich assay or to a standard antigen in a competitive assay. When the latter one was brought close to the electrode surface through antibody antigen interaction, the reaction product of one enzyme served as the reactant for the other one. With this, the background signal from the unbound enzyme label in the solution was negligible.

The most successful home use protein detection device is pregnancy test. The assay is based on lateral flow immune-chromatographic technique. The device is typically constructed with a nitrocellulose membrane with a sample addition pad and an absorbent pad at the two ends. A conjugate pad, which embeds monoclonal anti hCG antibody conjugated with dye label (*Eg*., gold nanoparticle and dye doped polystyrene micro/nanosphere), is sandwiched between the sample addition pad and the nitrocellulose membrane. Close to the absorbent pad end of the nitrocellulose membrane contains one test line and one control line with antibodies against hCG and the monoclonal antibody immobilized on it, respectively. When a urine sample is applied to the sample addition pad, it is transported toward the absorbent pad by capillary action. In the presence of hCG, the dye label is captured in both the test and control lines. Otherwise, only the control line is colored. This device cannot be classified as a biosensor as the result is read by naked eyes and no transducer is involved. In fact, it is possible to obtain a semi quantitative result by incorporating a reflectance based reader *Eg*.: Digital Pregnancy Test.

In recent years, efforts have been made in integrating the lateral flow immune-chromatographic technique with electrochemical detection system. McNeil *et al.* (2004) demonstrated an impedimetric measurement of the antibody antigen interaction occurred at the test line. Instead of a dye label, the mobile monoclonal antibody was labeled with Urease. Immediately after the capture of the monoclonal antibody Urease conjugate at the test line, a urea solution was allowed to flow through the test line to wash away the unbound materials and was, in the mean time, hydrolyzed by the Urease to effect a localized increase of pH. A pH-sensitive polymer coated electrode was positioned directly over the test line so that the pH change induced a breakdown of the polymer layer and thus a measurable change in the capacitance of the electrode. Besides enzyme label, Lin *et al.*, *(2007)* reported a highly sensitive assay utilizing quantum dot label. Regarding the electrochemical detection, the quantum dot label was first dissolved by a simple acidic treatment and the amount of cadmium ions was determined by stripping voltammetric measurement with a disposable screen printed electrode placed underneath the nitrocellulose membrane at the test line (Fig. 6.11). Both impedimetric and stripping voltammetric methods allow quantitative measurements to be made. Also, they are usually much more sensitive than reflectance based approach.

Aptamers, which are synthetic nucleic acid molecules that bind to non-nucleic acid targets *Eg*. Small molecules, Proteins and cells with high specificity and affinity, have emerged as a promising protein recognition element. They are obtained through an *in vitro* selection process known as systematic evolution of ligands by exponential enrichment (SELEX). When compared to antibodies, aptamers are much easier to synthesize and have higher stability, over the past few years, a number of aptamer based electrochemical protein detection schemes have been developed. The conformational change of aptamers upon target binding offers unique opportunities in achieving label free detection. Plaxco and co-workers immobilized a thrombin-binding aptamer to a gold electrode with its thiol group at the 5' end. The 3' end of the aptamer had a covalently attached methylene blue redox marker. In the absence of thrombin (Fig. 6.12), the aptamer assumed an unfolded state, thereby facilitating electron transfer (eT) between the redox marker and the electrode, whereas in the presence of thrombin, the aptamer became folded and eT was reduced (signal-off configuration). Other favorable features of this apta-sensor included the capability of separation free and real time measurements.

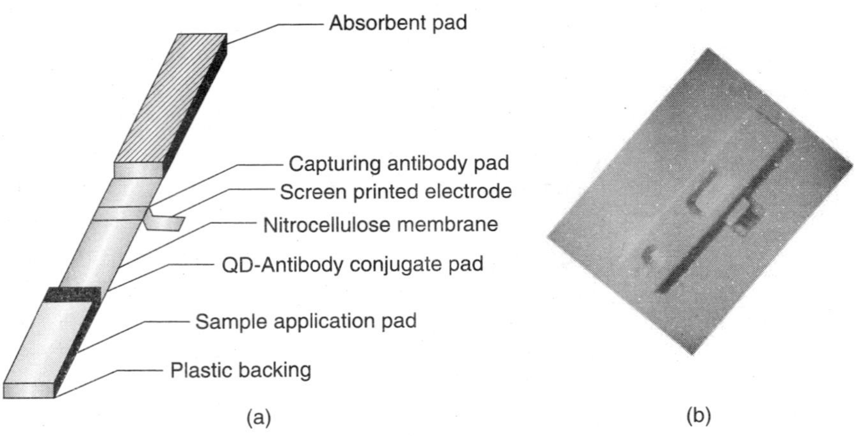

Fig. 6.11 *(a) Schematic illustration and (b) photograph of the disposable electrochemical immunosensor diagnosis device. (Source: From Liu et al., Anal. Chem.* ***2007****, 79, 7644-7653.)*

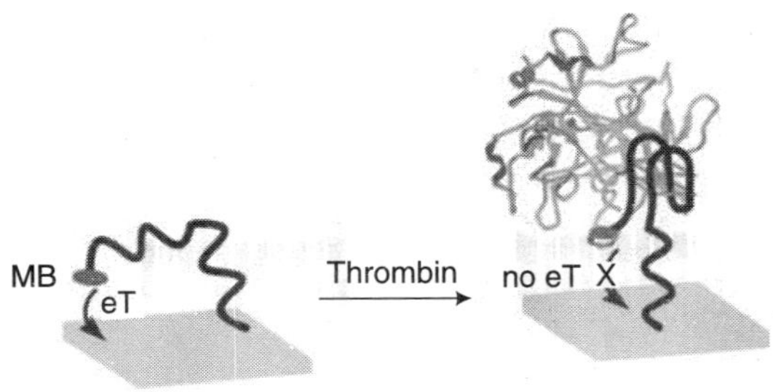

Fig. 6.12 *A schematic of the label-free electrochemical aptamer-based protein sensor. (Source: From Yi Xiao et al., Angew. Chem. Int. Ed. **2005**, 44, 5456-5459. Wiley-VCH, Verlag GmbH & Co. KGaA..)*

Microfluidic is a branch of MEMS that handles fluid flow in micro channels, offers many advantages to point-of-care protein detection. (1) The reduction in sample and reagent volumes (2) Attractive capability is the fully automated fluid control. *Eg.* The advanced microfluidics of i-STAT analyzer automates the washing of excess enzyme conjugate and enables the amperometric detection of cTnI in a sandwich ELISA format in just a few minutes. Besides washing function, microfluidics permits rapid separation of free antibody and antibody antigen complex in solution phase by microchip capillary electrophoresis.

Wang *et al.* (2001) developed a microfluidics device for performing electrochemical enzyme immunoassays (Fig. 6.13). The assay initiates with the mixing of an antibody-enzyme conjugate (Ab-E) and antigen (Ag) in an immuno reaction chamber (IRC), followed by electrophoretic separation of free (Ab-E) and antigen bound (Ag-Ab-E) antibodies. The substrate (S) of the enzyme was introduced close to the end of the separation channel and the product was measured amperometrically with an end-column three electrode system.

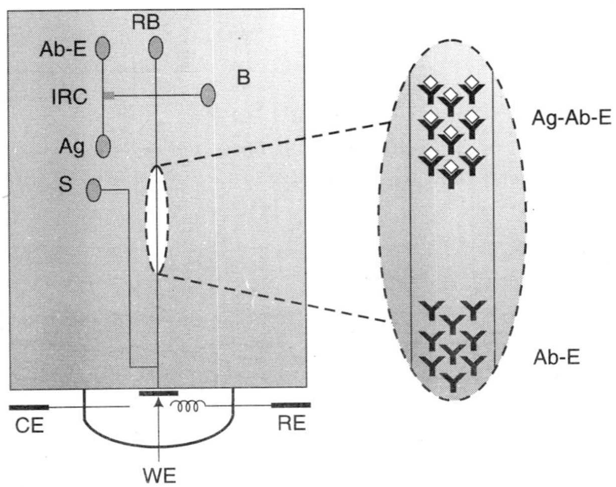

Fig. 6.13 *Schematic representation of a microchip-based electrochemical enzyme immunoassay method. (Source: From Wang et al., Anal. Chem. 2001, 73, 5323-5327).*

Recently the nanobiosensors are used for detection of proteins with ultra sensitive and multiplex mode, as well as label free and real time measurements. Lieber *et al.* (2005) fabricated a silicon-nanowire field effect device for multiplexed electrical detection of cancer markers. The conductance of the antibody functionalized silicon nanowire was strongly dependent on its surface charges. The conductance of a p-type (boron-doped) silicon nanowire increased instantaneously upon the specific binding of a negatively charged target protein (Fig. 6.14).

Carbon nanotube field-effect transistor is another important type of protein nanobiosensor. For the mobile charges of the silicon nanowire and carbon nanotube field effect transistor to pick up the surplus surface charges, the recognition event must occur within the electrical double layer. Because of this, aptamer (~2 nm) has been shown to be a better recognition molecule than antibody (~10 nm) in a carbon nanotube field effect transistor.

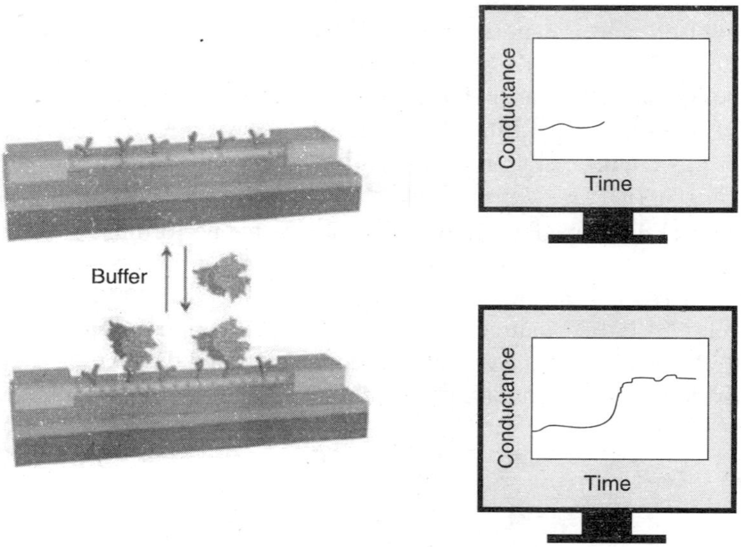

Fig. 6.14 *Schematic of the silicon nanowire sensor for label-free and real-time protein detection. (Source: From Patolsky et al., MRS Bulletin **2007**, 32, 142-148).*

6.4 UREA BIOSENSORS

Many clinical cases dealing with kidney and liver diseases require a fast and accurate measurement of urea from urine or blood samples. Biosensors are used for determination of urea from the samples because they are highly selective and simple to use such are called 'Urea Biosensors'. Urease enzyme used as a biological material in these biosensors, that catalyses urea into ammonium and hydrogenocarbonate through the following reaction:

$$(NH_2)_2Co + 2H_2O + H^+ \xrightarrow{\text{Urease}} 2NH_4^+ + HCO_3^- \qquad \text{... } \textbf{\textit{eq.-6.4}}$$

The development of biosensors based on pH-sensitive field effect transistors (pH-FETs) has been growing over the past decades due to their potential advantages, such as reduced size, sensitivity, integration of the information-processing circuit on the chip, low cost and the facility for creating multi-sensors. Micro-biosensors for urea determination have been prepared via urease immobilization on the surface of pH-FETs are one of the enzyme field effect transistors (ENFETs). The immobilization of enzyme on the surface of pH-FETs is a skilled process which is required technical expertise. Moreover, the conventional methods of enzyme deposition such as covalent binding, physical adsorption or cross-linking and enzyme entrapment has been widely used with high enzymatic activity and inorganic materials viz. oxides, clays, zeolites and sol gel are used for immobilization.

Clays are attractive materials for electrode functionalization of transducer surfaces because of their thermal stability, chemical inertia, well defined layered structure, ion exchange properties and low cost. Swelling cationic clays, known as Smectite clays, have been used in the elaboration of modified electrodes in particular enzyme electrodes. A layered double hydroxide (LDH)-type structure has an anion exchange property and can act as a passive filter allowing pre-concentration of analytes or redox mediators. Although LDH presents a lamellar structure and intercalation properties similar to cationic clays, such kinds of materials have never been exploited for the immobilization of biological macromolecules.

Urea ENFETs performance is based on the immobilization of urease in two synthetic clays, one is Laponite Si8 [Mg5.5Li0.5$H_4O_2$4] 0.7– Na 0.7+, a cationic clay and the other is layered double hydroxide (LDH) Zn3Al (OH)8Cl·$2H_2O$ referred to as [Zn-Al-Cl] an anionic clay. The use of sodium tetraborate in biosensor is a strong competitive inhibitor of urease, as an alternative to modulate the dynamic range of the clay.

6.4.1 Enzyme Immobilization

The Laponite (2.5 gL^{-1}) was dispersed overnight in deionized water under stirring conditions. The LDH colloidal suspension was prepared by the following procedure using deionized and decarbonated water.

Biomembranes are deposited on pH-FETs by using the following method. The pH-FET surface was treated with sulfochromic solution in order to obtain clean and hydrophilic surfaces which improve adhesion of the membrane. Then the sensor was thoroughly washed with ultra pure water and dried under nitrogen flow. A drop of a homogeneous mixture (2 μL) containing 1:1 Laponite/Urease (2.5 gL^{-1}) was deposited onto the sensitive area of the pH-FET. The device was placed in saturated glutaraldehyde vapor for the cross-linking of the membrane. After exposure to glutaraldehyde, the biomembranes were dried at room temperature for 30 min. Before measurements, the ENFETs were stored overnight at 4°C in a 5 mM phosphate buffer solution, 1 mM EDTA, pH 7.4 to equilibrate the membrane system.

A reference pH-FET (Laponite/BSA) was prepared using the same procedure by replacing urease by BSA, an inert protein. The cross-linking time and the volume of the mixture coated on the surface of the ENFETs were the same as those used for the Laponite/Urease biomembranes. For the rotating disk electrode experiments, 8 μL of the clay/enzyme mixture was deposited on a glassy carbon electrode disk (3 mm) and the enzyme macromolecules were cross-linked by glutaraldehyde vapor for 8 minutes and dried in air at room temperature.

6.4.2 Measurements

The biosensor response was measured at a controlled temperature (27 ± 0.3°C). The following conditions were used for the potentiometric measurements. The pH-FETs were connected to an amplifier system (ISFET) allowing the source voltage (V_s) to be measured while drain current (I_d) and drain voltage (V_d) remained constant. V_d was equal to 0.5 V and I_d was equal to 100 μA. The biosensor response as a function of temperature was studied by using a differential setup because pH-FETs are also sensitive to temperature. The linear sweep voltammetry was carried out in a conventional three-electrode cell, using a PAR 273 EG&G potentiostat coupled to a computer. The rotation speed of the electrode was controlled by a Tacussel EDI 101T/CTV 101T. The counter electrode was a Pt wire.

6.4.3 Optimization

pH-FETs were modified by simple adsorption of a Laponite/Urease 1:1 aqueous mixture (1 μL) followed by chemical cross-linking of the incorporated enzyme molecules by glutaraldehyde. The latter step was carried out at different time intervals and the calibration curves are obtained for the potentiometric response to urea from the resulting ENFETs. A decrease in the ENFET's sensitivity and maximum potential response is clearly observed with the increase in cross-linking time (Table 6.3). This may be described to a decrease in the activity of the immobilized urease or to an increase in steric hindrances in the coating related to the high reticulation degree of the entrapped enzymes.

Table 6.3 Analytical characteristics of ENFET biosensors based on Laponite for urea determination.

Laponite/ Urease ratio a	*Time (min) b*	*Max response (mV)*	*Sensitivity (V/M)*	*Linear range (mM)*	*Result*
1:1	8	69.5	53.22	0.005–0.55	0.9979
1:1	15	40.5	27.30	0.01–0.65	0.9974
1:1	25	22.3	13.59	0.03–0.55	0.9991
1:2	8	65.0	33.82	0.05–0.77	0.9992

It is observed from above data that the biosensor's performance increases with the increase of the enzyme/Laponite ratio from 1.1 to 1.2 and therefore produce a similar biosensor arrangement with double the amount of immobilized urease. The influence of the amount of Laponite/Urease mixture deposited on a pH-FET surface on biosensor performance, the Laponite/Urease ratio and the cross-linking time for all membranes being 1:1 and 8 minutes, respectively. The maximum response, at saturating substrate conditions, first increases with the increase in the amount of coating from 1 to 2 μL and hence with the amount of immobilized enzyme and then it seems to reach a constant value for thicker coatings. On the other hand, the response time of the biosensors increases strongly with the amount of coating, varying from 1.5 to 4 minutes when the deposited mixture increases from 1 to 4 μL, respectively. For this reason, 2 μL was chosen as the optimum deposition mixture for further experiments.

Several reference pH-FET sensors have been prepared and calibration curves were obtained with a modified pH-FET by Laponite containing an inert protein (BSA) instead of Urease. No response was observed in the presence of urea and this demonstrates that the potentiometric responses of ENFETs are due to the activity of urease and not to an accumulation of urea within the clay. Moreover, a Laponite/Urease (1:1, 2 μL) biomembrane elaborated without the glutaraldehyde cross-linking step appears to be totally inert in the presence of urea. This may be due to the complete release of enzyme into the buffer solution since the biosensors were stored overnight in aqueous solution before the measurements. Indeed, as previously observed with other enzyme/clay electrodes, the slow release of the enzyme, molecules incorporated within clay films, has implied the chemical cross-linking of the incorporated enzymes by glutaraldehyde. Likewise, depositing an equivalent amount of urease cross-linked with glutaraldehyde, without Laponite, produces a biosensor exhibiting a maximum response of 14% of that obtained with the equivalent biosensor containing Laponite and confirms the useful role of Laponite as an entrapment matrix or as a hydrophilic additive.

The maximum response for the ENFET based on layered double hydroxide (LDH) is 50% lower than that obtained with the corresponding ENFET based on Laponite while a less sensitive detection limit (5 μM) was recorded, highlighting the role of the clay structure (Table 6.4). In contrast to Laponite, the optimum LDH/enzyme configuration was obtained for a clay/enzyme proportion of 1:2. The latter enhances the maximum response (+40% vs. the initial LDH/Urease configuration) and the detection limit 3.5 μM, but the performance remains lower than that obtained with the Laponite/Urease biosensor. A similar evolution was observed for sensitivity values while different linear dynamic ranges were determined for both biomembranes (Table 6.4) and the response times is also different of these both biomembranes. For the ENFET based on Laponite, the response time varies between 1.5 and 4 minutes depending on the amount of Laponite/Urease mixture deposited on the pH-FET surface.

Table 6.4 Comparison of ENFETs based on Laponite and LDH for the determination of urea.

Biosensor composition	*Max. response (mV)*	*Sensitivity (V/M)*	*Detection limit (μM)*	*Linear range (mM)*	*Result*
Laponite/Urease (1:1)	93	110.6	2	0.005–0.5	0.9996
LDH/Urease(1:1)	47	18.2	5	0.06–1.18	0.9989
LDH/Urease (1:2)	72	37.26	3.5	0.04–1.37	0.9993

It is possible that the porosity of the membrane, combined with the characteristics of each clay, is the reason for such differences. The particle size of the LDH is larger than that of Laponite, 200 × 200 × 20 and 40 × 10 × 1 nm, respectively and the LDH/Urease biomembrane may be more porous than the Laponite/Urease. Thus, the substrate as well as the products of the enzymatic reaction diffuses faster through the LDH/Urease membrane, resulting in a decrease in the response time. Another large difference between biosensor responses is the signal/noise ratio. It appears that the biosensor response for ENFET based on Laponite is quite stable. In contrast, the noise grows as the substrate concentration increases for the ENFET based on LDH. Due to its high charge density, the LDH colloidal suspension is less stable and a slow precipitation can occur during the drying process resulting in a less homogeneous coating than with the well-delaminated Laponite gel.

6.4.4 Porosity

The mass transfer process through the clay/enzyme biomembranes onto the electrode surfaces can be investigated by using Rotating Disk Electrode (RDE) voltammetry. To avoid any interference with the charges of the clay sheets, the rotating disk voltammograms were recorded from Laponite/Urease and LDH/Urease modified electrodes in an aqueous solution of a neutral molecule (Hydroquinone) and compared to that obtained from a bare glassy carbon electrode.

The permeability (P_m) of the biomembranes was estimated using RDE experiments at different rotation rates. The results that describe the variation of steady state limiting current with the mass transport for a rotating disk electrode coated with an electro inactive membrane. The results obtained for a bare electrode and for an electrode coated with LDH/Urease and Laponite/Urease biomembranes. The estimated permeability of the LDH/Urease biomembrane was 1.4×10^{-2} cm/sec. and greater than that of the Laponite/Urease biomembrane (3.1×10^{-3} cm/sec.). This result is in agreement with the observed response time of the two biomembranes. Nevertheless, it is quite surprising that the less permeable membrane (Laponite) leads to the most sensitive biosensor. This may be attributed to the cationic-exchange properties of Laponite.

6.4.5 Influence of pH

Enzyme activity is known to be dependent upon pH and the pH effect on biosensor response was examined in a buffer solution for both clay/Urease biosensors at a fixed urea concentration. A maximum activity for both biomembranes (Laponite/Urease and LDH/Urease) was found at pH 7.5. These results are in agreement with previous ones reported for free enzymes or entrapped enzymes in different materials.

6.4.6 Thermal Stability

The effect of temperature on the biosensor response was also studied. The measurements for both biosensors were carried out between 13°C and 52°C. The ENFET based on Laponite presents a linear increase in sensitivity up to 35°C and then begins to lose its activity irreversibly for higher temperatures. This result is in agreement with that obtained for urease in solution or entrapped in another matrix. The ENFET based on LDH presents a wider stability range with maximum activity at 45°C.

The activation energy was estimated for both biomembranes through the *Arrhenius* equation. Plotting in k vs. 1/T produces a straight line whose slope is –Ea/R, inside the temperature range where enzyme activity increases. The activation energy values found for the Laponite/Urease and LDH/Urease biomembranes were very similar, 24.9 and 24.6 kJ/mol respectively. These results differ significantly from those reported for the free Urease, 12 kJ/mol, but are in agreement with those reported for this enzyme entrapped in different matrixes 20 kJ/mol.

6.4.7 Operational and Storage Stability

To estimate the operational stability of the biosensors, a continuous measurement of voltage was carried out for 90 minutes in a 0.2 mM urea solution for the Laponite/Urease biosensor and 0.6 mM urea for the LDH/Urease biosensor. For both biosensors, a small increase of the biosensor response is observed with time. This may be related to the variation of the local pH with time due to OH^- accumulation within the clay matrix. At the end of the experiment, the biosensors were washed with deionized water and their response was again measured. These results show that both biosensors have good operational stability, indicating that they can be used in measurements, lasting a long time.

Urease is very sensitive to the presence of heavy metals, a chelating agent 1 mM EDTA was added to the storage solution. The weak initial storage stability of ENFET based on Laponite may therefore be due to the more efficient action of EDTA in LDH than in Laponite, which is less porous. In addition, the anion-exchange properties of LDH should facilitate the percolation of the anionic EDTA unlike Laponite, which presents cation-exchange properties.

6.4.8 Sodium Tetraborate Effect

One limitation of the application of urea ENFETs is their linear domain, generally limited to low concentrations. In human blood, urea concentration varies between 6-8 mM for a healthy person. For a person with kidney dysfunction, urea concentration is higher up to 40 mM. The sensitivity strongly decreases with the sodium borate concentration from 0.125 to 0.5 mM, while the linear range remains almost constant (Table 6.5). The inhibition effect of tetraborate was also examined on the ENFET based on LDH. It indicated that greater influence of the tetraborate concentration on the linear dynamic range for the LDH/Urease biosensor. The LDH membrane presents two factors favorable to the increase of inhibitor concentration within the membrane: (1) Its permeability, which is higher than that of the Laponite membrane (2) its anion-exchange property allowing the accumulation of tetraborate.

Table 6.5 Role of Tetraborate on the Laponite/Urease and LDH/Urease biosensors.

Tetraborate (mM)	*Laponite/Urease*		*LDH/Urease*	
	Linear range (mM)	K_m^{app} *(mM)*	*Linear range (mM)*	K_m^{app} *(mM)*
0.00	0.005–0.5	2.8	0.035–1.4	2.4
0.125	0.025–1.8	4.4	0.045–7.2	10.1
0.25	0.030–1.8	6.1	0.060–9.5	37.2
0.50	0.035–1.8	10.0	0.100–12.0	61.7

Using the kinetic approach based on the Michaelis–Menten model, parameters such as the maximum rate (V_{max}) and the Michaelis constant (K_m) can be evaluated from a Line weaver–Burk plot of the calibration curves obtained with the Laponite/Urease and LDH/Urease biosensors in the presence and absence of tetraborate. The $K_m{}^{app}$ values for urea without any inhibitor determined for the Laponite/Urease and LDH/Urease biomembranes are quasi-identical (2.78 and 2.36 mM respectively) (Table 6.5). These results are in agreement with those (1–5 mM) reported in the literature for free Urease.

In the presence of the inhibitor, the Lineweaver–Burk plots present the same V_{max} value, reflecting a competitive inhibition of Urease by sodium tetraborate. However, for the ENFET based on Laponite, $K_m{}^{app}$ values do not change noticeably with the inhibitor concentration (Table 6.5). For the ENFET based on LDH, an important increase in $K_m{}^{app}$ values in the presence of increasing inhibitor concentration from 0.125 to 0.5 mM tetraborate corroborates. The results obtained for the Laponite/Urease and LDH/Urease biomembranes were 0.16 and 0.051 mM respectively. These results indicate that the enzyme/inhibitor affinity is stronger in the LDH/Urease than in the Laponite/Urease biomembrane highlighting the role of the clay matrix.

6.5 AMMONIA BIOSENSOR

6.5.1 Introduction

Ammonia oxidizing bacteria (AOB) used in biosensors for determination of NH_4^+ concentrations from the samples. The high sensitivity of bacterial ammonia oxidation to toxic substances may be used for the assay of toxicity. The biosensor consists of a Clark type Oxygen sensor with a layer of immobilized *Nitrosomonas europaea* cells in front of the sensor. It can be used in a fully oxygenated system, needs the addition of a buffering system to keep the pH constant and is linear up to only 70 µM NH_4^+.

Ammonia biosensor application for in situ NH_4^+ analysis is wastewater treatment, where nitrogen removing plants operate with alternating oxic and anoxic periods to first nitrify the ammonium to nitrite or nitrate and subsequently reduce this NO_x– to N_2. The need for measuring NH_4^+ in wastewater is largest under anoxic conditions, when NH_4^+ concentrations are highest and NH_4^+ biosensor that would work under anoxic conditions using the counter-diffusion principle. By placing the bacteria under a steady supply of oxygen from an internal oxygen reservoir, intended to optimize the growth conditions for improve the measuring characteristics of NH_4^+.

The biosensor was made accordingly at a macro scale with the following: tip (7 mm diameter), chamber housing, bacteria, internal O_2 supplying and sensing element (0.30 - 0.28 mm diameter). The oxygen micro-sensor was inserted into a tapered capillary with a tip diameter of 0.28 mm, so that the tip protruded about 0.1 mm in front of the capillary. The capillary was hereafter fixed in place with epoxy. Silicone was sucked into the capillary to a depth of 0.5 mm and excess silicone was wiped off with a piece of plastic foil. After curing, the assembly containing an air filled capillary was inserted into the bacterial chamber. The casing of the bacterial chamber was made from a 5 cm long piece of 7 mm glass tubing onto which 0.5 mm thick polyester front plate with a central 0.30 mm hole was fixed with epoxy. The 7 mm glass tube was fixed in place with epoxy so that the distance between the silicone rubber membrane of the O_2 supplying capillary and the front end of the polyester plate was 0.3 mm.

The AOB were grown in mineral salt medium with 5 mM Ammonia (NH_4^+) and 20 mM HEPES buffer at pH 7.8, 25°C. The cultures were harvested at the end of the logarithmic growth phase by centrifugation (22,000 × g, 20 min) and washed once with mineral salt medium without ammonia. The concentrated bacterial biomass was injected into the 0.3 mm hole of the bacterial chamber from the front end by use of a heat drawn 1 ml plastic syringe with 0.1 mm diameter tip of the plastic capillary. The ion permeable membrane at the tip of the sensor was sealed to the sensor; the membrane had a cut off value of 3500 Dalton. The bulk interior of the bacterial chamber was filled with the mineral salt medium without ammonia as used for suspending the culture. Finally, a few small granules of dry silica gel were placed in the shaft end of the conical air filled capillary that had a total air volume of about 0.3 ml and the capillary was sealed with epoxy. The epoxy seal to the air filled capillary was penetrated by a thin glass tube that could be broken to replace the air.

The ammonia biosensors were incubated for 2–3 weeks in mineral salt medium with 1 mM ammonia and 5 mM HEPES buffered at pH 7.5 to get the highest possible biomass of metabolically active bacteria. During this period the medium was changed twice a week. Different cultures of AOB have been used to fill the Ammonia biosensor *Eg. Nitrosomonas europaea*.

6.5.2 Principle

Ammonia biosensor diffuses Ammonia across the ion permeable membrane and at some depth in the bacterial chamber, it meets oxygen diffusing out from the gas reservoir inside the sensor. Due to the activity of the Ammonia oxidizing bacteria (AOB) the oxygen consumption will be proportional to the influx of ammonium and the oxygen concentration at the tip of the internal oxygen micro sensor will therefore change as a function of the external ammonia concentration. It can be shown that the signal from the internal oxygen sensor of such a biosensor is inversely proportional to the concentration of the analyte. Sensors made according to this counter diffusion principle have a high reading for low concentrations of the analyte.

6.6 HISTAMINE BIOSENSOR

Histamine biosensor's basic principle is the same as the principle of the glucose biosensor, the action of Diamine oxidase (DAO) that catalyzed the oxidative deamination of histamine to imidazoleacetaldehyde, hydrogen peroxide (H_2O_2) and ammonia (NH_3) (Fig. 6.15).

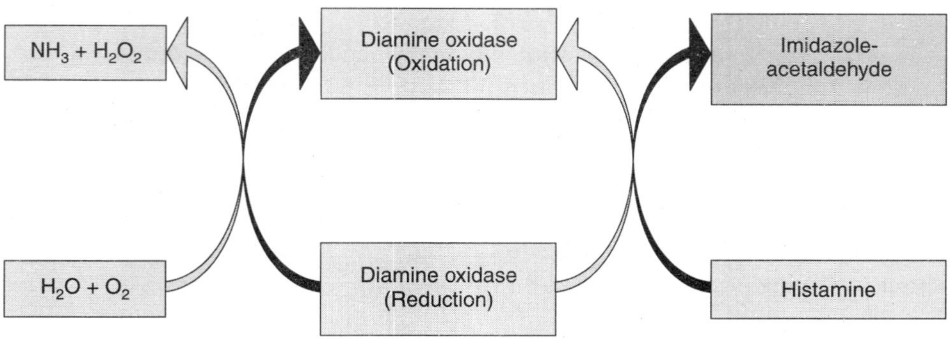

Fig. 6.15 *Oxidative deamination of histamine to imidazoleacetaldehyde, hydrogen peroxide and ammonia by Diamine oxidase.*

The reaction will become rate limited if either histamine or oxygen concentration is too low. Histamine concentration can be measured by monitoring either decrease in oxygen concentration or increase of hydrogen peroxide as the reaction proceeds.

For oxygen based histamine sensor, the oxygen is consumed during the reaction process. The oxygen concentration in the Diamine oxidase membrane will be a linear function of histamine concentration. The oxygen concentration can be measured by coupling the immobilized Diamine oxidase membrane to an electrochemical oxygen sensor. Since oxygen is also available in the sample, a similar reference oxygen sensor without the enzyme needs to be incorporated in the system. The signal current is then subtracted from the reference electrode and gives the result of histamine dependent difference current. The advantage of this type of sensor is that it has low electrochemical interference due to the use of a nonporous hydrophobic membrane. This membrane only allows gaseous molecules to reach the electrode and also can provide information on oxygen variations in the system. The immobilization of Diamine oxidase will help prolong the working lifetime of Diamine oxidase as catalase promotes the degradation of hydrogen peroxide to oxygen and water.

The hydrogen peroxide based histamine sensor has found wide applications in the development of sensor, especially an implantable version due to its simple sensor configuration that facilitates ease of miniaturization. Unlike oxygen, hydrogen peroxide is not present in the sample to be analyzed. However, it suffers from an intrinsic problem the interference from small endogenous analytes which may be electro active at the detection potential of hydrogen peroxide which is quite high (Azila, 2001).

The two types of sensors mentioned above are known as the first generation amperometric biosensors. The second generation of histamine biosensors makes use of mediators to shuttle electrons from the enzyme to the electrode instead of oxygen which are reversible, which had appropriate oxidation potentials and whose concentration could be controlled. If the system contains oxygen the biosensor become insensitive to histamine, thus only respond to changes in oxygen concentration. As oxygen remains in the system, the mediator must be able to compete effectively for the electrons.

The use of mediator in determining the content of histamine and other biogenic amines has been studied by Tombelli and Mascini (1998), compared to a single enzyme sensor and a flow system based on hydrogen peroxide generated by enzymatic reaction. A bi-enzyme FIA system with amperometric detection was used based on the following enzyme reactions: (i) Ferrocene carboxylic acid (Fc-COOH) as the mediator facilitating the electron transfer between the electrode and Horseradish Peroxidase (HRP) (ii) Pea Seedling Amine Oxidase (PSAO) catalyses the oxidation of the amine and subsequently the co-substrate, molecular oxygen is reduced to hydrogen peroxide. Hydrogen peroxide catalyzed with peroxidase using Fc-COOH as the mediator. The amperometric signal is monitored reductively at the electrode. Fig. 6.16 illustrates the basic principle of the electron transfer during measurements.

$$RCH_2NH_2 + O_2 + H_2O \xrightarrow{\textbf{Pea seedling amine oxidase (PSAO)}} RCHO + NH_3 + H_2O_2 \quad \textit{... eq. – 6.5}$$

$$H_2O_2 + Fe\ (II) \xrightarrow{\textbf{Horseradish Peroxidase (HRP)}} H_2O + Fe\ (III) \quad \textit{... eq. – 6.6}$$

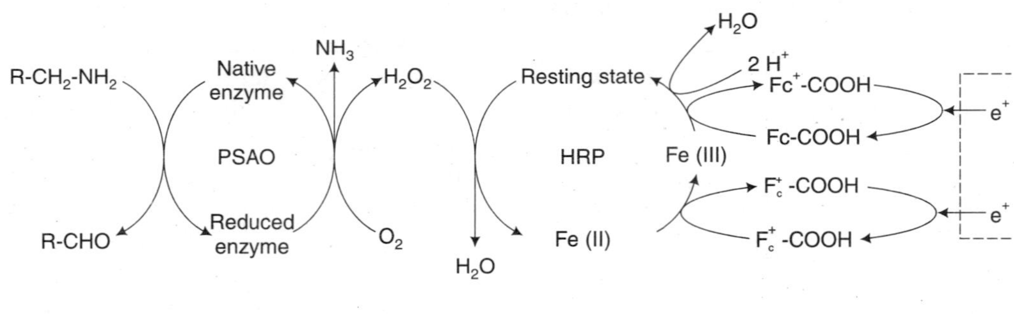

Fig. 6.16 *The basic principle of the electron transfer during measurements in Amperometric bi-enzyme system.*

Tombelli (1998) proposed second enzymatic reaction and mediator. Two distinct advantages are obtained: (a) It allows low potential detection (0.00 mV) of substrates hence avoiding interferences from electro active species which enhances the specificity of the assay (b) The assay sensitivity is enhanced by increasing the efficiency of the electrochemical detection.

The third generation Amperometric biosensors are based on the use of conducting organic salts or polymers. These conducting salts can be built into electrodes in three ways: (i) As single crystals (ii) as pressed pellets (iii) as a paste with graphite powder. The films are grown electrochemically and enzyme is entrapped in the membranes. The advantage of this system is that manipulation of the electro-polymerization can give a film that extends the linear range for substrate detection and reduces oxygen dependence.

6.7 LACTOSE BIOSENSOR

Sharma *et al.* (2004) designed and fabricated Lactose biosensor and to estimate lactose in milk (Fig. 6.17). The Lactose biosensor was based on Langmuir–Blodgett films in which the electrodes produced amperometric response at different concentrations of lactose. The lactose 5 biosensor was able to estimate the lactose concentration in few minutes compared to the tedious and time consuming methods such as spectrophotometry and polarimetry.

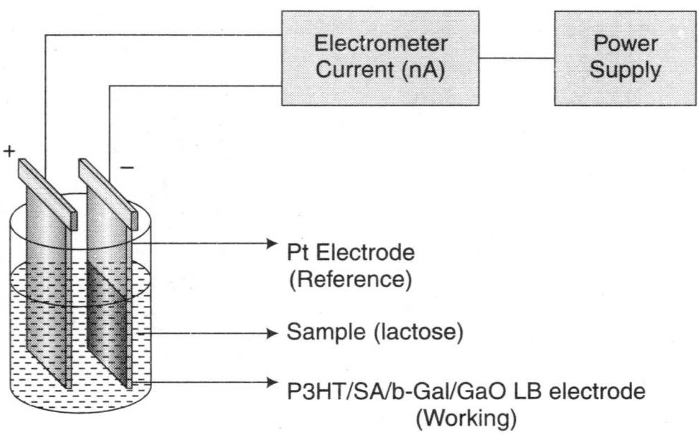

Fig. 6.17 *Schematic representation of lactose biosensor based on Langmuir Blodgett film as working electrode and platinum as reference electrode.*

6.8 METHANE BIOSENSOR

Methane biosensor consists of immobilized methanotrophic bacteria (*Methylomonas flagellata)* in contact with an oxygen electrode and these bacteria use methane as well as oxygen according to the following reaction.

$$CH_4 + NADH_2 + O_2 \rightarrow CH_3OH + NAD + H_2O \qquad \textit{... eq.-6.7}$$

Where NAD is nicotinamide adenine dinucleotide and $NADH_2$ is the reduced form of the coenzyme.

Oxygen consumption leads to a decrease in current, which is proportional to the methane concentration in the sample. Methane is an important intermediate in the global carbon cycle and it is thus of interest to measure its abundance and distribution in natural and man made systems. Two different techniques are used for measuring methane: (1) Membrane inlet mass spectrometry (MIMS): This method offers very robust and reliable measurements. However, a substantial membrane area is required to yield a sufficient signal size which results in a limited spatial resolution. Furthermore, it is difficult to apply the technique in field measurements as line power is required to run the mass spectrometer. (2) Micro scale biosensor (Fig. 6.18) for detection of methane, which exhibit micro scale characteristics with respect to size and stirring sensitivity (Damgaard and Revsbech, 1997).

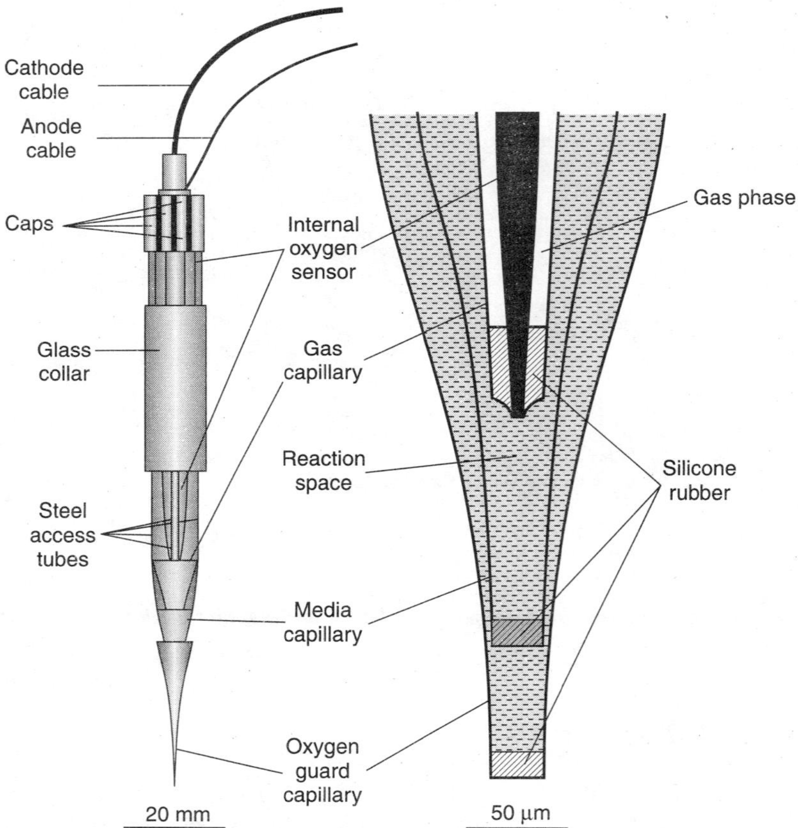

Fig. 6.18 *A Schematic representation of Methane biosensor.*

6.9 NITRITE MONITORING SENSORS

Biosensing system to detect nitrite in water using *Cytocrome cd1* Nitrite reductase immobilised in sol-gel or in Controlled Pore Glass (CPG). The increasing concentrations of nitrite and nitrate in groundwater, rivers and lakes bring serious risks for the public health (if present in drinking water nitrate and nitrite increase the risk of 'Methaemoglobinaemia' in infants or 'Intestinal Carcinogenic Nitrosamines' in adults) and the environment (may cause 'Eutrophication' in rivers). European Union sets the maximum admissible concentration of nitrite in drinking water at 0.1 mg/l. Laboratory techniques such as HPLC, potentiometric ion selective electrodes and colorometric assays are currently used to detect nitrite. The development of a very specific, rapid and on-site biosensing system may be an important tool for the nitrite monitoring and control in waters.

A nitrite biosensing system was developed using an enzyme Cytochrome cd1 nitrite reductase from the aerobic dinitrifier *Thiosphaera pantotropha* as the biological recognition molecule (Fig. 6.19). For the biological phase the Cytochrome cd1 was immobilized in bulk sol-gel or in CPG isothiocyana to derivatised. The biosensor is based on the measurement of the optical absorption of Nitrite reductase, since it shows spectroscopic changes when nitrite binds and oxidizes the enzyme.

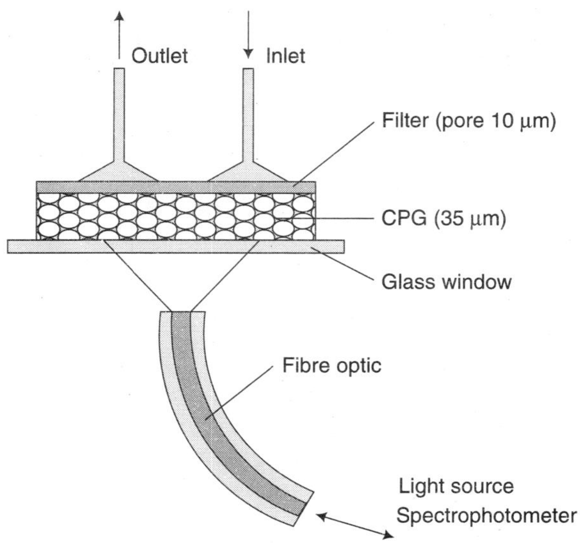

Fig. 6.19 *Nitrate monitoring sensor.*

The biosensors use Cytocrome cd1 immobilised either in sol-gel or in CPG allow the detection of nitrite in water. Detection limit of sodium nitrite achieved with these biosensors was 0.03mg/l [0.75×10^{-6} M] for cd1 immobilised in sol-gel and 0.04mg/l [0.93×10^{-6} M] for cd1 immobilised in CPG. The obtained detection limits were significantly lower than the nitrite concentration in water. The developed biosensors are potentially important tools for on-line nitrite monitoring and control in waters.

6.10 HYPOXANTHINE SENSORS

Hypoxanthine is measured on the basis of the reaction catalyzed by Xanthine Oxidase (XO). The pH change caused by uric acid is detected by using a Si Ion Sensitive Field Effect Transistor (ISFET). The oxidation of hypoxanthine to uric acid by Xanthine oxidase is initiated immediately after injection. The response to inosine has a time lag of 90 sec. after injection. This phenomenon is attributed to the three-step reaction. On the basis of this time lag, the sensor can determine inosine and hypoxanthine simultaneously.

To maintain quality, evaluation of freshness is important in the fish industry. When a fish dies, adenosine 5' triphosphate (ATP) decomposition in the fish meat occurs and adenosine 5' diphosphate (ADP) and adenosine 5' monophosphate (AMP) and related compounds are generated where IMP, HxR, Hx, X and U stands for inosine 5' monophosphate, inosine, hypoxanthine, xanthine and uric acid respectively. There comes the use of hypoxanthine and inosine sensor.

$$\text{ATP} \rightarrow \text{ADP} \rightarrow \text{AMP} \rightarrow \text{IMP} \rightarrow \text{HxR} \rightarrow \text{Hx} \rightarrow \text{X} \rightarrow \text{U} \qquad \text{... } \textbf{\textit{eq.}-6.8}$$

Where IMP, HxR, Hx, X and U stand for inosine 5' monophosphate, inosine, hypoxanthine, xanthine and uric acid, respectively. Consequently, Hx accumulation with an increase in storage time can be used as an indicator of fish meat freshness. Therefore, simple and rapid methods for the determination of Hx and HxR are required in the seafood industry.

Electrochemical Transducers

Chapter 7

7.0 ELECTROCHEMICAL SENSORS

Electrochemical sensors utilize chemical reactions which induce a current or voltage in the circuit between two electrodes, these electrodes are immersed in an electrolyte and this phenomenon was studied by Faraday in 1833.

Three generations of electrochemical cells are developed based on their influence of reaction process inside the cells. These are: (1) An electrochemical cell which generates electricity resulting from a spontaneous reaction occurring inside the cell is called a Galvanic cell (2) The reaction inside the cell is driven by an external source of current it is called an 'Electrolytic cell' (3) The third electrochemical cell is a cell in which the two reactant gases are supplied from outside to react inside the cell and produce a current called as 'Fuel cell'.

7.0.1 Principle

The electrochemical cell consists of a diffusion barrier, sensing electrode, counter electrode and electrolyte and an external circuit which connects the two electrodes. The diffusion barrier controls the diffusion of the reactants to the sensing electrode and it is the major component in the cell which influences the selectivity and response time of the sensor.

The sensing or working electrode is placed where the reactants have easy access to the electrode, whereas the counter electrode/reference electrode is placed way from the actual reactant so that no interference can take place between electrodes (Fig. 7.1). The electrode materials have a high catalytic activity towards electron transfer between the adsorbed reactant and the electrode material, low catalytic activity to other reactant species and they should have a high electrical conductivity and a high corrosion resistance to the electrolytes.

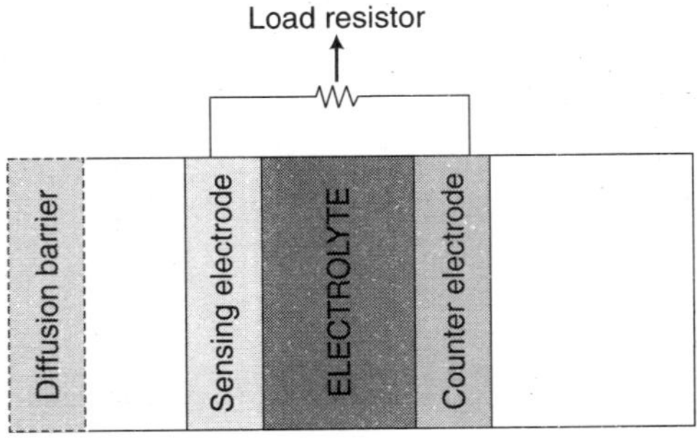

Fig. 7.1 *Schematic view of an electrochemical cell.*

Another important component of an electrochemical cell is the electrolyte solution, as with the diffusion barrier, it is to influence the selectivity of the cell by varying the electrolyte and choosing the appropriate one for the species of reactant. Some electrolytes are electrochemically inactive to the reactant (gas) whereas others are very active. This activity can very often be influenced by its pH value. *Eg.* Carbon dioxide electrochemically is active in an alkaline solution whereas in an acidic solution it is far less active. The electrolyte needs to be an ionically conductive solution, this can either be an aqueous or a solid solution. Another requirement of the electrolyte is that it should be in the liquid state over a wide range of temperatures (i.e. it should have a high boiling point and a low freezing point). The other component is the external circuit which supplies a voltage between the sensing and counter electrodes to measure the current output of the cell (Fig. 7.1).

Using an electrochemical approach to measure the reactants (gases), through two different methods: (1) The Amperometric method - A voltage is supplied to the circuit to supply electrons to the working electrode and the resulting Faradic current is measured and (2) The Potentiometric method - The potential between the two electrodes is measured under conditions where no current flows, therefore the voltage measured is a function of the reactant (gas) concentration.

7.0.2 Specifications

7.0.2.1 Sensitivity

The sensitivity of the electrochemical cell is dependent upon the surface area of the working electrode, permeability specifications of the membrane and temperature of the membrane. Electrochemical cells exhibit sensitivities ranging from the ppm level to the 100% level. They are being used in process control, in the medical field (*Eg.* incubators and ventilators) as well as in applications where high concentration up to 100% are being measured. The further criteria for lower detection limit is the fact that there is a resident base line current which follows even when there is no reactant (gas) at the sensing electrode. This baseline current can be the equivalent of a few tenths of a part per million of the reáctant (gas).

7.0.2.2 Response time

The response time of electrochemical sensor is in the range from seconds to minutes and is mainly dependent upon the thickness of the diffusion membrane (Fig. 7.2).

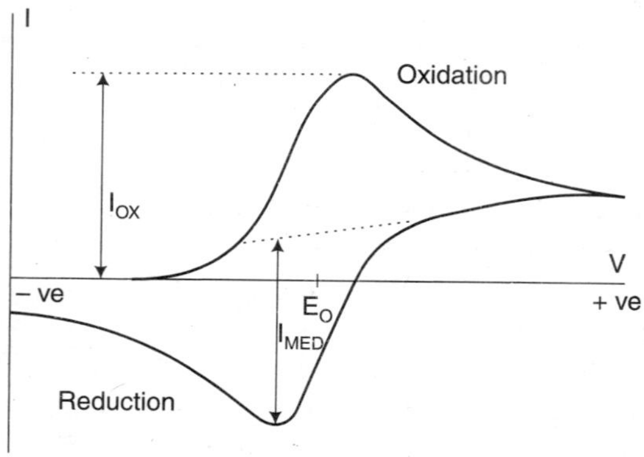

Fig. 7.2 *Cyclic Voltamogram for a typical Potentiometric sensor.*

7.0.2.3 Selectivity

The selectivity of electrochemical sensors is influenced by the choice of electrode and electrolyte solution. However, the diffusion barrier is often the major influence in the selectivity of the sensor. *Eg.* It is possible to select the type of molecule to pass the diffusion barrier by choosing an appropriate porous material with a specific micro-porous structure. The sensors are manufactured with stability of up to 1 year and a good reproducibility but some of the sensors show a slight decrease in sensitivity during the first month of their working period due to passivation of the electrode surface. After this passivation time most of the sensors exhibit good stability over a period of up to one year. The stability of the output signal can obviously be guaranteed for the test procedure, however in the long term the electrolyte and the electrodes get consumed during their use, in the sense the electrochemical cells need to be replaced regularly.

7.1 GLASS pH ELECTRODE

7.1.1 Introduction

pH is a measure of the concentration of hydrogens ions (= H^+) (= protons) in a solution. Numerically it is the negative logarithm of that concentration expressed in moles per liter (M). Pure water spontaneously dissociates into ions forming 10^{-7} M solution of H^+ (and OH^-). The negative of this logarithm is 7, so the pH of pure water is 7 (Fig. 7.3).

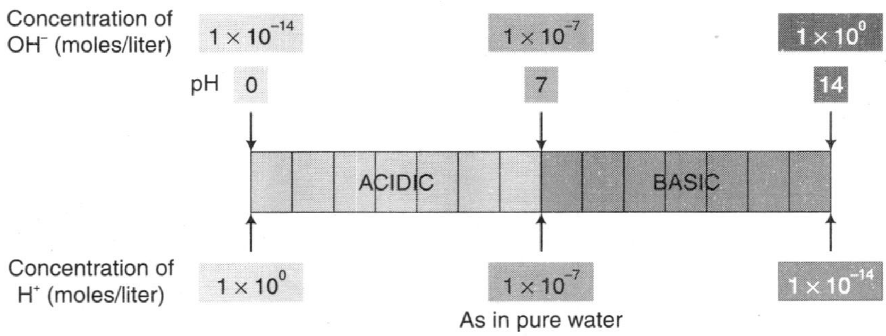

Fig. 7.3 *pH in pure water.*

Solutions with a higher concentration of H^+ than occurs in pure water have pH values below 7 are acidic. Solutions containing molecules or ions that reduce the concentration of H^+ below that of pure water have pH values above 7 are basic or alkaline (Fig. 7.3)

7.1.2 Construction

1. Bulb - a sensing part of electrode and made from a specific glass
2. AgCl precipitate - sometimes the electrode contains a small amount of AgCl precipitate inside the glass electrode
3. O.1M HCl – It is an internal solution
4. Silver chloride electrode or Calomel electrode – It is an internal electrode

5. Body of electrode - made from non-conductive glass or plastics.
6. Reference electrode
7. Junction - with studied solution usually made from ceramics or capillary with asbestos or quartz fibre.

Modern pH probe is a combination electrode (Fig. 7.4), which combines both the glass and reference electrodes into one body. The bottom of a pH electrode balloons out into a round thin glass bulb. The pH electrode is a dual glass tubes system, one inserted into another. The inner tube contains an unchanging saturated KCl and 0.1M HCl solution. Also, inside the inner tube is the cathode terminus of the reference probe. The anodic terminus wraps itself around the outside of the inner tube and ends with the same sort of reference probe. Both the inner tube and the outer tube contain a reference solution but only the outer tube has contact with the solution on the outside of the pH probe by way of a porous plug that serves as a salt bridge and this device is a 'Galvanic cell'. The reference end is essentially the inner tube of the pH meter, which cannot lose ions to the surrounding environment. The outer tube contains the medium, which is allowed to mix with the outside environment.

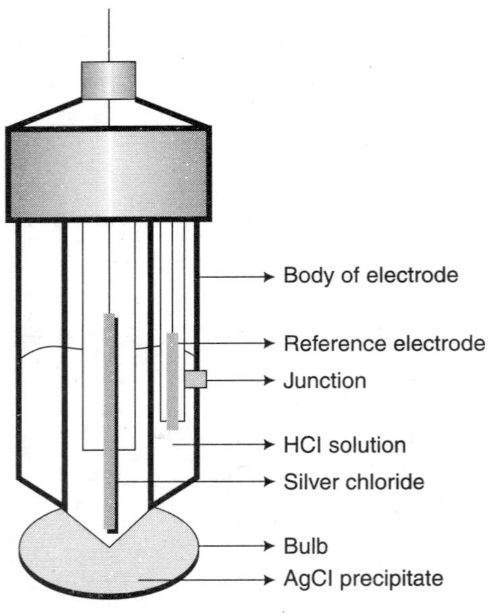

Fig. 7.4 *Schematic representation of pH glass electrode.*

The glass bulb on the bottom is coated both in and outside with a ~10 nm layer of a hydrated gel and is the measuring part of the electrode. These two layers are separated by a layer of dry glass. The silica glass structure is shaped in such a way that it allows Na^+ ions with low mobility. The metal cations (Na^+) in the hydrated gel diffuse out of the glass and into solution while H^+ from solution can diffuse into the hydrated gel. It is the hydrated gel, which makes the pH electrode an ion selective electrode. H^+ does not cross through the glass membrane of the pH electrode, it is the Na^+ which crosses and allows for a change in free energy. When an ion diffuses from a region of activity to another region of activity, there is a free energy change and this is what the pH meter actually measures. The hydrated gel membrane is connected by Na^+ transport and thus the concentration of H^+ on the outside of the membrane is 'relayed' to the inside of the membrane by Na^+.

All glass pH electrodes have extremely high electric resistance from 50 to 500 M Ohm. Therefore, the glass electrode can be used only with a high input-impedance measuring device like a pH meter, a high input-impedance voltmeter which is called an 'Electrometer'.

7.1.3 Principle

The most convenient and accurate way of determining pH is by using a glass electrode, developed in 1919. This device is one of the most basic equipment for biology laboratories. The pH electrode depends on ion exchange in the hydrated layers formed on the glass electrode surface. Glass consists of a silicate network amongst which are metal ions coordinated to oxygen atoms, and it is the metal ions that exchange with H^+. The glass electrode acts like a battery whose voltage depends on the H^+ activity of the solution in which it is immersed. The potential (E) due to H^+ is given by the equation:

$$E = 2.303\frac{RT}{F}\log\frac{[H^+]i}{[H^+]o} \qquad \text{... eq.-7.1}$$

where $[H^+]_i$ and $[H^+]_o$ are the molar concentrations of H^+ inside and outside the glass electrode.

Glass electrodes are particularly useful because of the lack of interference from the components of the solution. These electrodes are not contaminated by molecules in solution and if other ions are present they do not cause significant interference. However, at high pH they do respond to sodium.

A glass electrode consists of a thin, soft glass membrane that is situated at the end of a hard glass tube or sometimes an epoxy body. The glass electrode is an internal reference electrode of the silver/silver chloride type surrounded by an electrolyte of 0.1 M HCl (Fig. 7.5). This internal reference electrode gives rise to a steady potential. Thus, the varying potential of the glass electrode can be compared with a steady potential produced by an external reference electrode such as the standard Calomel electrode by joining the internal and external reference electrodes.

The external reference electrode can either be a separate probe or built around the glass electrode giving a combination electrode. If a combination electrode is used the level of the test solution must be high enough to cover the porous plug (liquid junction) but not as high as the level of the salt bridge solution (KCl) in the external electrode because it is essential for KCl to diffuse out slowly into the test solution (Fig. 7.6).

Whatever reference electrode is used, the measured voltage is the result of the difference between that of the reference and glass electrodes. There are other potentials present in the system, so called asymmetry potential, which is poorly understood but which is present across the glass membrane even when the H^+ concentration is the same on both sides. The potentials due to the Ag/AgCl and to the liquid junction to the reference electrode, which gives a potential because the K^+ and Cl^- do not diffuse at exactly the same rate and therefore generate a small potential at the boundary between the sample and the KCl in the reference electrode.

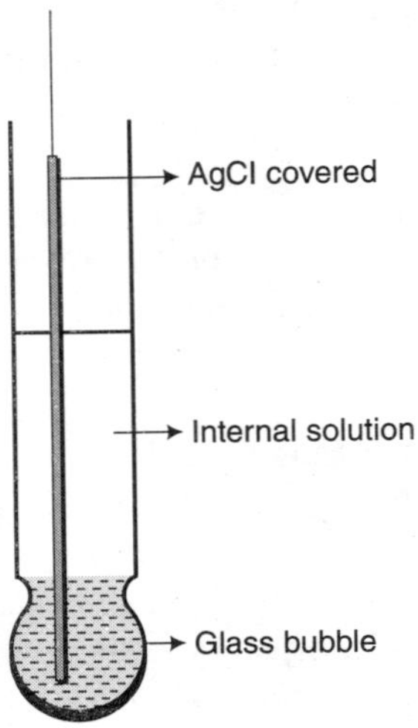

Fig. 7.5 *Glass pH electrode.*

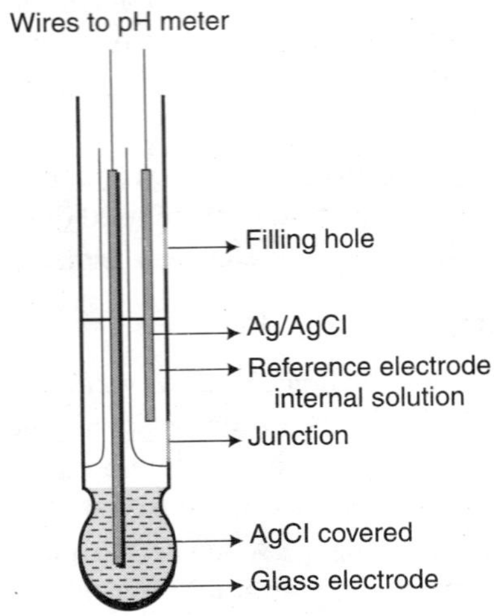

Fig. 7.6 *pH combination electrode.*

A pH electrode is used in conjunction with a pH meter. This records the potential due to the H^+ concentration but is designed to take little current from the circuit. A large current flow would cause changes in the ion concentration and hence pH changes; this is prevented by having a high resistance present.

7.1.4 Operation

It is important that the outer layer of glass on the glass electrode hydrated and immersed in solution. The thin glass membrane is fragile and care must be taken not to break or scratch it or to cause a build up of static electric charge by rubbing it. Many modern pH electrodes have a plastic casing surrounding the glass electrode to protect it from the damage. Gelatinous and protein-containing solutions should not be allowed to dry out on the glass surface as they would inhibit response.

The potential produced is temperature dependent and the effect is entirely predictable and can be compensated. Important is that the temperature compensation is applied accurately because of the accumulation of errors. The pH meter will thus have a temperature compensation dial that must be correctly set before the meter is calibrated.

7.1.5 Calibration

Calibration will necessitate the use of two solutions of widely differing pH. Usually calibration will be carried out with buffers pH 7.00, followed by pH 4.00 or pH 9.20. Once the pH electrode is calibrated it can be immersed in the test solution to be measured and pH will be estimated rapidly and accurately. Before measuring pH, first calibrate or standardize the electrode with two solutions of known pH. The different substances of standard solutions can be used for the electrode calibration process (Table. 7.1) most commonly used commercially available calibration buffers have pH of about 4.00, 7.00 and 9.20, at 25°C.

Table 7.1 The different substances of standard calibrating buffers.

S.No	*Substance(s)*	*Concentration*	*pH*
1	Potassium hydrogen phthalate- $KHC_8H_4O_4$	0.05000m	4.005
2	Disodium hydrogen phosphate- Na_2HPO_4	0.03043m	7.413
	Potassium dihydrogen phosphate- KH_2PO_4	0.008695m	
3	Disodium tetraborate- $Na_2B_4O_7$	0.01000m	9.180
4	Sodium hydrogen carbonate- $NaHCO_3$	0.02500m	10.012
	Sodium carbonate- Na_2CO_3 **(Used rarely)**	0.02500m	

Note: m stands for molality, M for molarity.

The calibration procedure depends on the type of pH meter model. First step is usually related to adjust the temperature at 25°C with knob. Now the electrode immerse into pH 7.00 buffer and rinse the electrode with distilled water before immersing it into new solution. Wash the electrode every time when it is moved from one solution to other solution to minimize the contamination. Check if the working part of the electrode is completely immersed in the buffer or solution, take care not to hit bottom of the beaker with the electrode and wait for the reading to stabilize (it takes few seconds usually, sometimes up to a minute).

Modern pH meter working in calibration mode often recognizes the buffer automatically and take necessary action by them. Sometimes pH readings will oscillate, if the oscillations are small try to find out the best position of the knob so that pH 7.00 is a mean displayed value. If the oscillations are large and erratic, they may be caused by faulty junction (check all), faulty cable (check them), faulty electrode (try other electrode) or faulty pH meter. Also, sometimes static electricity can be a reason of erratic readings considers changing clothes, grounding or shielding pH meter, cables and electrode.

The measurement of pH will also depend upon the type of solution. *Eg.* To measure pH value in acidic solutions, use pH = 4.01 buffer and to measure high pH value use pH = 10.00 buffer. Different pH meters may require slightly different operating procedures (use manual to proceed and how to maintain the electrodes).

7.1.6 Working Mechanism

When one metal is brought in contact with another, a voltage difference occurs due to their differences in electron mobility. When a metal is brought into contact with a solution of salts or acids, a similar electric potential is caused, which has led to the invention of batteries. Similarly, an electric potential develops when one liquid is brought in contact with another, but a membrane is needed to keep such liquids away from each other. A pH meter measures essentially the electrochemical potential between a known liquid inside the glass electrode (membrane) and an unknown liquid outside. Because the thin glass bulb allows mainly the responsive and small hydrogen ions to interact with the glass, the glass electrode measures the electrochemical potential of hydrogen ions or the potential of hydrogen.

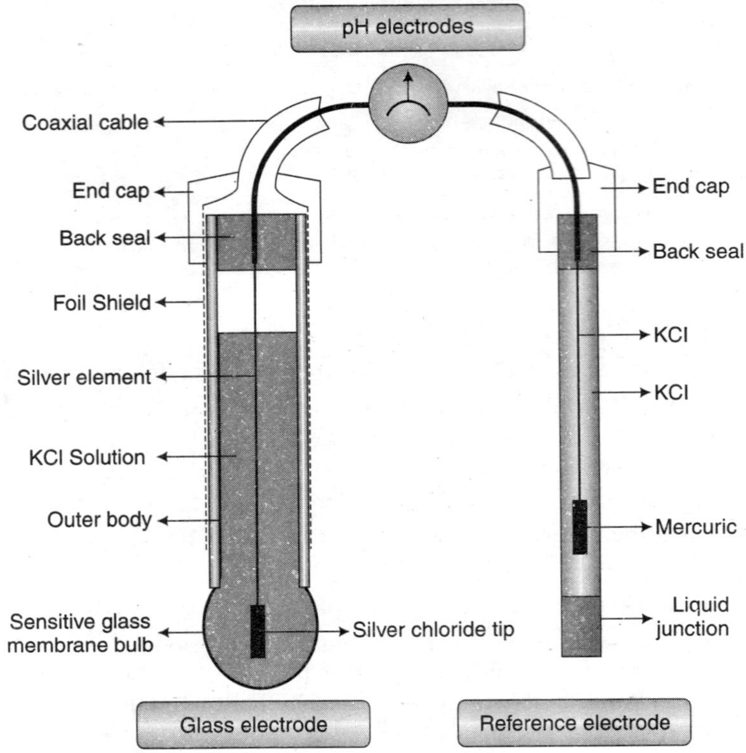

Fig. 7.7 *Schematic diagram of a Glass pH meter.*

The pH meter measures the electrical potential between the mercuric chloride of the reference electrode and its potassium chloride liquid, the unknown liquid, the solution inside the glass electrode, and the potential between that solution and the silver electrode. But only the potential between the unknown liquid and the solution inside the glass electrode change from sample to sample. So, all other potentials can be calibrated out of the equation.

The Calomel reference electrode consists of a glass tube with a potassium chloride (KCl) electrolyte which is in intimate contact with a mercuric chloride element at the end of a KCl element. It is a fragile construction, joined by a liquid junction tip made of porous ceramic or similar material. This kind of electrode is not easily 'poisoned' by heavy metals and sodium (Fig. 7.7).

The glass electrode consists (Fig. 7.7) of a sturdy glass tube with a thin glass bulb welded to it. Inside is a known solution of potassium chloride (KCl) buffered at a pH of 7.0. A silver electrode with a silver chloride tip makes contact with the inside solution. To minimise electronic interference, the probe is shielded by a foil shield, often found inside the glass electrode. Most modern pH meters also have a thermistor temperature probe which allows for automatic temperature correction, since the pH varies somewhat with temperature.

A pH meter depends on the types of electrode in use. Study the manufacturer's recommendations. It is better to keep the electrode moist, since moisturizing a dry electrode takes a long time, accompanied by signal drift. However, modern pH meters do not mind their electrodes drying out provided they have been rinsed thoroughly in tap water or potassium chloride.

7.1.7 Electrode Maintenance

Cleaning of the electrodes and their maintenance is as follows:

General

1. Soak in 0.1M HCl for 30 minutes.
2. Drain and refill the reference solution.
3. Soak the electrode in filling solution for 60 minutes.

Inorganic

1. Soak in 0.1M tetra sodium EDTA solution for 15 minutes.
2. Drain and refill the reference solution.
3. Soak the electrode in filling solution for 60 minutes.

Protein

1. Soak in 1% pepsin / 0.1M HCl for 15 minutes.
2. Drain and refill the reference solution.
3. Soak the electrode in filling solution for 60 minutes.

Grease and Oil

1. Rinse with detergent or Ethanol solution.
2. Drain and refill the reference solution.
3. Soak the electrode in filling solution for one hour. Electrode response may be enhanced by substituting a mixture of 1:1 pH 4 buffers and filling solution for the soaking solution.

Cleaning of the clogged junction -Pollution by sulfides

1. Use a solution of 8% thiocarbamide in 1 mol/L HCl.
2. Keep the electrode in the above solution till junction's color turns pale.

Pollution by silver chloride

1. Use concentrated ammonia solution.
2. Keep the electrode in the above solution for about 12 hours.
3. Rinse and put into pH 4 buffer for at least 60 minutes.

Other contaminations have to be removed by cleaning with distilled water, alcohol or mixtures of acids.

7.2 METAL OXIDE pH ELECTRODES

Glass pH electrodes can be easily broken and are not always the best choice for some measurement conditions. A stronger and more durable pH electrode can be made from solid metals that have metal oxide surfaces.

Glab *et al.*, 1989 described three preparation methods for metal oxide pH electrodes. (i) Iridium oxide is used for pH sensors, in addition to producing simple wire electrodes, iridium can also be sputtered onto other substrates. However, there are technical problems with producing a stable iridium oxide layer. (ii) Heating (800°C to 900°C) in a highly oxidizing flame is one method. Another popular technique uses heating in a flame after first immersing the tip in 1 M or higher molar solution of either sodium or potassium hydroxide. (iii) Palladium oxide has also been used as a pH sensor. A similar technique of flame heating after treating with a strong hydroxide solution is recommended over simple heating. Apparently both iridium and palladium can form single and double oxides, which complicate their pH sensitivity through multiple reactions. Iridium and palladium oxide pH sensors are reported to be insensitive to interference from ambient O_2.

Generally, if a metal (M)-metal oxide (M_xO_y) reaction is reversible, it can be represented as:

$$M_xO_y + 2y(e^- + H^+) \rightarrow {}_xM + {}_yH_2O \qquad \text{... eq.-7.2}$$

And the equilibrium potential will increase linearly with pH. However, it is also possible to have metal oxide-metal oxide reactions where additional reversible higher and lower valency oxide couples (M_xO_y, M_x 'O_y') are influencing the pH potential. A metal oxide-metal oxide reaction such as:

$$2MO_y + 2H^+ + 2e^- \rightarrow (2_y - 1)\, M_2O + H_2O \qquad \text{... eq.-7.3}$$

If there is a potential change of 59 mV/pH unit at 25°C, only one H_2O molecule is involved. Combined reactions can complicate the electrode potential when more than one form of metal oxide is present.

Further complications can arise if the metal oxide becomes hydrated. For example, a reaction such as:

$$MO_{[(x+n)/2}\, [H_2O]_{(y+m)} + ne^- + nH^+ \rightarrow MO_{(x/y)}\, [H_2O]_y + (m+n/2)H_2O \qquad \text{... eq.-7.4}$$

If the metal oxide is sparingly soluble, it may be possible to have steady state chemical equilibrium with adequate exchange current densities. However, if the oxide dissolves and diffuses away (corrosion), the electrochemical properties of the electrode will change with time. In this case, the electrode potential could also be sensitive to flow as convection influences the rate of oxide removal.

7.3 ANTIMONY pH ELECTRODES

The antimony (Sb) or antimony oxide (Sb_2O_3) electrode has been a viable alternative to the glass pH electrode, especially in highly corrosive environments (*Eg.* in hydrogen fluoride) where glass pH electrodes do not behave well. Antimony pH electrodes can be fabricated from crystalline or powdered material, or made by plating metal from solution. Monocrystalline electrodes have been found to be more stable than the poly crystalline electrodes. One of the major problems with antimony electrodes is related to the purity of the metal substrate. Instabilities can be caused by the anodic corrosion reaction and resulting surface pitting. The reaction is usually represented as follows:

$$2Sb + 3H_2O \rightarrow Sb_2O_3 + 6H^+ + 6e^- \quad \text{... eq.-7.5}$$

At the antimony anode while O_2 is reduced:

$$O_2 + 4H + 4e^- \rightarrow 2\,H_2O \quad \text{... eq.-7.6}$$

At the cathode, this can be an Ag/AgCl reference. Therefore, ambient O_2 can have a relatively large effect on the potential. The resulting equilibrium potentials for the two reactions are:

$$E_{Anode} = 0.15 - 2.303\frac{RT}{nF}\text{pH} \quad \text{... eq.-7.7}$$

and

$$E_{cathode} = 1.23 - 2.303\frac{RT}{nF}\text{pH} + \frac{RT}{4nF}\ln PO_2 \quad \text{... eq.-7.8}$$

The antimony electrode could be used as an O_2 sensor if the pH is maintained at a constant value. Glab *et al.,* 1989 have discussed the O_2 sensitivity of the antimony pH electrode, which appears to be much greater than some other metal oxide electrode types.

7.4 pH SENSITIVE MEMBRANES

Ammann 1986 has published a book on the use of neutral carrier ion selective membranes for electrodes, including those for pH measurements. The main component of the membrane is a complexing ligand which is an ionophore that is specific for H^+ or other ions. These ionophores can both form channels, or physically complex with individual ions and act as carriers. An ionophore for H^+, tri-n-dodecylamine is commercially available in formulations for use in liquid-filled pH sensitive microelectrodes or in solid pH sensitive polymer membranes. The ionophore may be incorporated into either polyvinyl chloride or silicone rubber membranes. After incorporation, the pH sensitive membranes can have shelf lives up to several years when stored in the dark at normal room temperatures. The membranes are highly specific, with minimal interferences from other ions.

7.5 OXYGEN ELECTRODE

7.5.1 Introduction

The dissolved oxygen (DO) is oxygen that is dissolved in water. The oxygen dissolves by diffusion from the surrounding air, aeration of water that has tumbled over falls and rapids and as a waste product of photosynthesis. From photosynthesis (in the presence of light and chlorophyll) process oxygen releases:

Carbon dioxide + water → Oxygen + Carbon-rich foods ... ***eq.-7.9***

CO_2 H_2O O_2 $C_6H_{12}O_6$

Fish and aquatic animals cannot split oxygen from water (H_2O) or other oxygen containing compounds. Only green plants and some bacteria can do that through photosynthesis and similar processes. Virtually all the oxygen we breathe is manufactured by green plants. A total of three-fourths of the earth's oxygen supply is produced by phytoplankton in the oceans.

If water is too warm, there may not be enough oxygen in it. When there are too many bacteria or aquatic animals in the area, they may over populate, using DO in great amounts. Oxygen levels also can be reduced through over fertilization of water plants by run-off from farm fields containing phosphates and nitrates (the ingredients in fertilizers). Under these conditions, the numbers and size of water plants increase. Then, if the weather becomes cloudy for several days, respiring plants will use much of the available DO. When these plants die, they become food for bacteria, which in turn multiply and use large amounts of oxygen and this depleting all the oxygen.

How much DO an aquatic organism needs depends upon its species, its physical state, water temperature (Fig. 7.8), pollutants present and more. Numerous scientific studies suggest that 4-5 parts per million (ppm) of DO is the minimum amount that will support a large, diverse fish population. The DO level in good fishing waters generally averages about 9.0 parts per million (ppm).

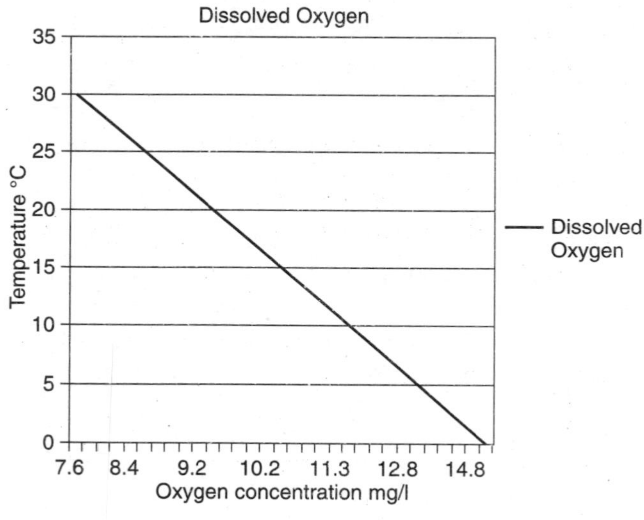

Fig. 7.8 *The effect of the temperature on the dissolved oxygen.*

Total dissolved gas concentrations in water should not exceed 110 percent. Concentrations above this level can be harmful to aquatic life. Fish in waters containing excessive dissolved gases may suffer from gas bubble disease however; this is a very rare occurrence. The bubbles block the flow of blood through blood vessels causing death. External bubbles (emphysema) can also occur and be seen on fins, skin and other tissue. Aquatic invertebrates are also affected by gas bubble disease but at levels higher than those lethal to fish.

Adequate dissolved oxygen is necessary for good water quality. Oxygen is a necessary element to all forms of life. Natural stream purification processes require adequate oxygen levels in order to provide for aerobic life forms. As dissolved oxygen levels in water drop below 5.0 mg/l, aquatic life is put under stress. Oxygen levels that remain below 1-2 mg/l for a few hours can result in large fish kills.

Dissolved oxygen is absolutely essential for the survival of all aquatic organisms. Moreover, oxygen affects a vast number of other water indicators, not only biochemical but esthetic ones like the odor, clarity and taste. Consequently, oxygen is perhaps the most well-established indicator of water quality.

A high DO level in a community water supply is good because it makes drinking water taste better. However, high DO levels speed up corrosion in water pipes. For this reason, industries use water with the least possible amount of dissolved oxygen. Water used in very low pressure boilers have no more than 2.0 ppm of DO, but most boiler plant operators try to keep oxygen levels to 0.007 ppm or less.

7.5.2 Electrode

The name Oxygen Electrode is just one of the many titles often used to describe the electrochemical sensor for oxygen. Other names include: 'Clark Cell', after its generally recognized inventor (1956) Leyland C Clark and 'Oxygen Membrane Polarographic Detector' (O_2-MPD), because of the mode of action of the electrochemical device.

The Oxygen electrode remains one of the most commonly used devices for measuring the partial pressure of oxygen (sometimes referred to as 'oxygen tension') in the gas phase or more commonly, dissolved solution. The application of Oxygen electrode in a wide variety of diverse subject areas including areas (1) Environmental studies (*Eg.* O_2-levels in natural waters) (2) Sewage treatment (vital in monitoring the progress of bacterial attack) (3) Alcohol production (O_2-levels in fermentation tasks need to be continuously monitored and controlled) (4) Medicine (invasive and non-invasive monitoring of a key physiological analyte).

The typical range of detection of O_2 of this device is from 10^{-4} atm (i.e. 0.01%) to 1 atm (i.e. 100%). The key to continuing supremacy of the Oxygen electrode over other electrochemical devices for O_2 detection is the utilization of a gas-permeable, ion-impermeable, membrane that separates the test system from the sensing electrode (the platinum cathode). This membrane prevents many problems of electrode passivation or poisoning that arise when the sensing electrode is placed in direct contact with the system (usually an aqueous solution) under test.

7.5.2.1 The cell

The Rank brothers' Oxygen electrode comprises two electrodes (1) The first is a small, diameter is 2mm, central platinum disc working electrode is the cathode and at this electrode, the O_2 diffusing through the membrane is reduced. (2) The second one is set in a well surrounding, this is a silver ring counter and reference electrode, it is about ten times larger in surface area than the Pt cathode. Conduction between the two electrodes is achieved using a 3M potassium chloride solution to saturate the paper tissue covering the two electrodes. On top of this is placed the key gas-permeable membrane, usually 12.7mm thick Teflon, sealed from the test sample in the incubation chamber by a silicone rubber 'O' ring (Fig. 7.9).

The controller is used to apply a voltage to the central platinum electrode that is sufficiently negative, with respect to the silver electrode, that all the oxygen diffusing through the membrane and reaching this electrode is reduced. The resultant current which flows between the two electrodes is proportional to the oxygen partial pressure in the test system, pO_2. The controller converts this current directly into a voltage and depending on the model will display this in units of percentage saturation. It can also be measured continuously using a x/t chart recorder or a data logger.

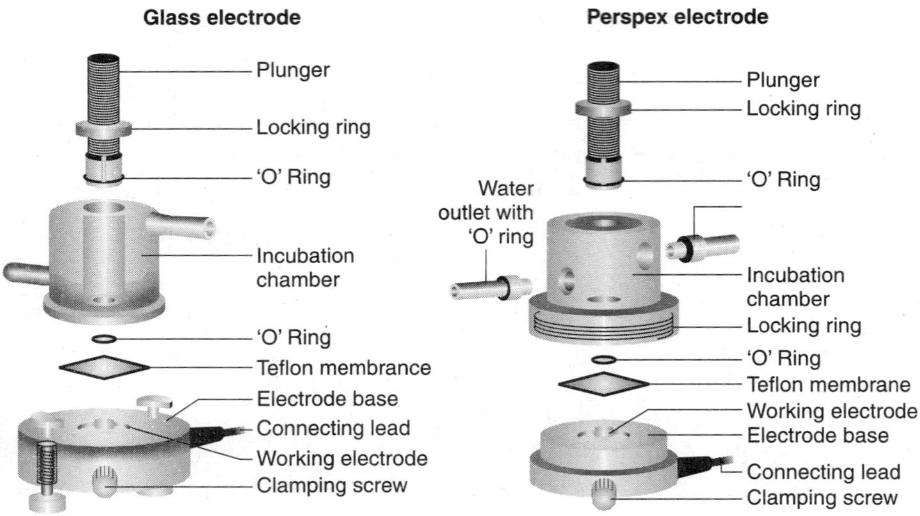

Fig. 7.9 *Mechanism of Dissolve Oxygen electrode. (Source: The Rank Brothers Oxygen Electrode, Issue 1a February 2002, Website: http://www.rankbrothers.co.uk/*

7.5.2.2 Principle

The platinum electrode is polarised at –0.6V with respect to the silver electrode, every oxygen molecule that reaches its surface from the test solution via gas permeable membrane, is reduced to water through the following reaction:

$$O_2 + 2H_2O + 4e^- \rightarrow 4OH^- \quad \text{... eq.-7.10}$$

For every reduction reaction there must be an oxidation and this occurs at the silver electrode as follows:

$$4Ag + 4Cl^- - 4e^- \rightarrow 4\ AgCl \quad \text{... eq.-7.11}$$

Thus the overall electrochemical process that occurs in an Oxygen Electrode is as follows:

$$4Ag + O_2 + 2H_2O + 4Cl^- \rightarrow 4AgCl + 4OH^- \quad \text{... eq.-7.12}$$

As the oxygen electrode is repeatedly used the bright silver ring electrode rapidly becomes 'tarnished'. Eventually an even coat of brown silver chloride forms on the silver electrode. The presence of this silver chloride layer is desirable (it stabilises the overall behaviour of the electrode) and should not be removed except if it grows very thick (after many months of use).

7.5.2.3 Setting up of the oxygen electrode

1. The small well containing the silver electrode is filled with the potassium chloride solution.
2. A 1.5 cm square piece of tissue paper is cut with a 2 mm hole in its centre and float this on the potassium chloride in the well ensuring that the hole is central above the platinum electrode.
3. The empty teat pipette is touched against the tissue paper and use it to suck off the excess electrolyte so that the paper is wet and clings to the surface of the electrode.
4. 1.5 cm square piece of Teflon membrane is cut and place it so that it covers both electrodes, ensuring that the platinum electrode is underneath the centre membrane and that there is no air bubbles trapped under the membrane.
5. The silicone rubber 'O' ring over the platinum electrode is pushed so as to hold the Teflon membrane in place when the plastic base and the incubation chamber are clamped together.
6. The electrode base is carefully clamped with the incubation chamber. It is important to ensure that the incubation chamber is not rotated on the base during clamping, as this will damage the membrane. The locking ring should be tightened by hand only. Over tightening may cause problems.
7. The electrode is connected to the controller and adjusts the polarizing voltage to 0.6V and stirring speed to a suitable level. Connect the water jacket of the incubation chamber to a constant temperature water bath and allow the sample temperature to stabilize. The electrode is now ready for calibration.

7.5.2.4 Maintenance

When the electrode is not in use for a few hours (*Eg*. overnight) it is best dismantled and the electrodes left to soak in distilled water. If the electrode assembly must be left intact, but non-operational for a few hours, it is best if the electrode is left on, but with the stirrer switched off. The platinum electrode needs to have a 'mirror' finish, any surface damage will affect the response of the electrode and it will thus need to be cleaned once every 5 to 7 days of use.

A suitable polish can be made by mixing thick slurry of 0.3 mm polishing alumina in distilled water. A piece of cotton wool can then be used to polish the platinum electrode until it is smooth, bright and clean (this should only take a few minutes). The silver electrode will need to have the layer of silver chloride removed and the surface polished every 2-3 months of use. A 10% ammonia solution on cotton wool can be used to remove the silver chloride layer.

7.6 BIOCHEMICAL OXYGEN DEMAND (BOD)

7.6.1 Introduction

Determination of biodegradable organic substances in wastewater has significant importance on the environmental pollution control. Traditional analysis of biochemical oxygen demand (BOD) involves 5 to 7 days incubation period and is thus ineffective for operation of activated sludge processes in wastewater treatment plants or determination of the concentration of biodegradable organic pollutants in polluted water. Because of the need for express method of biochemical oxygen demand (BOD) determination, the BOD biosensors have been developed. A biosensor is defined as a self contained integrated device capable of providing specific quantitative analytical information using a biological recognition element. The BOD sensors are easy to handle, portative, offer quick response and advantage is enable to evaluate the value of biochemical oxygen demand in 60 minutes.

Mathematical modelling of biosensors response provides several calibration methods and also gives a link among the design, fabrication and measurement procedure of the device. Most frequently, the biosensor output signal is analysed according to the stabilised initial and final values. It has been shown that the total oxygen uptake measurement needs 15-20 minutes and the recovery time between measurements can be as long as 3-4 hours. The alternative approach to the steady state analysis of a biosensors response is through the mathematical modelling of the dynamic behaviour of the output values.

7.6.2 BOD Sensor

A biosensor consisting of immobilised Yeast (*Trichosporon cutaneum)* and an oxygen probe was developed for BOD estimation. The BOD biosensor includes an oxygen electrode that consists of a platinum cathode and an aluminium anode bathing in saturated KCl solution and a Teflon membrane. Yeast cells are immobilized on a porous membrane and are trapped between the pores and the Teflon membranes. Oxygen consumption by the immobilized microorganisms causes a decrease in current until a steady state is reached. The BOD biosensor measures BOD at 3-60 mg/L and it is based on the diffusion limited amperometric oxygen sensor (the Clark-type oxygen sensor) and a biomembrane. The biomembrane consisted of mixed culture of microorganisms immobilised in 2% suspension of agarose in phosphate buffer. The agarose gel membrane was fixed on the top of the oxygen sensor polymer membrane with a special holder (Fig. 10).

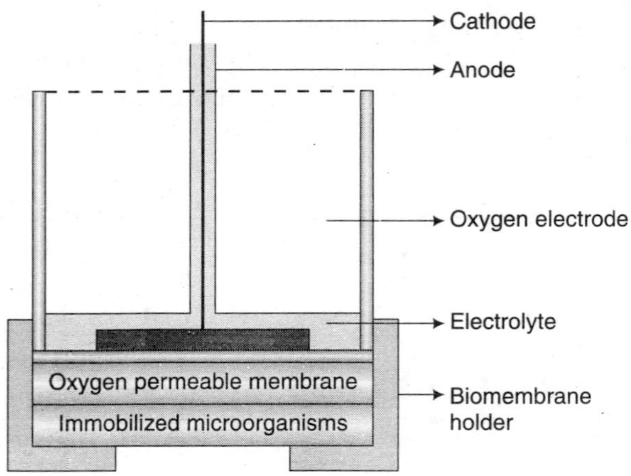

Fig. 7.10 *Schematic diagram of Biological Oxygen Demand (BOD) sensor.*

Microorganisms used in the biosensor were separated from the suspension of activated sludge. To accomplish a stable microbial consortium and improve mechanical properties of the biomembrane, the mixed culture of microorganisms was fed in laboratory scale process with synthetic waste water. The BOD_7 (7 days incubation) of this synthetic wastewater was 150 mg/L. The BOD_7 of this solution was 220±10 mg/L. The microbial consortium was adapted in this synthetic wastewater for 2 weeks and then 40 ml of suspension was centrifuged at 400 rpm for 15 minutes. The supernatant was discarded and the remainder was washed with distilled water and phosphate buffer solution (7.04 g/L $NaH_2PO_4H_2O$ and 17.55 g/L $Na_2HPO_412H_2O$) and centrifuged again. The remaining microorganisms were suspended in 4 ml of phosphate buffer and 1.5 ml of the final mix was immobilised with agarose.

The 2% suspension of agarose in phosphate buffer was heated to the boiling temperature, then cooled down to 45°C and 1.5 ml of prepared microorganisms suspension was added to it. The resulting mixture was spread on the polymer net of a particular thickness (0.5 cm) which, in order to gain a certain and even thickness of the layer containing microorganisms. It was then placed between two glass plates until the formation of a persistent layer of gel. The microbial membrane was kept for 5 days in phosphate buffer to reach stable condition. For the calibration of constructed biosensor the solution of D-glucose (150 mg/L) and L-glutamic acid (150 mg/L) (GGA) was used.

The measurements were carried out at 25°C in the thermostated measuring cell under continuous mixing with a magnetic stirrer and saturation of air oxygen with a micro compressor. The substrate solutions were added to the air saturated test media after the stabilized steady state output of the biosensor had been obtained. Sensor output signal was registered in a time interval of 1 second.

7.6.3 Calibration

The calibration of BOD sensor was performed by three different variables characterising the measurement process (1) Determination of the difference between the change of stabilized initial and final values of the biosensor response is a known method of biosensor calibration (2) A method of measurements, which offers much faster response (3) The determination of time constants for calibration of BOD sensor is that based analogously to dynamic method of output signal.

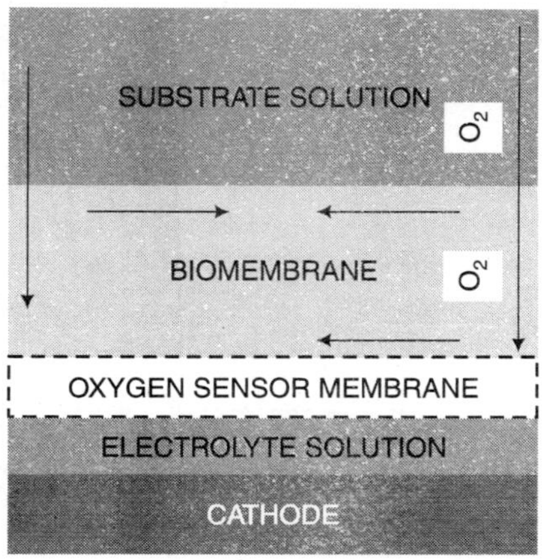

Fig. 7.11 *Characteristic processes of substrate and oxygen in a BOD sensor.*

The steady state response in the beginning of measurements shows the endogenous respiration of immobilised microorganisms. In a continuously mixed system dissolved oxygen diffuses from the aerated phosphate buffer to the biomembrane, where part of the oxygen is consumed by the microorganisms (Fig. 7.11). The remainder diffuses through the oxygen sensors polymer membrane and is detected by the Oxygen electrode.

When the substrate is added to the system, it diffuses into the biomembrane and is assimilated by microorganisms, resulting in the increase of the amount of oxygen consumed. As a result, the concentration of oxygen on the polymer membrane of sensor decreases leading to a decrease in biosensor output signal.

A typical biosensor curve characterising the response of the BOD sensor for a sample of substrate with a BOD of 88 mg/L is depicted. The value of change of the stabilised initial and final values of response depends on the substrate concentration added after the stable initial signal of biosensor has been reached. The time for one measurement in steady state regime takes approximately 12-15 minutes followed by 20-50 minutes recovery time depending on substrate concentration. The study of non-steady state response of biosensor output signal showed that to the maximum level of oxygen consumption by bacteria corresponds the maximum rate of change of output signal (V_m) which value can be determined by the course of output signal.

7.6.4 Non-Steady State Processes

The model is based on the main principles of working process of BOD sensors- substrate from the test media, into which the sensitive biosensor is immersed and is metabolised by microorganisms consuming oxygen in the biomembrane of the sensor. This process leads to the redistribution of oxygen fluxes and as a result, the concentration of oxygen decreases at oxygen sensors membrane leading to a decrease in the biosensor output signal. As in the non-steady state regime the flux of oxygen towards the cathode is not proportional to the oxygen concentration in biomembrane layer, the biosensor response, lag time and processes in biomembrane are expressed by transfer function:

$$h^*(t) = \exp\left(-\frac{t}{T_s}\right) - 2\sum_{n=1}^{\infty}(-1)^n \times \frac{T_d}{n^2T_s - T_d}\left[\exp\left(-\frac{t}{T_s}\right) - \exp\left(-n^2\frac{t}{T_d}\right)\right] \quad \text{... eq.-7.13}$$

where T_d and T_s are the time constants for the oxygen sensor and for the exponential decay of oxygen concentration at oxygen sensor membrane, respectively.

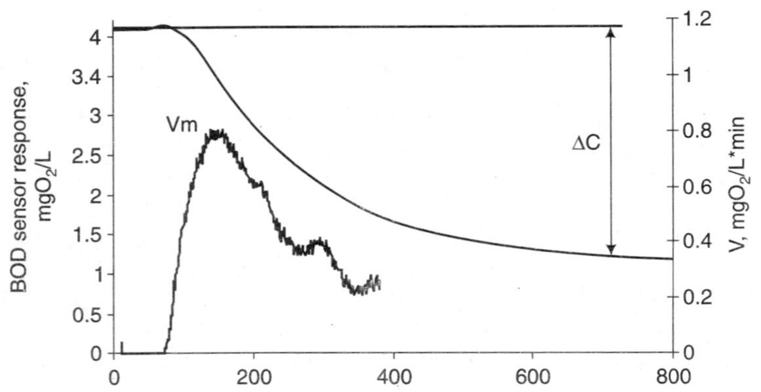

Fig. 7.12 *Analysis of output signal of a BOD sensor: ΔC is the change of the stabilised initial and final values of response; V is the rate of change of sensors response; V_m is the maximum rate of change of the sensor output. (Source: Siiri Velling, Kaja Orupold and Toomas Tenno (2007). BOD sensor for wastewater analysis—design and calibration methods. The Second Baltic Symposium on Environmental Chemistry, Kalmar, Sweden.)*

Transfer function of the oxygen sensor is characterised by time constant T_s which value shows the time needed to run all non-steady state processes of biosensor with maximum velocity (Fig. 7.12) (i.e. all the processes, diffusion and assimilation of substrate and oxygen and its electrochemical detection would have the maximum rate). So, that the T_s depends on the substrate concentration in the test media.

The biosensor response in the calibration solution was detectable in concentration range 10-90 mg/l in BOD units (Fig. 7.13). The sensor's response was analysed according to the steady state method and dynamic determination. The change of stabilised response dependency and dynamic values of biosensor output signal vs. BOD concentrations of calibration solutions were characterised by hyperbolic relationship (Fig. 4.14(a) and (b)). For the calibration, the linear parts of these dependencies can be employed. The hyperbolic relationships are characterised by theoretical maximum values of described variables (C_{max} or V_{max}) and by half-saturation coefficients that introduce the value of substrate concentration at which the variable under study has reached half of its maximum value. For the change of stabilised response and dynamic values of BOD sensor output signal the half-saturation coefficients are 18 mg/L and 27 mg/L, respectively. Different values of half saturation coefficients can be explained by the various factors affecting the calibration methods. The change between stabilised initial and final value of output signal depends mostly on the number, viability and respiration rate of microorganisms. Determination of dynamic variables is also influenced by several other factors—the characteristics and concentration of immobilisation material, the thickness of biomembrane and the rate of substrate and oxygen diffusion.

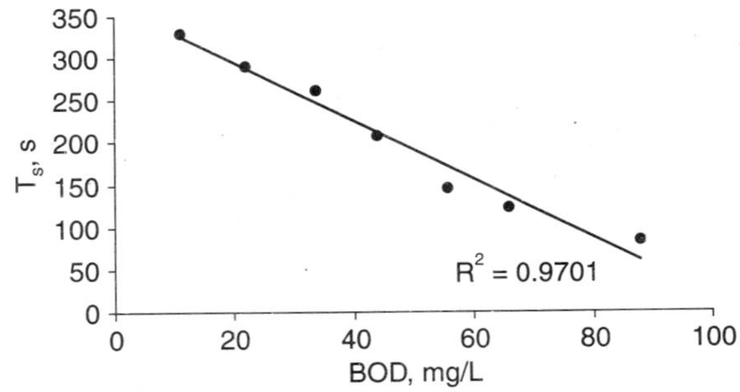

Fig. 7.13 *Normalised response of BOD sensor in calibration solution at various concentrations.*

The method of analysing the output signal of a biosensor thus provides a much broader measuring range and also shorter response and regeneration times. On the basis of these observations, it is obvious that the method of using the maximum rate of change is more advantageous for the practical applications of biosensors.

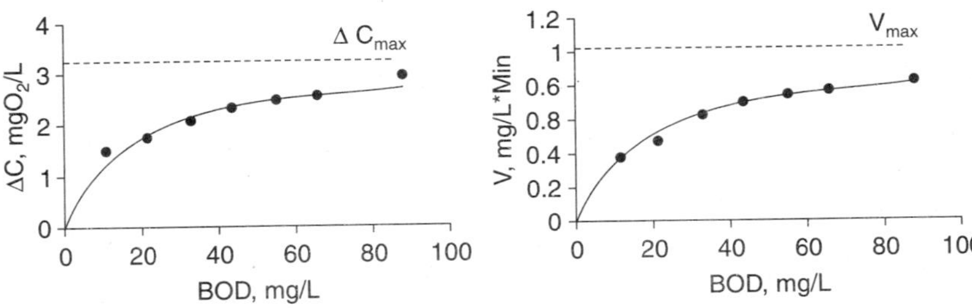

Fig. 7.14 *Dependence of the change of steady-state values (C) (a) The maximum rate of change (V_m) (b) Biosensors response on substrate concentration. The solid lines correspond to the theoretical curve obtained by the least squares curve fitting. (Source: Siiri Velling, Kaja Orupold and Toomas Tenno (2007). BOD sensor for wastewater analysis—design and calibration methods. The Second Baltic Symposium on Environmental Chemistry. Kalmar, Sweden.)*

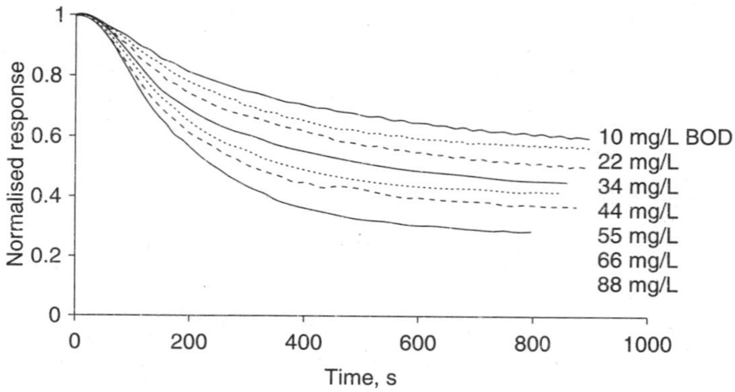

Fig. 7.15 *Dependence of time constant T_s on substrate concentration. (Source: Siiri Velling, Kaja orupold and Toomas Tenno (2007). BOD sensor for wastewater analysis—design and calibration methods. The Second Baltic Symposium on Environmental Chemistry. Kalmar, Sweden.)*

Mathematical modelling of the response of studied BOD sensor enabled researches to work out a new calibration method that is based on non-steady state processes in biosensor. The course of biosensor output signal can be characterised by time constants T_d and T_s that show the time lag and exponential decay of oxygen concentration at oxygen sensor membrane, respectively. The value of T_d that depends on the immobilisation matrix and thickness of the biomembrane should not vary much in the case of similar experimental conditions. The average value of T_d for constructed BOD sensor was 38 ± 5 seconds. The value of time constant T_s calculated depends on the substrate concentration as the increase of substrate concentration in test media causes the rise of oxygen consumption and therefore the amount of oxygen reaching the sensor membrane and oxygen electrode decreases. Linear dependence of T_s *vs* calibration solution concentrations (Fig. 7.15) can be used for BOD sensor calibration.

7.7 ION-SELECTIVE ELECTRODES

The pH measuring Glass electrode is an example for ion-selective electrodes (ISE). There are many different ISE's but it would be convenient to classify them in terms of the membrane used for partitioning.

7.7.1 Glass Membrane

Glass membrane originates with the hydrogen-ion selective electrode having well-behaving glass membrane that has high mobility for H^+ ion. Subsequently, electrodes have been developed for other cations such as Na^+ and K^+.

7.7.2 Inorganic Salt Membrane

These electrodes are based on inorganic halides and sulphides (*Eg*. silver salts, lanthanum fluoride and heavy-metal sulphides). These membranes have been produced from preparations ranging from whole crystals to dispersions in an inert matrix, such as polythene or silicon rubber. They are targeted at ions such as Ag^+, Cu^{++} and Pb^{++}.

7.7.3 Organic Membrane

The electrodes based on neutral carriers generally have the highest selectivity in this class. However, cation exchangers or complexing agents or anion exchangers have been successfully employed in electrodes with liquid or solid membranes, selective to cations or anions, respectively.

7.7.4 Gas Sensing Electrode

These electrodes are an extension of ion selective measurements to detection of gaseous analytes. Gas sensing probes are complete electrochemical cells incorporating with both the ion selective electrode and a reference electrode within the sensor. Assay of the target gaseous sample is not performed directly, but is related to a changing parameter (usually pH) which can be monitored by an ion-selective electrode.

Transducers

Chapter 8

8.0 TRANSDUCER

8.0.1 Introduction

The study of sensors is multidisciplinary and requires a broad knowledge of general physics, solid-state physics, electronics and semiconductor-manufacturing techniques. The main objective of the sensor is the information available in one energy form must be converted into the same or another energy form.

Transducer is a device which converts physical input quantities into an electrical output signal. It is used as a means of converting one form of energy into another form of energy (Table 8.1). Conversion of non-electrical signals into electrical ones is often effected for ease of processing.

The identification unit contains a device which reacts to the desired physical quantity and converts it into an electrical signal. Transducer converts a signal from one form to another form and most electronics circuits use both input and output transducers (Fig. 8.1).

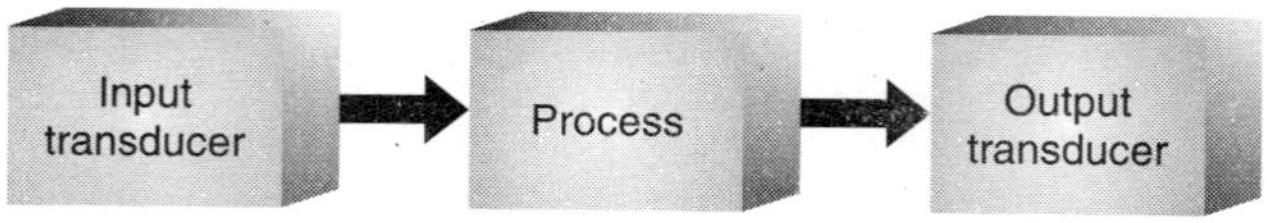

Fig. 8.1 *Electronics circuits with input and output transducers.*

The input transducer converts a quantity to an electrical signal (voltage) or to resistance (which can be converted to voltage) and these are also called sensors. *Eg.* LDR converts light to resistance, Thermistor converts temperature to resistance and Variable resistor converts angle to resistance.

If a machine or process must be supplied with new information, the output transducer must be able to detect it. In that case, often a type of actuator may be required to execute the right action. In measurement systems to be perceived by humans, we need a type of display to which our senses react, or can take action. So if we state that the output transducer is an actuator whether it activates a technical system or a living (biological) system. Output transducers convert an electrical signal to another quantity. *Eg*. lamp converts electricity to light, loudspeaker converts electricity to sound, motor converts electricity to motion and heater converts electricity to heat.

Table 8.1 Examples for energy transformation devices.

S.No.	*Energy source*	*Name of the device*
1	Radiant to electrical	Phototransistor
2	Mechanical to Electrical	Piezoresistance
3	Electrical to Mechanical	Piezoelectricity
4	Thermal to Electrical	Seebeck
5	Electrical to Chemical	Electroplating
6	Electrical to Electrical	Transistor
7	Electrical to Magnetic	Coil
8	Chemical to Electrical	pH
9	Chemical to Chemical	Chemical reaction

The range of possible transducers for each physical phenomenon is very extensive currently, more and more transducers are based on Hall effect, photo-detection, temperature sensors, piezo crystals etc. tend to have amplifiers, detectors and other signal processing circuits built in an integrated form as parts of their mountings. This eliminates the use of cables for carrying low level signals from the transducers to the processing circuitry, thereby minimizing noise and pick-up problems.

8.1 ELEMENTS

The following described elements are important in different type transducers.

8.1.1 Bonded Resistance Wire

One of the most useful of displacement transducers is the bonded resistance wire strain gauge. This gauge consists of fine metallic wire arranged in a grid. This wire is fine when the surface is strained, either in tension or compression its resistance changes and the changes in resistance are directly proportional to the displacement being measured.

8.1.2 Variable Resistance

Displacement may be measured by coupling the motion to the movable contact of a rheostat or potentiometer, through a light flexible cable wound onto a drum mounted on the rheostat shaft. Displacement thus rotates the shaft and changes the resistance in proportion to the displacement.

8.1.3 Capacitance Displacement Transducers

The basic idea is to detect the motion or distance to be measured so that changes the spacing between the plates of an air dielectric capacitance. This change affects the capacitance. In most applications the capacitance is of the order of 10 to 100 microfarads. The changes must be measured by the effect of the change in reactance (X) and in order to have a reactance of usable size, frequency of the order of megahertz must be used.

8.1.4 Velocity Transducers

There are various kinds of velocity transducers; the common type is the coil moving in a magnetic field as shown in Fig. 8.2. Voltage is induced in the coil as it is moved in the field of permanent magnet and induced voltage measures the velocity.

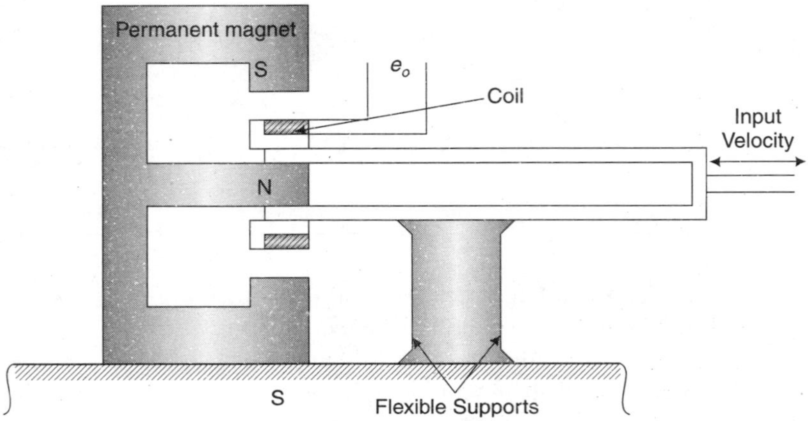

Fig. 8.2 *Schematic arrangement of moving coil in magnetic field for use as a velocity or Acceleration transducer.*

8.1.5 Acceleration Transducer

Linear acceleration may be measured by measuring the force due to acceleration. An arrangement by which acceleration is measured is shown in Fig. 8.3. Here the acceleration (a) moves the mass (m), so as to stress the bar (B), to bend or twist. The amount of stress is proportional to the acceleration. The stress can be conveniently measured by attaching bonded resistance wire strain gauge to the bar. Such a transducer has the advantage that it can be calibrated statically by measuring the force required to produce certain electrical output.

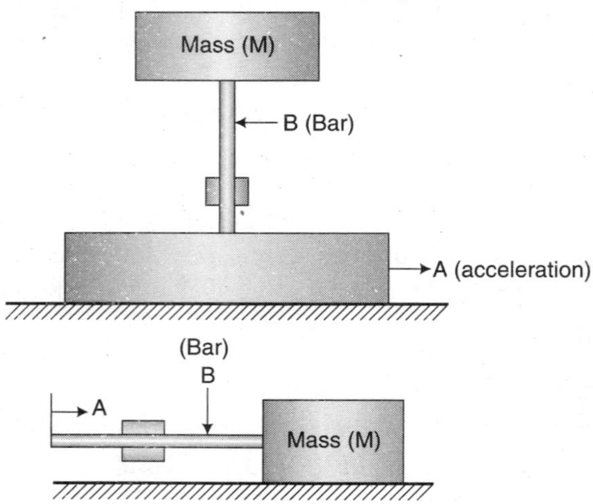

Fig. 8.3 *Acceleration measurement.*

8.2 SELECTION FACTORS

The selection of the right transducer for a given measurement application involves matching the transducer characteristics for the desired system performance and input requirements. The current availability of diverse technologies facilitates proper selection. The selection parameters have been detailed as follows:

8.2.1 Nature of Measurement

The measurement parameters are either in steady state, transient or dynamic, each of these groups falls into certain frequency ranges. The frequency response of a transducer determines its application limits. The sensitivity of a transducer is indicative of the amplification that will be required before the signal is processed. Accuracy, linearity, hysteresis and noise level are also the features which are considered while selecting a transducer.

8.2.2 Mechanical Input Characteristics

Some of the mechanical input characteristics are also one of the selection parameters to select the type or group of transducers which include linearity, mechanical hysteresis and viscous flow.

8.2.2.1 Linearity

In the case of static measurements, transducers should have linear relationship between input and output. This aspect is equally important for cyclic or transient signals as well as the dynamic response of the transducer.

8.2.2.2 Mechanical hysteresis

The factors like bearing friction, backlash, fractured or corroded components, plain dirt, etc., are responsible for adverse performance and phenomenon of hysteresis in the transducers. Mechanical hysteresis in transducers like springs, pressure capsules etc., is a manifestation of the imperfect response of the microscopic crystal grains, which gets integrated over the macroscopic dimension of the strained transducer element. The load/deformation diagram is a closed loop, the areas of which represent the energy dissipated into heat.

8.2.2.3 Viscous flow

The main difference between hysteresis and viscous flow is that the latter depends on time. The magnitude of viscous flow is in a material increase with the increasing load and temperature.

8.2.3 Loading Effect of Transducers

If the attachment or installation of a transducer in any way affects or changes the value of the parameter being measured, errors may be introduced which far exceed those inherent in the transducer itself. The loading characteristic is determined by the mass of transducer, exterior size of the transducer, and the geometric configuration of the transducer as well as the compliance of stiffness presented to the mechanical output. Some of the errors caused by improper physical loading are high frequency attenuation, waveform distortion and signal phase shift.

8.2.4 Environmental Considerations

Specifications should include the conditions under which a transducer is expected to give satisfactory output. There is no advantage using the unit in a severe environment if it cannot survive long enough to provide the needed data or will indicate the erroneous data. In modern research, the environment in which the transducers must operate is in the temperature range between 700° to 2700°C. In addition to high temperature requirements, transducers that are suitable for aircraft and space vehicles should be able to function under acceleration forces several times that of gravity. The critical aspects of transducer location are the temperature changes, shock, vibration and electromagnetic interference. General considerations for transducer environments include accessibility of transducer for adjustments and repair, simplicity of mounting and cable installations, convenient, size, weight, shape, resistance to corrosion, etc.

8.2.5 Transducer Capability

Transducer characteristics need to be investigated to the effects of overheating, over-range and power dissipation ratings etc., and have to withstand operating conditions taking into account.

8.2.6 Measuring System Compatibility

The compatibility of the transducer and the existing or proposed electronic or electrical systems of transmission, measurement or control must be considered before selecting the suitable transducer.

8.2.7 Economic Considerations

Some of the general factors involved in transducer selection are cost, availability, basic simplicity, reliability, low maintenance. The selection of a transducer for a given application is usually a compromise between the several factors listed above. Therefore, transducers should be selected which do not require excessive repairs or continuous calibration checking for the user's confidence in operations. Thus, the cost effectiveness is important while selecting a transducer.

8.3 TRANSDUCER MODELING

Transducer modeling is necessary to make accurate predictions of how a part or component might be inspected, prior to the actual building of that part. An ultrasonic transducer may be characterized by detailed measurements of its electrical and sound radiation properties. Such measurements can completely determine the response of any one individual transducer.

To develop research and commercial models that relates electrical inputs (voltage, current) to mechanical outputs (force, velocity) and vice-versa. These models can be very vigorous in giving accurate prediction of transducer response, but undergo from a lack of accurate modeling of physical variables inherent in transducer manufacturing. These electrical-mechanical response models must take into account the physical and electrical components (Fig. 8.4).

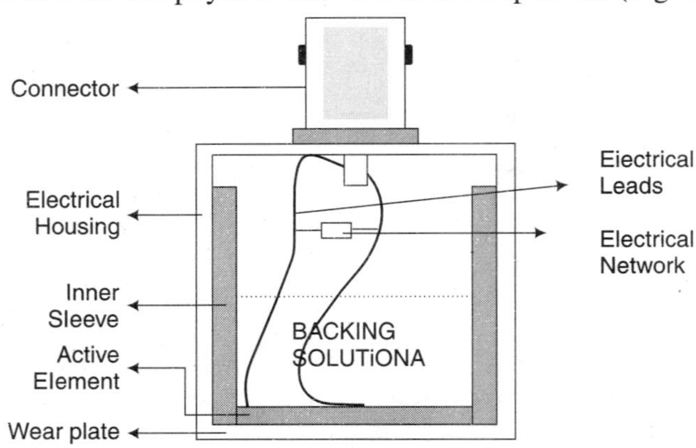

Fig. 8.4 *Physical and electrical components of a transducer.*

The Thompson-Gray Measurement Model which makes very accurate predictions of ultrasonic scattering measurements through liquid-solid interfaces, does not attempt to model transducer electrical-mechanical response. The Thompson-Gray Measurement Model approach makes use of reference data taken with the same transducer(s) to deconvolute electro-physical characteristics specific to individual transducers. The long term goal in ultrasonic modeling is to incorporate accurate models of the transducers themselves as well as accurate models of pulser receivers, cables, and other components that completely describe any given inspection setup and allow the accurate prediction of inspection signals.

8.4 TRANSDUCER TESTING-I

The transducer manufacturers have led in the development of transducer characterization techniques and have participated in developing the American Institute of Ultrasound in Medicine (AIUM) Standard Methods for Testing Single-Element Pulse-Echo Ultrasonic Transducers. The manufacturers perform characterizations according to Eritrean Standards Institution (ESI) and other industrial and military standards. Often, equipment in test labs is maintained in compliance with MIL-C-45662A Calibration System Requirements. As part of the documentation process, an extensive database containing records of the waveform and spectrum of each transducer is maintained and can be accessed for comparative or statistical studies of transducer characteristics.

Manufacturers often provide time and frequency domain plots for each transducer. The signals below were generated by a spiked pulser. The waveform image Fig. 8.5b on the left shows the test response signal in the time domain (amplitude versus time). The spectrum image on the right shows (Fig. 8.5a) the same signal in the frequency domain (amplitude versus frequency). The signal path is usually a reflection from the back wall (fused silica) with the reflection in the far field of the transducer.

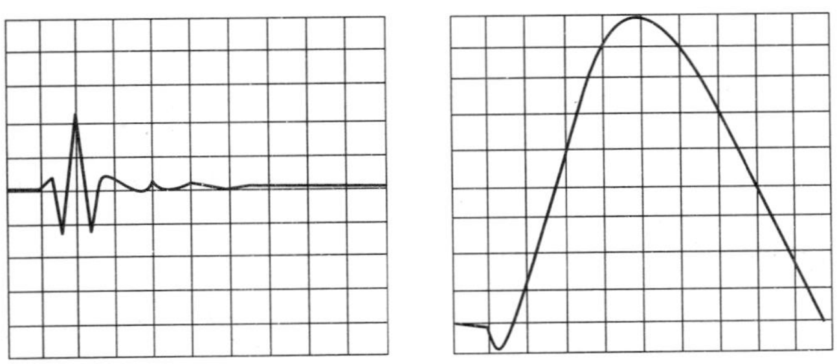

(a) Waveform (b) Spectrum: Linear

Fig. 8.5 *Waveform and spectrum of each transducer.*

Other tests may include the following:

1. **Electrical Impedance Plots** provide important information about the design and construction of a transducer and can allow users to obtain electrically similar transducers from multiple sources.

2. **Beam Alignment Measurements** provide data on the degree of alignment between the sound beam axis and the transducer housing. This information is particularly useful in applications that require a high degree of certainty regarding beam positioning with respect to a mechanical reference surface.
3. **Beam Profiles** provide valuable information about transducer sound field characteristics. Transverse beam profiles are created by scanning the transducer across a target (usually either a steel ball or rod) at a given distance from the transducer face and are used to determine focal spot size and beam symmetry. Axial beam profiles are created by recording the pulse-echo amplitude of the sound field as a function of distance from the transducer face and provide data on depth of field and focal length.

8.5 TRANSDUCER TESTING-II

The ASTM E1065 Standard Guide for Evaluating Characteristics of Ultrasonic Transducers, acoustic and electrical characteristics which can be described from the data, is obtained from specific procedures which are listed below:

8.5.1 Frequency Response

The frequency response may be obtained from shock excitation and sinusoidal burst procedures.

8.5.2 Relative Pulse-Echo Sensitivity

The relative pulse-echo sensitivity may be obtained from the frequency response data by using a sinusoidal burst procedure. The value is obtained from the relationship of the amplitude of the voltage applied to the transducer and the amplitude of the pulse-echo signal received from a specified target.

8.5.3 Time Response

The time response provides a means for describing the radio frequency (RF) response of the waveform. A shock excitation, pulse-echo procedure is used to obtain the response. The time or waveform responses are recorded from specific targets that are chosen for the type of transducer under evaluation, for example, immersion, contact straight beam, or contact angle beam.

8.5.4 Frequency Response

The frequency response of the above transducer has a peak at 5 MHz and operates over a broad range of frequencies. Its bandwidth (4.1 to 6.15 MHz) is measured at the -6 dB points, or 70% of the peak frequency. The useable bandwidth of broadband transducers, especially in frequency analysis measurements, is often quoted at the –20 dB points. Transducer sensitivity and bandwidth are chosen based on inspection needs.

8.5.5 Complex Electrical Impedance

The complex electrical impedance may be obtained with commercial impedance measuring instrumentation and these measurements may provide the magnitude and phase of the impedance of the search unit over the operating frequency range of the unit. These measurements are generally made under laboratory conditions with minimum cable lengths or external accessories and in accordance with specifications given by the instrument manufacturer. The value of the magnitude of the complex electrical impedance may also be obtained using values recorded from the sinusoidal burst.

8.6 TRANSDUCER SELECTING FACTORS

1. It should have high input impedance and low output impedance, to avoid loading effect.
2. It should have good resolution over the entire selected range.
3. It must be highly sensitive to desired signal and insensitive to unwanted signal.
4. Preferably small in size.
5. It should be able to work in corrosive environment.
6. It should be able to withstand for pressure, shocks, vibrations etc.
7. It must have high degree of accuracy and repeatability.
8. Selected transducer must be free from errors.

8.7 REQUIREMENTS

1. Smaller in size and weight.
2. High sensitivity.
3. Ability to withstand environmental conditions.
4. Low cost.

8.8 ACTUATORS

An actuator is a device that responds to a signal and develops an output in another form, usually mechanical movement of some sort. They are used extensively in diverse industrial applications like controlling flow or position, in photocopying machines, circuit breakers, tightening of textile screens, automobiles etc. An actuator is that portion of a final control element in a control system that furnishes the power to change and to maintain the valve plug position in response to a signal received from the automatic controller. In some applications the actuator may position elements other than a valve, such as a lever, damper, or pump speed governor.

Actuators are categorized as Mechanical, Pneumatic, Hydraulic and Electrical. Motor operated and solenoid actuator fall into electrical category. Combination forms include electro-hydraulic, electro-pneumatic actuators. The main types of actuators are used in automation to fluid processes, common in petroleum, chemical and allied manufacturing industries.

8.8.1 Types

The actuators are classified into two types (1) Control type and (2) Electrical type.

8.8.1.1 Control (valve) type

The control valve is extensively used for automatic control of system by varying the rate of flow of the fluid. The complete control valve consists of an actuator and a valve body. The actuator provides the power to vary the port opening in the valve body. These are classified on the basis of the means of actuation, notably pneumatic or electrically, the style of valve body design and the characteristics of the valve. These valves are extensively used for controlling fluid flow in any system.

8.8.1.2 Electrical type

Electrical actuators for control system fall into two basic categories – (1) Two position actuators include solenoid and relay actuation (2) Infinite-position actuators include reversible electric motor drives, saturable core reactors, Silicon Controlled Rectifiers (SCRs), variable-speed electric motor drives and electro-pneumatic converters.

Linear actuators are basically transducers which convert electrical signal into linear motion. Electro-magnetic actuators are on-off devices. *Eg.* Electrical brakes. Advent of electronics in Control Systems has made Electromagnetic actuators very important. Advantages of electromagnetic actuators are (1) Control is easy while the device is operating (2) Speed and characteristics can be controlled.

Electromagnetic actuators are gaining grounds because of the involvement of electronics and data processing. Thus sensors, transducers and actuators form a complete configuration to achieve close loop system in any control systems.

8.9 CLASSIFICATION

Transducers can be found both at the input as well as at the output stage of a measuring system. The input transducer is called the sensor, because it senses the desired physical quantity and converts it into another energy form. The output transducer is called the actuator, because it converts the energy into a form to which another independent system can react, whether it is a biological system or a technical system. So, for a biological system the actuator can be a numerical display or a loudspeaker to which the visual or aural senses react respectively. For a technical system the actuator could be a recorder or a laser, producing holes in a ceramic material. The results can be interpreted by humans.

Another important characteristic related to transducers implies the division of transducers into two categories, because some transducers require an auxiliary energy source to become operational. Hence a difference is made between modulating and self-generating transducers.

A modulating transducer is defined as a transducer which requires an auxiliary energy source to convert energy from one domain into another. *Eg.* Strain gauge and Linear Variable Differential Transformer (LVDT). This type of transducer requires an electrical energy source to become operational. The electrical current flowing in the strain gauge is modulated by a mechanical force which is converted into an electrical voltage change via a change of the resistive elements. As a consequence the input energy modulates the energy of the auxiliary source. At the output of a system a liquid-crystal display (LCD) is an example of a modulating transducer, because again an auxiliary energy source is required to perform the correct transduction.

A self generating transducer is defined as a transducer which requires no auxiliary energy source to convert energy. *Eg.* Solar cell, Thermocouple and Piezoelectric Crystal Transducer, and at the output one can find for instance, a heating element converting electrical energy into thermal energy without the need of an auxiliary energy source.

Transducers are also classified (Fig. 8.6) on the basis of Transduction (*Eg*. Capacitive, Inductive and Resistive), Material and technology (*Eg*. MEMS, Fibre Optic and Silicon Micro Technology), Primary Input Quantity (*Eg*. Temperature, Pressure and Flow) and various Applications (*Eg*. Environmental and Biomedical). Most important type of transducers is Resistive transducers which include Potentiometers, Thermistors, Light Dependent Resistor (LDR) and Strain Gages Resistance Temperature Detector (SGRTD).

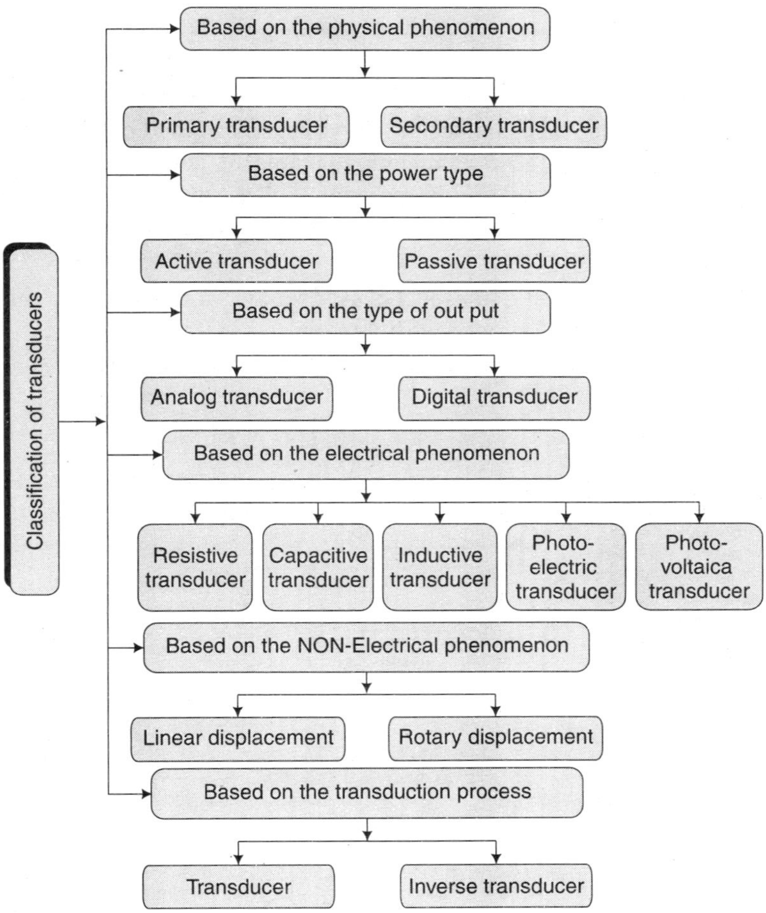

Fig. 8.6 *Classification of transducers.*

Some of the factors influencing the transducers include: Shape, Weight, Size, Sensitivity, Accuracy, Frequency response, Reliability, Availability, Cost, Impedance, Type of cable, Static characteristics, Ruggedness and Environmental compatibility.

8.9.1 Piezoelectric Transducers

Transducer is a device that converts energy from one form to another, at present, piezoelectric material is commonly used as a basic component of transducers. Piezoelectric devices are a very reliable and inexpensive means of converting electrical energy into physical motion and exhibit a high tolerance to environmental factors such as electromagnetic fields and humidity.

The active element is the heart of the transducer as it converts the electrical energy to acoustic energy and vice versa. The active element is basically a piece of polarized material (positively charged molecule and the negatively charged molecule) with electrodes attached to two of its opposite faces. When an electric field is applied across the material, the polarized molecules will align themselves with the electric field, resulting in induced dipoles within the molecular or crystal structure of the material. This alignment of molecules will cause the material to change dimensions and this phenomenon is identified as Electrostriction.

Piezoelectric element is a 'crystal' which delivers a voltage when mechanical force is applied between its faces and it deforms mechanically when voltage is applied between its faces. Because of these characteristics a piezoelectric element is capable of acting as both a sensing and a transmitting element. Permanently-polarized material such as quartz (SiO_2) or barium titanate ($BaTiO_3$) will produce an electric field when the material changes dimensions as a result of an imposed mechanical force. This phenomenon is known as the piezoelectric effect.

The active element of most acoustic transducers used today is a piezoelectric ceramic (Fig. 8.7), which can be cut in various ways to produce different wave modes. A large piezoelectric ceramic element can be seen in the image of a sectioned low frequency transducer.

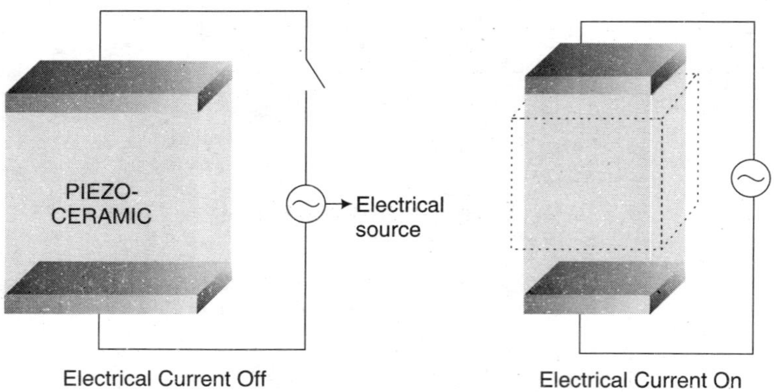

Fig. 8.7 *Piezoelectric ceramic electrical current Off and On process.*

Couplant is a material (usually liquid) (Fig. 8.8) that facilitates the transmission of ultrasonic energy from the transducer into the test specimen. Couplant is generally necessary because the acoustic impedance mismatch between air and solids (i.e. such as the test specimen) is large. Therefore, nearly all of the energy is reflected and very little is transmitted into the test material.

The couplant displaces (Fig. 8.8) the air and makes it possible to get more sound energy into the test specimen so that a usable ultrasonic signal can be obtained. In contact ultrasonic testing a thin film of oil, glycerin or water is generally used between the transducer and the test surface. When scanning over the part or making precise measurements, an immersion technique is often used. In immersion ultrasonic testing both the transducer and the part are immersed in the couplant, which is typically water. This method of coupling makes it easier to maintain consistent coupling while moving and manipulating the transducer and the part there of (Fig. 8.9).

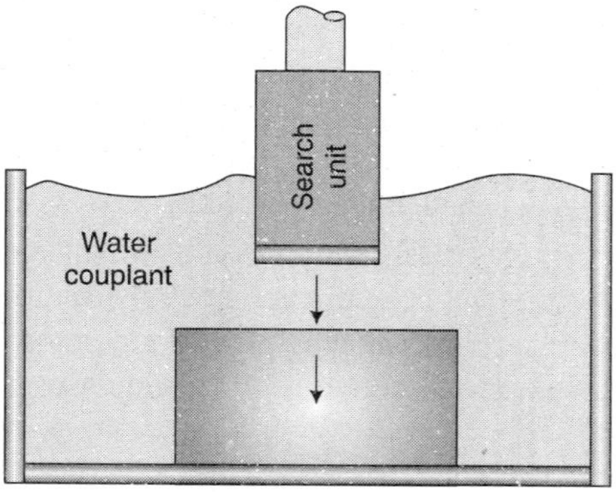

Fig. 8.8 *Schematic representation of piezoelectric couplant.*

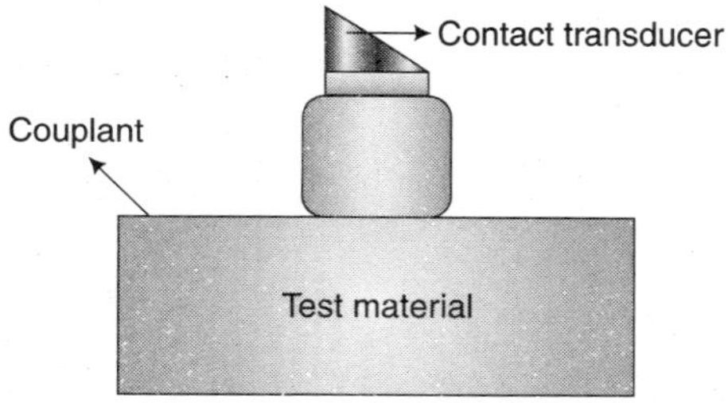

Fig. 8.9 *Piezoelectric couplant in testing process.*

Earlier the introduction of piezoelectric ceramics in the early 1950's, piezoelectric crystals made from quartz crystals and magnetostrictive materials were used. The active element is still sometimes referred to as the crystal by old timers in the NDT field. When piezoelectric ceramics were introduced, they soon became the dominant material for transducers due to their good piezoelectric properties and their ease of manufacture into a variety of shapes and sizes. They also operate at low voltage and are usable up to about 300°C.

The first piezoceramic used was barium titanate and that was followed during the 1960's by lead zirconate titanate compositions, which are now the most commonly employed ceramic for making transducers. New materials such as piezo-polymers and composites are also being used in some applications.

The thickness of the active element is determined by the desired frequency of the transducer. A thin wafer element vibrates with a wavelength that is twice its thickness. Therefore, piezoelectric crystals are cut to a thickness that is 1/2 the desired radiated wavelength. The high frequency contact transducers are not produced because the element is very thin and too fragile.

Piezoelectricity is a phenomenon in which positive and negative electric charges appear on opposite sides of some non-conducting crystals when subjected to mechanical pressure. The converse piezoelectric effect, electrostriction, is the property of some non-conductors, or dielectrics, that deform slightly under the application of an electric field. Piezoelectricity exists because some atomic lattice structures have an essential cell, a cubic or rhomboid atomic cage, and this cage holds a semi-mobile ion which has several stable quantum position states inside. Piezoelectric transducers have been conventionally used to convert electric signals into sound waves or other mechanical vibrations, or to convert mechanical vibrations into electric signals.

A piezoelectric transducer includes a vibrating piece which has on both its surfaces electrodes for providing an electric field to the vibrating piece. The piezoelectric transducer converts electric signals into mechanical vibrations or vice versa by utilizing the morphological change of a crystal which occurs on voltage application, or conversely by monitoring the voltage generated by a pressure applied on a crystal.

8.9.1.1 Characteristics

The transducer is a very important part of the ultrasonic instrumentation system. The transducer incorporates a piezoelectric element, which converts electrical signals into mechanical vibrations (transmit mode) and mechanical vibrations into electrical signals (receive mode). Many factors, including material, mechanical and electrical construction and the external mechanical and electrical load conditions, influence the behavior of a transducer. Mechanical construction includes parameters such as the radiation surface area, mechanical damping, housing, connector type and other variables of physical construction. As of this writing, transducer manufacturers are hard pressed when constructing two transducers that have identical performance characteristics.

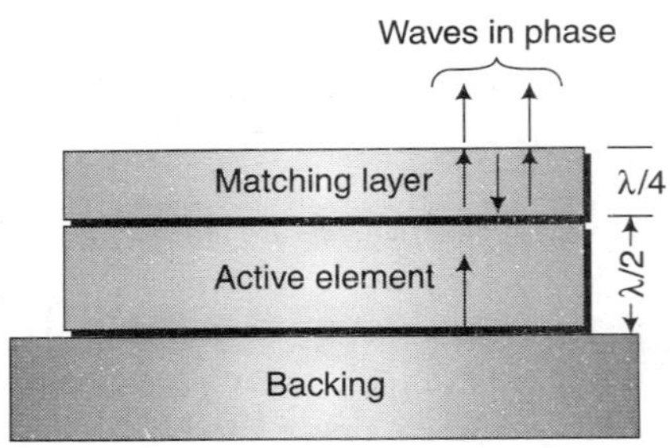

Fig. 8.10 *A cut away of a typical contact transducer.*

A cut away of a typical contact transducer is shown in Fig. 8.10. The piezoelectric element is cut to 1/2 the desired wavelength. To get as much energy out of the transducer as possible, an impedance matching is placed between the active element and the face of the transducer. Optimal impedance matching is achieved by sizing the matching layer so that its thickness is 1/4 of the desired wavelength. This keeps waves that were reflected within the matching layer in phase when they exit the layer (Fig. 8.10). For contact transducers, the matching layer is made from material that has acoustical impedance between the active element and steel, and also incorporates a wear plate to protect the matching layer and active element from scratching. Immersion transducers have a matching layer with acoustical impedance between the active element and water.

The backing material supporting the crystal has a great influence on the damping characteristics of a transducer. Using a backing material with impedance similar to that of the active element will produce the most effective damping. Such a transducer will have a wider bandwidth resulting in higher sensitivity. As the mismatch in impedance between the active element and the backing material increases, material penetration increases but transducer sensitivity is reduced.

8.9.1.2 Piezoelectric transducers series

a. 9000 SERIES TRANSDUCER

The 9000 Series Transducer is designed for operation in air at ultrasonic frequencies and it is a closed faced transducer with a rugged housing and construction that makes it ideal for harsh environments.

b. K – SERIES PIEZOELECTRIC TRANSDUCERS

The '*K*' Series Transducers offer the design engineer a unique blend of size, acoustical performance and environmental ruggedness when developing their ultrasonic sensor product. This line of transducers is specifically intended for operation in air at ultrasonic frequencies. These are used closed-faced, metal case construction with a resin filled back to ensure their integrity in antagonistic environments. The Pulse Transceivers incorporate internal mechanical damping techniques for improved temporal ring control and the high frequency transducers are coated with a special acoustic material for optimum efficiency and impedance matching to air.

c. L – SERIES PIEZOELECTRIC TRANSDUCERS

The '*L*' Series Transducers are designed with a unique blend of size, acoustical performance when developing their ultrasonic sensor product. These are specifically intended for operation in air at ultrasonic frequencies and all are used screen mesh open face construction housed in a metal case, offering increased sensitivity for non-hostile environments.

8.9.1.3 Applications

i. In particular, piezoelectric diaphragms have been employed as pressure sensors, in speakers for audio equipment, fluid ejection, fluid pumping and printing applications.

ii. Piezoelectric transducers commonly employ a ceramic piezoelectric element having electrodes on its oppositely polarized surfaces.

iii. A change of stress in the piezoelectric element produces a temporary potential difference across the electrodes.

iv. Piezoelectric ceramic transducers have found many important applications in adaptive structures for vibration control and acoustic noise suppression in modern space, civilian and military systems, such as launch vehicles, space platforms, aircraft, submarines and helicopters.

8.9.2 Electromagnetic Acoustic Transducers (EMATs)

One of the essential features of ultrasonic measurements is mechanical coupling between the transducer and the solid and it is achieved in one of two ways: (1) In immersion measurements, energy is coupled between the transducer and sample by placing both objects in a tank filled with a fluid, generally water. (2) In contact measurements, the transducer is pressed directly against the sample and coupling is achieved by the presence of a thin fluid layer inserted between the two. When shear waves are to be transmitted, the fluid is generally selected to have a significant viscosity.

Electromagnetic acoustic transducers (EMAT) act through different physical principles and do not need couplant. When a wire is placed near the surface of an electrically conducting object and is driven by a current at the desired ultrasonic frequency, eddy currents will be induced in a near surface region of the object. If a static magnetic field is also present, these eddy currents will experience Lorentz forces of the form

$$F = J \times B \qquad \text{... eq.-8.1}$$

Where F is the body force per unit volume, J is the induced dynamic current density and B is the static magnetic induction.

The most important application of EMATs has been in non-destructive evaluation (NDE) applications such as flaw detection or material property characterization. Couplant free transduction allows operation without contact at elevated temperatures and in remote locations. The coil and magnet structure can also be designed to excite complex wave patterns and polarizations that would be difficult to realize with fluid coupled piezoelectric probes. In the inference of material properties from precise velocity or attenuation measurements, using EMATs can eliminate errors associated with couplant variation, particularly in contact measurements.

Number of practical EMAT configurations are shown below (Figs. 8.11-8.15). In each, the biasing magnet structure, the coil, and the forces on the surface of the solid are shown in an exploded view. The first three configurations will excite beams propagating normal to the surface of the half-space and produce beams with radial, longitudinal, and transverse polarizations, respectively (Figs. 8.11-8.13). The final two use spatially varying stresses to excite beams propagating at oblique angles or along the surface of a component (Figs. 8.14-8.15). Although a great number of variations on these configurations have been conceived and used in practice, consideration of these three geometries should suffice to introduce the fundamentals.

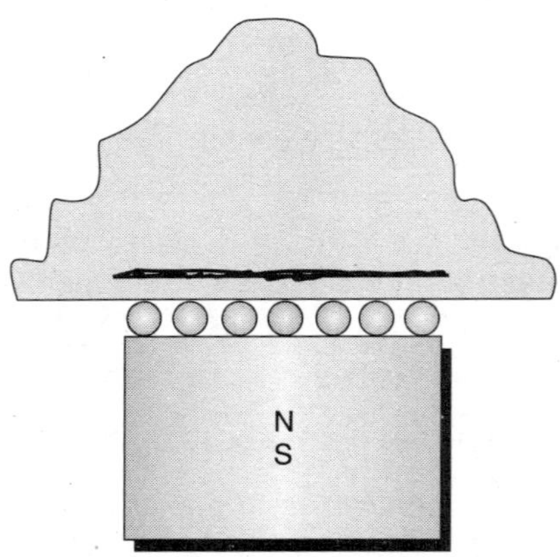

Fig. 8.11 *Cross-sectional view of a spiral coil EMAT exciting radially polarized shear waves propagating normal to the surface.*

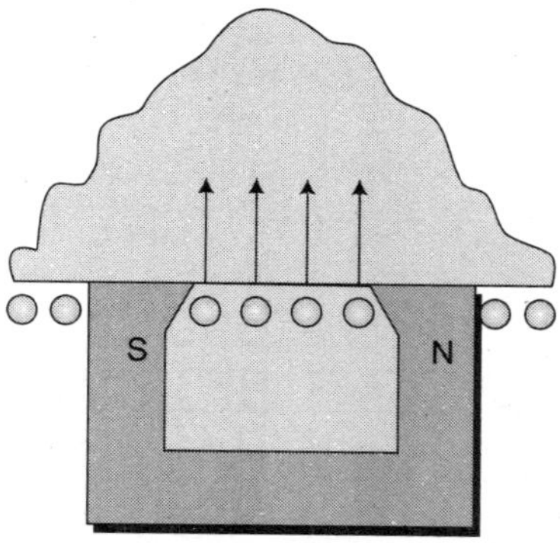

Fig. 8.12 *Cross-sectional view of a tangential field EMAT for exciting polarized longitudinal waves propagating normal to the surface.*

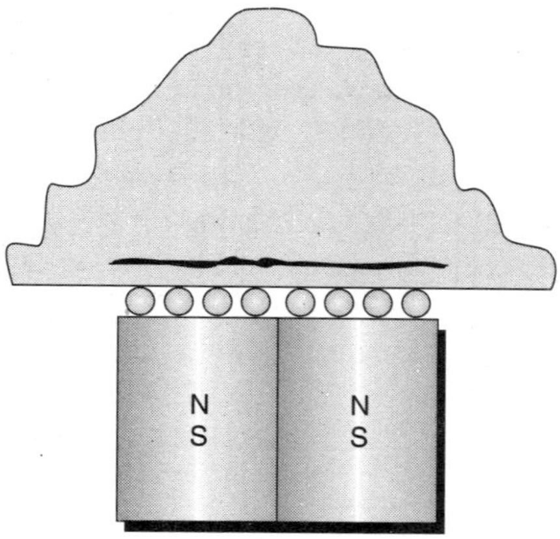

Fig. 8.13 *Cross-sectional view of a normal field EMAT for exciting plane polarized shear waves propagating normal to the surface.*

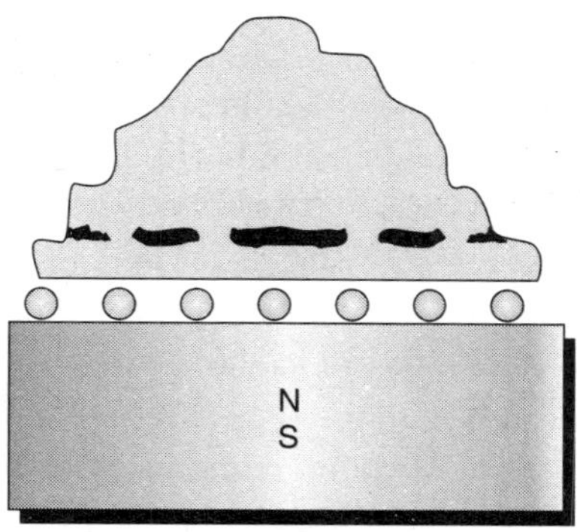

Fig. 8.14 *Cross-sectional view of a meander coil EMAT for exciting obliquely propagating L or SV waves, Rayleigh waves, or guided modes (such as Lamb waves) in plates.*

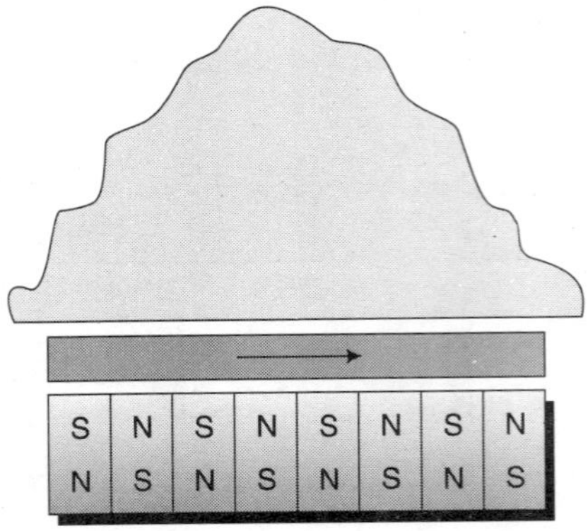

Fig. 8.15 *Cross-sectional view of a periodic permanent magnet EMAT for exciting grazing or obliquely propagating horizontally polarized (SH) waves or guided SH modes in plates.*

Practical EMAT designs are relatively narrow band and require strong magnetic fields and large currents to produce ultrasound that is often weaker than that produced by piezoelectric transducers. Rare earth materials such as Samarium-Cobalt and Neodymium-Iron-Boron are used to produce sufficient strong magnetic fields, which may also be generated by pulsed electromagnets.

The EMAT offers many advantages based on its couplant free operation. The advantages include the abilities to operate in remote environments at elevated speeds and temperatures, to excite polarizations not easily excited by fluid coupled piezoelectrics and to produce highly consistent measurements. These advantages are tampered by low efficiencies and careful electronic design is essential to applications.

8.9.3 Molecular Electronic Seismometers (MES)

The Russian Prince Golicyn developed the first seismometer at the beginning of the 20th century. It was a very simple instrument and looked like a stylus attached to a mass supported by spring. Later, electro or opto-mechanical converters were introduced which provide significant amplification of signals and allow for correction of the frequency response. Despite the enormous advances of science and technology in the 20th century, this traditional scheme of the seismometer hasn't changed for the most part. Pendulum based oscillating systems are delicate, expensive, precision instruments, which are usually installed and maintained by qualified personnel. These seismometers operate perfectly at high frequencies, but at the lowest frequencies the disadvantages of the pendulum based seismometers become extremely visible.

The instability of the natural frequency caused by temperature variations, inner convective flows, electromagnetic interactions between different mechanical parts of the seismometer and slight deviations of the angle of installation are lead to a dramatic loses of stability of the device.

Another approach to seismometer design is to simplify the oscillating system by restraining the pendulum and increasing the value of its natural damping. The increase of the damping plays the same role as the feedback mechanism that is usually used for expanding the frequency pass band of any recorder. Simultaneously, it is necessary to increase the conversion factor of the transducer, since the large value of damping leads to an essential decreasing of the sensitivity of oscillating system. This task has been successfully solved using the molecular electronic transducers (MET). It is a combination of an inertial mass, sensitive element and amplifier of the external acceleration with a power gain of 10^6 and large damping. The output current of an MET transducer is proportional to the velocity of the ground motion, since its transfer function is inversely proportional to the frequency. As for the mechanical system of the sensor, it represents a cylindrical tube filled with an electrolyte and provided with elastic membranes on the top and bottom of the tube (vertical sensor), or a toroid partially filled with electrolyte (horizontal sensor). In both cases, the electrolyte (10 – 15 g) plays a role of inertial mass and the hydraulic impedance of the transducer ~ (10^9 N*s/m^5) provides the necessary value of damping. Thus a three-component molecular electronic seismometer is light and compact since it does not require the use of large solid suspended inertial mass, handy since it contains no moving parts and inexpensive since no expensive materials or complex manufacturing processes are used.

MET needs power since it works as a vacuum triode. When a voltage source (~300 mV) is connected an electric current starts flowing between the anode and cathode and in the external bridge circuit. The value of this current depends on the charge concentration and the distance between the anode and cathode. The distance between the anode and cathode for the real transducer usually varies in the limits of 100 – 200 μm. The charge concentration may be varied randomly; as a result, the current variations are within the limits 10 μA – 1.5 mA. The value of this electric current is critical to the design of the support electronic circuit. The circuit design must balance a low self noise against the greater cost of components need to achieve it. Several electronic boards have been designed for different applications and it is possible to reduce seismometer's current to 100 μA and these have important in ocean bottom applications.

The design of MET seismometers makes them insensitive to the angle of installation, up ~ 5 deg. for vertical and 15 deg. for horizontal sensors. This is very important for the borehole version and it is a result of the fact that the liquid inertial mass (water in a cylinder) has a tendency to maintain a horizontal position.

The temperature range of operation is limited by freezing point of electrolyte. The low temperature version of MET seismometer uses electrolyte with some special agents and has freezing point about –55°C. It is very important that the MET technology based seismometers are extremely suitable for mass production.

8.9.4 Semiconductor Sensors

The semiconductor sensor type belongs to the class of solid state gas sensors and these are made with SnO_2, ZnO, and $BaSnO_3$ and used to detect carbon monoxide.

Semiconductor gas sensors measure the change in the conductivity of the material due to the presence of gas concentration in the environment. The principle of the conductivity change is the adsorption and desorption of the sensed gas on the surface of the semiconductor. According to the metal oxide have surface states where oxygen can be adsorbed in the form of oxygen ions (O^- or O^{2-}). Adsorption occurs because the atoms at the surface of the solid cannot satisfy their valency requirements and thus adsorb from the gas phase to reduce the surface energy of the solid. This process is called physisorption because the acting forces are the very weak van der Waals forces, in this process, no charge exchange has taken place and the semiconductor is still of low resistance. The energy states of the oxygen ions lie below the lower edge of the conduction band. In the next step, the oxygen ions bind electrons from the conduction band and it develops a depletion layer. Due to this, the semiconductor film becomes highly resistive and is sensitized for reducing gases in the environment. If there are reducing gases in the vicinity then the previously adsorbed oxygen reacts with the reducing gas. Leading to the release of some of the electrons into the conduction band of the semiconductor, thus the depletion layer diminishes and the conductivity of the semiconductor surface increases. The last step is the desorption of the reaction products from the surface making it available for the fresh adsorption of oxygen. When there is no reducing gas in the environment of the surface, the conductivity of the sample will return to a dynamic equilibrium which is given by:

$$\sigma = \sigma_0 + f.\ P^n \qquad \text{... eq.-8.2}$$

where σ_0 = Denotes the conductance of the sample in air, f = Constant which is specific for a certain type of semiconductor sensor and P^n = The partial pressure (respectively the concentration) of the reducing gas.

The equation indicates that when there is no reducing gas around the sensor it returns to its undisturbed state, this means it is a reversible process. The actual change in conductivity follows an exponential function, where the change in conductivity is given by an equation:

$$\sigma = A \exp(-E_A/kT)P^{1/N} \qquad \text{... eq.-8.3}$$

where σ is the conductivity, A is a constant, E_A is the activation energy for conduction, p is the partial pressure of the reducing gas and N is a constant which is determined by the type of material used.

For the two most widely used semiconductors in gas analysis, ZnO and SnO_2, the conductivity of the semiconductor film depends on the quantity of adsorbed oxygen on their surface and the adsorbed oxygen is a function of the concentration of the sample gas. Due to the fact that the sensing mechanisms are all surface correlated it is possible to change the features of a semiconductor sensor by changing the surface characteristics of the sensor. This is done by applying catalysts which change the behaviour of the surface physics in a 'controlled' manner. By using catalysts it is possible to increase the selectivity of the sensor for a certain gas, and to increase its long term reliability.

8.9.4.1 Catalysis

The selectivity and sensitivity of the sensor is influenced by changing the reaction of the sample gas with the oxygen, the actual release of electrons to the conduction band and desorption of the reaction product. Altering of these steps is achieved by using appropriate catalysts. A catalyst is a substance which increases the rate at which a chemical reaction approaches equilibrium and it is important to mention that the catalyst itself is not permanently affected by the reaction. The behaviour of the semiconductor sensor can be changed by varying the catalyst, the physical form of the oxide, the particle size and the distribution of particles on the surface. Using a catalyst, it is possible for the reaction to take an alternative route from the reactants to the product which is more energy efficient.

One possible effect of a catalyser is to change the sensitivity of the sensor. Here it is mainly the particle size and the distribution of the particles which have the dominant effect since they alter the effective area of possible adsorption sites for the oxygen. The influence of the amount of catalyst on the sensitivity is shown in Fig. 8.16 for a SnO_2 sensor which is doped with different amounts of palladium as a catalyser. The two graphs show the dependency of the sensitivity in relation to changing temperature and concentration respectively.

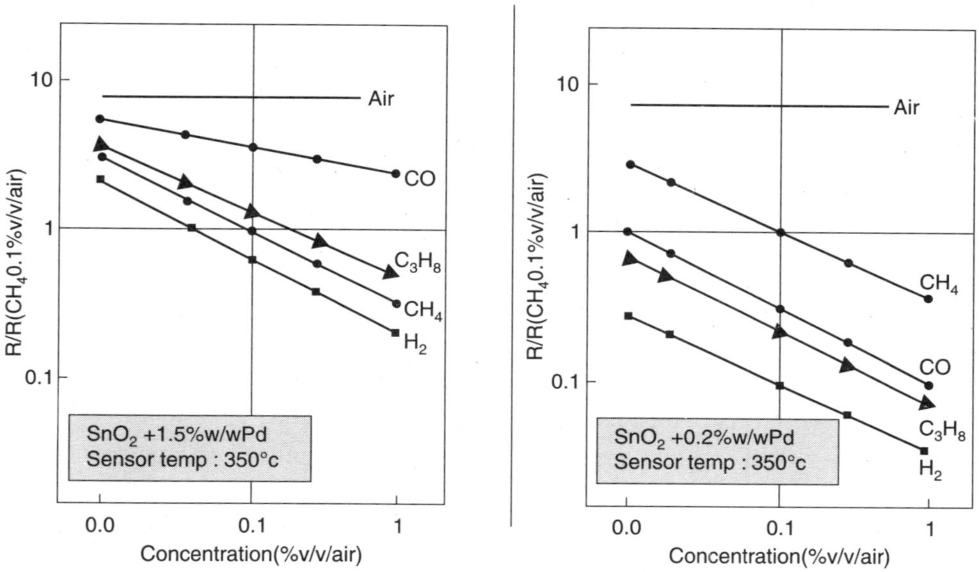

Fig. 8.16 *Graph of sensitivity of semiconductor sensors.*

The maximum sensitivity is dependent on the gas as well as on the oxides used. Other typical features of these sensors are that water vapour affects both the conductivity in air and the sensitivity to other gases.

8.9.4.2 Specifications

The parameters governing the performance of a semiconductor gas sensor are discussed with particular emphasize to the parameter specifications outlined.

1. Sensitivity

The sensitivity of semiconductor sensors for gases like H_2S, CO, NO_2 or C_2H_5OH are in the range of parts per million (ppm). As mentioned above, the sensitivity can be altered by using an appropriate catalyst which changes the number of adsorption sites for oxygen and the distribution of the sites on the surface of the semiconductor material.

2. Response time

The response time varies between milliseconds to minutes, for a typical semiconductor sensor the response time is typically in the order of minutes. The specification for a response time for a semiconductor sensor can be quite difficult, because these sensors exhibit quite regularly a fast initial response to a step change in concentration but have a very long "tail" before they reach their steady state.

3. Selectivity

In discussing the selectivity of semiconductor sensors, it is obvious that they show strong interference to other gases. This strong interference is due to their working principle. Theoretically any gas which is able to reduce the adsorbed oxygen on the solid surface is an interference factor and reduces therefore the selectivity of the sensor. However, there has been a lot of research on increasing the selectivity of the sensor by the use of different catalysts. This research however has not yet solved this problem.

4. Stability

The stability of these sensors is heavily dependent on the conditions at the surface and therefore on the environment it is held in. For instance, the sensitivity of the sensor changes due to the stability of the crystal grain size of the catalyst, which determines the area for effective adsorption sites. Furthermore, it is possible for the adsorbed oxygen to be bound strongly to the crystal lattice and hence be unable to desorb. A further reason for loss of sensitivity is surface poisoning which means that oxygen can no longer adsorb on the surface.

Temperature and humidity affect the stability of the above mentioned features in a major way because both of these effects change the resistivity of the oxides and therefore the conductance.

The main reason that this type of gas sensor has not yet been realized in large quantities for the quantitative measurement of gases is mainly due to the fact that there are too many parameters which influence the adsorption and desorption processes at the surface. Further, the basic mechanisms of catalysis and surface interactions are not yet well understood and the existing techniques for the better understanding of the technology are not used extensively. To achieve a better understanding of the geometrical, electric and dynamic processes on the sensor surface several techniques are used, for example, XRD (X-Ray Diffraction), SAM (Scanning Auger Microscopy), SEM (Scanning Electron Microscopy) and many more which are listed in. However, these are mostly used in a laboratory environment.

The semiconductor type gas sensor is not particularly suitable for the specific application of measuring the transfer factor of the lungs dynamically. It exhibits too slow a response time and suffers a strong negative interference when used in a gas sample which is saturated by water.

8.9.5 Conductometric Transducers

Enzyme reactions, which produce or consume ionic species, will, depending on the total ionic strength of the media, change the conductance/capacitance of the solution to a greater or lesser extent. Various planar inter digitated electrode configurations have been reported as conductometric transducers for biosensors. These have been used in combination with a variety of possible enzyme systems, which can confer specificity to this type of transducer.

Enzymes such as urease, which catalyze the production of ionic species, have been used in these devices. Nevertheless, because the conductance is sensitive to temperature, Faradaic processes, double layer charging and concentration polarization, differential methods with internal controls must be used. There are a variety of possible enzyme systems, which can confer specificity to this type of transducer. Urease-coated electrodes in combination with a control electrode coated with an inactive protein have been used to measure the decomposition of urea to ammonium, bicarbonate and hydroxyl ions.

The primary advantage of this technique is in the use of inexpensive, reproducible and disposable sensors. The main disadvantage is that ionic species produced must significantly change the total ionic strength to obtain a reliable measurement. This requirement increases the detection limit to unacceptable levels and results in potential interferences from variability in the ionic strength of the sample.

Thin film interdigitated planar conductometric electrodes have been used to measure heavy metals. Glucose oxidase, Alcohol oxidase, Butyril oxidase and Urease have been immobilized on transducer surfaces and used as bioactive elements for detection of Ag^+, Hg^{+2} and Pb^{2+}. Dynamic concentration ranges of 1-100 μM have been obtained for these devices.

8.9.6 Semiconductors

The semiconductor temperature sensor is an electronic device fabricated in other modern electronic semiconductor components such as microprocessors. Hundreds or thousands of devices are formed on thin silicon wafers and these sensors are available from a number of manufacturers. There are no generic types as with thermo couple and RTDs, although a number of devices are made by more than one manufacturer. The AD590 and the LM35 are the most popular devices, but over the last few years better alternatives have become available. These sensors share a number of characteristics: linear outputs, relatively small size, limited temperature range (–40 to +120°C typical), low cost and good. Often the semiconductor temperature sensors are not well designed thermally with the semiconductor chip not always in good thermal contact with an outside surface. Some devices are inclined to oscillate unless precautions are taken. Provided the limitations of the semiconductor temperature sensors are understood, they can be used in many applications.

The most popular semiconductor temperature sensors are based on the fundamental temperature and current characteristics of the transistor. If two identical transistors are operated at different but constant collector current densities, then the difference in their base emitter voltages is proportional to the absolute temperature of the transistors. This voltage difference is then converted to a single ended voltage or a current. The semiconductor temperature sensor is best suitable for embedded applications because they tend to be electrically and mechanically more delicate than most other temperature sensor types, however they do have valid application in many areas.

8.9.6.1 Types

The semiconductor temperature sensors can be grouped into five broad categories: voltage output, current output, resistance output, digital output and simple diode types.

i. Voltage output temperature sensors

These sensors provide a voltage output signal with relatively low output impedance, require an excitation power source and essentially linear.

Table 8.2 Examples of voltage output temperature sensors.

Sensor	*Out put*	*Tolerance (Range)*	*Package*	*Manufacture*
AD22100	22.5 mV/°C at tV 250 mV Offset	±2°C & ±14°C (−50 °C to +150°C)	TO-92 SO-8	Analog Devices
LM 135 LM 235 LM335	10 mV/°C	±2.7°C & ±9°C (−55°C to 150°C −40°C to 100°C)	TO-92 TO-46	National Semi Linear Tech.

The symptom is an offset in the sensors output - something which is not always obvious. It is wise to always include the manufacturer's recommended resistor - capacitor network close to the sensor.

ii. Current output temperature sensors

The current output sensors act as a high-impedance, constant current regulator typically passing 1 micro-amp per degree Kelvin and require a supply voltage of between 4 and 30 V.

Table 8.3 Examples of current output temperature sensors.

Sensor	*Out put*	*Tolerance (Range)*	*Package*	*Manufacture*
AD590	1μ A/°K	±5.5°C & ±10°C (−55°C to +150°C)	TO-52	Analog Devices
AD592	1μ A/°K	±1°C & ±3.5°C (−25°C to +105°C)	TO-92	Analog Devices

iii. Digital output temperature sensors

The digital temperature sensor is the first sensor to integrate a sensor and an analog to digital converter (ADC) on to a single silicon chip. In general, these sensors do not lend themselves for use with standard measuring devices because of their non standard digital interfaces. Many are designed specifically for the thermal management of microprocessor chips. A selection of representative devices given in Table 8.4.

Table 8.4 Examples of digital output temperature sensors.

Sensor	*Out put*	*Tolerance (Range)*	*Package*	*Manufacture*
LM95071	14 bit SPI	± 2°C (−45°C to +150°C)	SOT-5	National Semi Linear Tech.
LM56	2 comparators with suitable thresholds	± 3°C & ± 4°C (−40°C to +125°C)	SOP-8 MSOP-8	National Semi Linear Tech.

The Analog Devices parts employ a sigma-delta ADC that produces continuous pulse stream output with a mark space ratio, which is proportional to the temperature. This makes for easy interfacing to a microprocessor and also for isolating by optical or other means. The same signal could also be passed through a low pass filter to generate an analog voltage.

The Dallas DS2435 goes beyond that of a sensor plus ADC by providing simple data reduction using an eight bin time/temperature histogram with definable bin boundaries. It appears to have been specifically designed for battery management, but other application could include food transport monitoring, machine use monitoring. This sensor demonstrated the way of the future in sensor technology where sensor, ADC, memory and microcontroller are integrated to form an application specific task very cost effectively.

iv. Resistance output silicon temperature sensors

The temperature vs. bulk resistance characteristics of semiconductor materials allow the manufacture of simple temperature sensors using standard silicon semiconductor fabrication equipment (Table 8.5). This construction can be more stable than other semiconductor sensor, due to the greater tolerance to ion migration. However, other characteristics require that care must be taken in using these sensors.

Table 8.5 Examples of resistance output silicon temperature sensors.

Sensor	*Out put*	*Tolerance (Range)*	*Package*	*Manufacture*
KTY81	1K or 2K at 25°C	±1°C to ±12°C	SOD-70	Phillips
KTY82		(–55°C to +150°C)	SOT-23	
KTY83			SOD-68	

The silicon temperature sensor's resistance is given by the equation:

$$R = R_r\,(1 + a.\,(T - T_r) + \text{b}.\,(T - T_r)^2 - \text{c}.\,(T - T_i)^d) \qquad \textbf{... \textit{eq.}-8.4}$$

where R_r is the resistance at temperature T_r and a, b, c and d are constants. T_i is an inflection point temperature such that c = 0 for $T < T_i$.

The resistance of some of these semiconductor sensors is dependent on the excitation current (due to current density effects in the semiconductor) and the polarity of the applied voltage. As with other non-passive temperature sensors, self-heating can induce errors. There are a number of specialized cryogenic temperature sensors that use resistive semiconductor sensor elements made from silicon and germanium.

v. Diode temperature sensors

The ordinary semiconductor diode may be used as a temperature sensor (Fig. 8.17). The diode is the lowest cost temperature sensor and can produce more than satisfactory results if you are prepared to undertake a two point calibration and provide a stable excitation current.

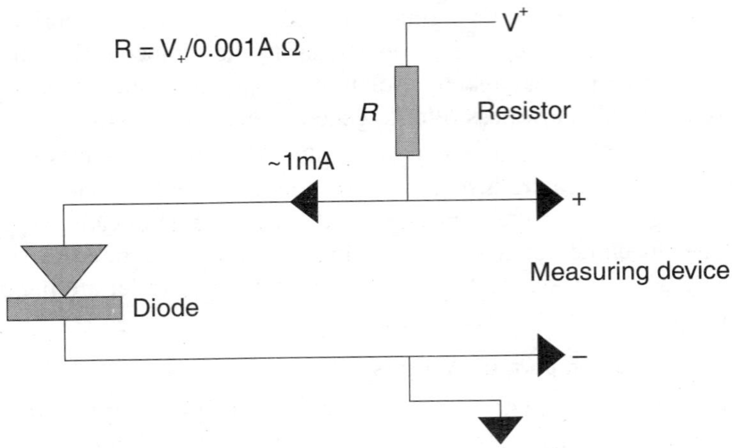

Fig. 8.17 *Diode as a temperature sensor.*

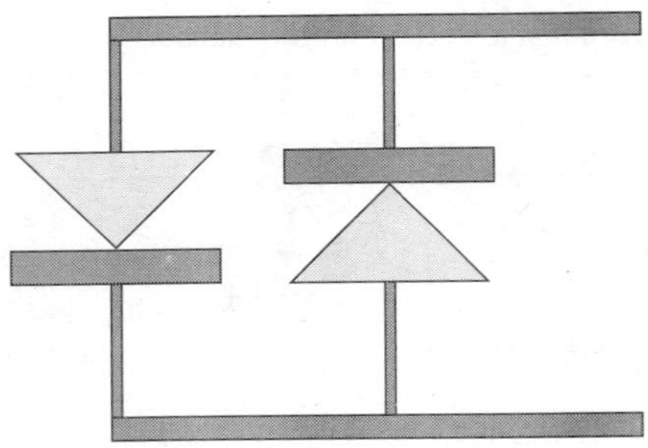

Fig. 8.18 *Back to back diodes.*

Almost any silicon diode is ok. The forward biased voltage across a diode has a temperature coefficient of about 2.3mV/°C and is reasonably linear. The measuring circuit is simple as shown to the right. The bias current should be held as constant as possible - using constant current source or a resistor from a stable voltage source. Without calibration, the initial error is likely to be too large, in the order of ±30°C and is largest of all the contact type temperature sensors. This initial error is greatly reduced if sensor grade parts are used.

One advantage of the diode as a temperature sensor is that it can be electrically robust-tolerant to voltage spikes induced by lightning strike. This is particularly true if power diodes (e.g. the common 1N4004) are used and a second back to back diode (Fig. 8.18) is used to limit power dissipation during high peak currents.

The transistor sensor is used in diode mode by connecting the base and collector together. If this is not done, the sensor is wired between base and emitter and the excitation current reduced by a factor of about 100. The result is a very low power, sensitive and linear sensor. The simplicity and performance of the sensor is under valued.

To improve the performance of the diode as a temperature sensor, two diode voltages (V_1 and V_2) can be measured at different currents (I_1 and I_2), typically selected to be about 1:10 ratio. The absolute temperature can be calculated from the equation:

$$T = (V_1 - V_2)/(8.7248 \times 10^{-5} \ln (I_1/I_2)) \quad \text{... eq.-8.5}$$

The result is in degree Kelvin (K). This is the method employed by most integrated circuit temperature sensors and explains why some output a signal proportional to absolute temperature.

8.9.7 Contact Transducers

Contact transducers are used for direct contact inspections, and are generally hand manipulated (Fig. 8.19). They have elements protected in a rugged casing to withstand sliding contact with a variety of materials.

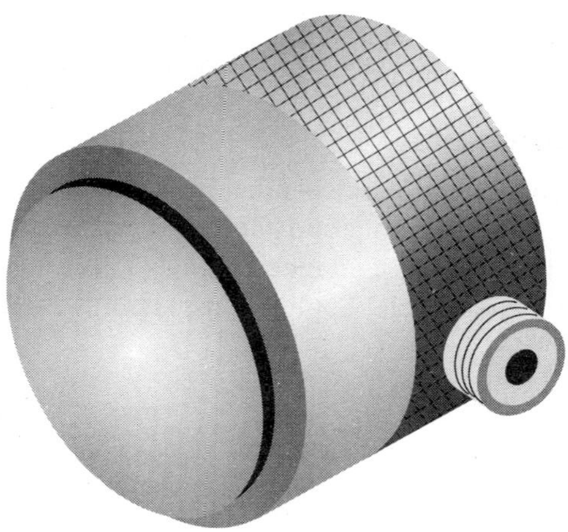

Fig. 8.19 *Contact transducers.*

These transducers have an ergonomic design so that they are easy to grip and move along a surface. They often have replaceable wear plates to lengthen their useful life. Coupling materials of water, grease, oils, or commercial materials are used to remove the air gap between the transducer and the component being inspected.

8.9.8 Immersion Transducers

Immersion transducers do not contact the component. These transducers are designed to operate in a liquid environment and all connections are watertight. Immersion transducers usually have an impedance matching layer that helps to get more sound energy into the water and, in turn, into the component being inspected (Fig. 8.20).

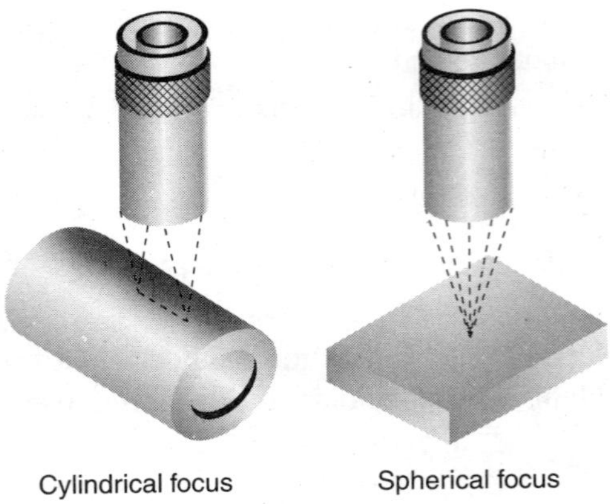

Fig. 8.20 *Immersion transducers.*

Immersion transducers can be purchased with a planer, cylindrically focused or spherically focused lens. A focused transducer can improve the sensitivity and axial resolution by concentrating the sound energy to a smaller area. Immersion transducers are typically used inside a water tank or as part of a squinter or bubbler system in scanning applications.

8.9.9 Dual Element Transducers

Dual element transducers contain two independently operated elements in a single housing (Fig. 8.21). One of the elements transmits and the other receives the ultrasonic signal. Active elements can be chosen for their sending and receiving capabilities to provide a transducer with a cleaner signal, and transducers for special applications, such as the inspection of course grained material.

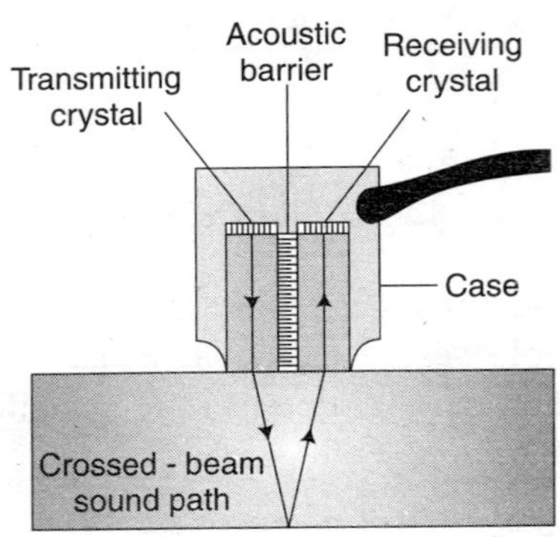

Fig. 8.21 *Dual element transducers.*

Dual element transducers are especially well suited for making measurements in applications where reflectors are very near the transducer since this design eliminates the ring down effect that single-element transducers experience (when single-element transducers are operating in pulse echo mode, the element cannot start receiving reflected signals until the element has stopped ringing from its transmit function). Dual element transducers are very useful when making thickness measurements of thin materials and when inspecting for near surface defects. The two elements are angled towards each other to create a crossed-beam sound path in the test material.

8.9.10 Delay Line Transduceres

Delay line transducers provide versatility with a variety of replaceable options.

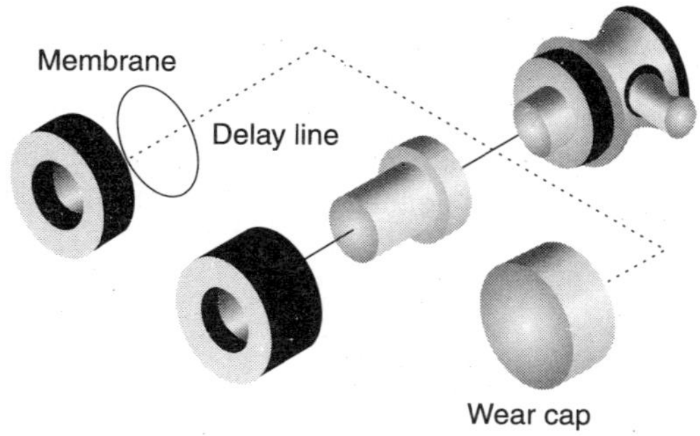

Fig. 8.22 *Delay line transducers.*

Removable delay line, surface conforming membrane, and protective wear cap options can make a single transducer effective for a wide range of applications. As the name implies, the primary function of a delay line transducer is to introduce a time delay between the generation of the sound wave and the arrival of any reflected waves (Fig. 8.22). This allows the transducer to complete its "sending" function before it starts its "listening" function so that near surface resolution is improved. They are designed for use in applications such as high precision thickness gauging of thin materials and delamination checks in composite materials. They are also useful in high-temperature measurement applications since the delay line provides some insulation to the piezoelectric element from the heat.

8.9.11 Align Beam Transducers

Angle beam transducers and wedges are typically used to introduce a refracted shear wave into the test material (Fig. 8.23).

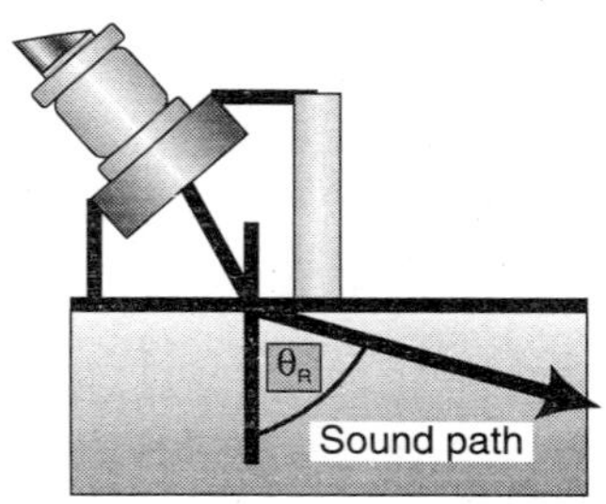

Fig. 8.23 *Align beam transducers.*

Transducers can be purchased in a variety of fixed angles or in adjustable versions where the user determines the angles of incidence and refraction. In the fixed angle versions, the angle of refraction that is marked on the transducer is only accurate for a particular material, which is usually steel. The angled sound path allows the sound beam to be reflected from the back wall to improve detectability of flaws in and around welded areas. They are also used to generate surface waves for use in detecting defects on the surface of a component.

8.9.12 Normal Incidence Shear Waves Transducers

Normal incidence shear wave transducers are unique because they allow the introduction of shear waves directly into a test piece without the use of an angle beam wedge. Careful design has enabled manufacturing of transducers with minimal longitudinal wave contamination. The ratio of the longitudinal to shear wave components is generally below –30 dB.

Optical Biosensors - I

Chapter 9

9.0 OPTICAL BIOSENSORS (OB)

9.0.1 INTRODUCTION

The last one decade has witnessed an exponential increase in the number and range of optical biosensors (OB) by using different optical phenomena. These biosensors are being developed on the reliability of optical methods and a number of optical transduction techniques. These optical biosensors are developed on the basis of linear optical phenomenon which includes absorption, fluorescence, phosphorescence, polarization, diffraction, interference, rotation and also non-linear phenomena. The abundance of a particular optical method depends on the nature of the application and desired sensitivities.

Fibre optic or optical fibre is the medium which is associated with the transmission of information as light impulses along with fibre matrix or plastic wire or glass material. Fibre optic wire carries much more information than copper wire and is far less subject to electromagnetic interference.

To increase the fibre optics adaptability this can be coupled with all optical techniques. The optical biosensors are involved in direct detection of analytes interest or indirect detection through optically labeled probes. These OBs are well established analytical tool for biomolecular interaction analysis. This biosensor consists of biological component, detector element and transducer associated with both components. The biological component is generally a protein target (bioreceptors) which is immobilized on the surface of the transducer. The OBs are coupled with microfluidics which introduces an analyte as well as separates unbound and bound molecules and is a powerful biophysical tool for characterization quantitatively.

There was a limitation on the ability to examine the living cells by using non-optical biosensors. Due to recent advancements in optical transduction technology in biosensors, that enable the discovery of specific types of cellular target dependent optical signals in cells, thus opening a new possibility for whole cell sensing in particular cell signaling analysis.

Optical biosensors have several advantages which include the high resolution, fast measurement with high sensitivity and reproducibility. The disadvantages of optical biosensor are the high cost of the apparatus and its large size.

9.1 SURFACE PLASMON RESONANCE

9.1.1 Introduction

Surface Plasmon Resonance is a physical phenomenon for which the electrons in thin gold film can be excited into collective oscillation by bombarding it with photons (light) of particular frequency. This phenomenon is analogous to resonant frequency in mechanical system and must have specific match between light frequency and momentum, and the characteristics of the gold. The gold will fail to exhibit the plasmon oscillation due to any mismatch between the wavelength of the light either too short (blue-shifted) or too long (red-shifted). The matching wavelength is also dependent on the permittivity of the material on the opposite side of the gold film (Fig. 9.1).

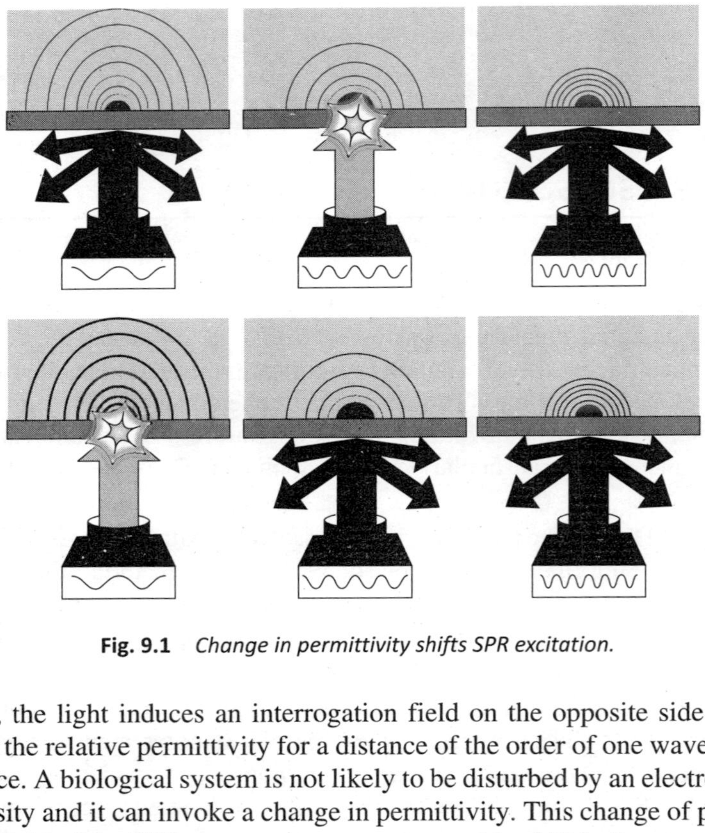

Fig. 9.1 *Change in permittivity shifts SPR excitation.*

In a thin film, the light induces an interrogation field on the opposite side of the gold. This field interrogates the relative permittivity for a distance of the order of one wavelength of the light striking the surface. A biological system is not likely to be disturbed by an electromagnetic field of low optical intensity and it can invoke a change in permittivity. This change of permittivity within the interrogation field of the SPR system can cause measurable shift in the wavelength required to excite the plasma state (Fig. 9.2).

Surface plasmon resonance (SPR) sensing is an exceedingly powerful and quantitative probe of the interactions. These interactions of variety of biopolymers with various ligands, biopolymers and membranes including protein:ligand, protein:protein, protein:DNA and protein: membrane. It provides not only for identifying these interactions and quantifying their equilibrium constants, kinetic constants and underlying energetics but also for employing them in very sensitive, label-free biochemical assays.

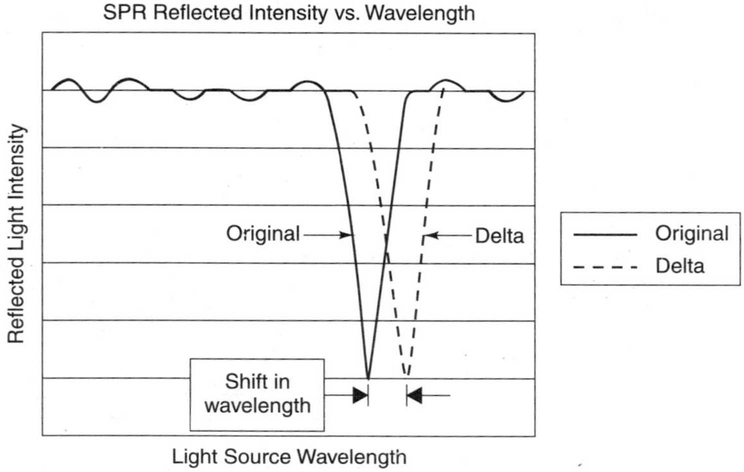

Fig. 9.2 *Depiction of measurable shift in SPR curve.*

In a typical SPR biosensor, the interactant (ligand or biomolecule) is immobilized on SPR active gold coated glass slide which forms one wall of a thin flow cell. The other interactant in an aqueous buffer solution is induced to flow across this surface by injecting it through thin flow cell channel (Fig. 9.3). Selected light passes through the prism and is focused on the slide, which reflects the light from the gold and the same passes back through the prism to the detector and can be measured. Changed in reflectivity versus angle or wavelength gives a signal that is proportional to the volume of biopolymer bound near the surface. A flow cell allows solutions above the gold surface to be rapidly changed, when light (visible or near infrared) is shined through the glass slide and onto the gold surface at angles and wavelengths, so it is called as Surface Plasmon Resonance (SPR).

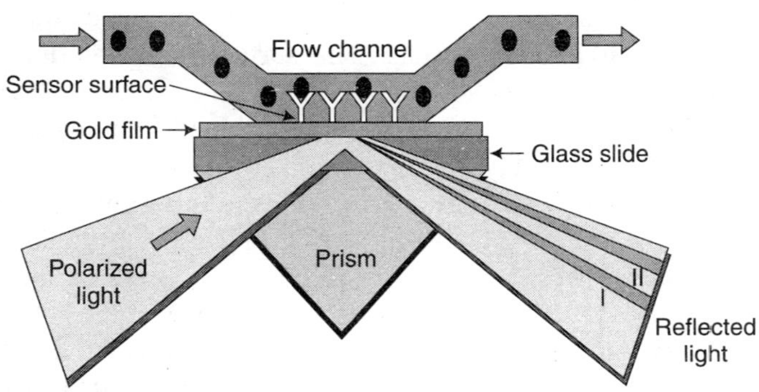

Fig. 9.3 *Basic components of an instrument for SPR biosensing.*

The high sensitivity of the optical response is due to the efficient and collective excitation of conduction electrons near the gold surface. The observation and quantifications of binding between the solution phase interactant and immobilized interactant by monitoring reflectivity changes (Fig. 9.4). An advantage of SPR is its high sensitivity without any fluorescent or other labeling of the interactants.

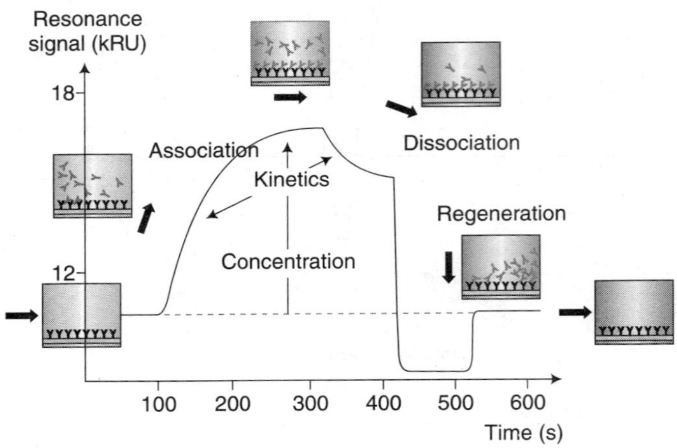

Fig. 9.4 *A typical SPR biosensing experiment- showing the optical response vs. time.*

Based on signal transduction in biosensors in a typical experimental set up, incident light is reflected from the internal face of prism in which the external face has been coated with a thin metal film. At critical angle, the intensity of the reflected light is lost to the creation of resonant oscillation in the electrons at the surface of the metal film. The critical angle is dependent on the refractive index of the material present on the metal surface; this method has been used to measure the binding of antibodies to antigens immobilized at the sensor surface. This might be visualized as a multi layer configuration consisting of quartz (prism surface), metal, antigen and antibody and the reflected light from the inside surface of the prism does not pass through the sample.

The Surface Plasmon Resonance (SPR) based devices use an optical method to measure the refractive index near (~300 nm) a sensor surface. This surface forms the floor of a small flow cell (20-60 nL) through which an aqueous solution (running buffer) passes under continuous flow (1-100 μL. min^{-1}). In order to detect an interaction the ligand is immobilized onto the sensor surface and its binding partner (the analyte) is injected in aqueous solution (sample buffer) through the flow cell under continuous flow. As the analyte binds to the ligand, the accumulation on the surface results in an increase in the refractive index. This change in refractive index is measured in real time and the result plotted as response or resonance units (RUs) vs. time (sensorgram). Background response is also generated, if there is a difference in the refractive indices of the running and sample buffers. This background response must be subtracted from the sensorgram to obtain the actual binding response. The background response is recorded by injecting the analyte through control or reference flow cell which has no ligand or an irrelevant ligand immobilized to the sensor surface.

Resonance Unit (RU) represents the binding of approximately 1 pg protein/mm^2 and needs a binding analyte >50 pg/mm^2. It is very difficult to immobilize sufficiently high density of ligand onto surface to achieve this level of analyte binding. The sensor surfaces with a 100-200 nm thick carboxymethylated dextran matrix attached were already commercialized. Much higher levels of ligand immobilization are possible by adding third dimension to the surface however, these levels of ligand has two drawbacks (i) High ligand density, the rate at which the surface binds the analyte may exceed the rate at which the analyte can be delivered to the surface (mass transport) (ii) Related problem is that, following dissociation of the analyte, it can rebind to the unoccupied ligand before diffusing out of the matrix and being washed from the flow cell. Although the dextran matrix may exaggerate these kinetic artifacts (mass transport limitations and rebinding) they can affect all surface binding techniques.

9.1.2 Immobilisation of Ligand

The most challenging step while setting up the surface plasmon resonance (SPR) device experiment is immobilization of one of the ligands to the sensor surface without disrupting its activity. This SPR ligand immobilization process can either be direct by covalent coupling or indirect through capture by covalently coupled molecule. The major advantage of direct covalent immobilization is that it can be used for any ligand/protein provided, it is pure (>50%) and has isoelectric point >3.5 (pI). However, it has three important drawbacks (a) because they usually have multiple copies of the functional group that mediates immobilization, proteins are coupled heterogeneously and sometimes at multiple sites (b) Direct coupling often decreases or completely abrogates binding to analyte (c) Directly coupled proteins are difficult to regenerate.

Indirect immobilization has the disadvantage that it can only be used for proteins that have a suitable binding site or tag for the covalently coupled molecule. However, it has four important advantages, which make it the method of choice in most cases (i) Proteins are seldom inactivated by indirect coupling (ii) The protein need not be pure. It can be captured from a 'crude' sample (iii) All the molecules are immobilised in a known and consistent orientation on the surface (iv) Using appropriate buffers it is often possible to dissociate selectively the non-covalent ligand/analyte bond, thereby enabling the ligand surface to be reused, is termed 'regeneration'.

9.1.2.1 Covalent coupling

The covalent coupling of direct immobilization method for immobilization of a protein/ligand onto the surface of the SPR with the following procedures:

1. Selection of coupling chemistry
2. Preparation of protein
3. Optimization of the pre-concentration
4. Coupling of protein – Amine coupling
5. Establishment of regeneration conditions
6. Adjustment of the immobilization conditions

9.1.2.1.1 Selection of coupling chemistry

In this approach, three main types of coupling chemistry utilize viz., amine (*Eg.* Lysine), thiol (*Eg.* Cysteine) and aldehyde (*Eg.* Carbohydrate) functional groups on glycoproteins. All covalent coupling methods utilize free carboxymethyl groups on the sensor chip surface; hence they can be used for any of the sensor chips. The methods of choice for the protein is to be immobilized, its surface has to be exposed to disulphide or a free cysteine and ligand thiol coupling. Failing this, amine coupling should be tried in the first instance. If amine coupling inactivates the protein (as assessed by ligand or MAbs binding), aldehyde coupling can be attempted, provided that the protein is glycosylated.

9.1.2.1.2 Preparation of protein

The protein should be pure with high level of activity because direct coupling is relatively indiscriminate. If the protein solution is contaminated by any other proteins or other solutions, the level of protein binding decreases. 5-10 μg/ml of protein is required for SPR biosensors. The protein is coupled with primary amine groups or sodium azides when the solutions have strongly buffered or high salt concentrations.

9.1.2.1.3 Optimiztion of pre-concentration

The purpose of pre-concentration is to concentrate the protein to very high levels (>100 mg/ml) within the dextran matrix, thereby driving the coupling reaction. Without pre-concentration for higher concentrations of protein would need to be injected to get equivalent levels of coupling. Pre-concentration is driven by an electrostatic interaction between the negatively charged carboxylated dextran matrix and positively charged protein. To this end the protein is diluted into a buffer with a low ionic strength (to minimize charge screening) with a pH below its isoelectric point (pI) (to give the protein net positive charge). However amine coupling is most efficient at a high pH because activated carboxyl groups react better with uncharged amino groups. Thus the highest pH compatible with pre-concentration is determined empirically. Electro statically bound protein should dissociate rapidly and completely when injection of running buffer because the proteins net positive charge will decrease. The electrostatic interactions screened by the high ionic strength of the running buffer. Incomplete dissociation suggests that the interaction was not purely electrostatic because of the binding, low pH and denatured form of the protein.

9.1.2.1.4 Coupling of protein–amine coupling

Amine coupling follows four standard steps which involved in the coupling of the proteins. (1) The first step is to activate the carboxymethyl groups with N-hydroxysuccinimide (NHS), thus creating a highly reactive succinimide ester which reacts with amine and other nucleophilic groups on proteins. (2) The second (coupling) step is to inject the protein in pre-concentration buffer, thereby achieve high protein concentrations and driving the coupling reaction. (3) The third (blocking) step, blocks the remaining activated carboxymethyl groups by injecting very high concentrations of ethanolamine. The high concentration of ethanolamine also helps to elute any non-covalent bound material. (4) The final regeneration step is optional when a protein is coupled for the first time it is advisable not to include any regeneration step.

9.1.2.1.5 Establishment of regeneration conditions

If the protein is included and the protein has poor activity it will not be clear whether covalent coupling or regeneration was responsible for disrupting the protein. The structural integrity of the protein should be evaluated before regeneration. Once regeneration conditions have been established these can be added on to the coupling protocol. Regeneration allows surfaces to be reused many times, saving both time and money. However establishing an ideal regeneration conditions are time consuming and in many cases impossible. Thus it may be more cost-effective to opt for imperfect or no regeneration, using new sensor surfaces instead.

9.1.2.1.6 Adjustment of the immobilization conditions

Since the level of ligand coupling achieved is unpredictable it is usually necessary to modify the initial protocol in order to achieve the desired immobilization level. It is also often necessary to create several surfaces on the same chip with different levels of coupling. The best way to achieve different levels of coupling is to change the duration of the activation step by varying the volume. The level of coupled ligand varies in proportion with the duration of the activation step. Thus, a twofold reduction in activation period will usually lead to a twofold reduction in coupling. It is possible using an option in the inject command to couple ligand simultaneously in multiple flow cells varying only the length of activation step.

9.1.3 Ligand Capture or Non-Covalent Immobilisation

Non-covalent or indirect immobilization of a ligand is called ligand capture, it depends on two requirements: (a) It must be possible to obtain or create a sensor surface that can capture a ligand and it should be possible to regenerate this surface, so that repeated capture is possible (b) The ligand needs to have a suitable binding site or modification that allows it to be captured.

9.1.4 Activity of Immobilized Ligand

It is important to evaluate the functional integrity of the immobilized protein and this can be achieved by using protein which binds to correctly folded ligand. An ability to bind its natural ligand is reassuring evidence that an immobilized ligand is functionally intact. Monoclonal antibodies (MAbs) usually bind to 'discontinuous epitopes' on a protein they are excellent probes of protein structure. It is preferable to check the binding of several MAbs, including ones that bind to the same binding sites or failing this, the same domain of the natural ligand. MAbs are particularly useful if the natural binding partner has not been identified or a candidate binding partner being assessed for an interaction with the ligand. It is important to know that whether the immobilized ligand is 'active' but also what proportion is active.

9.1.5 Analyte

The extent to which the analyte needs to be characterized depends on the nature of the experiment. Quantitative measurements require that the analyte is very well characterized and of the highest quality. In contrast, this is less important for qualitative measurements.

9.1.5.1 Purity, activity and concentration

In order to determine affinity and association rate constants it is essential that the concentration of the injected material is known with great precision. Accurate is the measurement of total concentration of analyte, to use spectrophotometer to measure the optical density of a solution of pure protein, usually at 280 nm (OD_{280}). In order to calculate the concentration from the absorption at OD_{280} two additional measurements are required: (1) It is necessary to determine the extinction coefficient. This is calculated from the primary sequence and it is best determined directly by amino acid analysis of a sample of the protein with a known OD_{280}. (2) It is important to assess what proportion of the purified protein is 'active', i.e. able to bind to the ligand. This can be done by depletion experiments in which the ligand-coated sepharose beads are used to deplete the analyte from solution. MAbs coated sepharose beads can be used instead if ligand coated beads are impractical. If all the analyte can be depleted in this way it is 100% active.

9.1.5.2 Valency

A single binding site (monovalent) for each molecule is required to measure the affinity and kinetics. If the protein has a single binding site it is only necessary to show that it exists as a monomer in solution and this can be achieved by size exclusion chromatography or analytical ultracentrifugation. It is important to emphasize that these analytical techniques will not detect the presence of very low concentrations of multivalent aggregated material. If the dissociation of bound analyte is monophasic (mono-exponential) multivalent binding can be ruled out. If dissociation is bi-exponential with >10 fold difference in the two k_{off} values, multivalent binding is likely.

9.1.5.3 Refractive index effect and control analytes

Injection of an analyte over the surface has two reasons: (1) This helps rule out non-specific binding. (2) It controls for any refractive index artefacts. These occur when the background signal measured during the injection of an analyte sample differs between flow cells. Clearly such an artefact will create problems for affinity measurements. It can occur whenever there is a substantial difference between surfaces. *Eg.* If very different levels of material are immobilised on each surface because the immobilized material displaces volume, the volume accessible to the injected analyte sample will differ between flow cells. If the analyte sample has a higher refractive index than the running buffer larger background signal seen from the surface with less immobilized material. This artefact is greater when (i) there are big differences in the levels of immobilized material (*Eg.*>2000 RUs) (ii) the background signal is very large (*Eg.* >1000 RUs) (iii) the binding response is much smaller (<10%) than the background response. These conditions are common when measuring very weak interactions because high concentrations of analyte are injected and with low molecular weight analytes which give a very small response. A refractive index artefact can be detected by injecting a control solution with a similar refractive index to the analyte sample. If the control solution gives the same response in both flow-cells a refractive index artefact can be excluded. Refractive index artefacts are most easily avoided by taking care to immobilize the same amount of total material on both the control and ligand sensor surfaces.

9.1.6 Applications

Surface Plasmon Resonance biosensors are used in detection of various analytes/targets of different samples from various sources/fields and few applications related to SPR biosensors mentioned below.

9.1.6.1 Evaluation of macromolecules

Most laboratories studying biological problems at the molecular or cellular level need to produce recombinant proteins. It is important to be able to show that the recombinant protein has the same structure as its native counterpart. With the possible exception of enzymes, this is most easily done by confirming that the protein binds its natural ligands. Because such interactions involve multiple residues, which are usually far apart in the primary amino acid sequence, they require a correctly folded protein. In the absence of natural ligands Monoclonal antibodies (MAbs) that are known to bind to the native protein are an excellent means of assessing the structural integrity of the recombinant protein. The commercialized SPR biosensors are used to evaluate the binding of recombinant proteins to natural ligands and MAbs. Setting up an assay for any particular protein is very fast and the data provided are highly informative.

9.1.6.2 Equilibrium measurements

Equilibrium analysis requires multiple sequential injections of analyte at different concentrations at different temperatures. Because this is very time consuming and it is only practical to perform equilibrium analysis on interactions that attain equilibrium within 30 min. The time it takes to reach equilibrium is determined primarily by the dissociation rate constant or k_{off} and an interaction should reach 99% of the equilibrium level within 4.6/k_{off} seconds. High affinity interactions (K_D < 10 nM) usually have very slow k_{off} values and are therefore unsuitable for equilibrium analysis. Conversely, very weak interactions (K_D >100 μM) are easily studied. The small sample volumes required for injections (<20 μl) make it feasible to inject the very high concentrations (>500 μM) of protein required to saturate low affinity interactions. Equilibrium affinity measurements are highly reproducible and this feature, the very precise temperature control make it possible to estimate binding enthalpy by van't Hoff analysis. This involves measuring the (often small) change in affinity with temperature.

9.1.6.3 Kinetic measurements

The real time binding data is used to analyse the binding kinetics and has some of the important limitations to kinetic analysis. Largely because of mass transport limitations it is difficult to measure accurately k_{on} values faster than about $10^6\ M^{-1}s^{-1}$. This upper limit is dependent on the size of the analyte. Faster k_{on} values can be measured with analytes with a greater molecular mass. This is because the larger signal produced by a large analyte allows the experiment to be performed at lower ligand densities and lower ligand densities require lower rates of mass transport. For different reasons measuring k_{off} values slower than $10^{-5}s^{-1}$ or faster than ~1 s^{-1} is difficult because it is easy to use and easy to generate kinetic data. However obtaining accurate kinetic data is a very difficult, time consuming, binding kinetics understanding and the potential sources of artefact.

9.1.6.4 Analysis of mutant proteins

To visualize the capture of proteins from crude mixtures onto the sensor surface and the SPR is convenient for analyzing mutants generated by the site-directed mutagenesis. Mutants can be expressed as tagged proteins by transient transfection and then captured from crude tissue culture supernatant using an antibody to the tag, thus effectively purifying the mutant protein on the sensor surface. It is simple to evaluate the effect of the mutation on the binding properties viz., affinity, kinetics and thermodynamics of the immobilized protein. This is used for quantifying the effect of mutations on the thermodynamics and kinetics of weak protein/ligand interactions.

9.1.7 Disadvantages

9.1.7.1 High throughput assays

At a time only one sample can be analyzed and each analysis takes 5-15 minutes. Automation does not solve this problem because the sensor surface deteriorates over time and reuse. Air bubbles in the microfluidic system are affecting the measurements of kinetics of analytes.

9.1.7.2 Concentration assays

It is not suitable for measurement of the sample concentrations because these require the analysis of many samples in parallel, including the standard curve. Another problem is the optimal sensitivity and long equilibration periods of the assays.

9.1.7.3 Studying small analytes

The SPR measures the mass of material binding to the sensor surface, very small analytes (M_r <1000) give very small responses. The recent improvements in signal to noise ratio have made it possible to measure binding of such small analytes. However high surface concentration of active immobilized ligand (~ 1mM) is needed and this is difficult to achieve. At high ligand densities accurate kinetic analysis is not possible because of mass transport limitations and rebinding. Equilibrium analysis is possible with very small analytes and under optimal conditions.

9.2 FIBRE OPTIC BIOSENSORS

9.2.1 Introduction

Recent developments in fibre optic sensors have two important scientific advanced components: laser and modern low cost fibre optics. Fibre optics have become an important part of sensor technology and their use as a probe or as a sensing element which is increasing in clinical, pharmaceutical, industrial and military applications. Main advantages of fibre optic biosensors have excellent light delivery, long interaction length, low cost and ability not only to excite the target molecules but also to capture the emitted light from the targets.

Fibre optics transmits light on the basis of the principle of Total Internal Reflection (TIR). Fibre optic biosensors are analytical devices in which fibre optic device serves as transduction element and it is to produce a signal that is proportional to the concentration of chemical or biochemical to which the biological element reacts. Fibre optic biosensors are based on the transmission of light along with silica glass fibre or plastic fibre optic to the site of analysis. These biosensors can be used in combination with different types of spectroscopic technique such as Absorption, Fluorescence, Phosphorescence and Surface Plasmon Resonance (SPR).

Fibre optic biosensors can be classified into two different categories: (1) Intrinsic sensors, where interaction with the analyte occurs within an element of the fibre optic (2) Extrinsic sensors, in which the fibre optic is used to couple light usually to and from the region where the light beam is influenced by the measured. These fibre optic biosensors are useful in routine tests, patient home care, surgery and intensive care as well as emergency situations.

9.2.2 Absorbance Measurements

Simple fibre optic biosensors use absorbance measurements to determine changes in the concentration of analytes that absorb given wavelength of light. It works by transmitting light through fibre optical to the sample and the amount of light absorbed by the analyte is detected through the same fibre. Fibre optics is immobilized with suitable biological material and produces the analyte that absorbs the light.

Wolthuis *et al.* (1992) developed fibre optic pH sensor and fibre optic oxygen sensor for medical applications. The fibre optic pH sensor used an absorptive indicator compound with long wavelength absorption peak near 625 nm and change in absorption over the pH range 6.8 to 7.8 is linear. Return light signal is split into short and long wavelength components with dichroic mirror; the respective signals are detected by photodiodes and their photocurrents are used to form a ratiometric output signal. This sensor system resolution is 0.01 and response time 30 to 40 seconds. Advantages of this sensor are good performance, low cost and solid state instrumentation. The fibre optic oxygen sensor's viologen indicator (viologens are toxic bipyridinium derivatives of 4, 4'-bipyridyl) becomes strongly absorbent after UV stimulation and then returns to the transparent state. The rate of indicator return to transparency is proportional to the oxygen concentration. The solid state sensor system has performance comparable to oxygen measurement techniques and applicable for both *in-vitro* and *in-vivo* oxygen measurements.

Surface plasmon resonance (SPR) is an optical phenomenon caused by charge density oscillation at the interface of two media with dielectric constants of opposite sign *Eg.* metal and dielectric. In this way, by tuning the plasmon resonance to a wavelength for which the outer medium is absorptive, a significant variation of the spectral transmittance of the device is produced as a function of the concentration of the analyte. With this mechanism, selectivity can be achieved without the need of any functionalization of the surfaces or the use of recognizing elements, which is a feature for any kind of chemical sensor or biosensor. Doubly deposited uniform waist tapered fibres are well suited for the development of these new sensors.

9.2.3 Reflectance Measurements

The free radicals are implicating primary or secondary in the initiation or progression of many diseases which are characterized by an inflammatory reaction. Naughton *et al.* (1993) was developed fibre optic sensor, immobilized with nitrophenol and used for the continuous monitoring of OH^- radicals. Nitrophenol was immobilized onto XAD-7 methacrylate beads and these were attached to the polymethylmethacrylate fibre optic. This reflectance based sensor incorporated OH^- radicals, sensitive chromophore and decrease in its reflectance spectrum upon by reactive oxygen derived radical attributable to the formation of nitrocatechol. Nitrocatechol generated from the attack of OH^- radical on nitrophenol exhibits strong absorption band in the visible region of the electromagnetic spectrum (λ_{max} = 510 nm).

Based on the principle of multiple reflections in white light interferometry, an optical fibre immobilized with protein molecule either an antigen or antibody is adopted as a high sensitivity biosensor to achieve the real time monitor of the immunoassay *Eg.* Rabbit-IgG with anti-rabbit-IgG immuno reaction. Optical thickness (biolayer 0.70 nm) can be detected by measuring the reflected spectrum phase shifting between the two interfaces of biolayer. This method had advantages such as simple in structure, low cost, high sensitivity and reliability compared with other optical sensing methods.

SPR biosensor has been widely used for the last decade to analyze the affinity between ligand and analyte. The SPR condition is usually achieved in the attenuated total reflection (ATR) geometry by using prism coupling optics. This optical setup is usually bulky and incompatible with micrometer sized probes. At present the Localized Plasmon Resonance (LPR) is being developed in the place of conventional SPR in view of the fact that LPR is quite compatible with such microsensors. The

LPR of gold appears as an absorption band at about 550 nm in the presence of thin molecular layers on gold. The LPR produces a large enhancement of the electric field around small gold particles or roughness of a gold surface, so that it is applicable in a highly sensitive affinity sensor of size less than micron. Kajikawa *et al.* (2004) was proposed simple and sensitive optical sensing method for biological applications.

Gold behaves as dielectric with large extinction index under blue or violet light. The presence of transparent surface layer on gold produces large decrease in the reflectivity of the gold surface due to multiple reflections in the surface layer, this phenomenon is called Anomalous Reflection (AR). AR is applicable to affinity biosensors based on fibre optics, so that inexpensive and disposable micrometer sized biosensor probes are possible. The application of AR to real time measurements of the adsorption process of Octadecanethiol on gold and the affinity of Streptavidin to biotin labeled monomolecular layer on gold.

Misiakos *et al.* (2004) described an optical affinity sensor based on a Monolithic Optoelectronic Transducer (MOT) which integrates on silicon die thin optical fibres (silicon nitride) along with self aligned Light Emitting Diodes (LED) and photo detectors (silicon p/n junction). The LEDs are optically coupled to the corresponding photo detectors through silicon nitride fibres. Specially designed spacers provide for the smooth bending of the fibre at its end points and toward the light source and the detector ensuring high coupling efficiency. The transducer surface is hydrophilized by oxygen plasma treatment, silanized with (3-aminopropyl) triethoxysilane and bioactivated through adsorption of the biomolecular probes.

The use of micro-fluidic module allows real time monitoring of the binding reaction of the gold nanoparticle labeled analytes with the immobilized probes. Their binding within the evanescent field at the surface of the optical fibre causes attenuated total reflection of the wave guided modes and reduction of the detector photocurrent. The biotin Streptavidin model assay was used for the evaluation of the analytical potentials and detection limit 3.8 pM - 13 pM. Gold nanoparticle labeled Streptavidin developed for continuous and stopped flow assay modes. The detection sensitivity was improved by silver plating of the immobilized gold nanoparticles and detection limit of 20 fM was obtained after 20 minutes of silver plating.

Recently, fibre optic biosensor was introduced which could directly detect biological interaction such as immunoreactions of antigens and antibodies without destroying the biolayer. The test is based on the theory of multilayer-reflection principle in white light interferometry. When immunoreactions occur, the reflected spectrum phase shifts and this could be detected by means of reflected spectrum phase shifting or by biolayer thickness changing. Continuously detecting of thickness changing on a fractional nanometer scale with sub second repetition times is allowed in this system. The detecting system has high sensitivity, precision, speed, cost effective and reliability.

9.2.4 Fluorescence Measurements

Fluorescence techniques provide sensitive detection of biomolecules. Since fluorescence intensity is proportional to the excitation intensity even weak signals can be observed. In last decade reagent less fibre based biosensors have been developed and these are capable of detecting changes in cell behaviour, metabolism and cell death when exposed to toxic agents. Fluorescence measurements are not used as often as absorbance and reflectance with enzyme optical fibre based biosensors, as it is not common for enzyme reactions to produce fluorescent products or intermediates. Majority of the fluorescence techniques employs fluorescent dye(s) to indirectly monitor the formation or consumption of transducer. Several enzyme catalysed reactions involve the production or consumption of fluorescent species. NAD^+ (Nicotinamide adenine dinucleotide) dependent reactions catalysed by group of dehydrogenase enzymes and in general the dehydrogenase reaction occurs as follows:

$$\text{Substrate} + \text{NAD (P)}^+ \xrightarrow{\text{Dehydrogenase}} \text{Product} + \text{NAD (P) H}^+ \qquad \textbf{... \textit{eq.}-9.1}$$

NADH or NADPH is produced which can be detected using fibre optics through its fluorescence at 350 nm and 450 nm. Several optical sensors have been produced by integrating different dehydrogenase or oxidoreductase enzymes with optical fibres.

Anderson *et al.* (2000) developed fibre optic based biosensor which integrates an array of optical and electrical components together with long fused silica fibres and proteins for detection of analyte in solution. The optical fibre core near the distal end is tapered and coated with either antibodies or DNA binding proteins. Assays are performed by flow of a solution containing the fluorescently tagged ligand molecules over the coated fibre. Within seconds, analyte recognition occurs and fluorescence signal is transmitted back up the fibre. Applications for this biosensor include clinical diagnostics, pollution control and environmental monitoring.

The application of DNA as selective recognition element in biosensor design is a new and electrifying area in analytical chemistry. The ssDNA was covalently immobilized onto quartz optical fibres for development of an optical biosensor for DNA. The non-optimized configuration was able to detect femtomolar (fM) concentrations of cDNA with an analysis time of less than 1 hour. This analysis indicates that biosensors which can selectively detect genetic material from biological samples with advantages of low detection limits, reasonable analysis times, highly stable biorecognition elements and regenerability.

Fibre optic biosensors based on fluorescence detection often used combination of fluorescent bioreceptor associated with an optical transducer. Fluorescent biosensors may also be obtained by immobilizing whole cells on the surface of a sensor layer. This bioactive layer is usually placed in front of the tip of an optical fibres bundle to generate fluorescence signal. The optical fibres are required to send the excitation radiation to the fluorescent bioelement and transmit the fluorescence radiation up to fluorimeter.

Algal cells are deposited on the surface of an opaque support and they are placed at a little distance to the optical fibres bundle to allow the fluorescence radiation emitted by the illuminated area to be collected properly. In order to improve the simplicity and reliability of fluorescence based biosensors, optically translucent supports are used because their optical properties enable detection of fluorescence emitted by the *Algal* cells even if they are entrapped in the bulk of the translucent support. Silica matrix has properties like optical translucence, biocompatibility and chemical inertness. The design flexibility of sol–gel technique and ease of fabrication can fulfil to create the support with structural and chemical features that could be compatible with biomaterials. While immobilization of enzymes and whole cells in sol–gel is a well known technique in biosensor application, it is not yet used for the construction of vegetal cell optical biosensors. When an enzyme is immobilized, substrate is required and fluorescent indicator may be added to enable optical detection of the analyte. Recently, Nguyen-Ngoc and Tran-Minh (2007) investigated translucent matrix obtained from a sol–gel process for entrapment of fluorescent biomaterials in order to improve the fluorescent biosensor fabrication. Vegetal cells as bioreceptors have been tested for herbicides determination.

Examples for Fibre optic biosensors based on fluorescence measurements:

Eg. 1: **Ethanol vapor** - Immobilizing alcohol oxidase (AOD) onto a tip of a fibre optic oxygen sensor with tube ring using an oxygen sensitive ruthenium organic complex, reaction unit for circulating buffer solution applied to the tip of the device; after experiment in liquid phase, the sniffer device was applied for gas analysis using a gas flow measurement system with a gas generator; optical device applied to detect the oxygen consumption induced by AOD enzymatic reaction with alcohol application.

Eg. 2: **Glucose** - Biochemical assay based on homogeneous singlet/singlet energy transfer affinity assay; sensor probe indirectly measures glucose concentrations from the level of fluorescence quenching caused by the homogeneous competition assay between tetramethylrhodamine (TRITC) labeled concanavalin A (receptor) and Fluorescent Iso-thio cyanide (FITC) labeled Dextran (ligand).

Eg. 3: **Glutamate** - Enzyme layer composed of glutamate dehydrogenase (GDH) and glutamate pyruvate transaminase (GPT) to produce NADH at the tip of a fibre-optic probe; GDH catalyzes the formation of NADH and GPT drives the GDH reaction by removing a reaction product and regenerating glutamate; optimal response obtained at pH 7.4 tris-HCl buffer at 25°C.

Eg. 4: **Ethanol and Urea** - Increase of NADH fluorescence during the enzymatic conversion of ethanol by using an enzyme system of ADH and AlDH; for urea based on a pH sensor measuring the change in fluorescence of the pH indicator fluorescein; both sensors are suitable for controlling fermentation processes when ethanol and urea are produced or consumed.

9.2.5 Chemiluminescence Measurements

Molecular assembly which conjugated polymer is interfaced with a photodynamic protein described by Ayyagari *et al.* (1995). The conjugated polymer is functionalized with biotin and chemically grown off a glass surface. The Streptavidin derivatized protein immobilized on polymer matrix through the strong biotin-Streptavidin interactions. The assembly built on the surface of an optical fibre or on the inside walls of a glass capillary form an integral part of a biosensor and used for the detection of environmental pollutants such as organo phosphorus based insecticides. The protein in the system can be replaced by any biological macromolecule of interest.

Chen *et al.* (1996) immobilized multilayer enzyme assembly on a fibre optic surface *Eg.* A multilayer of an enzyme alkaline phosphatase was immobilized by chemical cross linking on fibre optical surface. Chemiluminescence, Ellipsometry and Surface plasmon resonance were used to characterize the structure and activity of the assembly. A chemiluminescence based fibre optic biosensor developed for the detection of organophosphorus based pesticides. Fibre optic based biosensor on the electro-chemiluminescence of luminol for glucose and lactate flow injection analysis by Marquette and Blum (1999). The electro-chemiluminescence of luminol was generated using glassy carbon electrode polarized at +425 mV vs. platinum pseudo-reference electrode. After optimization of the reaction conditions and physicochemical parameters influencing the sensor response, the measurement of hydrogen peroxide could be performed in the range 1.5 pmol-30 nmol. Glucose oxidase or lactate oxidase were immobilized on polyamide and collagen membranes. With collagen as the enzymatic support, the detection limits for glucose and lactate were 60 pmol and 30 pmol respectively, whereas with the enzymatic polyamide membranes the corresponding values were 150 pmol and 60 pmol.

Chemiluminescence biosensing system for antioxidants was developed by Palaoran *et al.* (2000) based on luminol and hematin co-immobilized on a cellulose membrane disc. The concentration of the antioxidant was quantified through the measurement of the inhibition of the chemiluminescence emitted when hydrogen peroxide was introduced into the reagent phase. The minimum detectable concentration was 100 μM and the response time was < 60 seconds.

The enhanced chemiluminescence (CL) reaction of the luminal, H_2O, Horseradish Peroxidase (HRP) with immobilized HRP by using micro-encapsulation in sol-gel matrix used to develop biosensor for the measurement of p-iodophenol, p-coumaric acid, 2-naphthol and hydrogen peroxide. The detection limits obtained for p-iodophenol, p-coumaric acid and 2-naphthol were 0.83 mM, 15 nM and 48 nM respectively. Direct enzyme immobilization onto the end of the optical fibre permits the construction of a remote enhanced CL biosensor, this biosensor is used to detect hydrogen peroxide and its detection limit is 52.2 μM.

Magrisso *et al.* (2006) described computerized multi-sample temperature controlled luminometer for a fibre array based biosensor to monitor circulating phagocyte activity. It can perform integral measurements of chemiluminescence emitted from up to six samples containing less than 0.5 μl, whole blood while the samples and detector do not change their position during the measurement cycle. The optical fibres in this luminometer are used as both light guides and solid phase sample holders. The instrument is designed to simplify the assessment process of both the extracellular and the intracellular parts of the phagocyte emitted chemiluminescence using the same system. This technology is being used in a wide range of analytical luminescence applications in biology, biophysics, biochemistry, toxicology and clinical medicine.

9.2.6 Bioluminescence Measurements

Biosensor associates bioactive sensing layer with a suitable transducer giving usable output signal and the selective molecular recognition of the target analyte can be achieved with various kinds of affinity enzyme systems. The biocatalysts are generally immobilized on an artificial support placed in close contact with the transducer. Covalent immobilization is preferable, because it prevents the leakage of enzymes from the support and this can now be easily achieved with the availability of commercially preactivated membranes. Sensing layers prepared with such membranes were used for the development of amperometric as well as luminescence based fibre optic biosensors. In addition to the target analyte, enzymatic reactions generally involve one or two other substrates, which must be added to the reaction medium when an enzyme based biosensor is operated. The sensing and transducer associated with the sensing layer determines the sensitivity of an enzyme based sensor. Due to the unusual nature of one of the reaction products i.e., light which can be detected at a very low level, bioluminescence and chemiluminescence reactions associated with an optical transduction can be used to design highly sensitive biosensors.

The reproducible and easy immobilizations of receptors on sensor surfaces are prerequisite for the development of receptor based fibre optic biosensors. Using a fused silica fibre as the transducer binding processes of luminescently labeled ligands can be monitored by evanescent wave sensor technology. The vesicle fusion technique was used for the immobilization of membrane bound receptors in order to preserve their binding specificity and activity by embedding them in an environment similar to a lipid bilayer. Using the binding of fluorescently labeled Streptavidin to biotinylated lipid layer as a model system for receptor ligand interaction, good sensitivity, combined with low non- specific binding were observed.

The characteristics and performance of biosensors mainly depend on the properties of the bioactive layer associated with the transducer. Two approaches are developed by Blum *et al.* (1995): (a) Reagent less fibre optic biosensor with co-substrates embedded in the vicinity of the immobilized enzyme. (b) Compartmentalized layer with sequential enzymatic reactions for improved signal detection. The reagent less biosensor focused on the bacterial bioluminescent system with two enzymes, Oxidoreductase and Luciferase for NADH detection. The compartmentalized enzyme layers, sequential bienzymatic system Lactate oxidase-peroxidase is chosen as a model. The hydrogen peroxide produced by the Lactate oxidase reaction serves as a co-substrate for the chemiluminescence reaction of luminol catalyzed by Peroxidase. Compartmentalization of the bioactive layer is obtained by immobilizing the two enzymes separately on different membranes stacked at the sensing tip of the fibre optic sensor. When using such a design, a 20-fold increase of the sensor response for lactate is obtained compared with a sensor including Lactate oxidase and Peroxidase randomly co-immobilized on the same membrane.

Recently, biotin was covalently coupled with alginate in an aqueous-phase reaction by carbodiimide mediated activation chemistry to provide a biotin-alginate conjugate for subsequent use in biosensor applications. The synthetic procedure was optimized with respect to pH of the reaction medium (pH 6.0), the degree of uronic acid activation (20%) and addition of the reagents. The new biotin alginate conjugate was used for the encapsulation of bioluminescent reporter cells into microspheres.

Ionic and colloid gold influences on luminous bacteria *Photobacterium phosphorum* (B7071) bioluminescence were studied by Gruzina TG *et al* (2005). It was shown that both forms of gold inhibited bioluminescence of the studied bacteria depending on their concentrations and incubation time with cells. The approaches for the creation of biosensor systems based on luminous bacteria and semiconductor structures or optical fibres are proposed.

Eltoukhy *et al.* (2006) described bioluminescence detection lab-on-chip consisting of a fibre-optic face plate with immobilized luminescent reporters/probes that is directly coupled to an optical detection and processing system-on-chip fabricated in a 0.18 μm process. The lab-on-chip used for determining gene expression using reporter gene assays, determining intracellular ATP and sequencing DNA.

On the other hand, hydrophobic phosphorescent Pt-porphyrins have been used for the development of luminescent polymer films designed for fibre-optic oxygen sensors. Luminescent and quenching characteristics of several Pt-porphyrins incorporated into polymer matrices have been studied to optimize the preparation of sensitive coatings for fibre-optical sensors. The films thus obtained have been used for fibre-optical oxygen monitoring in solutions. More recent, a simple system for enzymatic flow-injection analysis of metabolites is described by Ovchinnikov *et al.* (1999) which is based on the phosphorescence lifetime based detection of molecular oxygen using phase-modulation techniques and a simple instrument phosphorescence phase detector equipped with a fibre-optic probe. The phase detector is connected to the oxygen sensor membrane and allows real-time continuous monitoring of the phosphorescence phase shift. This parameter is related to the phosphorescence lifetime of the oxygen probe, therefore giving a measure of the dissolved oxygen concentration, and its changes as a result of the enzymatic oxidative reaction with the substrate. The sensor membrane is positioned in a compact integrated flow-through cell and exposed to the flow stream. Using glucose as a test analyte and glucose oxidase enzyme, two different sensor setups were tested: (1) the membrane type biosensor in which the enzyme is immobilized directly on the oxygen sensor membrane. (2) The microcolumn type biosensor in which the enzyme is immobilized separately, on a microparticle sorbent (controlled pore glass) and put into a microcolumn with the oxygen sensor membrane placed at the column outlet. In either case a new type of oxygen sensitive material was used, which provides a number of advantages over the existing materials. In this material the oxygen-sensitive coating was applied on a microporous scattering support, the latter comprised a layer of cellulose particles on polyester support.

9.2.7 Refractive Index

Akkin *et al.* (2002) described a fibre based optical biosensor which is capable of detecting ultra small refractive index changes in highly scattering media with high lateral and longitudinal spatial resolution. The system is dual channel phase sensitive optical low coherence tomography system that measures relative optical path length differences between the orthogonal modes of the polarization maintaining fibre.

Tubb *et al.* (1997) described new design of optical fibre surface plasma wave chemical sensor. The basic sensor consists of single mode optical fibre with thin layer of silver evaporated onto the tapered section. The gradually changing diameter of the fibre along the taper and the variation in silver depth around the taper result in a distributed coupling between the guided mode of the fibre and the surface plasma wave. As a result, the coupling to the surface plasma wave occurs over an enlarged spectral range. The device shows good sensitivity to refractive index and the change of refractive index (5×10^{-4}) being detectable.

Long period grating (LPG) technology based biosensor is used for the detection of large molecules (*Eg.* proteins) and small molecules (*Eg.* pesticides). The LPG sensor is spectral loss optical fibre based system that provides direct detection of large molecules by using an antigen or antibody modified hydrogel without the need for secondary amplification. The binding of the specific target results in mass increase that produces a localized refractive index change around the LPG region and thus a spectral shift in wavelength is observed. The magnitude of the observed shift can be correlated to the target concentration. *Eg.* HIV protein (p24) was directly detected at 1 ng/ml with a specific signal that was 5-7 times that of the system noise. A direct and indirect competitive assay was demonstrated with the target atrazine. The sensitivity of the two competitive assay formats was in the range of 10-50 ng/ml.

Optical sensors based on the excitation of Surface plasmon (SP) to hold great potential for biomolecular interaction analysis and detection of biological analytes. The surface plasmon resonance (SPR) sensors developed into portable sensing devices capable of operating in the field. An optical fibre SPR sensor based on polarization maintaining fibres and wavelength modulation described by Piliarik *et al.* (2003). It is demonstrated that this design provides superior immunity to deformation of optical fibres of the sensor and thus allows for more accurate SPR measurements under realistic operation conditions. This fibre optic SPR sensor is able to resolve the refractive index changes as low as 4×10^{-6} under moderate fibre deformations.

Lin *et al.*, (2007) developed side polished multimode fibre sensor based on SPR as the transducing element with halogen light source. The SPR fibre sensor is side polished until half the core is closed and coated with 37 nm gold thin films by d.c. sputtering. The SPR curve on the optical spectrum is described by an optical spectrum analyzer and can sense a range of widths in wavelengths of SPR effects. The measurement system using the halogen light source is constructed for several real time detections that are carried out for the measurement of the index liquid detections for the sensitivity analysis. The sensing fibre is demonstrated with a series of refractive index liquids and a new type of the fibre-optic micro-sensor for SPR was created on the basis of the fabrication technology of optical fibre probes in near field scanning optical microscopy (NSOM). The newly developed SPR microsensors were prepared by coating a gold metallic film on the chemically etched single mode fibre containing a conical core. They were applied to the real time monitoring of the refractive index (RI) of transparent liquids flowing in the microfluidic device.

9.2.8 Applications

Fibre optic biosensors are used for a variety of sensing applications. The small physical dimensions and the ease with which they can be used for multiplexed analyte detection make them ideal platforms for sensing. Bio-analytical micro systems based on miniaturized biosensing elements could find wide applications in DNA analysis, drug discovery, medical diagnostics and environmental monitoring as well as in protection against bioterrorism. Miniaturization, portability, multi analyte potential and interfacing with electronic functions are critical elements for biosensing devices to meet the demands in these fields.

Optical fibres have been used to develop sensor based on nucleic acids and cells. Sensors employing DNA probes have been developed for various genomics applications and the detection of microbial pathogens. Live cell based sensors used for monitoring of environmental toxins and fundamental studies on populations of individual cells.

Advances in nanotechnology have recently led to the development of fibre optics based nanosensor systems having nanoscale dimensions suitable for intracellular measurements. The possibilities to monitor *in-vivo* processes within living cells could dramatically improve understanding of cellular function, thereby revolutionizing cell biology. Fibre optic sensors provide significant advantages for *in situ* monitoring applications due to the optical nature of the excitation and detection modalities. Another advantage of fibre optic sensors is the small size of optical fibres, which allow sensing intracellular/intercellular physiological and biological parameters in micro environments.

The combination of nanotechnology, biology and photonics opens the possibility of detecting and manipulating atoms and molecules using nano-devices which have the potential for a wide variety of medical uses at the cellular level. The nano-probes were fabricated with optical fibres pulled down to tips with distal ends having sizes of approximately 30 nm-50 nm. The nanoscale size of this new class of sensors, allows for measurements in the smallest of environments. One such environment that has evoked a great deal of interest is that of individual cells. Using these nano-biosensors, it has become possible to probe individual chemical species in specific locations throughout a cell.

Biosensors utilizing immobilized antibodies have become increasingly useful due to the specificity of antigen–antibody interactions and the fact that the biosensors can be used where typical cuvette based spectroscopic measurements are not practical. Of particular usefulness are fibre optic based biosensors, as the fibre upon which the antibody is immobilized serves as a conduit for measuring optical changes occurring during the antigen–antibody binding event.

Fluorescence based biosensors provide advantages over other forms of biosensors in that the high signal-to-background discrimination associated with fluorescence is coupled to the antigen binding event. Biosensors utilizing fluorescence may also take advantage of evanescent wave excitation, where excitation light propagating through the fibre excites fluorophores within close proximity (100 nm) to the fibre surface. Excitation via this technique results in fluorescence from labeled antigens only, when the antigens bound by antibodies are within the evanescent wave. Thus, background signals from bulk solution contribute little or nothing to the total measured fluorescence, in contrast to cuvette based measurements where signal from unbound antigen may dominate. Evanescent wave biosensors are also extremely useful for conducting analyses in opaque media, heterogeneous solutions or colloidal suspensions.

9.3 MYOGLOBIN BASED OPTICAL BIOSENSOR

Myoglobin is a protein, made up of α-helices joined by short loops and it is a water-soluble globular protein. The tertiary (3-D) structure consists of 8 α-helices which fold to make a compact globular protein. The folding occurs in such a manner that almost all of the hydrophilic (polar) groups are on the outside of the protein, facing the aqueous environment. The hydrophobic group's arte almost all inside the protein and the hydrophobic effect plays a large role in maintaining the stability of the folded protein. The α-helices mostly lie with one side of the helix facing the interior of the protein and the other facing the outside. For this reason the helices are amphipathic with the side facing the interior having amino acids with hydrophobic side-chains and the side facing to outside having polar side-chains.

Myoglobin based optical biosensor originated from an LED/photo resistor based colorimeter coupled with immobilized myoglobin and this is suitable for introductory chemistry, environmental chemistry, instrumental analysis and biochemistry.

Metalloporphyrin containing proteins such as myoglobin are intensely colored and their absorbance wavelength maxima are sensitive to the interaction with ligands such as CN^-, CO, O_2, N_3^- and F^-. In these, specifically the cyanide compound releases total about 1.5 million lb from steel mills and metal heat treating industries. The Maximum Contaminant Level Goals (MCLG) for cyanide is 0.2 ppm (mg/l). The cyanide ion combines with iron in Cytochrome a_3 or Cytochrome oxidase and also has a high affinity for the ferric iron in Methemoglobin.

Myoglobin was entrapped by mixing it to final concentration of 1 mg/ml in a 10% acrylamide/1.67% bisacrylamide and 10 mM tris HCl pH 7.4, giving final volume of 15 ml. After thorough mixing, 0.05 ml of 10% (w/v) ammonium per sulfate and 0.01 ml of TEMED were added. 2 ml of this mixture goes into glass vials with rubber septa (0.50–0.75 ml each). The septa had been modified by punching a hole in each just large enough to insert an LED. Two small needles were inserted in the septa next to the LED. Green LEDs (Radio Shack, $V_F = 2.1$ V) were used as the light source. The LED voltage was maintained at 2 V using a 15 turn, 10 kΩ trimpot (potentiometer) and 9 V batteries. The detector was Radio Shack, CdS photo resistor. Photo resistor was glued to the bottom of each vial by using a thin layer of five minute epoxy. The photo resistor was connected to the 9 V batteries in series with a 1 kW resistor. Because the resistors in series form a voltage divider, the voltage can be measured across the fixed resistor. Voltage is proportional to the light intensity impingent on the photo resistor. The entire assembly was then wrapped with black electrical tape. The "dark" voltage (V_{dark}) was measured with the LED off. The cell was filled with water via syringe connected to one of the needles. With the LED lit, the voltage measured is the 100% transmittance (V_{ref}). The water was removed from the cell and the cell was then filled with the cyanide containing sample solutions. The voltage was again measured, V_{sample}. Absorbance was calculated using the following relationship:

$$A = -\log T = -\log\left(\frac{I}{I_o}\right) = -\log\left(\frac{V_{sample} - V_{dark}}{V_{reference} - V_{dark}}\right) \quad \text{... eq.-9.2}$$

Cyanide is toxic and contact with acid liberates poisonous gas. Acrylamide is a suspect carcinogen and neurotoxin. The cyanide containing solutions should be collected and oxidized prior to disposal. This is accomplished by adjusting the pH to about 10 followed by treatment with excess commercial bleach (sodium hypochlorite). The solution pH is then neutralized.

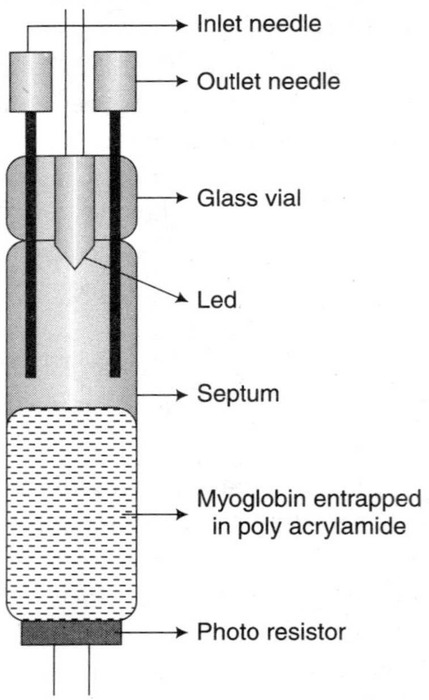

Fig. 9.5 *Schematic representation of Myoglobin based biosensor. (Source: Construction of a Simple Myoglobin—Based Optical Biosensor, Todd A. Wells, The Chemical Educator, 2007).*

Myoglobin based biosensor (Fig. 9.5) biological material myoglobin was immobilized (entrapment method) in a polyacrylamide (hydrogel). Finally, the transducer should be chosen so that the chemical interaction of the biological element with the sample is converted into an electrical signal that can be processed and displayed. Because the absorbance of myoglobin red shifts is upon cyanide binding a simple LED-based colorimeter was chosen as the transducer.

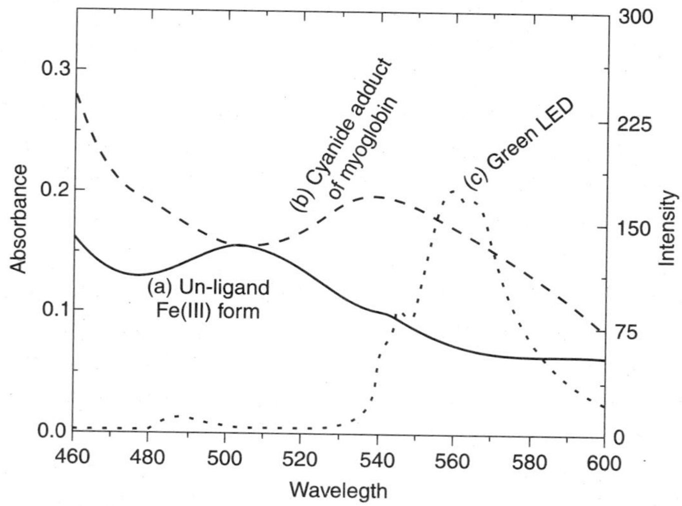

Fig. 9.6 *Comparison between the absorbance spectra of myoglobin and myoglobin with cyanide and the emission spectrum of the green LED. (Source: Construction of a Simple Myoglobin—Based Optical Biosensor, Todd A. Wells, The Chemical Educator, 2007)*

The green LED used for the biosensor had a maximum emission wavelength of 560 nm, which overlaps significantly more with the cyanide adduct of myoglobin than the un-ligand Fe (III) form and is shown in Fig. 9.6. A plot of absorbance versus cyanide concentration is shown in Fig. 9.7. The sensor has a linear response in over the concentration range. Care must be taken to allow the sample solution to equilibrate with the polymer matrix. Faster equilibration can be achieved by using a thinner polyacrylamide myoglobin element; however, it may be necessary to increase the myoglobin concentration.

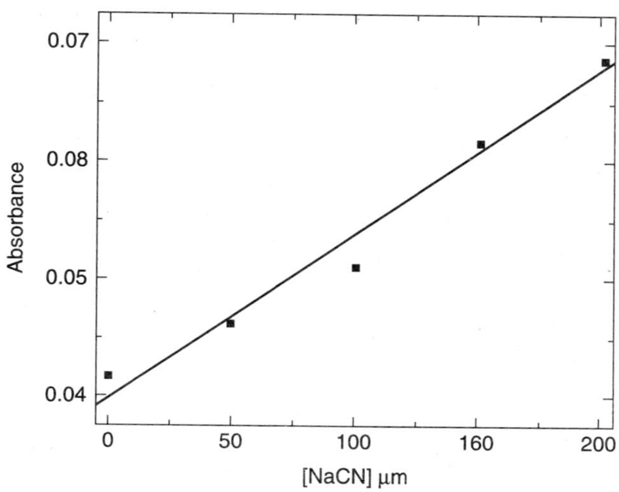

Fig. 9.7 *The myoglobin based biosensor with increasing cyanide concentrations. (Source: Construction of a Simple Myoglobin—Based Optical Biosensor, Todd A. Wells, The Chemical Educator, 2007)*

A single needle inserted through the septum may be used instead of separate inlet and outlet needle. The liquid transfer may be accomplished through needle and not by cap removal because this may cause a change in the distance between the LED and photo resistor. The light level measured by the photo resistor does vary with the separation distance.

9.4 AUTOMATED BIOSENSOR

9.4.1 Introduction

The fibre optic biosensor has developed over the last decade into a portable, automated four channel sensor and the assay consisting of sandwich fluoroimmunoassay. Antibodies are immobilized onto the surface of an optical fibre. When sample flows over the fibre probes, immobilized antibody captures any analyte present and the amount of analyte bound is determined. The binding of fluorescent tracer antibody to the bound analyte forms 'fluorescent complex' or 'sandwich'. The biosensor monitors this complex formation by evanescently exciting surface bound fluorophores with diode laser. The optical probe captures portion of the emitted fluorescence, which returns back up the fibre to the photodiode detector. Since, the excitation intensity and efficiency of fluorescence recovery falls exponentially with distance from the fibre probe surface, the system is highly discriminatory for the surface bound fluorophores.

9.4.2 RAPTOR

The RAPTOR is the only fibre optic biosensor commercially available that is designed to withstand the rigors of field testing. The system employs separate 635 nm diode laser to excite each of the four fibre optic probes. To match this excitation wavelength, the fluorescent dye Cy5 is used to tag the tracer antibody. Since the sensor performs two step sandwich assay, fluorescent tracer reagent is not contaminated by sample and negligible signals are obtained from optical probes when negative samples are analyzed. Hence, both the fibre probes and fluorescent reagent can be reused until positive is obtained. This makes the RAPTOR ideal for continuous monitoring applications. If a positive sample is detected, the user need only replace the coupon which holds the four capture antibody coated optical probes. The optical probes are injection molded polystyrene which permits inexpensive mass production. These probes are easily coated with antibody by passive adsorption overcoated with stabilizing solution and dried while being glued into the coupon thereby providing an extended shelf life. Each coupon is marked with a bar code that tells the sensor the appropriate assay protocol and the identity of the antibody on each of the probes. To use, the coupon is inserted into the holder on the RAPTOR and sample and reagents are introduced through blunt tipped needles, which connect to the coupon upon closure of the top door. From its initial design, the RAPTOR was made to be field hardened and very simple to use, even for an operator with no technical background. It can be battery operated and accepts samples introduced either manually or automatically from an attached air particle collector.

Data analysis is also completely automated which removes any subjective bias of the user. After the data are collected and analyzed using preloaded software, with user modifiable settings, the results are displayed on the LCD screen of the instrument in a simple display specifying the assays performed and the instrument's assessment of the results. While the results are displayed in qualitatively, easy-to-read format (negative, suspect, positive, high positive), quantitative data can be downloaded through serial port to a computer for further analysis.

9.4.3 Preparation of Probe and Coupon

Capture antibodies were immobilized onto fibre optic probes by passive adsorption. Injection molded, polystyrene fibre optic probes were first blackened at their distal ends to prevent reflection of excitation light. Probes were then placed into capillary tubes (100 µl, cut to 4 cm length) prefilled with 36 µl of the appropriate antibody solution (100 µg/ml IgG in 0.1 M sodium carbonate buffer at pH 9.6). After overnight incubation at 4°C, unbound IgG was rinsed off the probes by brief immersion in distilled water and the probes were incubated for 15-30 minutes in immunoassay stabilizer solution. The probes were then mounted and glued into the disposable coupons.

9.4.4 Preparation of CY5 Labeled Antibodies

Cy5 conjugated antibodies were prepared by reacting 3 mg of protein (1 mg/ml) in 50 mM sodium tetraborate, 40 mM NaCl, pH 9.0 with one vial bis-functional Cy5-reactive dye for 30 minutes at room temperature in the dark. Subsequently, labeled protein was separated from unincorporated dye by size-exclusion chromatography. Dye to protein ratios ranged from 2 to 4 Cy5 molecules per IgG.

9.4.5 Raptor Design and Operation

Insertion of the assay coupon into the RAPTOR coupon compartment aligns all necessary optical paths and engages all fluidic connections required for sample analysis. A pneumatic pump moves buffer, air, fluorescent reagent or sample within the system. Serpentine channels in the coupon provide a common path across the probe surfaces. Bubble detectors, which monitor liquid-to-air interfaces, control introduction of sample and fluorescent reagent.

During the two step sandwich immunoassay, sample is flowed over the four fibre optic probes mounted in the assay coupon. Antigen present in the sample binds to the fibre optic probe coated with capture antibody specific for that antigen; unbound material is washed away by brief rinse with Poly Butylene Teraphalate (PBT). Fluorescently labeled antibody is next introduced and binds to the antibody-antigen complexes on the probe surface, completing the sandwich assay. This tracer reagent is maintained at a suitable temperature in an onboard thermal storage module and recovered after each assay cycle, allowing multiple sequential analyses to be performed. Excitation light from four 5.0 mW laser diodes (635 nm) within the RAPTOR is focused into the fibre optic waveguides. An evanescent wave is created along each probe, exciting the fluorescent emission of specifically bound Cy5-labeled antibodies. The portion of the fluorescence captured by the optical probe is collimated by the probe's molded lens and focused onto a photodiode using a ball lens, chosen for its light-gathering power and short focal length. A long pass dichroic filter (665 nm) rejects reflected laser light. If each fibre optic probe has been coated with capture antibodies with differing specificities, one sample is interrogated for four different analytes simultaneously.

9.4.6 Assay Procedure

Each assay consists of multiple steps, which are performed automatically by the RAPTOR. Immediately after loading the coupon into the instrument, prior to analyte challenge, the RAPTOR automatically initiates a five minute baseline protocol. The probes are briefly rinsed with Poly

Butylene Teraphalate (PBT) (for hydration) and are then incubated for 90 seconds with fluorescent tracer antibody. After tracer reagent is returned to its compartment, the probes are rinsed with PBT and an initial background wash value is determined. This background rate represents the rate of nonspecific binding of the fluorescent antibody to the probe surface and must be determined before samples are analyzed.

To analyze a sample, it is first loaded into the sample port using 1 ml syringe equipped with a blunt-tipped needle or a transfer pipette. After the coupon is rinsed briefly with PBT, the sample is flowed into the coupon and is incubated with the probes for seven minutes. A subsequent wash with PBT eliminates unbound material; the RAPTOR automatically flushes the sample port with PBT at the same time as the PBT wash, thus preventing sample carryover. After that, the coupon is cleared with air to prevent dilution of the incoming tracer reagent. Cy5 antibody reagent is flowed into the coupon and incubated for 90 seconds to interrogate the amount of antigen bound to the probes. The tracer reagent is returned to its reservoir for reuse and a final PBT wash completed. The entire standard assay cycle, including all wash steps, is completed in 10 minutes.

The rate of fluorescence increase during incubation with the Cy5 antibody is calculated ('assay rate') and a final reading is taken to determine the increase in fluorescence due to Cy5 antibody bound during the entire assay cycle (wash delta). While one parameter is sufficient for strongly positive samples, use of both factors greatly reduces the number of false positives obtained. Samples are considered positive if both the values are greater than the means of the background values plus three standard deviations. The assays are fast (3-10 minutes) and sensitive (a limit of detection is 1-10 ng/ml, 50-5000 cfu/ml, depending on the antibody: analyte combinations).

RAPTOR's have sensitivity and reliability for the field operations. The RAPTOR fibre optic biosensor is well suited for testing for pathogens and toxins and can be used as an in-line monitor or to analyze discrete samples. Only minimal sample preparation (*Eg.* dilution or coarse filtration) is required. Disposable coupons can be made with optimum assay combinations, allowing multiple samples to be analyzed for multiple pathogens without changing probes or reagents.

9.4.7 Applications

The RAPTOR and the fibre optic biosensors from which it evolved have been used to detect and quantify a variety of hazardous substances in numerous sample matrices; which includes detection of explosives, toxins, bacteria, viruses, in groundwater, soil extracts, air extracts, meat homogenates and clinical fluids. The limit of detections for several different analytes is shown in Table 9.1.

Table 9.1 Limit of detection for toxins and pathogens using a 10-minute assay performed with RAPTOR.

Type of Analyte	*Name of the Pathogen*	*Detection limit*
Toxin	Cholera toxin	1 ng/ml
Bacterial surface protein	*Yersinia pestis* F1	10 ng/ml
Gram positive bacterium (vegetative form)	*Bacillus anthracis*	50 cfu/ml
Gram positive bacterium (spore)	*Bacillus globigii*	5×10^4 spores/ml
Gram negative bacterium	*Brucella abortus*	7×10^4 cfu/ml

Optical Biosensors - II

Chapter 10

10.0 GUIDED WAVE OPTICAL BIOSENSORS

10.0.1 Introduction

Photonic and electronic micro- and nanosensors are emerging as very attractive devices to be employed in various fields such as medicine, microbiology, particle physics, automotive, environmental safety and defense. At present, great efforts have been made in development of the sensors by using the actuator monolithic integration, micro-electro-mechanical system (MEMS) and micro-opto electro-mechanical system (MOEMS). Optical sensors have considerable attention because of their immunity to electromagnetic interference, good compactness, robustness and high compatibility with fibre networks, but also because they usually exhibit shorter response time and higher sensitivity with respect to MEMS/MOEMS devices.

Sensors used for biochemical applications (*Eg*. biomolecules concentration measurement, DNA molecules investigation and pH estimation) must reach some specific performance requirements which are significantly different from those typical for other sensors. In particulars, biosensor has to be contextually highly sensitive and selective to the analyte being detected, to be biocompatible and immune to external disturbances as either pressure or temperature changes. Using photonic technologies, it is possible to fabricate very compact, highly performing and low cost biosensors.

10.0.2 Effective Index Change In Guiding Structures

In an optical guiding structure shift of propagating optical mode effective index can be induced by two different physical phenomena (Fig. 10.1): (i) The change of thickness (equal to ρ) of an adlayer constituted by adsorbed or bound molecules which are transported (by convection or diffusion method) from the gaseous or liquid medium, that serves as waveguide cover medium and to cover medium/guiding film interface. (ii) The change of refractive index (Δn) of the homogeneous liquid that serves as waveguide cover medium (Fig. 10.1).

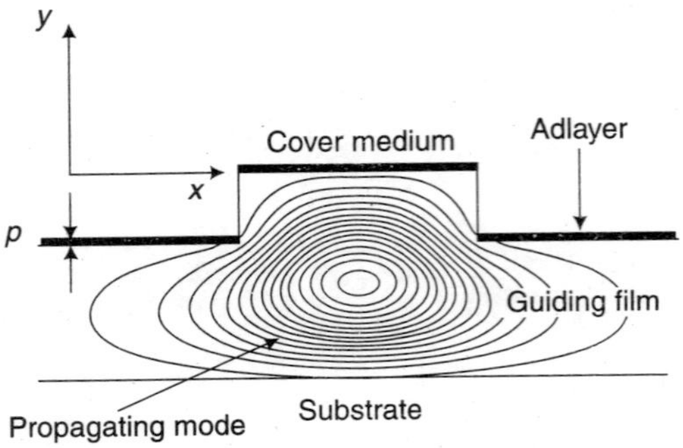

Fig. 10.1 *Effective index change in integrated optical sensors. (Source: Guided-Wave Optical Biosensors, Sensors ISSN 1424-8220 @ 2007 by MDPI, www.mdpi.org/sensors)*

10.0.3 Integrated Optical Biosensors

Integrated optical sensors adopt waveguide to confine optical power and are based on evanescent field sensing. Although light is confined within the guiding film of the waveguide, there is a part of the guided light that travels through a region extending outside into the medium surrounding the waveguide and can interact with the environment. This kind of optical sensors exhibits advantages with respect to other optical sensors such as compactness, high sensitivity and possibility of mass production.

In large number of integrated optical biosensors, the analyte presence either produces shift in waveguide cover medium refractive index (homogeneous sensing) or determines the thickness increase of a molecular layer deposited on guiding film or cover medium interface (surface sensing). Both phenomena affect the propagating optical mode effective index which is measured in different ways according to specific biosensor architecture.

10.0.4 Interferometer

Interaction between the sample and the optical signal propagating in the sensor produces change of optical mode effective index and then in its phase. To convert this phase shift in an amplitude change, interferometer architectures are commonly used. Among these architectures, Mach-Zehnder approach assures high sensitivity and mostly used. In these kinds of integrated optical biosensors (Fig. 10.2) input optical signal is split by Y-junction in two signals. These two signals propagate in the reference and sensing arms (in this last arm the interaction between the sample and the optical signal takes place). After the propagation in these arms, the two optical signals accumulate a phase shift $\Delta\varphi$ and the shifted phase optical signals interfere by an output Y-branch. If input power pin is unequally divided by input Y-junction between the two arms (the sensing arm receives an optical power equal to $Y_m P_{in}$ and the reference arm receives an optical power equal to $(1 - \gamma_1)$. P_{in} and output power coming from output Y-junction is the sum of optical power coming from sensing arm multiplied by γ_2 and optical power coming from reference arm multiplied by $(1 - \gamma_2)$, the ratio between the output and input powers is shown in Fig. 10.2.

$$\frac{P_{out}}{P_{in}} = \gamma_1 \gamma_2 + (1 - \gamma_1).(1 - \gamma_2) + 2\sqrt{\gamma_1\gamma_2(1 - \gamma_1)(1 - \gamma_2)}.\cos(\Delta\varphi) \quad \text{... eq.-10.1}$$

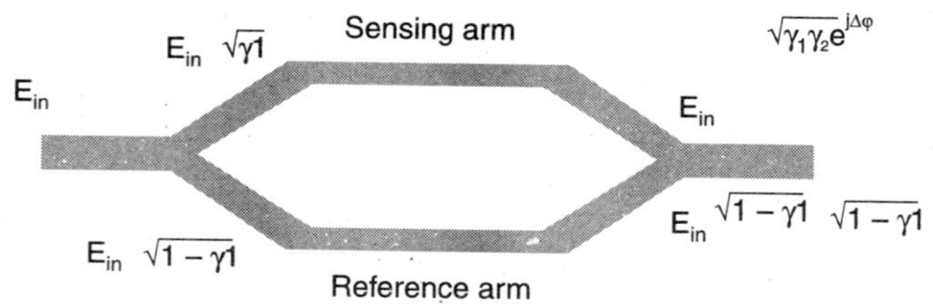

Fig. 10.2 *Mach-Zehnder interferometer architecture. (Source: Guided-Wave Optical Biosensors, Sensors ISSN 1424-8220 @ 2007 by MDPI, www.mdpi.org/sensors)*

Figure 10.3 shows the contour plot of $\frac{P_{out}}{P_{in}}$ ratio (for $\Delta\varphi = 0$) with respect to γ_1 and γ_2 and observed that $\frac{P_{out}}{P_{in}} = 1$, if $\gamma_1 = \gamma_2$. The perfect matching is obtained for $\gamma_1 = \gamma_2 = 1/2$. In fact, as it is shown in Fig. 10.3, only when $\gamma_1 = \gamma_2 = 1/2$ the sensor exhibits maximum extinction ratio.

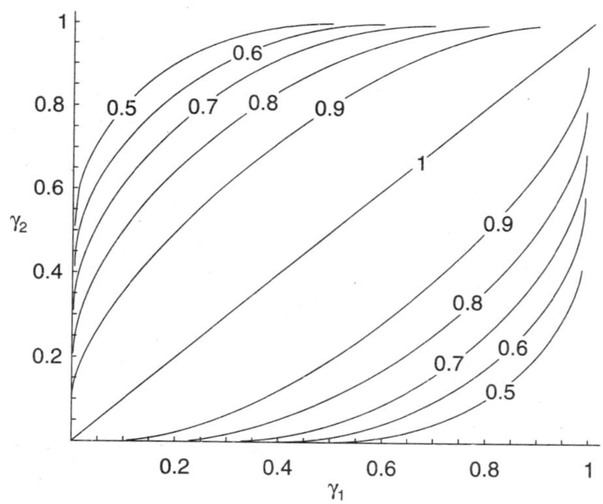

Fig. 10.3 *Contour plot of output/input power ratio (P_{out}/P_{in}) vs. γ_1 and γ_2. (Source: Guided-Wave Optical Biosensors, Sensors ISSN 1424-8220 @ 2007 by MDPI, www.mdpi.org/sensors)*

The integrated optical sensors is based on Mach-Zehnder interferometers (MZI), to detect different chemical species as organic compounds and proteins. Technologies employed for fabrication of these sensors are prevalently CMOS compatible (guiding film in silicon, silicon nitride or silicon oxynitride) but glasses and III and V semiconductor compounds technology have been developed. The sensitivity 'S' of sensors based on Mach-Zehnder interferometer can be expressed, assuming that $\Delta\varphi$ varies (Fig. 10.4) in a narrow range around $\Pi/2$, as:

$$S = \frac{2\pi}{\lambda} LS_w \quad \text{... eq.-10.2}$$

where S_w is the waveguide sensitivity.

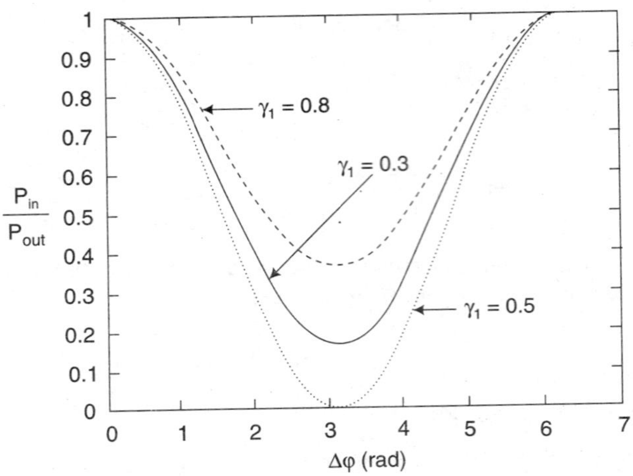

Fig. 10.4 *Output/input power ratio vs. $\Delta\phi$ for different γ_1 values. (Source: Guided-Wave Optical Biosensors, Sensors ISSN 1424-8220 @ 2007 by MDPI, www.mdpi.org/sensors)*

In Mach-Zehnder optical biosensor has been fabricated, these are capable to monitor antigen-antibody immunoreactions and detection limit is 7×10^{-6} (with respect to detectable change in cover medium refractive index). A detection limit equal to 1.5×10^{-6} has been obtained by a Mach-Zehnder optical biosensor fabricated using a waveguide produced in a BGG36 glass substrate. In glass based MZI, a graded index waveguide is commonly adopted. This kind of guiding structure supports optical modes having a weak evanescent field and so exhibits a reduced sensitivity. To improve glass waveguide sensitivity to adopt a channel-planar composite optical waveguide (COWG) to realize MZI sensing arm has been developed. A channel planar COWG consists of a single mode channel waveguide locally overlaid with a tapered film (Fig. 10.5). This kind of guiding structure permits to enhance evanescent field and consequently waveguide sensitivity.

Other configurations adopted for interferometer integrated optical biosensors are those based on Young interferometers. Using a four-channel Young interferometer, the simultaneous and independent monitoring of tree antigen-antibody reactions has been achieved. In this device the generated interference pattern is recorded by a CCD camera and analyzed by Fast Fourier Transform (FFT) algorithm and detection limit of 8.5×10^{-8} has been obtained.

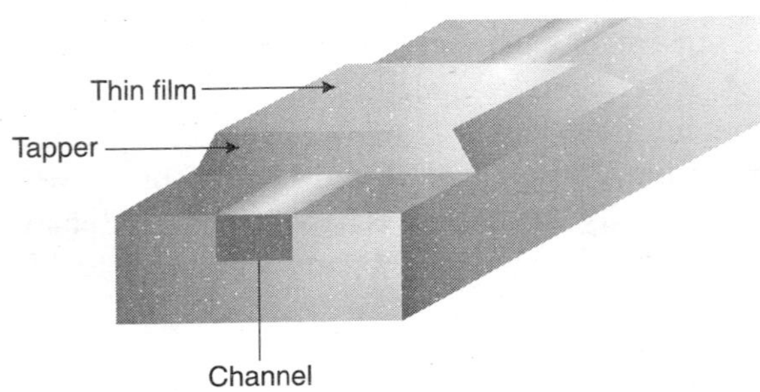

Fig. 10.5 *Schematic diagram of a channel-planar composite optical waveguide. (Source: Guided-Wave Optical Biosensors, Sensors ISSN 1424-8220 @ 2007 by MDPI, www.mdpi.org/sensors)*

10.0.5 Anti-Resonant Reflecting Optical Waveguides

Anti-resonant reflecting optical waveguide (ARROW) is a five-layer guiding structure where light is confined within the core layer by total internal reflection at the cover medium or core interface and by anti-resonant reflection (99.96%) due to the presence of the two cladding layers underneath the core (Fig. 10.6).

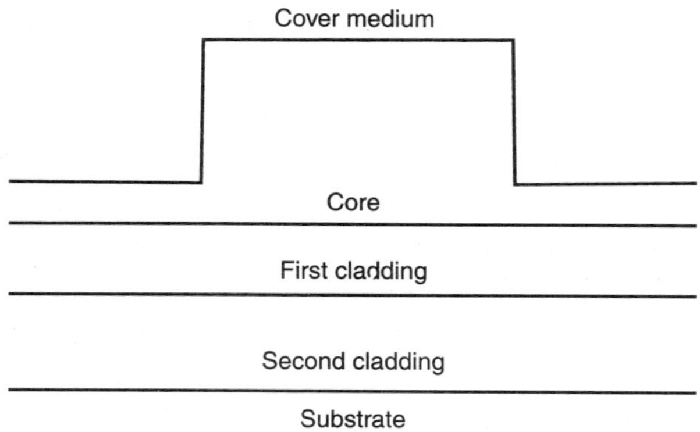

Fig. 10.6 *ARROW waveguide structure. (Source: Guided-Wave Optical Biosensors, Sensors ISSN 1424-8220 @ 2007 by MDPI, www.mdpi.org/sensors)*

ARROW is a leaky guiding structure with virtual mono mode behaviour, where higher order modes are filtered out by loss discrimination due to the low reflectivity assured to these modes by the two cladding layers. The refractive indices and thicknesses of the cladding layers have to be properly designed for the working wavelength to assure high reflectivity and having good guiding characteristics. ARROWs are usually fabricated in a CMOS compatible technology and the materials commonly used are silicon for the substrate, silicon nitride for the first cladding, silicon oxide for the core and second cladding layers.

ARROW structures exhibit low losses, permit larger dimensions than conventional total internal reflection based waveguides (micrometers instead of nanometers) enable to confine optical power in low index material (silicon oxide) and exhibit good sensitivity and these features are quite attractive for sensing applications in high sensitive and low cost integrated optical sensors. In Mach-Zehnder interferometer adopting an ARROW as guiding structure has been used to fabricate 15 mm long optical biosensor with detection limit of 2×10^{-5}.

10.0.6 Hollow Waveguides

Hollow waveguides permit to simultaneously confine the optical power and the solution to be analyzed in the waveguide core constituted by a low-index liquid substance. In the waveguides, the optical signal is confined inside the core (refractive index - 1.33) by two cladding layers designed to produce the anti resonant reflection (Fig. 10.7).

In liquid core, hollow waveguide has been adopted to realize an integrated sensor capable to detect refractive index changes in liquid substance constituting the core. This sensor exhibits minimum detectable change in core medium refractive index equal to 9×10^{-4}.

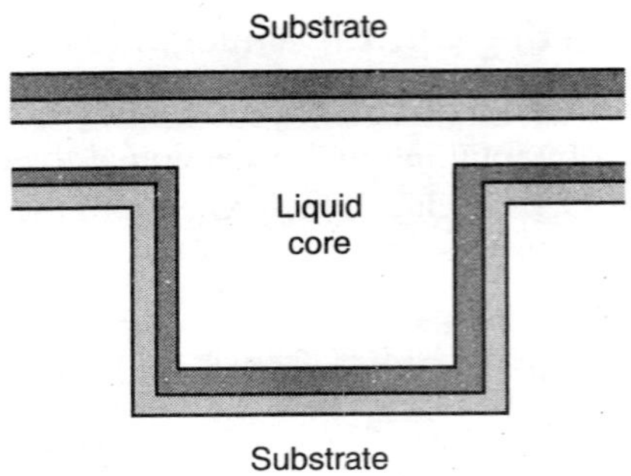

Fig. 10.7 *Hollow waveguides cross section. (Source: Guided-Wave Optical Biosensors, Sensors ISSN 1424-8220 @ 2007 by MDPI, www.mdpi.org/sensors)*

10.0.7 Surface Plasmon Resonance

Integrated optical biosensors exploiting surface plasmon resonance (SPR) are constituted by a single mode waveguide produced. *Eg.* Thermal ion-exchange in glass substrate. A small region of waveguide is coated with a thin metal film and the metal is coated with a part of the sensor can be supported to the surface plasmon waveguides (SPW) and forms the interaction region with the solution containing the analyte. The structure incorporates Y-junction used as 3 dB power splitter. A quasi-TM mode is excited in the input waveguide. After passing through the 3 dB power divider, this mode couples across the step discontinuity between the metal coated and uncoated waveguide regions exciting the surface plasmon wave (SPW). After propagating through the metal coated region of the sensor, SPW excites the optical mode in the uncoated output waveguide (Fig. 10.8).

The integrated optical biosensor for Simazine detection has been designed and fabricated. This sensor exhibits a minimum detection of 1×10^{-4} with a detection limit equal to 0.1 μg/l for Simazine concentration measurements. A minimum detectable change in cover medium refractive index equal to 1.2×10^{-6} is achievable by the SPR based integrated biosensor. This sensor is able to detect 2 nm of human chorigonadotropin present in 1 ml of 1% bovine serum albumin solution.

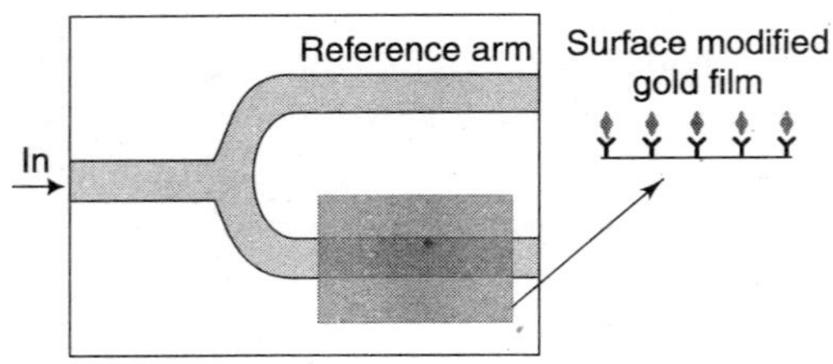

Fig. 10.8 *Integrated optical biosensor exploiting the SPR scheme. (Source: Guided-Wave Optical Biosensors, Sensors ISSN 1424-8220 @ 2007 by MDPI, www.mdpi.org/sensors)*

In planar structure (Fig. 10.9) the propagation constant of SPW propagating at the interface between metal and dielectric medium. In order to optically excite the SPW and the propagation constant of light in the dielectric medium has to be increased. This enhancement can be done using either attenuated total reflection (ATR) method or grating couplers.

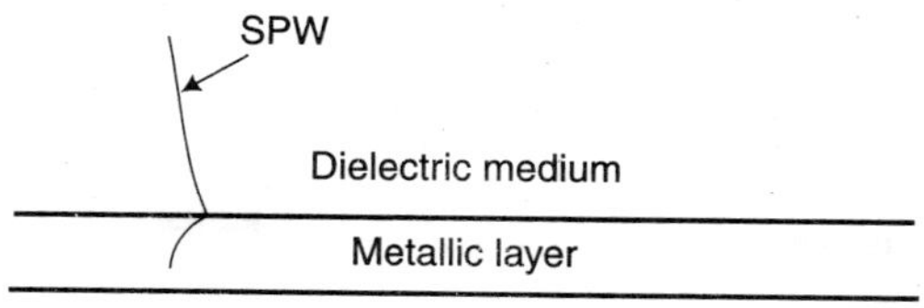

Fig. 10.9 *Planar structure in which SPW propagates. (Source: Guided-Wave Optical Biosensors, Sensors ISSN 1424-8220 @ 2007 by MDPI, www.mdpi.org/sensors)*

10.0.8 Silicon Slot Waveguides

Recently sub-micrometer slot waveguides develop by using silicon on insulator (SOI) technology (Fig. 10.10) and emerging as very attractive guiding structures for integrated optics. The guiding structure is constituted by two Si wires very close to each other and supports a quasi-TE mode highly confined in the low index gap region between the two Si-wires (Fig. 10.11).

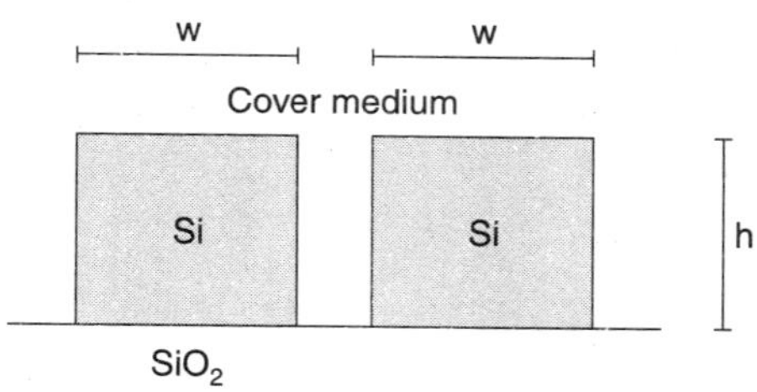

Fig. 10.10 *SOI slot waveguide structure. (Source: Guided-Wave Optical Biosensors, Sensors ISSN 1424-8220 @ 2007 by MDPI, www.mdpi.org/sensors)*

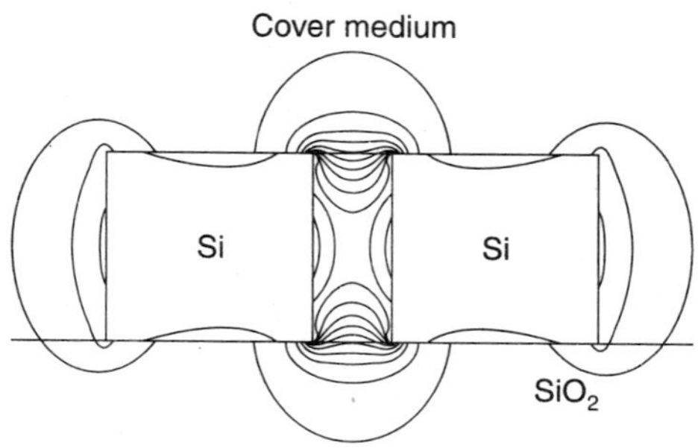

Fig. 10.11 *Quasi-TE mode profile in silicon slot waveguide. (Source: Guided-Wave Optical Biosensors, Sensors ISSN 1424-8220 @ 2007 by MDPI, www.mdpi.org/sensors)*

Variety of optical devices have been realized by adopting slot waveguides including, micro-ring resonators, optical modulators, electrically pumped light emitting devices, directional couplers, all-optical logic gates and beam splitters. High values (around 65%) reached by confinement factor in cover medium when quasi-TE modes propagates in a slot waveguide, suggest to adopt this guiding structure for ultra-compact and ultra-sensitive integrated optical biosensor fabrication. Finite

Element Method (FEM) based investigation of silicon slot waveguide sensitivity has demonstrated that this guiding structure can exhibit sensitivity values significantly larger than those shown by other very sensitive waveguides (such as Si-wire waveguides).

10.0.9 Integrated Optical Micro Cavities

Micro-ring and micro-racetrack resonators are emerging as very attractive building blocks for micrometer scale photonics, essentially because of their compactness and versatility. In fact, a great number of optical functions related to telecommunication systems such as filtering, modulation, amplification and switching can be performed using guided wave micro-resonators. Micro-resonator advantages in sensing applications are related to the device size reduction by some orders of magnitude which do not compromise the device sensitivity, because the large photon lifetime within the resonator at the resonance provides an equivalently long interaction length to achieve a detectable phase shift.

Light coupled to micro-resonators is confined within the structure by total internal reflection forming high quality factor resonant modes. Any interaction with the evanescent tail of the optical field (usually the change in waveguide cover medium refractive index) affects the guided mode and thus changes the resonance behaviour of the cavity (Fig. 10.12). This change in resonance can be detected with very high sensitivity by optimizing the micro cavity design and the method of observation.

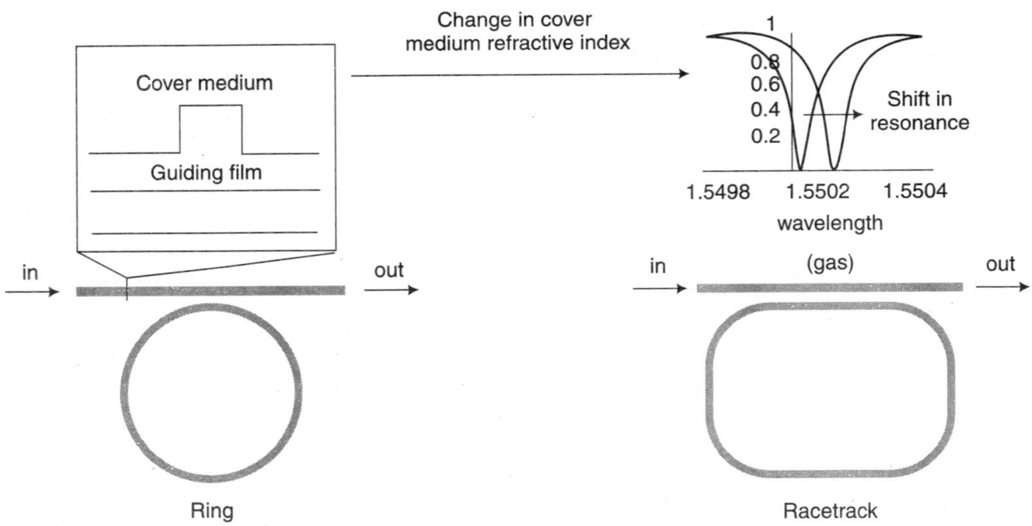

Fig. 10.12 *Micro cavity-based architectures for integrated optical sensors. (Source: Guided-Wave Optical Biosensors, Sensors ISSN 1424-8220 @ 2007 by MDPI, www.mdpi.org/sensors)*

Any micro-racetrack resonator can be theoretically investigated by using the schematic diagram as shown Fig. 10.13 (micro-ring resonator can be seen as a micro-racetrack resonator in which the resonator straight section has a length LS equal to zero and so it can be described adopting the same formalism).

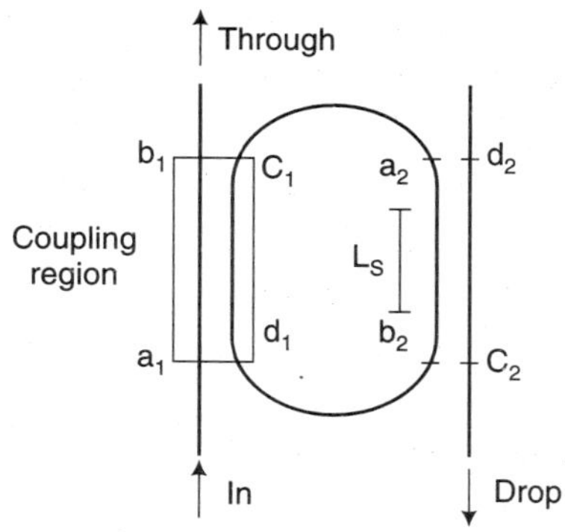

Fig. 10.13 *Resonator racetrack structure schematic representation. (Source: Guided-Wave Optical Biosensors, Sensors ISSN 1424-8220 @ 2007 by MDPI, www.mdpi.org/sensors)*

The capability of a micro-ring resonator in SOI technology adopting a Si-wire waveguide and having a radius of 3 μm to sense the change in ammonia concentration have been theoretically demonstrated, whereas a design strategy to optimize Si_3N_4/SiO_2 micro-ring based integrated optical sensors have been proposed.

A micro racetrack resonator fabricated using polymeric materials, in which the resonance slope has been enhanced by the introduction of two partially reflecting elements (Fig. 10.14) have been adopted for biochemical sensing. Exploited configuration produces a fan resonant line shape that greatly enhances the device sensitivity. Proposed sensor exhibits a minimum detectable concentration change of glucose solution around 0.024% or 24 mg/dl.

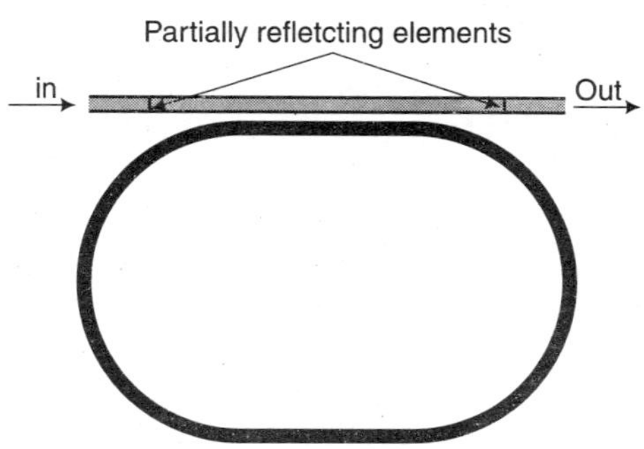

Fig. 10.14 *Micro racetrack resonator biosensor with sensitivity enhanced by introducing two partially reflecting elements. (Source: Guided-Wave Optical Biosensors, Sensors ISSN 1424-8220 @ 2007 by MDPI, www.mdpi.org/sensors)*

A lowering of detection limit related to glucose concentration measurement is obtainable by increasing the micro-cavity quality factor Q. The Q value around 20,000 has been estimated for the micro-ring resonator proposed (fabricated using polymeric materials), so this sensor exhibits a detection limit equal to 0.915 mg/dl in glucose concentration measurement (homogeneous sensing) and equal to 250 pg/mm^2 in Spavidin molecule detection (surface sensing). Finally, the capability of this sensor to detect shifts of propagating mode effective index around 10^{-7} has been also proved. Vertical coupling between straight waveguide and micro-ring resonator is employed in other biosensors (Fig. 10.15). Micro-ring quality factor is around 12,000 and detection limit, in terms of minimum detectable change in effective index is equal to 1.8×10^{-5}.

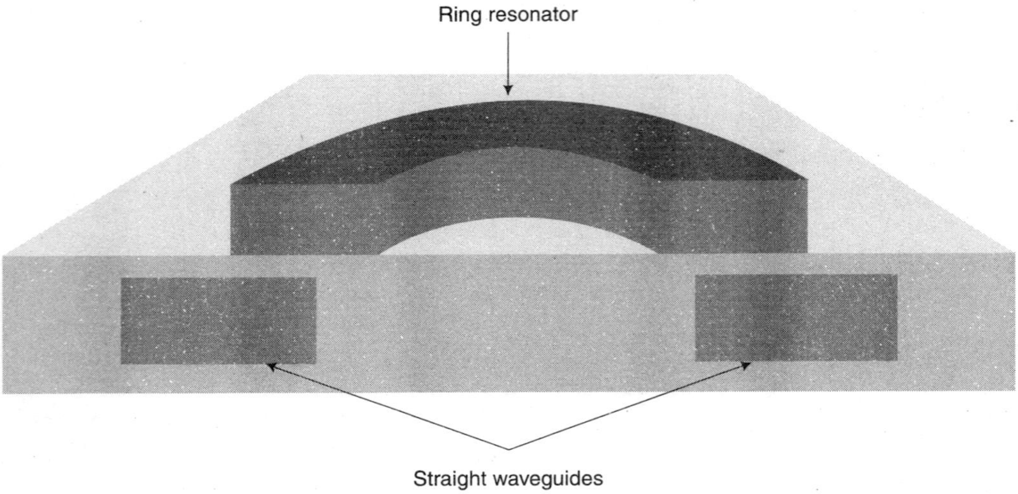

Fig. 10.15 *Vertical coupling between micro-ring and straight waveguides. (Source: Guided-Wave Optical Biosensors, Sensors ISSN 1424-8220 @ 2007 by MDPI, www.mdpi.org/sensors)*

Adopting a Si_xN_y/SiO_2 waveguide, a racetrack resonator having a radius of 2 mm has been fabricated for protein detection. This sensor is capable to detect very low concentrations of Avidin in phosphate buffered saline solution (detection limit is equal to 0.1 nM). Adopting a micro disk resonator fabricated using a CMOS-compatible technology (Si_3N_4/SiO_2 waveguide) and having a radius of 15 μm measurement of a change in cover medium refractive index as low as 10^{-4} was experimentally demonstrated.

10.1 OPTICAL BIOSENSOR FOR PROBING CELL SIGNALING

10.1.1 Introduction

Optical biosensors are well established analytical tool for biomolecular interaction analysis and to examine living cells. Mainly in cell signaling analysis, the advancements in instrumentation of optical biosensors particularly large scale screening platforms enable the discovery of specific types of cellular target dependent optical signals in cells. Three types of optical biosensors are being used for biomolecular interaction analysis and their principles are discussed below:

1. Surface Plasmon Resonance: The receptor (Y) is covalently coupled to the derivatized surface of gold thin layer, which is deposited onto glass substrate (Fig. 10.16). The incident light on the gold layer is directed by a prism. The reflectance of the gold surface is a function of incident angle. The resonance angle shifts when target molecules (Y) in solution bind to the immobilized receptors.

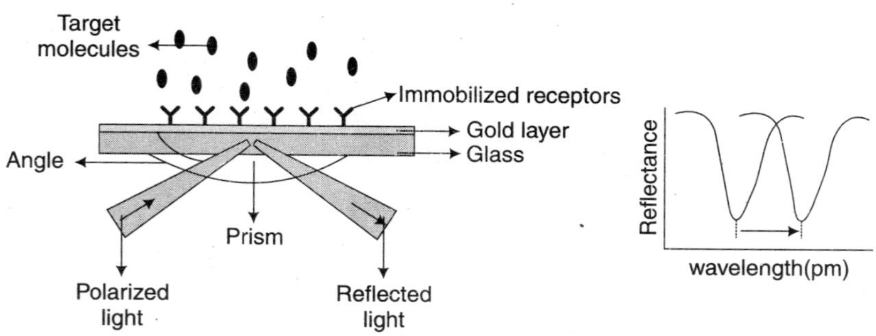

Fig. 10.16 *Principle of Surface Plasmon Resonance optical biosensor for biomolecular interaction analysis. (Source: Non-Invasive Optical Biosensor for Probing Cell Signaling Ye Fang, sensors, ISSN 1424-8220, © 2007 by MDPI, www.mdpi.org/sensors)*

2. Resonance Wave Guide biosensor: The coupling of incident light into the waveguide is achieved by the diffraction grating (Fig. 10.17). The intensity of reflected light is a function of the resonant wavelength. The binding of target molecules in a sample to the immobilized receptors results in shift in the resonant wavelength.

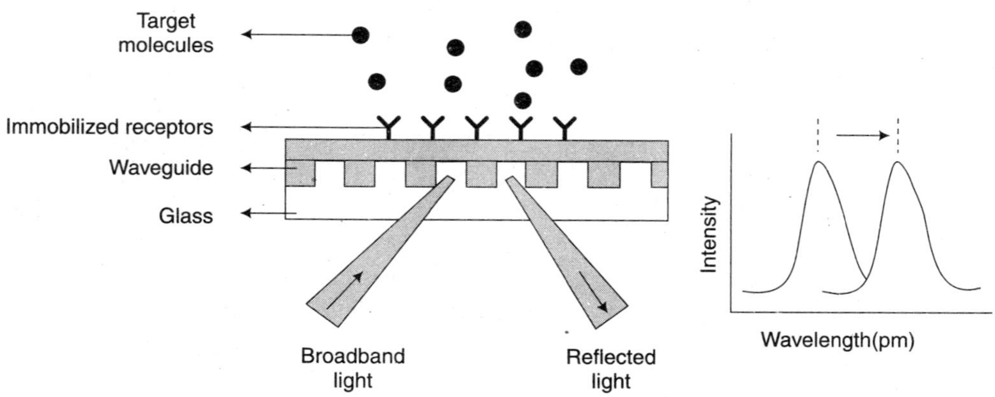

Fig. 10.17 *Principle of Resonance Wave Guide optical biosensor for biomolecular interaction analysis. (Source: Non-Invasive Optical Biosensor for Probing Cell Signaling Ye Fang, sensors, ISSN 1424-8220, © 2007 by MDPI, www.mdpi.org/sensors)*

3. Interferometric biosensor: A broadband light resource is directly incident at the solution surface interface (Fig. 10.18). Two types of reflected light waves originate from the two reflecting surfaces (i) the interface with the optical layer and (ii) the surface of bioreceptors, interact each other, leading to interference patterns. The binding of biomolecules to the bioreceptors alters the inference pattern.

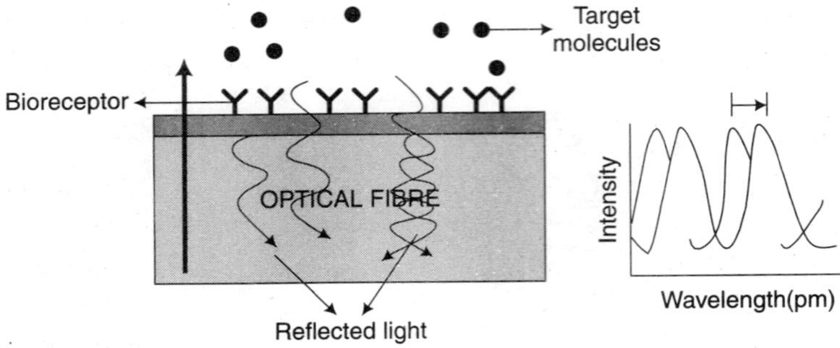

Fig. 10.18 *Principle of Interferometric optical biosensor for biomolecular interaction analysis. (Source: Non-Invasive Optical Biosensor for Probing Cell Signaling Ye Fang, sensors, ISSN 1424-8220, © 2007 by MDPI, www.mdpi.org/sensors)*

10.1.2 Theory

All optical biosensors commonly measure changes in local refractive index at or very near the sensor surface and which are applicable for cell sensing. In principle, they can utilize an evanescent wave to characterize ligand induced changes in cells. The evanescent wave is an electromagnetic field produced by the total internal reflection of light at solution surface interface which typically extends short distance (~hundreds of nanometers) into the solution with a characteristic depth, termed 'penetration depth' or 'sensing volume'.

Ramsden *et al.* (1994) described theory to study cell adhesion and spreading on the surface of a Resonant Wave Guide (RWG) biosensor. This theory assumes 3-layer waveguide system wherein the cells undergo shape changes with well defined geometrical parameters and steps upon adhesion and spreading on the biosensor surface. Such model allows both the number of cells at the surface and parameter characterizing their shape to be accurately determined. Recently, developed theoretical and mathematical models that describe the parameters and nature of optical signals measured in living cells in response to stimulation with ligands. These models, based on three-layer waveguide system in combination with known cellular biophysics, provide link from ligand induced optical signals to several cellular processes mediated through a receptor.

To probe cell signaling, cells are brought to in close contact with the sensor surface through cell culture. The biosensor measures an averaged response of cells located at the area illuminated by the incident light. The cell layer of highly convergence is used in order to achieve optimal assay results, because of the large dimension of cells compared to the short penetration depth (~100-200nm) of biosensor. Configuration biosensor is non-conventional three-layer system, which includes substrate, waveguide film with grating structure and cell layer (Fig. 10.19). Ligand induced change in effective refractive index (i.e., the detected signal) is directly proportional to the change in refractive index of the bottom portion of cell layer:

$$\Delta N = S(C)\, \Delta n_c \qquad \text{... eq.-10.3}$$

Where $S(C)$ is the sensitivity to the cell layer, and Δn_c the ligand induced change in local refractive index of the cell layer sensed by the biosensor. Because the refractive index of given volume within the cell is determined by the concentrations of biomolecules such as proteins, Δn_c is directly proportional to ligand-induced change in local concentrations of cellular targets or molecular assemblies within the sensing volume. Considering the exponentially decaying nature of the evanescent wave extending away from the sensor surface, the ligand induced optical signal is given by:

$$\Delta N = S(c)\ \alpha d \sum_i \Delta C_i \left[e^{\frac{-Z_i}{\Delta Z_c}} - e^{\frac{-Z_{i+1}}{\Delta Z_c}} \right] \qquad \text{... eq.-10.4}$$

Where ΔZc is the penetration depth into the cell layer, α the specific refraction increment (about 0.18/mL/g for proteins), z_i the distance where the mass redistribution occurs and d is an imaginary thickness of a slice within the cell layer. Here the cell layer is divided into an equal-spaced slices in the vertical direction. Equation indicates that the ligand-induced optical signal is the sum of mass redistribution occurring at distinct distances away from the sensor surface, each with an unequal contribution to the overall response. Furthermore, the detected signal, in terms of wavelength or angular shifts, is primarily sensitive to mass redistribution occurring perpendicular to the sensor surface. Because of its dynamic nature, it is also referred to as dynamic mass redistribution (DMR) signal.

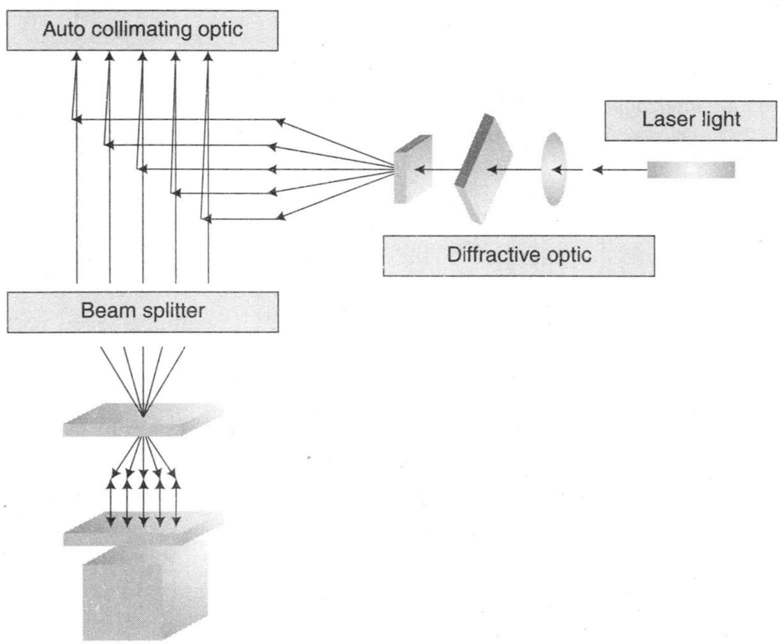

Fig. 10.19 *Angular interrogation detection system of RWG biosensor for high throughput applications. (Source: Non-Invasive Optical Biosensor for Probing Cell Signaling Ye Fang, sensors, ISSN 1424-8220, © 2007 by MDPI, www.mdpi.org/sensors)*

Two detection schemes of RWG biosensors are used for high throughput applications: (1) Angular interrogation system–a laser light is split into an array of 7 × 7 light beams, such that each illuminates a biosensor located within one well of 96 well SBS standard micro plate. The reflected lights are received with a CCD camera as an image of 7 × 7 resonant bands (Fig. 10.19). (2) Wavelength interrogation system–a linear array of 16 illumination/detection heads is used to scan a 384 well micro plate, one column at a time. In each bundled head, there is an optical fibre to send a laser light to illuminate a biosensor located in each well and a second optical fibre to receive the reflected light (Fig. 10.20).

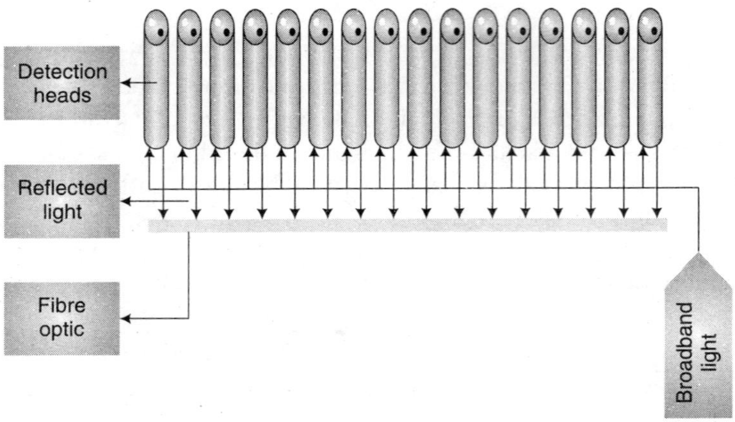

Fig. 10.20 *Wavelength interrogation detection system of RWG biosensor for high throughput applications. (Source: Non-Invasive Optical Biosensor for Probing Cell Signaling Ye Fang, sensors, ISSN 1424-8220, © 2007 by MDPI, www.mdpi.org/sensors)*

In the angular interrogation system (Fig. 10.19) the total internal reflection is observed at a critical angle of incidence at the interface of the waveguide film and the cell layer. This system consists of launch system to generate an array of light beams such that each illuminates a RWG biosensor with a dimension of ~200 μm × 3000 μm and CCD camera-based receive system to record changes in resonant angle of the reflected light. The array of light beams is obtained using a beam splitter and diffractive optical lenses. This system allows an array of 7 × 7 biosensor wells in a 96 well micro plate to be simultaneously sampled at every 3 seconds. This system does not have any moving parts, except of a liquid handling system for compound addition.

In the wavelength interrogation system (Fig. 10.20), the total internal reflection is achieved at resonant wavelength of broadband light source illuminated at the waveguide. This system consists of a detector with temperature controlled environment, an external liquid handling system and 384 well standard SBS microtiter plates. The detector system is centered on integrated fibre optics to measure any wavelength shift of the reflected light. A broadband light source, generated through fibre optic and collimating lens, is used to illuminate a small region of the biosensor through the bottom of micro plate. A detection fibre for recording the reflected light is bundled with the illumination fibre. A series of illumination/detection heads are arranged in a linear fashion, so that reflection spectra are collected from all 16 wells within the same column of a 384 well micro plate simultaneously. The whole plate is scanned so that each sensor is addressed multiple times and each column is addressed in sequence. The temperature controlling unit is built-in to minimize any temperature fluctuations.

10.1.3 Probing Cell Signaling with Optical Biosensor

Cells rely on multiple cellular pathways or machineries to process, encode and integrate the information received. Unlike the affinity analysis with optical biosensors that specifically measures the binding of analytes to protein target, living cells are much more complex and dynamic. This section discusses the cellular mechanisms that provide the basis for cell signaling studies with optical biosensors as well as the applications of optical biosensor in cell signaling.

10.1.3.1 Cell-surface interactions

To study cell signaling, cells are brought in contact with the surface of a biosensor which can be achieved through cell culture. These cultured cells are attached onto the biosensor surface through three types of contacts: focal contacts, close contacts and extracellular matrix contacts, each with its own characteristic separation distance from the surface. As a result, the basal cell membranes are generally distant away from the surface by ~10-100 nm. Since the penetration depths or sensing volumes of most optical biosensors are in the range of 100-200 nm, these biosensors are able to sense the bottom portion of cells.

Cells, in many cases exhibit surface dependent adhesion and proliferation. In order to achieve robust cell assays the biosensor surface may require coating to enhance cell adhesion and proliferation. On the other hand the surface properties could have direct impact on cell biology, surface bound ligands may influence the response of cells, while the mechanical compliance of substrate material which dictates how it will deform under forces applied by the cell, is influential. Together with the culture conditions (time, serum concentration, confluency, *etc.*), cellular status obtained may be distinct from one surface to another and from one condition to another. Thus, special attention to control cellular status are necessitated for developing biosensor-based cell assays.

10.1.3.2 Cells under un-stimulated conditions

Cells are dynamic objects with relatively large dimensions – typically tens of microns. Even without stimulation, cells constantly undergo micro motion dynamic movement and remodeling of cellular structure, as observed in tissue culture by time lapse microscopy at the subcellular resolution as well as by bio-impedance measurements at the nanometer level.

Under un-stimulated conditions the cells generally give rise to an almost net-zero DMR response, as examined with RWG biosensor. This is partly because of the low spatial resolution of optical biosensors, as determined by the large size of the laser spot and the long propagation length of the coupled light. The biosensor typically measures an averaged response of large population of cells located at the light incident area. Although cells undergo micro motion at the single cell level, the large populations of cells examined give rise to net-zero DMR response in average. Furthermore, it is known that intracellular macromolecules are highly organized and spatially restricted to appropriate sites in mammalian cells. The localization of proteins is tightly controlled in order for cells to regulate the specificity and efficiency of proteins interacting with their proper partners, to spatially separate protein activation and deactivation mechanisms and thus to determine specific cell functions and responses. Thus, under un-stimulated conditions, the local mass density of cells within the sensing volume can reach an equilibrium state, thus leading to a net-zero optical response.

10.1.3.3 Cellular responses upon stimulation

Living cells have exquisite abilities to sense and respond to exogenous signals. Cell signaling was originally thought to function via linear routes where an environmental cue would trigger a linear chain of reactions resulting in a single well defined response. However, evidence show that cellular responses to external stimuli are much more complicated. It has become apparent that the information of the cells received is processed and encoded into complex temporal and spatial patterns of phosphorylation and topological relocation of signaling proteins. The spatial and temporal targeting of proteins to appropriate sites is crucial to regulate the specificity and efficiency of protein-protein interactions, thus dictating the timing and intensity of cell signaling and responses. Pivotal cellular decisions, such as cytoskeletal reorganization, cell cycle checkpoints and apoptosis, depend on the precise temporal control and relative spatial distribution of activated signal-transducers. Thus, cell signaling mediated through a cellular target such as G protein-coupled receptor (GPCR) typically proceeds in an orderly and regulated manner and consists of a series of spatial and temporal events, many of which lead to changes in local mass density or redistribution in local cellular matters of cells. These changes or redistribution when occurring within the sensing volume can be followed directly in real time using optical biosensors.

10.1.4 Applications of Optical Biosensors in Cell Signaling

Optical biosensors can follow receptor signaling in real time and offer a kinetic optical signal in a ligand-dependent manner. A ligand-induced DMR signal is an integrated and quantifiable response of living cells, containing contributions of many downstream signaling events. Chemical biology studies, based on the modulation profiles of various modulators on ligand-induced DMR signal, suggest that DMR signal can be treated as a unique readout for systems cell biology of receptor signaling. Such approach has been applied to map the signaling pathways and network interaction of epidermal growth factor receptor (EGFR), as well as to study the cellular functions of cholesterol in quiescent A431 cells. The EGF-induced DMR signal is saturable to EGF concentration and sensitive to the activity of EGFR tyrosine kinase, Methyl Ethyl Ketone (MEK) and dynamin. These optical biosensor cell assays link the EGF-induced DMR signal of quiescent A431 cells mainly to the Ras/mitogen-activated protein kinase pathway, which primarily proceeds through MEK and leads to cell detachment.

The G-protein coupled receptors (GPCR) signaling is sophisticated receptor may couple to more than one G protein subtype and interact with other signaling molecules and scaffolding proteins such as arrestins. The DMR signal imitates the complexity of GPCR signaling by bradykinin B2 receptor signaling in A431. Bradykinin, a vasoactive nonapeptide, is implicated in the regulation of many physiological and pathophysiological responses, including vascular permeability, inflammation, pain and neurotransmitter release. Endogenous bradykinin B2 receptor in A431 cells is mitogenic or anti-mitogenic, dependent on the cellular states as well as the doses of bradykinin. In proliferating states, low doses (<100nM) of bradykinin trigger Gs-mediated signaling, while high doses (>100nM) of bradykinin favors Gq-signaling. On the other hand, bradykinin between 0.5nM and about 100nM mediated both Gs and Gq-pathways in quiescent A431 cells obtained with 0.1% fetal bovine serum for about 20 hours. A quiescent cell, blockage of protein kinase-C activity only partially attenuates the initial P-DMR, but potentiates the retarded event. Conversely, blockage of protein kinase-A activity increases the initial P-DMR event, but suppresses the retarded event.

These results suggest that the activation of B2 receptors leads to both Gq and Gs signaling and both pathways negatively cross-regulated each other.

DMR signal is a novel physiological response of living cells and can be used to study ligand pharmacology. When a ligand is specific to a receptor expressed in a cell system, the ligand-induced DMR signal is also receptor-specific, dose dependent and saturable. Furthermore, the DMR signal exhibits expected desensitization patterns. Furthermore, the DMR signal also maintains the fidelity of GPCR ligands, similar to those obtained using conventional technologies. The biosensor can distinguish full agonists, partial agonists, inverse agonists, antagonists and allosteric modulators.

10.2 SU-8 MICRO BIOSENSOR

10.2.1 Introduction

Because of the huge demand for smaller biosensors, capable of transforming a biological reaction into a measurable signal, integrated optic sensors attracted considerable interest, taking into account their advantages such as small size, easy to handle, possibility to obtain integrated analyzing microsystmes. Biosensors are also becoming more important devices for the determination of the mechanisms and kinetics of biological events.

The properties of light propagation can be modified in different ways due to variations of absorptive, refractive or luminescent properties of the biological layer. There are many techniques used to detect properties of biosamples: luminescence, interference, Surface Plasmon Resonance (SPR) depending on the way that the bio-layer manifests itself during the interaction with light field. The SPR method for the detection of thin biological films outputs a single value comprising two parameters describing the biological layer: its thickness and its refractive index.

The fabrication and operation of robust integrated optical refractometers, suitable for precise measurements of small changes without ambiguity, a sensing principle is based on mono-mode planar optical waveguides which are integrated using an interferometric architecture: a Mach-Zehnder Y splitter waveguide with far field interference. The shift in the effective refractive index will be induced by the biological sample immobilized on the surface of the core or the waveguide, in the open window which allows the interaction with the surrounding medium and subsequently will be sensed at the output signal. The main idea is to obtain cheap miniaturized sensors that can be integrated with the light source and field detector with a high working accuracy. The main interest of the sensor is the detection of DNA genome and enzyme. The basic structure of the sensor has the architecture represented in Fig. 10.21. As it can be observed, a polymer (typically Poly (methyl methacrylate) (PMMA) or other) with a lower refractive index, than SU-8 covers all sensors surface except the sensing area. The sensing area presents an opening in the cladding at the

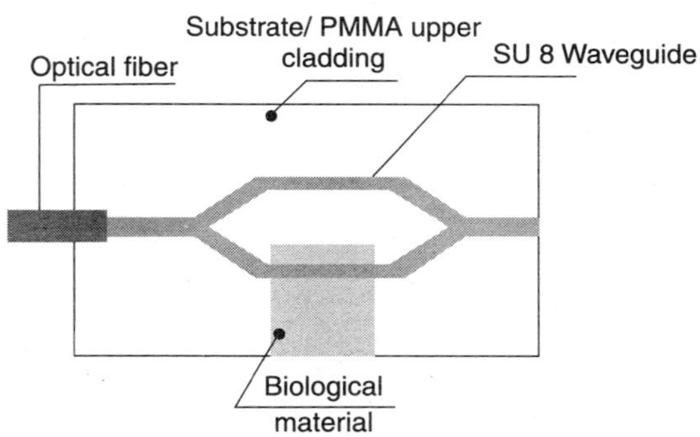

Fig. 10.21 *Schematic presentation of the Mach Zehnder biosensor. (Source: Su-8 Micro-Biosensor based on Mach-Zehnder interferometer, D. Esinenco et al., Rev. Adv. Mater. Sci. 10, 2005)*

top of the waveguide. Through this window, the biomaterial can directly be in contact with the surface of the waveguide.

The purpose is to determine more precisely the interaction between the wave front and the bio-specimens, which results mainly from the main modes and propagation principles of light through the waveguide. Bio-interaction also changes the effective refractive index and absorbance coefficient of the layer at a certain frequency.

The structure has to be designed in such a way that the profile of the electric field components to extend above the waveguide to surface. In this way the waveguide lets out an evanescent wave, which is sensitive to materials deposited on the top. This structure is obtained by thermally grown SiO_2 coating with a thickness of 1.6 mm - 1.7 mm. A positive photo resist (PMMA) was used as the upper cladding, with a lower refractive index than SU-8. Strip optical waveguides were defined using photolithographic techniques. The dimensions of the SU-8 core were: thickness 0.9 μm, width 1-12 μm and a refractive index of n = 1.58 - 1.59.

Special attention should be given to the waveguide top surface relief and cladding. In SU-8 photo-resist, a deformation near the edges was obtained (Fig. 10.22) maybe due to the hydrophobic properties of the resist. In some cases, this kind of surface non-uniformities could be helpful in more precise quantitative detection.

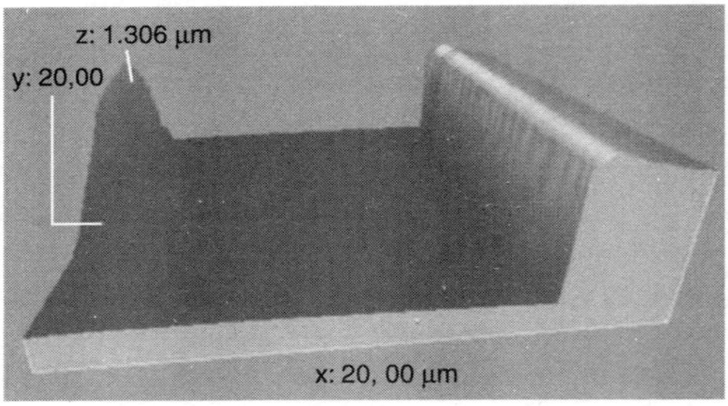

Fig. 10.22 *AFM profile of an SU-8 waveguide edge. (Source: Su-8 Micro-Biosensor based on Mach-Zehnder interferometer, D. Esinenco et al., Rev. Adv. Mater. Sci. 10, 2005).*

Deformed edges extend the Transverse Electric (TE) (Ey in the current case) mode outside the waveguide, increasing its sensibility. Typically, thin gold nano-particles layer can be used to bind complementary DNA strands and to amplify bio-sample reaction with the evanescent field. The biomolecules must be some how aligned or selected. The hybridization process DNA strands leads to 'in plane' amino acids aligning, resulting amino acids planes parallel to the binding surface. When DNA molecules bind to the waveguide top surface, they form an amino-layered cladding. Each of the cladding planes, during the interaction with the evanescent field, can act as a floating refractive index layer. The TE and Transverse Magnetic (TM) mode are affected depending on the 'biocladding' composition. TE and/or TM modes interference analysis should give enough information about the qualitative and quantitative aspect of the bio-samples. The differences in refractive index values between the reference arm and the sensing one result in a phase shift between the two arms that can be read weather at the end of the waveguide with an integrated photo detector or in the far field using CCD and numeric analyses.

10.2.2 Simulation

Optiware Finite-Difference Time-Domain (FDTD) tools were used in order to optimize the size of the design and to improve the Match-Zehnder interferometer parameters. The first step was to establish the thickness of the SU-8 optical waveguide in order to have low losses and monomode propagation. Fig. 10.23 presents the results of the simulation: the normalized intensity of light vs. the SU-8 core thickness. As it can be observed, for the value less than 1 μm the losses become higher. The light intensity is strongly dependent on the waveguide thickness up to a certain value. As anticipated, increasing thickness shows less attenuation on superior modes, which makes it a multimodal waveguide.

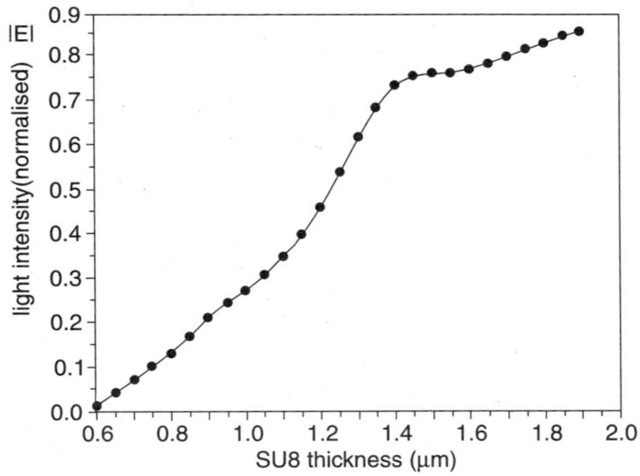

Fig. 10.23 *Output SU-8 waveguide power function of the core thickness. (Source: Su-8 Micro-Biosensor based on Mach-Zehnder interferometer, D. Esinenco et al., Rev. Adv. Mater. Sci. 10, 2005).*

For mono mode propagation, the thickness of the guiding layer has to be reduced, so that a value between 1.1 - 1.5 μm was chosen in order to respect both conditions. The second step was the design of Y beam splitter. Low losses can be obtained only if small separation angles are used. Fig. 10.24 shows the simulation results.

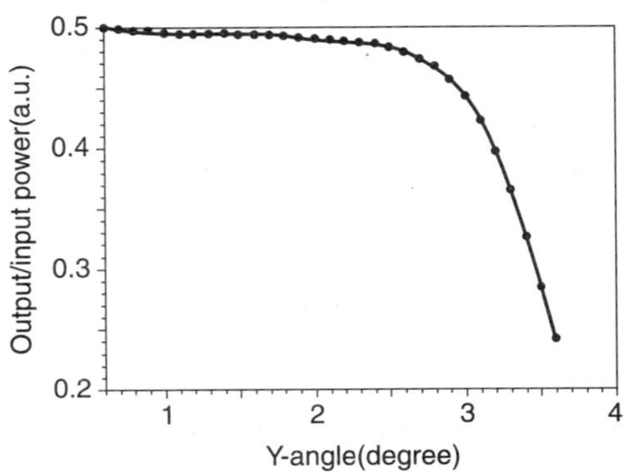

Fig. 10.24 *Light intensity losses in the Y-splitter versus angle value. (Source: Su-8 Micro-Biosensor based on Mach-Zehnder interferometer, D. Esinenco et al., Rev. Adv. Mater. Sci. 10, 2005).*

A number of computations and measurements for the interferometric sensor were carried out. The simulation results showed that very small refractive index changes in the cladding of one arm, cause a notable light intensity change at the merging point of the two arms that can be read at the end of the Mach-Zehnder interferometer. In addition, the fundamental transverse mode for the designed configuration was simulated. The corresponding results are illustrated in below Fig. 10.25. The test structure of the Mach-Zehnder biosensor was coupled to an optical fibre, for an electro-magnetic wavelength of 635 nm. Both the SU-8 waveguide itself without any biolayer or cladding and the final sensor structure were tested.

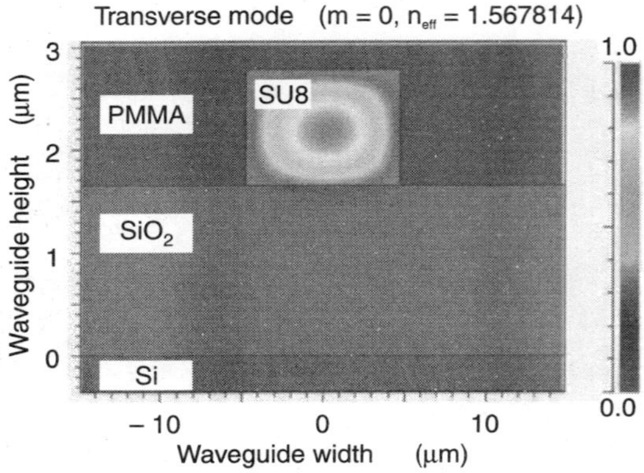

Fig. 10.25 *Field distribution of the fundamental mode of the SU-8 Waveguide. (Source: Su-8 Micro-Biosensor based on Mach-Zehnder interferometer, D. Esinenco et al., Rev. Adv. Mater. Sci. 10, 2005).*

10.3 CHEMILUMINESCENCE BIOSENSORS

10.3.1 Introduction

Chemiluminescence is the process whereby energy from a chemical reaction is released directly as light without the involvement of heat or flame. Chemiluminescence is not similar to fluorescence (neon signs), phosphorescence (glow in the dark toys) or incandescence (light bulbs) but most similar to bioluminescence. Chemiluminescence is the production of light from a chemical reaction in excess of the black body radiation expected from that body and it is often referred to as 'cold light'. Chemiluminescence involves the production of an electronically excited species from a number of reactants which goes on to release visible light in order to revert to its ground state energy.

A chemoluminescent reaction carried out in an Erlenmeyer flask producing large amount of light. It is the emission of luminescence (light) with limited emission of heat as the result of chemical reaction. Given reactants A and B with an excited intermediate [◊],

$$[A] + [B] \rightarrow [\lozenge] \rightarrow [\text{Products}] + \text{light} \qquad \textbf{... } \textit{eq.}\textbf{-10.5}$$

Eg. if $[A]$ is luminol and $[B]$ is hydrogen peroxide in the presence of suitable catalyst:

$$\text{luminol} + H_2O_2 \rightarrow \text{3-APA}[\lozenge] \rightarrow \text{3-APA} + \text{light} \quad \text{... eq.-10.6}$$

where 3-APA is 3-aminophthalate, 3-APA[◊] is the excited state fluorescing as it decays to a lower energy level. The decay of the excited state [◊] to lower energy level is responsible for the emission of light. One photon of light should be given off for each molecule of reactant, or Avogadro's number of photons per mole. In actual practice, non-enzymatic reactions seldom exceed 1% Q_C, quantum efficiency. Standard example of chemiluminescence in the laboratory setting is the luminol test. When chemiluminescence takes place in living organisms the phenomenon is called bioluminescence.

10.3.2 Types

Based on the type of phase (liquid/solid/gas/other) reactions chemiluminescence based biosensors can be classified into the following types:

10.3.2.1 Liquid – phase reactions

a. Luminol in an alkaline solution with hydrogen peroxide in the presence of iron or copper or an auxiliary oxidant, produces chemiluminescence (see *eq.*-10.5).
b. Pure oxygen (O_2) can also emit light. If solutions of 30% hydrogen peroxide and 5% alkaline sodium hypochlorite (NaClO) are mixed, red light is emitted. It is barely visible, for this reason a sensitizer is included to boost light emission in terms of brightness and intensity.
c. Other liquid phase chemilumiscence reagents are also used viz., Peroxyoxalates, Aryl oxalates, Acridiniumesters and Dioxetanes.

10.3.2.2 Gas - phase reactions

The oldest known chemoluminescent reaction is that of elemental white phosphorous oxidizing in moist air, producing a green glow. This is actually a gas-phase reaction of phosphorus vapor, above the solid, with oxygen producing the excited states $(PO)_2$ and HPO. Another gas phase reaction is the basis of nitric oxide detection in commercial analytic instruments applied to environmental air quality testing. Ozone is combined with nitric oxide to form nitrogen dioxide in an activated state.

$$NO + O_3 \rightarrow NO_2\,[\lozenge] + O_2 \quad \text{... eq.-10.7}$$

The activated NO_2 [◊] luminesces broad band visible to infrared light as it reverts to a lower energy state. A photomultiplier and associated electronics counts the photons which are proportional to the amount of NO present. To determine the amount of nitrogen dioxide (NO_2) in a sample (containing no NO) it must first be converted to nitric oxide (NO) by passing the sample through a converter before the above ozone activation reaction is applied.

The ozone reaction produces a photon count proportional to NO which is proportional to NO_2 before it was converted to NO. In the case of a mixed sample containing both NO and NO_2, the above reaction yields the amount of NO and NO_2 combined in the air sample, assuming that the sample is passed through the converter.

If the mixed sample is not passed through the converter, the ozone reaction produces activated NO_2 [◊] only in proportion to the NO in the sample. The NO_2 in the sample is not activated by the ozone reaction. Though inactivated NO_2 is present with the activated NO_2 [◊], photons are only emitted by the activated species which is proportional to the original NO. The final step is subtracting NO from (NO + NO_2) to yield NO_2.

Chemiluminescence biosensor based on supported lipid layer incorporated with ganglioside GM1 was developed for the detection of cholera toxin (CT). The planar supported lipid membrane was prepared as biosensing interface via spontaneous spread of ganglioside-incorporated phospholipid vesicles on the octadecanethiol-coated gold surface. The specific interaction of multivalent CT by ganglioside GM1 molecules enables the biosensor to be implemented via a sandwiched format using a liposome probe functionalized with GM1 and horseradish peroxidase (HRP). Then, the presence of the target CT could be determined via the HRP-catalyzed enhanced chemiluminescence reaction.

The developed strategy offers several unique advantages over conventional biosensors in that it allows for an easy construction and renewal of the sensing interface. The small background signal is due to low non-specific adsorption of serum constituents on the lipid membrane and effective immobilization of multiple biocatalytic amplifiers and recognition components via common phospholipid reagents. The developed biosensor gives chemiluminescence signal in linear correlation to CT concentration within the range from 1 pgml/l) to 1 ngm/l with readily achievable detection limit of 0.8 pgml/l.

The chemiluminescence biosensor measured 25 × 75 × 6.5 mm in dimension and was produced in an analytical laboratory. The sol-gel method is used to co-immobilize Horseradish Peroxidase (HRP) and luminol in the micro reactor and to immobilize Uricase in the enzymatic reactor. The main characteristic of the biosensor was to introduce air as the carrier flow instead of the more common solution carrier for the first time. The uric acid was determined by a chemiluminscent (CL) reaction between the hydrogen peroxide produced from the enzymatic reactor and luminol under the catalysis of HRP in the micro reactor. The linear range of the uric acid concentration was 1 to 100 mg l^{-1} and the detection limit was 0.1 mg l^{-1}.

Chemiluminescence biosensor with alcohol dehydrogenase and tris (2, 2'-bipyridyl) ruthenium (II) immobilized in sol-gel hybrid material.

An ethanol biosensor based on electro generated chemiluminescence detection was developed. Electro generated chemiluminescence reagent tris (2, 2'-bipyridyl) ruthenium (II) and alcohol dehydrogenase were immobilized in the same sol-gel hybrid film. The copolymer poly vinyl alcohol with 4-vinylpyridine and cation exchanger Nafion were incorporated into sol-gel film to provide the micro-environment for retaining the activity of enzyme and immobilize tris (2, 2'-bipyridyl) ruthenium (II). The intensity of electrogenerated chemiluminescence increased linearly with ethanol concentration from 2.5×10^{-5} to 5.0×10^{-2} M and detection limit was 1.0×10^{-5} M. The prepared biosensor exhibited high sensitivity, wide linear range and good stability.

10.4 FLUORESCENT BASED BIOSENSORS

Higher energy radiation is absorbed by species and stored for long enough that the molecules have collided and lost vibrational energy in collisions. When the excited molecule reemits, the photon is of less energy depending on the species and is therefore of lower frequency. *Eg.* White clothes-detergent manufacturers put fluorescents in with their mixture to make clothes really 'whiter than white' as they absorb UV radiation in sunlight and emit it in the visible region causing the net light reflected from clothes to be higher than expected and it is normally emitting from singlet state.

Fluorescence phenomena are related to an energetic transition from an excited state to lower energy level producing photon emission. Energy levels in an organic molecule are shown by Jablonskii diagram (Fig. 10.26). In this diagram, $S_0,\ldots, S_n$ and $T_1,\ldots, T_m$ represent the discrete electronic energy levels of a molecule ($S_1,\ldots, S_n$ are excited singlet states and $T_1,\ldots, T_m$ are excited triplet states). At equilibrium, molecules have the lowest possible energy S_0. When a molecule absorbs electromagnetic energy, it moves to an excited energy level (that usually is a singlet state). The energy level reached by the molecule absorbing electromagnetic energy depends on the incident radiation wavelength.

The intensity of the absorption (fraction of ground state molecules promoted to the electronic excited state) depends on the intensity of the excitation radiation (number of photons) and the probability of the transition with photons of the particular energy used. To characterize the intensity of an absorption band, a term often used is the oscillator strength *f*, which may be defined from the integrated absorption spectrum by the relationship:

$$f = 4.315 \times 10^{-19} . f\ \varepsilon_V . \, d\, V \qquad \text{... eq.-10.8}$$

where ε_v is the molar extinction coefficient at the frequency V.

A molecule in an excited state tends to reach a lower energy level. Relaxation of the molecule from an excited state may occur in different manner. If relaxation takes place by emission of a photon without any change in spin multiplicity (from a singlet state to another singlet state), the transition is known as fluorescence. Emitted photon wavelength depends on the loss of energy by the molecule in the relaxation process.

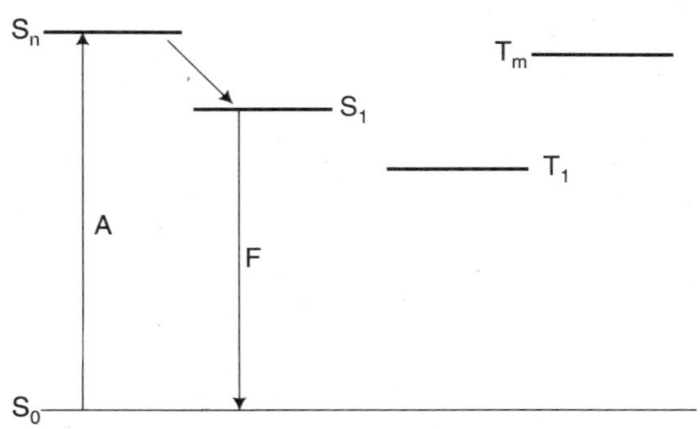

Fig. 10.26 *Jablonskii diagram. (Source: Guided-wave optical Biosensors, Sensors 2007, MDPI).*

Fluorescence process can be characterized by its lifetime, defined as the time T required for the emission to decrease to 1/e its original intensity, following a pulse excitation. Fluorescence lifetimes of organic molecules are of the order of 10^{-9} to 10^{-7} seconds. Polarization is another physically observable property of the luminescence; it is caused by unique symmetries and orientations of electric moment vectors and wave functions involved in electronic transitions. The electric dipole moment determines the direction and with charge is displaced in a molecule undergoing an electronic transition. It is possible to study the polarization of the transition using polarized light to excite and detect luminescence. The degree of polarization is related to the angle between the absorption vector and the emission vector (Fig. 10.27).

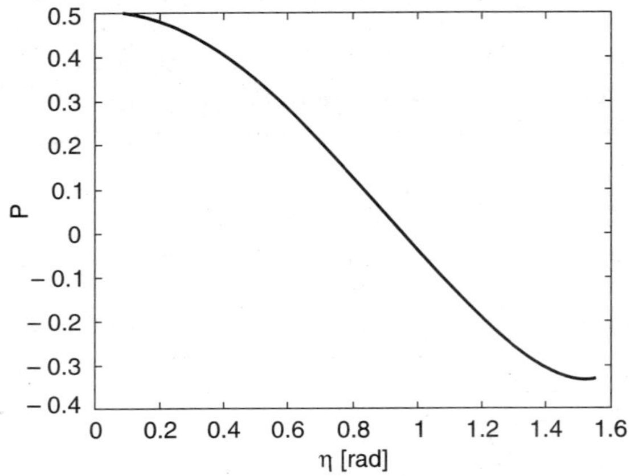

Fig. 10.27 *Degree of polarization vs. η. If η = 0°, absorption and emission vectors are parallel and P = 0.5. If η = 90°, absorption and emission vectors are perpendicular and P = –0.33. (Source: Guided-wave Optical Biosensors, Sensors 2007, MDPI).*

The principal scheme of typical fluorescence based biosensor is shown in Fig. 10.28. In this sensor, the optical signal provided by a laser propagates in an optical fibre until fibre section on which antibodies are immobilized. The optical radiation produced by antigen fluorescence is sent back to the detection system. This sensor can be used to detect the carcinogen benzo[a]pyrene (BaP) or to distinguish BaP and benzopyrene tetrol (BPT).

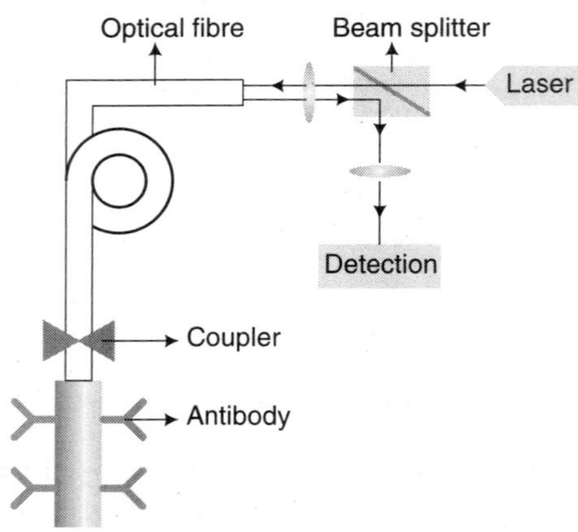

Fig. 10.28 *Typical fluorescence-based photonic sensor. (Source: Guided-wave Optical Biosensors, Sensors-2007, MDPI).*

Fluorescence based optical sensors are used in biochemical analyses of other chemical compounds and *in vivo* and *in vitro* analyses of tissues and cells (for example, neoplastic cells). Moreover, fluorescence is widely used in DNA sequencing (individuation of nucleotides sequence constituting DNA molecules), because it enables to reach a high readout speed.

Most biological matter is fluorescent, which includes three aromatic amino acids and proteins containing them, the green fluorescent protein (GFP), nucleic acids, flavine nucleotides and NADH, whilst NAD^+ and most saccharides and lipids are non-fluorescent. In addition, most colored matter in nature including coumarins, flavones, anthocyans and chlorophylls (but with the notable exception of hem) is fluorescent.

10.4.1 Fluorescent Probes

Probes, in contrast to labels are expected not to be inert but rather to respond to their microenvironment or to chemical species. Probes responding to chemical species (such as an ion, pH or oxygen) may also be referred to as indicators.

10.4.2 Direct Fluorescent Sensors

Numerous bioanalytical assays are based on the fact that NADH is fluorescent, while NAD^+ is not. All enzymatic reactions based on NAD/NADH are amenable to fluorescence analysis and this is widely exploited in practice, even though NADH has to be excited at around 350 nm which can cause substantial background fluorescence from other biomatter. On the other side, most assays are performed in the kinetic mode so that it is the relative signal change (above the background) that is measured rather than the total intensity (including the background) and this reduces interferences a lot.

The co-enzyme FAD is another strongly luminescent species but has found less wide applications because both the oxidized (FAD) and the reduced form (FADH2) display fluorescence so that they are less useful for monitoring the course of a biochemical reaction. Their excitation is at around 450 nm, and fluorescence peaks at 512 nm. Both NADH and FAD have been shown to be useful for purposes of chemical sensing using immobilized reagents and, in some cases, using fibre optics waveguides. However, their practical applications have been confined mainly to conventional cuvette tests so far.

10.4.3 Indirect Fluorescent Sensors

10.4.3.1 Chemosensors

There is a tremendous interest in sensing (monitoring) clinically important species such as oxygen, the proton (pH), carbon dioxide, glucose and urea. These are major metabolites or substrates which cannot be visualized directly via luminescence spectroscopy. Hence, use is made of respective indicators that can be added to the system of interest and which allow indirect visualization of the species of interest.

Optical sensing of oxygen is almost exclusively accomplished on the basis of the quenching effect which it exerts on numerous luminophores. Both luminescence intensity and decay time may be measured and the latter is being preferred for reasons of system stability, precision and accuracy. The pH of blood and related biological matter is determined via classical pH indicators such as the fluorescein which are immobilized on solid supports (or inside polymeric matrices) and undergo a pH-induced spectral change. Numerous schemes exist that are based on measurement of luminescence intensity, decay time or energy transfer efficiency. Fibre optic sensors are used for the measurement of oxygen, pH and carbon dioxide (with ~10 μm thin tips) in algae and in the deep sea.

10.4.3.2 Enzymatic biosensors

The above chemosensors may also be used to monitor the course of enzymatic reactions during which oxygen or protons are consumed or produced. A typical example being the oxidation of glucose by the enzyme glucose oxidase according to:

$$\text{glucose} + \text{oxygen} \rightarrow \text{gluconic acid} + \text{hydrogen peroxide (1)} \qquad \textit{... eq.-}\mathbf{10.9}$$

It can be seen that such a reaction can be monitored kinetically in various ways, namely by measuring the consumption of oxygen (via the above fluorescent oxygen sensor), the formation of the acid (via the above pH sensor), or the formation of hydrogen peroxide (via a respective sensor). Since 2 - 4% of Caucasians suffer from *Diabetes mellitus*, there is a substantial need for such sensors.

Enzymatic reactions can be inhibited by toxic species including organophosphates and heavy metals. This is the basis for so called inhibition biosensors which respond to environmentally harmful species like herbicides, warfare agents, and heavy metals including lead and cadmium.

10.4.3.3 Cellular biosensors

Fluorescent biosensors may also be obtained by immobilizing whole cells on the surface of a sensor layer. *Eg.* Immobilization of yeast on the surface of a luminescent oxygen sensor led to a fibre optic sensor for measurement of biological oxygen demand. This is a parameter for the loading of water with organic waste. Yeast and certain bacteria digest the organic fraction and thereby consume oxygen and this is detected by the luminescent oxygen sensor.

Living Biosensors

Chapter 11

11.0 MICROBIAL BIOSENSORS

11.0.1 Introduction

Biosensor is an analytical device that combines biological sensing element with transducer to produce signal which is proportional to the analyte concentration. This signal results from change in protons concentration, release or uptake of gases, light emission and absorption brought about through the metabolism of the target compound by the biological recognition element. The transducer converts this biological signal into a measurable response such as current, potential or absorption of light which is amplified, processed and stored.

Biomolecules such as enzymes, antibodies, receptors, organelles and microorganisms as well as animal and plant cells or tissues are used as biological sensing elements. Microorganisms are recommended to detect a wide range of chemical substances, amenability to genetic modification and broad operating pH and temperature range, making them ideal as biological sensing materials. To construct various types of living biosensor devices, microorganisms are incorporated with diversity of transducers such as amperometric, potentiometric, conductometric, colorimetric, luminescence and fluorescence.

Enzymes are the most widely used biological sensing elements in the fabrication of biosensors. Purified enzymes have very high specificity for their substrates or inhibitors, their application in biosensors construction may be limited by the tedious, time consuming and costly enzyme purification, requirement of multiple enzymes to generate the measurable product or need of cofactor/coenzyme. Many enzymes and co-factors that co-exist in the cells have the ability to consume and hence detect large number of chemicals; however, this can compromise the selectivity. They can be easily manipulated and adapted to consume and degrade new substrate under certain cultivating conditions. The progress in molecular biology or recombinant DNA technology has opened endless possibilities of tailoring the microorganisms to improve the activity of an existing enzyme or express foreign enzyme or protein in the host cells and make the microbes excellent biosensing elements.

11.0.2 Immobilization of Microorganisms

The basis of a microbial biosensor is in close contact between microorganisms and transducers. Fabrication of microbial biosensor requires immobilization on transducers with a close proximity. Since, microbial biosensor response, operational stability and long-term use are a function of the immobilization strategy used, immobilization technology plays a very important role and the choice of immobilization technique is critical. Microorganisms can be immobilized on a transducer or support matrices by chemical or physical methods.

11.0.2.1 Chemical methods

Chemical methods of microbe immobilization include covalent binding and cross-linking. Covalent binding methods rely on the formation of a stable covalent bond between functional groups of the microorganisms' cell wall components such as amine, carboxylic or sulphydryl and the transducer such as amine, carboxylic, epoxy or tosyl. To achieve this, all cells are exposed to harmful chemicals and harsh reaction conditions, which may damage the cell membrane and decrease the biological activity. To overcome this drawback is still a challenge for immobilization through covalent binding, this method has therefore not been successful for immobilization of viable microbial cells.

Cross-linking involves bridging between functional groups on the outer membrane of the cells by multifunctional reagents such as glutaraldehyde and cyanuric chloride, to form a network. Because of the speed and simplicity, the method has found wide acceptance for immobilization of microorganisms. The cells may be cross-linked directly onto the transducer surface or on a removable support membrane, which can then be placed on the transducer. The ability to replace the membrane with the immobilized cells is an advantage of the latter approach. While cross-linking has advantages over covalent binding, the cell viability and/or the cell membrane biomolecules can be affected by the cross-linking agents. Thus cross-linking is suitable in constructing microbial biosensors, where cell viability is not important and only the intracellular enzymes are involved in the detection.

11.0.2.2 Physical methods

Adsorption and entrapment are the two widely used physical methods for microbial immobilization. Because these methods do not involve covalent bond formation with microbes and provide relatively small perturbation of microorganism native structure and function, these methods are preferred when viable cells are required. Physical adsorption is the simplest method for microbe immobilization. Typically, a microbial suspension is incubated with the electrode or an immobilization matrix, such as alumina and glass bead, followed by rinsing with buffer to remove un-adsorbed cells. The microbes are immobilized due to adsorptive interactions such as ionic, polar or hydrogen bonding and hydrophobic interaction. However, immobilization using adsorption alone generally leads to poor long-term stability because of desorption of microbes.

The immobilization of microorganisms by entrapment can be achieved by either retention of the cells in close proximity of the transducer surface using dialysis or filter membrane or in chemical/biological polymers/gels such as alginate, agarose, chitosan, collagen, polyacrylamide, poly vinyl alcohol, poly ethylene glycol and polyurethane. A major disadvantage of entrapment immobilization is the additional diffusion resistance offered by the entrapment material, which will result in lower sensitivity and detection limit.

11.0.3 Classification of Microbial Biosensors

Microbial biosensors are classified into electrochemical, optical and others based on the type of transducers.

11.0.3.1 Electrochemical microbial biosensor

Electrochemical microbial biosensors are classified into three types: amperometric, potentiometric and conductometric.

11.0.3.2 Amperometric microbial biosensor

Amperometric microbial biosensor operates at fixed potential with respect to a reference electrode and involves the detection of the current generated by the oxidation or reduction of species at the surface of the electrode.

Amperometric microbial biosensors have been widely developed for the determination of biochemical oxygen demand (BOD) for the measurement of biodegradable organic pollutants in aqueous samples. The conventional standard method for the determination of BOD measures the microorganisms' oxygen consumption or respiration over a period of 5 days and is reported as BOD5. While BOD5 is a good indicator of the concentration of organic pollutants in water, it is extremely slow and hence not suitable for process control. To address this limitation, several BOD biosensors based on amperometric oxygen electrode transducer modified with microorganisms degrading or metabolizing organic pollutants have been reported. The microbial strains used as biological sensing element include *Torulopsis candida, Trichosporon cutaneum*, *Pseudomonas putida*, *Klebsiella oxytoca* AS1, *Bacillus subtilis*, *Arxula adeninivorans* LS3, *Serratia marcescens* LSY4, *Pseudomonas* sp., *P. fluorescens*, *P. putida* SG10, *Thermophilic bacteria*, *Hansennula anomala* and yeast. Because, any given strain provides a narrow substrate spectrum, single strain BOD-biosensor has limitations in analyzing the complex samples. This bottleneck can be alleviated by employing a mixture of two or more microorganisms to broaden the substrate and hence analyte spectrum with a stable performance.

The first commercial BOD biosensor was fabricated by Nisshin Denki in 1983 and development in BOD biosensors has focused on identifying different microorganisms that can determine BOD of a specific waste. *Eg.* A miniaturized oxygen electrode based on thick-film screen-printing was recently developed to replace the bulky Clark dissolved oxygen electrode transducer. The widely used thick-film screen-printing technique was used to print the platinum-working electrode, Ag/AgCl reference electrode and platinum auxiliary electrode of the amperometric oxygen electrode on an inert substrate. The oxygen electrode was then modified with *A. adeninivorans* LS3 by entrapment in polycarbamoyl sulfonate (PCS) gel and successfully applied for rapid (100 seconds) and stable (up to 2 months) BOD determination. Similarly, to extend the dynamic range of the BOD sensor, which in the case of dissolved oxygen electrode is limited by the solubility of oxygen in the sample, a ferricyanide mediated microbial biosensor using a novel yeast strain for BOD measurement was developed. Recently, an amperometric transducer array featuring four individually addressable platinum electrodes was constructed and modified with two microbial strains with different substrate spectra for the measurement of BOD and poly cyclic aromatic hydrocarbons (PAH) simultaneously.

Besides BOD biosensor, amperometric microbial biosensors have also been applied for measurement of several other chemicals. Because of its importance in fermentation industry and clinical toxicology, microbial biosensors for ethanol have garnered the second most research attention after BOD. Different microorganisms metabolizing ethanol such as *Trichosporon brassicae*, *Acetobacter aceti*, *Candida vini*, *Gluconobacter suboxydans*, *C. tropicalis*, *Aspergillus niger*, *Saccharomyces ellipsoideus*, *G. oxydans* and *Pichia methanolica* have been immobilized on oxygen electrode to fabricate ethanol biosensors. While these biosensors possess good sensitivity and stability, they usually have poor selectivity. Thus, there is a great interest to develop selective microbial ethanol biosensor. An improved selectivity for ethanol determination in presence of glucose was achieved by replacing oxygen with ferricyanide as the electron acceptor mediator for *G. oxydans* immobilized on a glassy carbon electrode by cellulose acetate membrane which also restricted the availability of glucose to the cells by size exclusion.

Sugars are important ingredients of different media and sensors. Microbial biosensors for sugars have ranged from the simple modification of Clark and micro-fabricated oxygen electrode with *S. cerevisiae* and *E.coli* K12 mutants, respectively, to modification of graphite electrode with *G. oxydans* in conjunction with hexacyanoferrate (III) as a mediator. Phenol and substituted phenols have received considerable attention in waste analysis program due to their high toxicity to mammals, humans and plants. A variety of amperometric microbial biosensors have been reported for these EPA priority chemicals. *p*-Nitrophenol (PNP) degrading bacterial *Arthrobacter* JS 443 and *Moraxella* sp. isolated from PNP contaminated sites have been immobilized on oxygen and carbon paste electrodes using polycarbonate membrane and Nafion, respectively and by directly mixing in the carbon paste to fabricate biosensor for PNP. Other microbial biosensors for phenols include *Rhodococcus erthropolis* modified Clark oxygen electrode for 2,4-dinitrophenol (2, 4-DNP) and *P. putida* DSM 50026, a well-known phenol degrading microorganisms, modified thick-film and screen printed graphite electrodes for phenols.

Neurotoxic organophosphate (OP) compounds have found wide applications as pesticides and insecticides in agriculture and as chemical warfare agents in military practice. Amperometric biosensors based on genetically engineered *Moraxella* sp. and *P. putida* with surface expressed organophosphorus hydrolase (OPH) have been developed for sensitive, selective and cost-effective detection of OPs. These biosensors relied on the amperometric detection of PNP generated from hydrolysis of OP compounds by surface-displayed OPH or oxygen consumed and electrochemically active intermediates formed during the further mineralization of the PNP by the cells.

The inhibition of bacterial respiration and hence the decrease of oxygen consumption rate has been utilized to fabricate cyanide biosensor. Whole-cell biosensors consisting of dissolved oxygen electrode modified with *Nitrosomonas europaea*, *Thiobacillus ferrooxidans*, *Saccharomyces cerevisiae* and *Pseudomonas fluorescens* were reported for batch and continuous cyanide monitoring.

Other amperometric microbial biosensors based on monitoring of cell respiration include biosensor for surfactants, representing a widespread group of organic pollutants using surfactant degrading bacteria, hydrogen peroxide by coupling immobilized living *Acetobacter peroxydans* and for acetic acid using *Fusarium solani*.

Over the last two decades, the microbiologically influenced corrosion (MIC) of metallic materials has received great attention. A stable, reproducible and specific microbial biosensor was developed for monitoring MIC of metallic materials in industrial systems based on *Pseudomonas* sp. isolated from corroded metal surface and immobilized on acetyl-cellulose membrane at oxygen electrode.

A linear relationship between the biosensor response and the concentration of sulfuric acid (the most corrosive inorganic acid involved in microbial corrosion) was established. The biosensor response time was 5 minutes and was dependent on many parameters such as pH, temperature, corrosive environment and immobilized cell loading. The same group also used *Acetobacter* sp. to develop amperometric microbial biosensor for monitoring microbiologically influenced corrosion caused by fungal species.

Another application of amperometric microbial biosensors is the detection of heavy metal ions for environmental control. A microbial biosensor to detect Cu^{2+} by an amperometric method has been developed using recombinant *S. cerevisiae* containing plasmids with Cu^{2+} inducible promoter fused to the *lacZ* gene. In the presence of Cu^{2+}, the recombinant strains are able to utilize lactose as a carbon source and lead to the oxygen consumption change, which can be detected by using oxygen electrode. A cadmium-responsive promoter from *E. coli* was fused to a promoter less *lacZ* gene and then the galactosidase activity was monitored using screen-printed electrode in the presence of cadmium. This whole-cell biosensor could detect nanomolar concentrations of cadmium on-line or in-site within minutes.

11.0.3.3 Potentiometric microbial biosensors

Conventional potentiometric microbial biosensors consist of an ion-selective electrode (pH, ammonium and chloride) or a gas-sensing electrode (pCO$_2$ and pNH$_3$) coated with an immobilized microbe layer. Microbe consuming analyte generates a change in potential, resulting from ion accumulation or depletion, measure the difference between working electrode and reference electrode, and the signal is correlated to the concentration of analyte. A wide detection range is possible due to a logarithmic relationship between the generated potential and the concentration of analyte. However, this method requires a very stable reference electrode, which may be a limitation of these transducers. A few examples of biosensors based on potentiometric transducers are summarized in Table 11.1.

Table 11.1 Potentiometric microbial biosensors.

S.No	*Name of Transducer*	*Name of Microorganism*	*Name of Target*	*Detection limit*
1	pH electrode	*Flavobacterium* sp.		0.025–0.4 mM
		Recombinant *E coli*	Organo-phosphates	2 µM
		Recombinant *E coli*		3 µM
		Recombinant *E coli*	Penicillin	1–16 mM
2	Flat pH electrode	Recombinant *E coli*	Penicillin	5–30 mM
3	WP2 LAPS	*E coli*	Tryptophan	0–12 µM
4	NH_4^+ ion selective electrode	*Bacillus* sp.	Urea	0.55–550 µM
5	Oxygen Electrode	*S. ellipsoideus*	Ethanol	0.02–50 mM
		S. cerevisiae	Sucrose	3.2 µM

The simplest potentiometric microbial biosensor is based on the modification of ion selective electrode. Several microbial biosensors based on modification of glass pH electrode with genetically engineered *E. coli* expressing organophosphorus hydrolase intracellularly and on the outer surface of cells and wild-type OP degrading bacteria *Flavobacteium* sp. The principle of detection is based on the detection of the protons released by OPH catalyzed hydrolysis of OP and correlating to the concentration of OPs. Similarly, recombinant *E. coli* harboring the plasmids encoding for Lactamase and Penicillinase synthesis immobilized on pH electrode using gluten and acetyl cellulose membranes entrapment, respectively were developed for penicillin monitoring. A new type of solid state silicon-based light Potentiometric sensor for monitoring hydrogen ion was integrated to the auxotrophic bacteria *E. coli* WP2 (requiring tryptophan for its growth) to fabricate a potentiometric microbial assay for tryptophan.

The pH electrodes are most widely applied as ion selective electrodes in microbial sensors. The other ion selective electrode utilized for the development of microbial sensors is ammonium ion selective electrode coupled with Urease yielding from *Bacillus* sp., isolated from soil to develop a disposable microbial biosensor for monitoring urea in milk. Similarly, a chloride ion selective electrode was modified with Trichloroethane (TCE) degrading bacterium *Pseudomonas aeruginosa* (JI104) for TCE monitoring in batch and continuous modes in wastewaters.

A potentiometric oxygen electrode with immobilized *S. ellipsoideus* was also successfully used to produce a microbial biosensor for the determination of ethanol with an extended response range. Based on the same format, sucrose biosensor based on an immobilized *S. cerevisiae* was also developed.

11.0.3.4 Conductimetric biosensor

Many microbe catalyzed reactions involve a change in ionic species and the associated change is a net change in the conductivity of the reaction solution. Even though the detection of solution conductance is non-specific, conductance measurements are extremely sensitive. Recently, a single used microbial sensor based on conductivity variation was developed to investigate the effect of both species and concentration or osmolarity of anions on the metabolic activity of *E. coli*. This hybrid sensing system combines physico-chemical and biological sensing and greatly increases the ease with which comparative data could be assimilated.

11.0.3.5 Microbial fuel cell type biosensor

Microbial fuel cells (MFCs) have been studied as a BOD sensor for a long time. Since Karube *et al.* (1977) reported a BOD sensor based on MFC using the hydrogen produced by *Clostridium butyricum* immobilized on the electrode, a variety of MFC BOD sensors with use of electron-mediator have been developed. Even though the addition of mediators in these biosensors can enhance the electron transfer, these biosensors have poor stability because of the toxicity of mediators. Recently, mediator-less microbial fuel cells have been exploited to fabricate novel BOD sensors for continuous and real time monitoring.

11.0.3.6 Optical microbial biosensor

The modulation in optical properties such as UV Visible absorption, bio- and chemiluminescence, reflectance and fluorescence brought by the interaction of the biocatalyst with the target analyte is the basis for optical microbial biosensors. Optical based biosensors offer advantage of compactness, flexibility, resistance to electrical noise and a small probe size.

11.0.3.6.1 Fluorescence biosensor

Fluorescence spectroscopy has been widely applied in analytical chemistry. It is a sensitive technique that can detect very low concentrations of analyte because of the instrumental principles involved. At low analyte concentrations, fluorescence emission intensity is directly proportional to the concentration. Fluorescent materials and green fluorescent protein (GFP) have been extensively used in the construction of fluorescent biosensor.

11.0.3.6.2 Green fluorescence protein (GFP) based biosensor

Like bioluminescent reporter *lux* gene, *gfp* gene coding for the green fluorescent protein (GFP) has also been widely applied as reporters and fused to the host gene that allows reporter activity to be examined in individual cells. Because GFP is very stable and not known to be produced by microorganism indigenous to terrestrial habits, it provides great advantage and flexibility when evaluating reporter activity. The primary disadvantage of GFP as a reporter protein is the delay between protein production and protein fluorescence.

The GFP based microbial biosensor has been shown to be useful in assessing heterogeneity of iron bioavailability on plant. In this sensor, ferric iron availability to cells was assessed by quantifying the fluorescence intensity of cells containing a plasmid-borne transcriptional fusion between an ion-regulated promoter and *gfp*. Recently, Wells *et al.* (2005) developed an ultrasensitive biosensor for arsenite by using laser-induced fluorescence confocal spectroscopy to measure arsenite stimulated enhanced green fluorescent protein synthesis of genetically engineered *E. coli* bioreporter cell, which has an inherent single-molecule detection capability. A recombinant soil bacterium *Sinorhizobium meliloti* has been constructed by fusing the *gfp* gene to the *melA* promoter, which is induced on exposure to galactose and galactosides. Using this fusion strain, a biosensor was developed to determine the concentration of galactosides. Similarly, *gfp* reporter gene has also been used to develop biosensors for various applications, such as detecting bio-available toluene and related compounds and *N*-acyl homo serine lactones in soil, measuring water availability in a microbial habitat and monitoring cell populations. With the development of DNA recombinant technologies and understanding to microbes, this type of biosensor will become an increasingly more powerful technique.

11.0.3.6.3 O_2-Sensitive fluorescent material based biosensor

Besides green fluorescent protein, other fluorescent materials have also been used in the construction of microbial biosensor. The biosensors consisted of either a layer of oxygen-sensitive fluorescent materials that are made up of sea water microorganisms immobilized in poly vinyl alcohol, sol–gel matrix and an oxygen fluorescence quenching indicator with linear range of 4–200 mg/l or an immobilized *P. putida* membrane attached to an optical fibre sensor for dissolved oxygen with detection limit of 0.5 mgL^{-1}.

11.0.3.6.4 Bioluminescence biosensor

Bioluminescence is associated with the emission of light by living microorganisms and it plays a very important role in real time process monitoring. The bacterial luminescence *lux* gene has been widely applied as a reporter either in an inducible or constitutive manner. In the inducible manner, the reporter *lux* gene is fused to a promoter regulated by the concentration of a compound of interest. The concentration of the compound can be quantitatively analyzed by detecting the bioluminescence intensity. In the constitutive manner, the reporter gene is fused to promoters that are continuously expressed as long as the organism is alive and metabolically active. This kind of reporter is good for evaluating the total toxicity of contaminant. Both types of reporters have been shown to be useful for biosensor development.

Heavy metal mediated toxicity in the environment is dependent on bio-available metal concentrations. Bioluminescent microbial biosensors have been extensively investigated to monitor bio-available metal. *Ralstonia eutropha* AE2515 was constructed by transcriptionally fusing *cnr*YXH regulatory genes to the bioluminescent *lux*CDABE report system to fabricate a whole cell biosensor for the detection of bio-available concentration of Ni^{2+} and Co^{2+} in soil. Several optical biosensors consisting of bacteria that contain gene fusion between the regulatory region of the *mer* operon (*mer*R) and *lux*CDABE have been developed to quantitatively response to Hg^{2+}. The *mer* promoter is activated when Hg^{2+} binds to *Mer*R, then result the transcription of the *lux* reporter gene and subsequent light emission. Bio-available copper in soil is also monitored by using engineered *P. fluorescens* through mutagenesis of *P. fluorescens* containing copper-induced gene and *Tn5::lux*AB promoter probe transposon.

In order to monitor nutrients in an aquatic ecosystem, a biosensor for monitoring phosphorus bio-availability to *Cyanobacteria* (*Synechococcus* PCC 7942) was developed. The reporter strain *Synechococcus* harbors, the gene coding the reporter protein luciferase under the control of an inducible alkaline phosphatase promoter, which can be induced under phosphorous limitation and shows improvement to conventional phosphorus detection methods. Bioluminescent microbial biosensors using the inducible reporter gene have also been developed for the measurement of bio-available naphthalene and halogenated organic acids.

The environmental problems caused by industrial and agricultural pollution have increased the demand for the development of pollutant and toxicity detection methods. The fusion of reporter genes to promoters that are induced when cells are stressed by toxic chemicals are one promising approach that has been used to fabricate biosensor for such application. Recombinant *E. coli* bearing *fabA lux* fusion and plasmid pUCD607 containing the full *luxCDABE* cassette has been constructed as biosensors for water pollutant detection. The online pollutant and toxicity test using bioluminescence based biosensors was proved to be sensitive and reliable. *Lux* marked rhizobacterium *P. fluorescens* has been developed to evaluate the pollution-induced stress, which influences rhizobacterium carbon flow based on the fact that bioluminescence output of biosensor is directly correlated with metabolic activity and reports on carbon flow in root exudates. Furthermore, *lux*-marked whole cell biosensors for the evaluation of interactive toxicity of chlorophenol and toxicity assessment of wastewater treatment plant and these biosensors responded to test pollutants/ toxicity rapidly.

Genotoxicants is a class of hazards, which can cause DNA damage. An optical-fibre bioluminescent microbial sensor to detect the DNA damage hazard-mitomycin C by the induction of a selected promoter and the subsequent production of bioluminescent light through a recombinant *lux* reporter was reported. Bioluminescence production was shown to be dose dependent *E. coli* containing plasmid-borne fusion of the *recA* promoter-operator region to the *Vibrio fischeri lux* genes has also been reported for genotoxicant detection. When the recombinant *E. coli* strains are challenged with DNA damage hazards, they increase their luminescence. Furthermore, this study was expanded by investigating and demonstrating the luminescence response of these strains to ultraviolet radiation, which can cause DNA damage. Another *lux*-based *Psedumonoas aeruginosa* biosensor was fabricated to quantify bacteria exposure to UV radiation in biofilm.

11.0.4 Colorimetric Biosensor

A sensitive biosensor based on color changes in the toxin sensitive living cells of fish was reported. In the presence of toxins produced by the microbial pathogens, the cells undergo visible color change and the color changes in a dose-dependent manner. This cell-based biosensor's potential application in the detection and identification of virulence activity associated with certain air, food and water-borne bacterial pathogens. A simple fibre-optic based microbial sensor to detect organophosphates based on the absorbance of P-nitrophenol (PNP) formed from the hydrolysis of organophosphates by the genetically engineered *E. coli* expressing organophosphorus hydrolase on the cell surface. This biosensor can be easily extended to other organophosphates such as Coumaphos through the monitoring of its hydrolysis product Coumarin.

A colorimetric whole cell bioassay for the detection of common environmental pollutants benzene, toluene, ethyl benzene and xylene (BTEX), found at underground fuel storage tanks, using recombinant *E. coli* expressing toluene dioxygenase and toluene dihydrodiol dehydrogenase was reported. The bioassay was based on the enzyme catalyzed conversion of the BTEX components to their respective catechols followed by the reaction with hydrogen peroxide in presence of Horseradish Peroxidase (HRP) to colorimetric products that can be monitored at 420 nm.

11.0.5 Other Types of Microbial Biosensors

Besides electrochemical, optical and colorimetric microbial biosensors, there are few other types of biosensors reported recently.

11.0.5.1 Sensors based on baroxymeter

As a new application, baroxymeter has been developed as a portable wastewater direct toxicity assessment device based on manometric bacterial respirometry. Respirometry was measured as the pressure drop in the head space of a close vessel due to oxygen uptake by the microorganism in contact with sample. This microbial pressure sensor showed good reproducibility and comparable responses with other reported methods.

11.0.5.2 Sensors based on infrared analyzer

A new method was reported for monitoring inhibitory effects in wastewater treatment plants based on continuous measurement of the microbial respiration product CO_2. Activated sludge microbes are used as the biological elements and their respiratory activity is inhibited by the presence of toxic compounds, resulting a decrease in CO_2 concentration which was analyzed by using a CO_2 infrared analyzer. Based on the measurement of CO_2 concentration in the off gas produced during degradation of carbon compound by microbial respiration activities, a microbial biosensor was developed to monitor the extent of organic pollution in wastewater both off-line in a laboratory and online in a wastewater treatment plant.

11.1 CELL BASED BIOSENSORS (CBBs)

11.1.1 Introduction

Biosensors with biological precognitive components, such as organisms or biological materials, are used to detect the parameters concerned with the physiological process or the biochemical process. According to precognitive components, there are three classes of biosensors: (1) Molecular biosensors (based on antibodies, nucleic acids, enzymes or ion channels) (2) Cellular biosensors and (3) Tissue biosensors. As the sensitive elements are concerned, the former is immobilized biological components and the latter two are based on the organisms themselves. Thus molecular biosensors are generally designed to respond only to target molecules and have high sensitivity. However, because of these characteristics, relevant molecules with similar function may not be detected by molecular biosensors. By using living cells as sensitive elements, cell based biosensors overcome the limitations and may detect unexpected molecules. Nowadays, with the development of micro-fabrication technology of semiconductor, miniaturization of analytic technology has offered a powerful tool for analysis of the cellular microenvironment.

Cell based biosensors can detect the analytic information quantificationally. In other words, it can verify the presence and the concentration of some substances. *Eg.* by using the CBBs with some special type receptors, the sensitivity of the sensors for certain agonist can be deduced by the receptor ligand combination constant. So the concentration of agonist can be detected easily. The more important fact is that cell based biosensors are able to measure the functional information. *Eg.* CBBs can be used to analyze the effect of pharmaceutical compound on a given physiological system. Sometimes, people make CBBs as detectors to study the mechanism of action of second messenger.

In general, cell based biosensors can keep living cells under constant observation to study cellular physiological action when cells are subjected to stimulus. From the point of view of biology, CBBs can explore the cellular state, function and the basic life action. From the point of view of analytic, CBBs can evaluate the analytic function. But there are many complex problems when living cells were treated as the primary sensor, including the selection, the culture and the maintenance of living cells. In addition, the coupling of living cells and the secondary sensor constitutes one of the challenges. However, cell based biosensors can perform real time bioassays dynamically and rapidly and have numerous applications ranging from biomedicine to environmental detection.

11.1.2 Applications

11.1.2.1 Monitoring cellular microenvironment

Real time detection *in-vivo* contributes to understand comprehensively the cellular physiology. Sensors are required to measure and analyze the extracellular and intracellular microenvironments quantificationally. Ion sensitive microelectrode (ISME) is one of the best tools for detecting concentration of some free ions in cells. Integrating various ISMEs, we can observe simultaneously many ions such as NH_4^+, Cl^{-1}, Na^+ and Mg^{2+}. Another useful method is fluorescence imaging, which can detect changes of ion concentration and expression of proteins concerned with transfer of cellular signaling. Combined with fluorescent probe, the laser confocal microscope can attain the precision and the legibility that have never been attained before, which makes it possible to detect concentration of various ions such as free Ca^{2+} and H^+.

Changes of extracellular metabolism products such as ions, large biomolecules, depend on the intracellular physiological state. Thus, deduce the intracellular physiological state by detecting the metabolism products. In 1990s, using silicon based technology cytosensor was invented, which is now called a micro-physiometer. It is a system for performing bioassay, which integrates living cells with Si-based biosensors. Acidic products of energy metabolism acidify cellular environments and the microphysiometer measures the rate of proton excretion from cells using a light addressable potentiometric sensor (LAPS). Similar to LAPS, ion-sensitive field effect microelectrode arrays are developed to measure the pH close to the nearby surface and the global extracellular acidification rate at the same time. Furthermore, O_2 electrode sensors and CO_2 electrode sensors are able to detect the consumption of O_2 and the production of CO_2 during the glycolytic cycle.

Large number of experiments on ligand receptor proves that the combination of receptor and agonist increases the rate of extracellular acidification (ECAR). Microphysiometer is able to estimate the effect of chemotherapy by ECA. So it can be used to evaluate and screen the drugs. Hafner *et al.* (2000) developed the special application of this kind of sensors in sub-organelle, such as G protein coupled receptors, receptors with intrinsic catalytic activity, ligand gated ion channel, receptors linked with cytoplasmatic tyrosine kinase. The sensor technology for monitoring extracellular environment has made great advance. If there is a system consisting of the units for detecting extracellular environment and intracellular environment, it is possible to understand the trans-membrane principle of biological substance and the effect mechanism of drugs on cell.

11.1.2.2 Cellular electrophysiological characteristics

Living cells grown *in vitro* are particularly attractive as potential detector elements. The electrophysiological signals of excitable cells depend on the cellular functional information. Excitable cells, such as neural cells, myocytes, evoke action potential when subjected to stimulus. Patch clamp can be used to detect the cellular membrane potential directly, but the measurement is limited to a fixed recording site, so it cannot be used to measure the coupling of cells. Furthermore, the invasive nature of intracellular recording, as well as the voltage sensitive and optical sensitive dyes, limits the utility of standard electrophysiological measurements and optical approaches. As a result, planer microelectrode arrays are developed and have become powerful tools, which offer a non-invasive and long-term approach to the measurement of excitable cells action potential or communication between neural tissues.

Fig. 11.1 illustrates a multi-electrode array used in recording potentials from a chick cardiac monolayer. Microelectrode array was fabricated with 36 microelectrode sites that were 10 μm in diameter. Standard thin film photolithographic techniques were used to implement the array of gold microelectrode sites and leads. The simultaneous record from a chick cardiac monolayer is shown in Fig. 11.2. The spike delay between proximal microelectrode sites allowed quantification of cardiac propagation velocity. It was reported that the propagation was approximately 100 mm/second. Krause *et al.* (2000) developed an extended gate electrode arrays for extracellular signal recordings which were arranged in a matrix of 8 × 8 with active areas ranging from 6 to 30 μm in diameter. An electronic amplification circuitry based on commercial Junction Field Effect Transistors (JFET) was used, where the gold sensor fields acted as extended gate electrodes (EGE) of the transistors, which provides a new approach for long term extracellular recording systems *in vitro*.

The measurement of activity of a collection of neurons in response to various substances allowed detecting the action potential of neurons cultured on them. Because of morphological changes of neurons and slices, long term recording will lead to drift and blur, which can be minimized by signal processing methods. Neuron-silicon equivalent circuit technology was developed to analyze the aberrant signals. Vassanelli *et al.* (1998) constructed a model to detect the weak signals from rat embryonic neurons and illustrated that the variation of extracellular potential resulted from the variation of ion conductance where the electrodes contact the cellular membrane. Vast extracellular recording signals are useful in pharmaceutical screening.

Fig. 11.1 *Microelectrode arrays, with 36 microelectrode sites that are 10 μm in diameter. (Source: Cell-based biosensors: Towards the development of cellular monitoring, Wang Ping et al., Chinese Science Bulletin Vol. 47 No. 22 November 2002)*

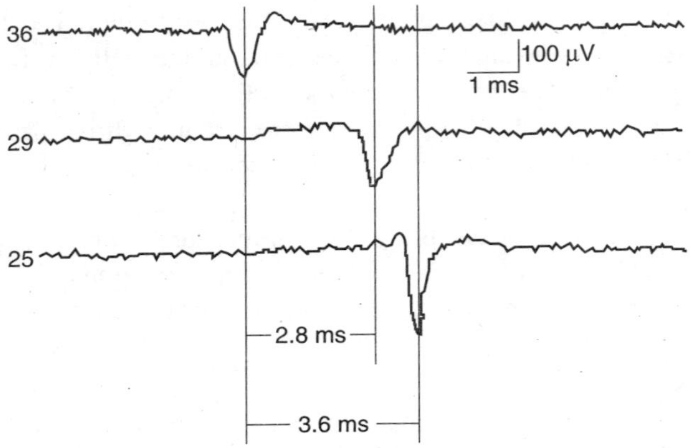

Fig. 11.2 *Recording from a monolayer of embryonic day 11 chick myocardiocytes after 23 d in vitro. (Source: Cell-based biosensors: Towards the development of cellular monitoring, Wang Ping et al., Chinese Science Bulletin Vol. 47 No. 22 November 2002)*

11.1.2.3 Detecting the cellular mechanical behavior

Generally, cells respond to broad classes of effectors and alter the mechanical behavior or properties of living cells. For instance, some kinds of cells may move in response to stimulus of hormone and viral infection of other kinds of cells may change in cytoskeletal activity. Thus, the mechanical behavior of cells has potential for sensor applications and offers a novel method to study the mechanical behavior and the response of living cells.

Antonik *et al.* (1997) fabricated cantilevers to detect the mechanical behavior, which were typically 100-300 μm long, 20-40 μm wide and 0.4 0.8 μm thick, with resonant frequency in the range of 5-100 kHz and spring constants of 0.001-0.5 N/m. Cells can respond mechanically either actively in the form of movement or passively by changing their visco-elastic properties. By growing Madin Darby Canine Kidney (MDCK) cells directly on one side of AFM-type cantilever, mechanical activity of the cells can be cross-examined by monitoring the deflection of the cantilever with a conventional optical lever detection system (Fig. 11.3). The experimental data explains that these integrated cell cantilevers were capable of detecting the response of MDCK cells to different toxins, with response time of the order of several seconds.

The cell electrode interface impedance depends on adhesion, spreading, and motility of the cells cultured on the microelectrodes. Giaever *et al.* (1993) designed the impedance sensors that can observe the morphological changes of mammalian cells continuously. Theoretically, the impedance technique is capable of dynamic measurements of cellular movement at the nanometer order, a resolution above that of conventional measurement methods. Impedance measurements have been used to assess the effect of nitric oxide on endothelin induced migration of endothelial cells and subtle changes in cell morphology of fibroblasts treated with prostaglandin E2.

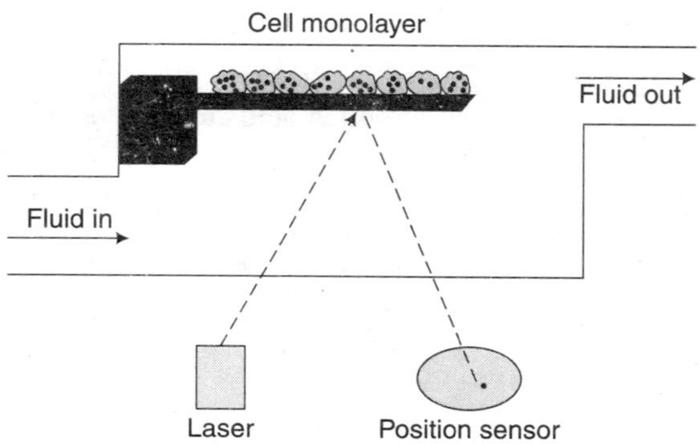

Fig. 11.3 *Biosensor based on micromechanical interrogation of living cells. (Source: Cell-based biosensors: Towards the development of cellular monitoring, Wang Ping et al., Chinese Science Bulletin Vol. 47 No. 22 November 2002)*

Stimulating a cell with an effector begins a cascade of metabolic events that attempts to adapt the cell to its new environment. It is the integration of these reactions that determines the overall response of the cell, but any single reaction or group of reactions in the chain can be interrogated and used as the basis for sensing by the biosensor based on micromechanical interrogation of living cells.

11.1.2.4 Measuring cellular specific behavior

There is another family of cellular measurement systems that are used to detect specific features of certain types of cell or receptor ligand systems. *Eg.* Some bacteria that are naturally bioluminescent or have been engineered with a gene for a Luciferase that is tied to a promoter sensitive to a given substance are used in biosensors based on bioluminescence. Generally, this kind of biosensor is applied in detecting heavy metals. The problem that has to be solved is how to control the factors that affect the amount of light emitted, such as oxygen concentration, pH, composition of the medium and the number of cells present.

The chromatophores are pigmented cells responsible for the brilliant and changeable skin colors of many cold blood animals, including fish, amphibians and reptiles. These cells have been employed in biosensors because many types of biologically active agents change their appearance. There are several advantages of this kind of biosensors: high specificity and accuracy.

11.1.3 Trends

11.1.3.1 Multi-parameter detection

There are many kinds of cell based biosensors with different principles and applications, but most of them have single function, i.e. they are able to detect certain substance. Getting more information about the multifunctional cellular processing of input and output signals in different cellular plants is essential. Baumann *et al.* (1999) developed the cell monitoring system (CMS) and this CMS integrated O_2, temperature, pH, optical fibre and mobilized enzyme sensors. The former two guaranteed the standard living environment of cells and the others offered the measurement of extracellular environment.

In 1992, micro-physiometer developed for performing bioassays, which was able to detect indirectly the physiological changes of living cell, according to the concentration changes of extracellular H^+ that was the rate of extracellular acidification (ECAR). Although, the micro-physiometer has been applied widely, it is also required to improve the concentrations of extracellular ions, such as Na^+, K^+, Ca^{2+}, it may change the cell physiology.

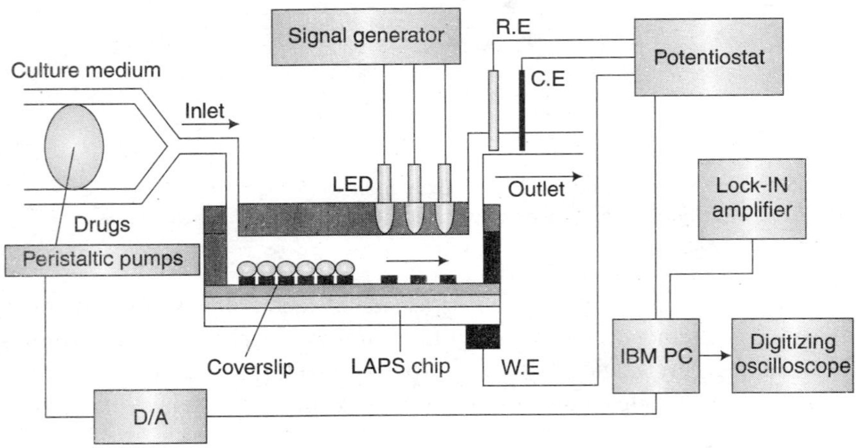

Fig. 11.4 *The schematic drawing of microphysiometer based on multi-light addressable potentiometric sensor (MLAPS). (Source: Cell-based biosensors: Towards the development of cellular monitoring, Wang Ping et al., Chinese Science Bulletin Vol. 47 No. 22 November 2002)*

The fundamental of multifunction microphysiometer based on MLAPS is as follows: The surface of the light addressable potentiometric sensor (LAPS) is deposited with different sensitive membranes (Sensitive to Na^+, K^+, Ca^{2+}) by silicon micro-fabrication technique and the Poly Vinyl Chloride (PVC) membrane technique. In the microphysiometer, different sensitive membranes are illuminated in parallel with N light sources at different frequencies, the response amplitudes of each frequency component can be measured on-line by parallel processing algorithm. The instrument (Fig.11.4) is equipped with a fluid control system consisting of a chamber, a pump and a valve, the latter two control the fluid (drug or medium) flow through the chamber at periodic intervals in order to keep cells living, analyzed the ion channel of suckling mouse's myocardial cell under the effects of western medicines (Phenobarbital sodium, penicillin sodium). The result proves that the microphysiometer can be utilized to monitor the micro-environmental changes of normal or abnormal cells under the effects of various drugs, including extracellular various ions and metabolic product, estimate the drug effect (stimulator or inhibitor) and evaluate and explore the drugs.

The trend of cell based biosensors is to combine multifunctional microphysiometer with ion-sensitive ultra-microelectrode and laser scanning confocal microscopy or optical nano-fibre. Then the integrated system can be used to study the extracellular and intracellular environments synchronously to explore the physiological mechanism deeply and to investigate the specificity of Agonist. Changes of extracellular and intracellular ion concentration relate to the changes of ion channels. The latter can be regarded as a special kind of cellular surface receptor. So, microphysiometer will play an important role in drug discovery of cellular surface receptor structure. According to these receptors to design regulator, to the regulate cellular function and to discover the novel drugs.

11.1.3.2 Biosensors based on Bio-MEMS

With the development of semiconductor micro-fabrication the detectors are small in size, *Eg.* Photo-sensors, amplifiers, discriminators and logic circuitry all built into one chip which is so called the Biological Micro-Electro Mechanical Systems (Bio-MEMS) and these are combined with micro-sampling and micro-separating techniques. Advantages of Bio-MEMS cell-based biosensors are to reduce the volume of sample, automatic sampling, to minimize the number of cells required for single cell, to improve sensitivity, to enhance the resolution of time and space. Microelectrode array (MEA) fabricated by the traditional micro-process technology has been applied widely. While integrating micro-pump, micro-valve, signal process circuit and MEA into one silicon device. This device is capable of automatic selection of the samples, flow of medium and to process the signals. Baxter *et al.* (1994) fabricated an 8 channel cell metabolism detecting system by silicon micromechanical technique which can realize high throughput drug screening (Fig. 11.5).

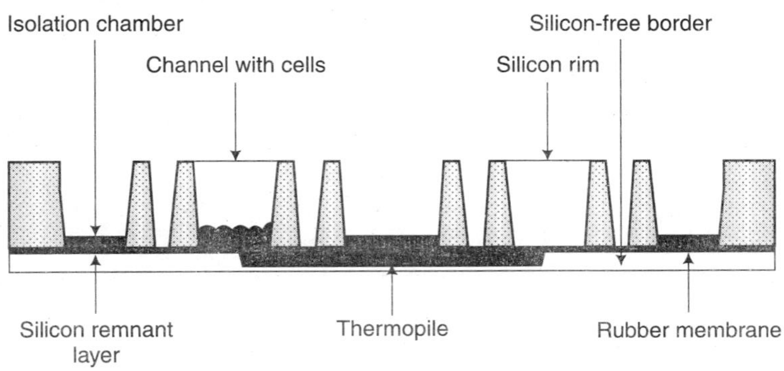

Fig. 11.5 *A high-throughput silicon microphysiometer. (Source: Cell-based biosensors: Towards the development of cellular monitoring, Wang Ping et al., Chinese Science Bulletin Vol. 47 No. 22 November 2002)*

Semiconductor material has its particular characteristic of cellular physiology and the characteristic allows inventing more types of cell based biosensors. Verhaegen *et al.* (2000)

associated cellular energy metabolism with seebeck effect and invented a microphysiometer to detect cellular energy delivery (Fig. 11.5). They fabricated two individual thermally insulated micro-chambers that were connected with the hot end and cold end of thermopile on silicon wafer with the standard integrated circuit micro-process technique. The thermopile consisted of 666 aluminum/ polysilicon thermocouples. Power and temperature sensitivity of the sensor are 23 V/W and 130 mV/K respectively. Both chambers were added with 100 μl medium. Then, the primary chamber housed 106 eukaryotic cells cultured to confluence, the balanced chamber housed no cell at all. The thermal equilibration time of the device was about 40 min. Basal metabolism was measured to be 330 pW/cell. Stimulation of transport mechanisms by reducing bath osmolality by 50% (add more 100 μl medium to both chambers) increases metabolism by 40 pW/cell. Stimulation of transport mechanisms by adding the Oxytocin hormone 1μl, 0.2 U/ml increased metabolism to 436 pW/cell.

11.1.3.3 Portable biosensors

Cell based biosensors are having great success in development of some commercial products. The use of cell-based biosensors outside the laboratory has been limited due to many issues including preparation of the sample, maintenance of the biological environment and integration of the electronics for data collection and analysis. Kovacs *et al.* (2001) developed an integrated silicon-poly dimethyl Siloxane cell cartridge that provided electrical and fluidic interconnects and environmental control. The cell cartridge also contained a functional CMOS silicon chip that incorporated a digital interface, temperature control system, micro-electrode sensors and analog signal buffering. Additionally, it supported separate cell populations in two 10 μl sealed chambers that had independent fluidic channels for sample injection. The data acquisition module offered 32 channels to fulfill the amplification and filtration of action potential and to send real time data streaming for computation. So this system is a field-portable cell-based biosensor platform. In addition, long-term recording of action potential required enormous storage space and special algorithm for signal process such as non-linear systems analysis and short time series analysis.

11.1.3.4 Genetically engineered biosensors

Cell based biosensors utilize whole cells to detect biologically active agents. Although cell-based biosensors have shown success in detecting the presence of biological agents, efforts to classify the type of agent based on functional activity have proven difficult because multiple biochemical pathways can lead to the same cellular response. These complicated interrelationships between cellular pathways and external environmental conditions can be considered 'cross-talk' noise. Cell based biosensors are composed of two transducers, the cells which are the primary transducer that converts the presence of a bioactive substance into a cellular signal and the secondary transducer that detects the cellular signal and outputs an electrical signal. Modification of the secondary transducers and signal process methods are difficult to eliminate the 'cross-talk' noise. With the development of molecular biology, people attempt to remove the 'cross-talk' noise based on modifying the primary transducer. The primary transducer is engineered to have either an increased sensitivity to agents that act on the pathway of interest or a decreased sensitivity to agents that act on pathways that interfere with the pathway of interest. Both situations can dramatically reduce the effect of 'cross-talk' noise due to interfering pathways. In practice, using the techniques of molecular biology to genetically engineer cells to have the desired response does modification of the primary transducer. A gene coding for a critical element in the biochemical pathway of interest is either inserted or removed from the cell genome, creating cells with the appropriate enhanced or depressed response to agents that act on the engineered pathway. *Eg.* The genetically engineered cell-based biosensors operate by assaying for an agent's ability to differentially activate two populations of cells, wild-type (WT) cells and cells genetically engineered to lack a specific receptor, knockout (KO) cells. Any biological agent that targets the knocked out receptor will evoke a response in the WT but not in the KO. This approach provides the benefits of an assay for specific functional activity while simplifying signal analysis.

Aravanis *et al.* (2001) designed the genetically engineered cell based biosensors for functional classification of agents. The primary transducers used in the genetically engineered cell based biosensors were mouse cardiomyocytes cultured on micro-electrode arrays. This approach successfully used a modified primary sensor to translate the 'cross-talk' noise into common mode noise that could be rejected. However, there are limitations to the types of receptors that can be deleted using current genetic engineering techniques. These cells were harvested from a genetically engineered animal. The deletion of many types of cellular receptors would result in a lethal mutation during embryonic development, preventing the harvesting of the differentiated cell type. Additionally, receptor deletions that are fatal for an entire animal might not impair single cell viability, which will decrease the precision of functional analysis, even lead to abhorrent result. However, the method provides for useful new assays in a variety of situations such as pharmaceutical screening.

11.2 WHOLE-CELL BACTERIAL BIOSENSORS

11.2.1 Introduction

Bacteria can be used as biosensors to detect the toxicity of a variety of environmental media including soil, sediment and water by coupling the bacteria with the transducers that convert cellular response into detectable signals. These bacterial biosensors are engineered by pairing a reporter gene that generates signal with contaminant sensing component that responds to chemical or physical change such as exposure to a specific analyte. When the biosensor is exposed to such a change, the sensing component stimulates the reporter gene through a biochemical pathway in the cell. The reporter gene then produces a measurable response such as emitting visible light, which is indicative of the degree of chemical or physical change. Several biosensors have been developed that indicate toxicity of any chemical or physical change and new biosensors are being developed to respond to particular analytes. Such biosensors have been developed for sensing heavy metals and metalloids including arsenic, cadmium, mercury and lead.

The microbial biosensors measure the bioavailable concentration of contaminants. To test the measurements made by biosensors, a chelating agent known to decrease bioavailability of lead was added to a lead solution. Measurements of the lead solution containing chelating agents were taken and compared to measurements of the lead only solution. A decrease in the biosensors luminescence matched a decrease in bioavailable concentration of lead in the solution. So, those biosensors are sensitive to the bioavailable fraction of the contaminant and their luminescence reflects the bioavailable concentration.

Biosensors have been further tested by comparing their results with the results from chemical analysis of arsenic contaminated samples. Chemical analyses such as sequential extraction procedure can be used to determine the total arsenic, water soluble arsenic and acid soluble arsenic content of the sample. The acid and water soluble contents are used to determine the mobility of arsenic in an attempt to clarify true exposure potential. The results from a biosensor correlated moderately with the water soluble arsenic determined by chemical analysis but not with the acid soluble or total arsenic content. Therefore, the acid soluble or water soluble arsenic content does not completely represent the bioavailable content of arsenic.

11.2.2 Reporter Genes

A reporter gene encodes for a mechanism that produces a detectable cellular response. It determines the sensitivity and detection limits of the biosensor. Specific characteristics are needed for the reporter gene to be used in a biosensor. The gene must have an expression or activity that can be measured by using simple assay and reflects the amount of chemical or physical change. Also, the biosensor must be free of any gene expression or activity similar to the desired gene expression or activity that is being measured. Ensuring that the biosensor is free of any similar gene expression or activity prevents misinterpretation of the response and guarantees the measurement directly reflects the desired chemical or physical change.

Several reporter genes meet the necessary requirements and are frequently used. The most commonly used reporter gene is the *luc* operon from the firefly (*Photinus pyralis)* and it has been widely used as a monitor of gene expression and a reporter in 'bacterial biosensors'. The *luc* operon produces the enzyme Luciferase, capable of generating the luminescence of the firefly. The activation of the gene results in the transcription or reading of the *luc* operon, which causes the cell to produce Luciferase. The enzyme Luciferase spurs a chemical reaction that produces CO_2 and visible light by catalyzing the oxidation of its substrate, D-luciferin which binds to the enzyme active site. The visible light produced can be measured with a variety of instruments including a luminometer and optical fibres. The reaction has a high sensitivity level and a broad dynamic range but is complicated by requirements such as the addition of the substrate, an aerobic environment and ATP as a source of energy. However, this reporter gene has great versatility and can be mutated to produce enzymes that express a range of colors from green to red, which can then be independently controlled for multianalyte assays.

Other organisms have similar genes capable of emitting light. A close homolog to the firefly is the Click beetle (*Pyrophorus plagiophthalamus*), whose Luciferase is capable of producing a wide range of colors and the click beetle gene is being used in biosensors. However, the *luc* operon from the firefly has proven better suited for biosensors than the *luc* operon from bacteria such as *Vibrio harveyi* and *Vibrio fischeri*. The firefly *luc* operon is simply more efficient at converting chemical energy to light.

A photo protein emits light when excited with light of a specific wavelength range. A photo protein, green fluorescent protein (GFP) and its encoding gene from the Jellyfish (*Aequorea victoria)* have been used in biosensors. The production of GFP in the jellyfish results in the emission of a green fluorescence that can be measured. Therefore, the GFP system can be used as a reporter gene for biosensors, although it has a lower sensitivity compared to the *luc* operon. The GFP system allows real time detection without the addition of substrates and without disrupting the cells metabolism. The GFP does not rely on internal reducing equivalents being produced by the cell which may mean that this reporter gene is not as sensitive to the growth or nutritional status of the biosensor. The ease of detection and the minimal metabolic cost to the host cells will lead to an increase in the use of GFP in biosensors.

Another advantage of GFP is its ability to alter its stability and spectral properties through structural alterations and thus produce mutants with improved fluorescence intensity, thermo-stability and chromophore folding. GFP is a very stable protein, which means it can accumulate in the cell over time which results in background fluorescence. Mutants can be developed with less stable protein to improve the biosensor by reducing the background fluorescence. Measurements can be taken of the number of proteins produced or their fluorescence if the background fluorescence is considered. Results are more stable if the number of proteins is measured instead of their fluorescent activity because their activity and the number of proteins are not dependent on the performance of each protein.

11.2.3 Contaminant Sensing Components

Microorganisms have developed systems for detoxifying or excreting toxic substances. These systems can be used as the contaminant sensing component of the biosensor by detecting the substance for which it is designed to detoxify or excrete. The contaminant sensing component is combined with reporter genes to create biosensors that can identify toxic substances at very low levels. When the contaminant-sensing component detects the substance, it triggers the reporter gene which produces the luminescent enzyme. The sensing component determines the specificity of the biosensor. Some microorganisms have been developed genes for heavy metal resistance, which tend to be specific towards a particular metal instead of a general mechanism for all heavy metals. Thus, biosensors can be developed to detect specific metals and the genes have been found for resisting lead, nickel, mercury, chromium and arsenic. The pathway developed by bacteria to manage arsenic is used as the contaminant sensing part of the biosensor. A gene located in plasmid, encodes the mechanism of the pathway and has been cloned to use a biosensor. Almost all the bacteria have homologous genes with slight variation in the number of components involved in the mechanism.

In gram negative bacteria, the arsenic resistance gene remains inactive with the absence of As (III) in the cell due to the binding of the *ars* operon repressor protein to the promoter region of the gene. As (III) activates the system by binding to the repressor protein and release the promoter region for transcription. The released promoter region is transcribed to produce various components of the mechanism such as *ars*B and an arsenite translocating protein that serves as a trans-membrane efflux channel. This protein functions either chemi-osmotically, without an energy source or by ATP hydrolysis when coupled with *ars*A, an arsenite specific ATPase. *ars*C, the enzyme arsenate reductase, is also transcribed to reduce As (V) to As (III), since As (V) cannot pass through the *ars*B/*ars*A pump. *ars*D is a regulatory protein for additional control over the expression of the system and *ars*R is a transcriptional repressor. The mechanism varies slightly in gram positive bacteria, which lack *ars*A and *ars*D. Understanding the mechanism for arsenic resistance is necessary in order to develop an appropriate biosensor and to better understand its response.

11.2.4 Arsenic Sensing Biosensors

Various biosensors have been developed and tested on a research level for detecting bioavailable arsenic. To develop the biosensor, the arsenic resistant gene and the reporter gene are cloned and inserted onto one plasmid, which is then inserted into host bacteria. All arsenic sensing biosensors are triggered by arsenic, the analyte, entering the biosensor and activating the transcription of the resistant gene, which is followed by the transcription of the reporter gene. The entire resistant gene is not needed, so many biosensors use the beginning components such as the promoter. It is able to recognize the arsenic and begin the transcription of the plasmid that contains the reporter gene. The transcription of the reporter gene produces proteins, which glow in direct correlation to the amount of arsenic entering it.

Both gram negative and gram positive arsenic resistance systems have been utilized in biosensors as the arsenic sensing component. Constructed biosensors have used either Luciferase or GFP as the reporter gene coupled with various combinations of arsenic resistance mechanism components. Various strains of bacteria have served as the host bacteria. Many combinations of reporter genes, resistance gene components and host strains can be incorporated into a biosensor. Ji and Silver (1992) studied to determine which combination is most advantageous for regulation of the *ars* operon in *Staphylococcus aureus* plasmid p1258. The plasmid was inserted into various strains of bacteria to determine the pattern and causes of induction. Corbisier *et al.* (1993) gene fusion studies through the coupling of the *lux*AB gene encoding Luciferase from *V. harveyi* and the arsenic operon from *Staphylococcus aureus* plasmid p1258 and achieved the results. The recombinant plasmid was tested in two host strains, *E.coli* and *S.aureus*, which differed in their patterns of induction. Both biosensors were induced by arsenite, arsenate and bismuth while the *S.aureus* biosensors were also induced by antimonite. Arsenite was the strongest inducer for the *E. coli* biosensors and generated the maximum light emissions at a concentration of 10 µm.

Another arsenic biosensor was created by Tauriainen *et al.* (1999) using the *luc* operon from the firefly, which is more efficient at converting chemical energy to light energy (luminescence). The *luc* operon was coupled with the *ars* promoter and the *ars*R gene from *E. coli* plasmid R773 and inserted into two strains of *E. coli*. It demonstrated greater sensitivity than its predecessor, pTOO21, which is used the *ars* operon from *S.aureus* plasmid p1258. Unlike the previous plasmid, the recombinant plasmid pTOO31 is able to detect both As (III) and As (V) with equal strength. One strain of *E. coli* MC1061, proved to be a better host for the plasmid pTOO31. It has a detection range of 33 nm to 1 mM for As (III) and a range of 33 µm to 33 mM for As (V). Luminescence can also be moderately induced by antimony and cadmium. *E. coli* MC1061 (pTOO31) was used to test the bioavailability of arsenic in contaminated soil samples. The sensor reacted to As (III) at concentrations 100 times lower than As (V). The sensitivity was 0.1 µm for As (III) and improved to 0.5 µm for As(V) when cultivated in the nutrient rich Luria Bertani medium.

S.aureus, B.subtilis, E. coli and *P.fluorescens* all perform equally well as the host strain when testing for As (III) using either plasmid pTOO31 or pTOO21. Overall, *E. coli* tends to be the most sensitive; it detects levels as low as 100 nm and works well for both As (III) and As (V). However, *E. coli* may not be the best suited host strain for testing environmental samples as well as laboratory samples, since it is not native to soil. *P.fluorescens,* a soil bacterium may be better suited for testing soil and sediments. *P.fluorescens* also has advantage over *E. coli* due to the low amount of substrate needed to reach the maximum luminescence.

Two arsenic sensing strains were developed using the *luc*GR gene encoded for luciferin from the click beetle. The promoter and a*rs*R were used from plasmid p1258 and plasmid R773 as the arsenic sensing components for plasmid pTPT21 and pTPT31, respectively. While their detection limits were similar when used in either *E. coli* or *P. fluorescens* pTPT31 performed better than pTPT21. For *P.fluorescens* with plasmid pTPT31 induction was shown to increase steadily between 10 nm and 10 μm with toxic inhibition seen at 100 μm. While *E. coli* with plasmid pTPT31 strains have detected levels of 100 nm, induction increased steadily between 10 nm and 1 μm with a partial drop at 10 μm. The maximal induction for both strains was seen from 100 nm to 10 μm, when plasmid pTPT31 was compared to plasmid pTOO31 and no significant difference was found.

Roberto *et al.* (2002) developed an arsenic biosensor by coupling a*rs*R, a*rs*D and their promoters with the gene encoding for GFP from the marine jellyfish (*A. victoria).* The recombinant plasmid was inserted in *E. coli,* which created a biosensor that produced GFP in the presence of arsenic. This biosensor is capable of detecting both As (III) and As (V) with a range of 1 to 10,000 ppb (0.01 to 100 μm). A cell density of 109 cells/ml was necessary to exceed the background fluorescence. An incubation period of 10 to 12 hours was needed to reach a steady state of fluorescence, although growing cells to log phase before exposing them to arsenic reduced the incubation period to 2 hours.

11.2.5 Methods for Using Biosensors

The method for testing samples using bacteria as biosensors is generally the same for all strains of bacteria, with slight variations in induction time and substrate addition. Once the desired biosensor has been selected, the cultures are grown overnight, usually in a Luria-Bertani medium, which is sometimes supplemented with Kanamycin at 37 EC. The next day a new dish with 50 ml of new medium is inoculated with 0.5 ml of the overnight culture. The new culture is grown at 37 EC to an optical density of 0.8 to 1.2 at 600 nm. After the cells are washed, they are suspended and then diluted in M9 medium at a concentration of 1×10^6 cells per measurement.

The biosensor culture is then exposed to the contaminated sample as water samples, soil water extractions or soil suspensions. The procedures for properly exposing and reading the biosensors need to be refined for all types of medium including water, soil and food. Arsenic sensing biosensors have not been tested in the field and therefore have not been thoroughly tested with environmental media of soil, sediment and water. For a better understanding of arsenic bio-availability from heterogeneous medium, it is important that the biosensor can be exposed to the arsenic without any alteration of its form in the medium, in order to ensure accurate information about its bio-availability. The exposed final culture is then incubated for 60 to 90 minutes (depending on the host strain) as a mixture of one part cells to one part metal solution. Once the culture has been incubated, luminescence or fluorescence can be measured. The quantification of light emissions is one of the most sensitive means of detection. This can be performed using electrochemical transducers, where cells are immobilized onto an electrode and amperometric or potentiometric techniques measures the cellular response. Optical transducers can also be used to measure absorbance or luminescence or fluorescence by immobilizing the cells onto optical fibres.

11.2.6 Measuring Luminescence or Fluorescence by Luminometer

A luminometer or fluorimeter can be used to measure the intensity of the luminescence or fluorescence by counting the number of photons emitted by a cell. An individual cell response is not simply on or off, but rather a sliding scale with greater intensity correlated to higher concentration of arsenic within the cell. The mixture of arsenic solution and cells is placed in a well plate, which is inserted into the luminometer. The background luminescence is measured for the first 5 seconds.

A substrate is added to induce luminescence that is associated with arsenic. Induction can take 2 to 12 hours, depending upon the host strain used. Luminescence can be induced in *P. fluorescens* in 2 hours when incubated at 22 EC, while *E. coli* takes 4 hours. Induction time improves for *E. coli* when it is incubated at 37 EC. Green fluorescent protein (GFP) sensors not grown to log phase actually have the greatest induction time at 10 to 12 hours before peak luminescence is reached.

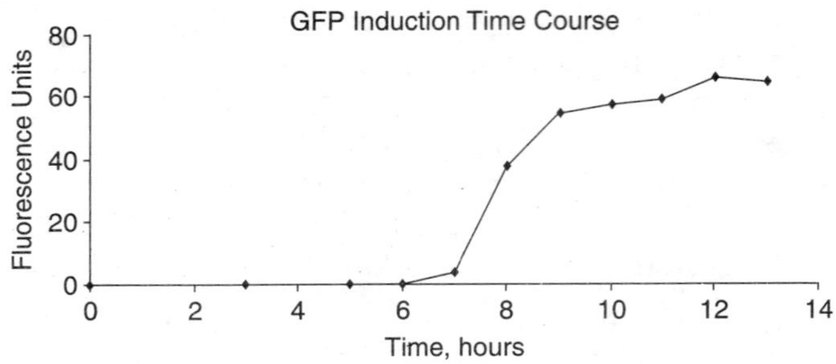

Fig. 11.6 *Plot of Induction. (Roberto et al., 2002)*

The measurements are taken periodically until the luminescence peaks without adding more substrate. The time needed to reach this peak varies with the bacterial strain and reporter gene used in the biosensor. A plot can be generated of the biosensors response to the sample as fluorescence over time. *Eg*. Fig. 11.6 shows the results of one biosensor using GFP as the reporter gene when exposed to a sample contaminated with arsenic. An induction coefficient is calculated by dividing the peak luminescence value of the sample by the luminescence value of a culture without arsenic.

11.2.7 Advantages and Disadvantages

Whole cell bacterial biosensors have several advantages over conventional methods of detecting contaminants at a site. The greatest advantage is the ability of biosensors to detect the bio-available fraction of the contaminant, as opposed to the total concentration. Knowing the bio-available fraction allows for a more accurate assessment of the site and the potential risks involved. The biosensors create a clear picture by providing physiologically relevant data in response to the contaminant. This response (usually luminescence) is quick and easy to measure, resulting in real time data.

Advantages

1. To measure the bio-available fractions
2. Inexpensive and less labor intensive
3. More sensitive and produces real time data
4. Suitable for field work

Disadvantages

1. Short lifetime
2. Lack of genetic stability

3. Unknown rate of Type I and II errors
4. Limited understanding of applicability to higher organisms
5. Performance dependent on environmental condition

It is concluded that the biosensors are fast, less expensive and less labor intensive than other traditional methods such as atomic absorption spectrometry, inductively coupled atomic electron spectrometry and sequential extraction procedure. The results obtained from biosensors are compatible and comparable to chemical analysis, while being free of chemical extractions and analytical procedures. Biosensors can also be more sensitive than chemical methods. The detection limit of biosensors depends on its design but it has been demonstrated as low as 7 ppb for As (III) and 35 ppb for As (V). They are particularly for field work or *in situ* analysis, since the procedure for using biosensors does not involve the bulky, fragile equipment or specialized training that most analytical methods require.

Disadvantages also exist when using biosensors instead of traditional methods. A major disadvantage is the limited understanding of the biochemistry involved. It can only be assumed that the fraction of arsenic bio-available to bacteria is equivalent to the fraction of bio-available to higher organisms. The response made by the biosensor is specific to its membrane and uptake procedures. The uptake of arsenic through phosphate pumps and aqua-glyceroporins has been documented in a variety of organisms from bacteria to humans, indicating the relevance of the biosensors to higher organisms. The details of bacterial response to most contaminants including arsenic are understood, but the response of higher organisms is still unclear. It is therefore difficult to transpose information gathered through bacterial biosensors and apply it directly to higher organisms. Discoveries are being made surrounding the responses to arsenic for higher organisms including humans. Further, the advances in biosensor research may clarify the meaning and applicability of the results for use in ecological and human health risk assessment.

Bacterial biosensors perform best under conditions similar to their natural environment. Unfortunately, little is understood about the relationship between the microorganisms and their natural environment. It is difficult to determine the effects that certain procedural conditions have on the performance of biosensors. The temperature, pH, incubation time, medium and reagents all can have effects on their performance that can only be understood with further research and testing. Problems with performance include luminescence without the presence of the analyte or no luminescence in the presence of the analyte. Such Type I and Type II errors are not identifiable and the rate of committing such errors is unknown. Without fully understanding the biochemistry and specificity of each sensor such errors cannot be identified or predicted.

11.2.8 Potential

Despite their disadvantages, whole cell bacteria biosensors do have the potential to become a common tool for site characterization. A few challenges, such as longevity, reproducibility and linearity must be overcome in order to fully implement the use of biosensors in the field. As these challenges are met, the applicability of biosensors may increase. Biosensors can be freeze dried so they are easier to handle and store. Freeze dried cells are more reagent like and they are suitable for field work because they do not need to be carefully stored and cultured. Freeze dried cells maintain their usability but do have lower induction coefficients and sensitivities. Cells can be reconstituted by incubating them in water and then placing them in a medium in order to regain sensitivity.

Lysis of whole cell bacterial biosensors has also been done in an attempt to make them easier to use. By separating the recombinant plasmid that contains the reporter gene and the arsenic sensing gene from the rest of the cell, the biosensor becomes more like a reagent without the need for cell cultivations. This reduces the day-to-day variation in the genetic makeup of the biosensor, as well as the time needed to perform the test. Using cell lysate, the plasmid without the remaining cellular components is more convenient and easier to use in the field as an *in vitro* method for determining arsenic bioavailability. Despite the increased ease of use for cell lysate, there are disadvantages. While results from tests using whole cell and cell lysate biosensors correlated well, the results varied more between replicates and samples for the cell lysates biosensor. The whole cell biosensor created more reproducible results for all environmental samples tested. This may be due to the fact that the natural environment and ingredients for the mechanism provided by the whole cell. The reaction is simplified and efficiency is improved when the whole cell is used.

Further improvements on the applicability of biosensors include the development of a multi-analytical biosensor. By genetically labeling the bacteria used in the biosensor, a multi-analytical biosensor can be developed to test more than one contaminant at one time. Multiple strains of bacteria, which test for different specific contaminants, could be combined into one biosensor. Each strain of bacteria can have another gene incorporated into its plasmid. The gene encodes for a protein, such as the Cyan Fluorescent Protein (CFP), which emits a particular color. Measurements can be taken of the luminescence of each strain, which is identified by its individual color. A similar effect can be produced using mutated reporter genes that produce distinct colors ranging from green to red. Each mutant can be coupled with a specific contaminant-sensing gene to develop a multi-analytical biosensor.

11.3 YEAST BASED BIOSENSORS

Kubiak and Wang (1989) described a yeast modified carbon paste electrode that is sensitive to primary alcohols. Dry Red Star active yeast granules were mixed with graphite powder and mineral oil. Carbon pastes containing 2,5 and 10% (w/w) yeast were tested. The modified carbon pastes were either packed into a glass tube (3 mm opening) with the amplifier connected by a copper wire or into the electrode cavity of a thin-layer flow detector. Measurements were obtained for different alcohol concentrations in the presence of 1 mM NAD^+ and potassium hexacyanoferrate (III) in a well stirred 0.05 M phosphate buffer (pH:7.4). Potassium hexacyanoferrate was used as a redox mediator, permitting the transfer of electrons for the oxidation of NADH to occur at a lower potential. Steady state current responses to 0.5 mM ethanol at a potential of 0.6 V were reported to be 298, 405 and 590 nA respectively for the test solutions with 2, 5 and 10% yeast contents. In the flow injection tests, the 95% response times ranged from 7-10 seconds. The yeast-carbon paste electrode sensitivity and linearity were examined for different alcohols. The order of sensitivity for the following primary alcohols was ethanol > 1-propanol > 1- butanol > 1-amyl alcohol. The electrodes were not sensitive to methanol or to the secondary alcohols (2- propanol and 2 – butanol). The Line Weaver-Burk plot for 1-amyl alcohol was not described by simple Michaelis-Menten kinetics. For low alcohol concentrations below the apparent K_m current concentration relationships were approximately linear. The alcohol content of several commercial beverages, including vodka, red and white wines and beer were measured and found to be in good agreement with the labeled values.

Biochip

Chapter 12

12.0 BIOCHIP

12.0.1 Introduction

A biochip is a collection of miniaturized test sites (microarrays) arranged on a solid substrate that permits many tests to be performed at the same time in order to achieve higher output and speed. The development of biochip is a major thrust of the rapidly growing biotechnology industry, which encompasses a diverse range of research efforts including genomics, proteomics, computational biology and pharmaceuticals. Advances in these areas provide new methods for sorting out the complex biochemical processes occurring inside cells, with the larger goal of understanding and treating human diseases. At the same time, the semiconductor industry has been increasingly developing the technology of microminiaturization. The inclusion of these two fields in present days has enabled biotechnologists to begin the packing bulky sensing tools into smaller and smaller spaces onto so called biochips. These chips are essentially miniaturized laboratories that can perform hundreds or thousands of simultaneous biochemical reactions. Biochips enable researchers to quickly screen large numbers of biological analytes for a variety of purposes. *Eg.* For diagnosis of diseases, detection of the bioterrorism agents, etc.

12.0.2 History

The development of biochips has a long history; one of the first chemistry based sensors was the glass pH electrode invented by Hughes (1922). Measurement of pH was accomplished by detecting the potential difference developed across a thin glass membrane selective to the permeation of hydrogen ions; this selectivity was achieved by exchanges between H^+ and SiO sites in the glass. The basic concept of using exchange sites to create permeation, at present selective membranes was used to develop other ion sensors. *Eg.* A K^+ sensor was produced by incorporating Valinomycin into a thin membrane. The first true sensor (oxygen sensing electrode) is developed by Leland Clark (1956). The first true biosensor (sensor with biological material) is developed by Clark *et al.* (1962) which is utilized by glucose oxidase molecules embedded in a dialysis membrane. The enzyme functioned in the presence of glucose, to decrease the amount of oxygen available to the oxygen electrode, thereby relating oxygen levels to glucose concentration. These biosensors became known as enzyme electrodes and are being used for measurement of various targets.

In 1953, Watson and Crick discovered double helix structure of DNA molecules and set the stage for genetics research that continues to the present day. In 1977, the development of sequencing techniques by Gilbert and Sanger, to directly read the genetic codes that provide information for the synthesis of protein. And this research showed how hybridization of complementary single oligonucleotide strands could be used as a basis for DNA sensing. Two additional developments enabled the technology used in modern DNA based biosensors (1) In 1983, Kary Mullis invented the polymerase chain reaction (PCR) technique for amplifying DNA concentrations and this was made possible the detection of small quantities of DNA in samples. (2) In 1986, Hood *et al.,* developed a method for label the DNA molecules with fluorescent tags instead of radiolabels, thus enabling hybridization experiments to be observed optically.

In 1980, the rapid technological advances in biochemistry and semiconductor fields, led to the large scale development of biochips in the 1990's. At this time, the biochips were largely a platform technology which consisted of several separate, yet integrated components. Figure 12.1 shows the structure of a typical biochip platform. The actual sensing a component or a chip is just one piece of a complete analysis system. Transduction must be done to translate the actual sensing event (DNA binding, oxidation/reduction) into a format understandable by a computer (voltage, light intensity and mass), which then enables additional analysis and processing to produce a readable output. The multiple technologies needed to make a successful biochip from sensing chemistry, to micro-arraying, to signal processing, making the barrier to entry steep. The first commercial biochip (Gene Chip) was introduced by Affymetrix and it contain thousands of individual DNA sensors for use in sensing defects or single nucleotide polymorphisms (SNPs), in genes such as p53 (a tumor suppressor) and BRCA1 and BRCA2 (to breast cancer). The chips are produced by using the microlithography techniques used in the fabrication of integrated circuits.

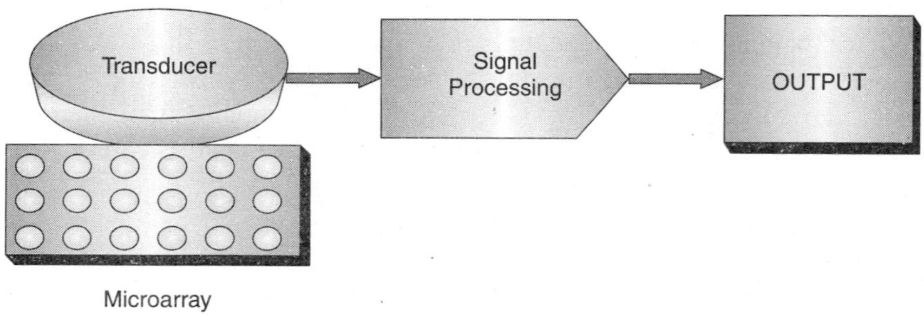

Fig. 12.1 *The structure of a typical biochip platform.*

Today, large multiplicities of biochip technologies are either in development or being commercialized. Numerous advancements continue to be made in sensing, that enable new platforms to be developed for new applications i.e., (a) Cancer diagnosis through DNA typing is just one market opportunity (b) Number of industries currently desire the ability to screen for a wide range of chemical and biological agents, with purposes ranging from testing public water systems for disease agents to screening airline cargo for explosives (c) Pharmaceutical companies wish to screen drug candidates against target enzymes. To achieve these, DNA, RNA, proteins and living cells are being employed as sensing mediators on the biochips. Various transduction methods can be employed including surface plasmon resonance, fluorescence and chemiluminescence. The particular sensing and transduction techniques chosen depend on factors such as price, sensitivity and reusability.

USES

(i) Forensics
(ii) Food industry
(iii) Identity testing
(iv) Medical diagnostics
(v) Pharmaceutical investigation
(vi) Transplantation screening
(vii) Water and environmental testing

12.1 FABRICATION PROCESS

The microarray is an important component of a biochip platform. Typically, these are deposited on a flat substrate, which may either be passive (*Eg.* silicon or glass) or active, consisting of a micromechanical devices that perform signal transduction process. The surface chemistry is used to covalently bind the sensor molecules to the substrate medium. The fabrication of microarrays is non-trivial and is a major economic and technological difficulty that may ultimately choose the success of the biochips. The main manufacturing dispute is the process of placing each sensor at exact position on the substrate. To achieve this dispute, micro-printing systems are used to place the sensor material on the chip surface. Because each sensor is unique, only a few spots can be placed at a time. The low throughput nature of this process results in high manufacturing costs.

Fodor *et al.* (1991) developed a unique fabrication process in which a series of microlithography steps is used to combinatorially synthesize thousands of unique, single stranded DNA sensors on a substrate, one nucleotide at a time. One lithography step is needed per each base type; thus, a total of four steps are required per nucleotide level. Although this technique is very powerful, in which many sensors can be created simultaneously, it is currently only feasible for creating short DNA strands (15-25 nucleotides). Reliability and cost factors limit the number of photolithography steps that can be done.

Most of the microarrays are consist of a Cartesian grid of sensors. This approach is used chiefly to map or encode the coordinate of each sensor to its function. Sensors in these arrays typically use a universal signaling technique. *Eg.* Fluorescence. These arrays must be made using a serial process (sequential steps) to ensure that each sensor is placed at the correct position.

In Random fabrication process, the sensors are placed at arbitrary positions on the chip and it is an alternative to the serial method. In this approach, large batches of identical sensors can be produced; sensors from each batch are then combined and assembled into an array. A non-coordinate based encoding scheme must be used to identify each sensor. The beads are placed randomly in the wells of an etched fibre optic cable and each bead was uniquely encoded with a fluorescent dye. However, this encoding scheme is limited in the number of unique dye combinations.

12.1.1 Magnetoresistive Biochip Microarray

Most of the lab-on-a-chip devices are biochips that incorporate smart passive microfluidics with embedded on chip power sources and integrated biosensor arrays and used in a number of biochemical analysis operations such as clinical analysis & diagnostics and DNA analysis & Proteomics analysis.

New technologies have been developed for DNA recognition or microorganism or identification of cell. Biological targets are labeled with magnetic-nanoparticles and local oscillating magnetic fields guide the target biomolecules over immobilized biological probes placed over magneto resistive sensors. Upon biomolecular recognition and washing, the sensors detect the fringe fields created by the magnetic labels of the remaining targets.

The system is based on an array of magnetic sensors (tunnel junction or spin valve), with a thin film diode or transistor associated with each sensor to read out data. Two main blocks can be identified in this system: (i) the biochip platform, where addressing and read out of the biosensor array, automatic sensor calibration and local temperature control are performed (ii) a hand-held analyzer capable of multi-parameter identification and detection, based on the analysis of the data collected by the biochip platform. The hand-held analyzer, based on a Personal Digital Assistant (PDA) or a laptop computer, includes a communication module with the biochip platform not only for collecting data but also for controlling its general operation. A prototype complete system is being developed with off-the-shelf electronic components, and after this phase it is intended to integrate the major part of the biochip platform, make it even more portable, cost-effective and easy to use.

12.1.2 Structure

The biochip platform comprises an array of 256 sensing elements, which are to be functionalizing with biological probes or to be used as electronic and biological references. A matrix element includes a magneto-resistive sensor (Spin-Valve or Magnetic Tunnel Junction) in series with a Thin Film Diode (TFD), together with an unshaped current line adjacent to the sensor. In addition, on-chip resistor structures will be used to affect local sensing element temperatures.

The biochip is fabricated by using microfabrication technique. Spin-valves (SVs) and Magnetic Tunnel Junctions (MTJs) are deposited on Si or Al_2O_3 substrates by an ion beam deposition system or magnetron sputtering and are defined by direct write laser lithography and ion milling. The U-shaped spin-valve sensors of dimensions of 2.5 μm × 130 μm show a maximum change in resistance with the applied magnetic field or magneto-resistance ratio (MR) of ~ 8% and a sheet resistance of 20 Ω/□. On the other hand, MTJs of dimensions of 2 μm × 10 μm show a tunneling magneto-resistance ratio (TMR) of ~ 40% and a resistance-area product ($R \times A$) of 10 $k\Omega\cdot\mu m^2$. Sensor dimensions are chosen to provide a linear sensor response to the applied field, which is necessary for quantification of biomolecular recognition. Furthermore, sensor dimensions also take into account the diode characteristics such that a maximum response is obtained for magnetic nanoparticle detection.

Silicon Thin Film Diode (TFD) is fabricated by using chemical vapor deposition and are patterned by reactive ion etch with dimensions of 100 μm × 100 μm or 200 μm × 200 μm. Diodes show a reverse-bias current density, J_0, of 9.22×10^{-14} A/μm^2 and a characteristic diode factor, n of 1.23, for the diode response $J = J_0 \exp\left[\frac{eV}{nkT}\right]$. The saturation current density for these diodes is

10^6 times larger than the reverse bias current density. Sensor and diode leads and u-shaped current lines are made by evaporated aluminum and lift-off and a sputtered Al_2O_3 layer is used to separate distinct metal layers. In particular, current lines are tailored such that magnetically labeled targets are focused over surface immobilized probes using a combination of AC and DC magnetic fields and are detected with the magneto-resistive sensors. In addition, heating elements are deposited adjacent to the matrix elements to effect and control local temperatures, such that biomolecular interaction can be promoted or hindered depending on the temperature. This stringency method may be particularly important when distinction between single DNA mismatches is required. Finally, the biochip is passivated with Al_2O_3 and SiO_2 layers to protect the chip against chemical corrosion and to provide a suitable surface for probe functionalization.

12.1.3 Architecture

The architecture is used to develop a modular and portable system for DNA recognition or microorganism. The corresponding programmable electronic interface circuits are implanted in a small electronic card, which is pluggable in or can communicate with a hand-held device. Three main blocks can be acknowledged as the architecture of magnetoresistive biochip system: (i) The biochip platform, which is formed by the sensor array, the auxiliary sensors used as references and the heater and carrier circuits (ii) A digital Microcontroller/Digital Signal Processor (MC/DSP) with an associated Analog to Digital Converter (ADC) and Digital to Analog Converter (DAC), which is used to calibrate the sensors, for controlling the read out of the sensors and the heater and carrier operations (iii) The PDA for multiparameter analysis and to the viewer interface, also equipped with an Input/Output (I/O) module for communicating with the digital MC/DSP, through a standard bus or by performing wireless communication. The biochip is fed with the biological fluid by the fluid dispenser/flux controller using micro-valves and pumps controlled by the MC/DSP.

12.1.3.1 Biochip platform

In each site of the sensor array, high precision measurements of two main physical properties have to be performed: resistance and temperature. In order to get sensor and electronics scalability (aiming up to 256 sensors) the system will be organized in a matrix structure. The sensor addressing is done by using a commutating matrix of integrated Thin Film Diodes (TFDs). Each sensor, whose resistance value is represented by S_{xy}, is associated with a TFD D_{xy}, each Thin Film Diode (TFD) and the corresponding magnetoresistive sensors are connected in series and this series circuit is driven by a programmed current, provided through a Digital to Analog Converter (DAC). A current mirror-circuit provides a current with equal value for similar circuits, D_r and S_r, placed on a specific location of the chip; these circuits are used to provide reference values. The current flows through row and column multiplexers, according to the address of the element to access, establishing only a close circuit at a time. This allows the use of a single DAC and only one amplifier.

The Thin Film Diodes (TFD) has two main functions that correspond to two different modes of the circuit operation: (i) only the diode corresponding to the particular addressed element of the matrix is forward conducting while all others are reverse-biased (ii) the voltage-temperature characteristic of these diodes is used for sensing the temperature of the diode site. These temperature sensors are calibrated through the MC/DSP, by generating pulses of current modulated in width, Pulse Width Modulation (PWM), to control resistive heaters placed around the chip. The calibration is made in DC, by amplifying the voltage at the terminals circuit. The gain of the amplifier is programmed and the output analog signal is applied to the input of an ADC in order to have the digital values available on the MC/DSP. Each of the temperature sensors is calibrated at setup time by experimentally computing the junction parameters of the equation relating voltage with temperature.

For measuring the changes in the variation of the resistance values in a magnetoresistive sensor, an alternate magnetic field is created by a coil placed close or below the chip. By using an AC analysis, we solve the difficult problem of measuring small variances on the resistance, typically less than 5% for an average value of 1 kΩ. Moreover, approximately half of the voltage across the circuit is due to this resistive sensor, since the other half is due to the voltage to the forward-biased diode. Therefore, by only applying the alternate component of the voltage at the terminals of the amplifier we can considerably increase the gain of the amplifier to get reading with more resolution. Furthermore, taking advantage of an amplifier with high a Common-Mode Rejection Rate (CMRR), the reference signal is placed in the negative input terminal providing the reference for the resistance variation by amplifying the difference of the input values.

The carrier circuit generates local fields to guide the target biomolecules over immobilized biological probes. The superposition of the DC magnetic field created by the current lines with a low frequency, few Hz, AC external field leads to attractive magnetic forces concentrating the magnetic labels in the inner region of the U-shaped line. This in turn allows much faster (few min) hybridization rates between biological targets and the immobilized probes, when compared with diffusion controlled processes. This technique is also used to control the number of labels over each sensor area. The operation of the biochip platform and the required digital signal processing are implemented in the MC/DSP. The main characteristics of this device and its main features are discussed in the next section.

12.1.3.2 Microcontroller or digital signal processor

The MC/DSP is the programmable core of the platform, operating stand-alone but with an interface for communicating with a hand-held analyzer. Its main tasks are: to control the operation of the reading out circuits, to calibrate the sensors and to implement the signal processing and the temperature control algorithms. For example, algorithms have been developed to characterize each cell from the point of view of magnetic and temperature sensitivities and nonlinearities; these algorithms are implemented in the MC/DSP and are executed when a reset occurs. On the other hand, the signals are mainly processed in digital domain (*Eg.* filtering) by the MC/DSP. The control signals to operate external micro-valves opening or closing the individual micro-chambers and pushing the fluids around are also generated by the MC/DSP. At the end, the automatic individual sensor calibration will allow the counting of labels (and therefore target biomolecules) over each sensor, once the output voltage is measured comparatively to the reference sensor.

12.1.3.3 Hand-held analyzer

The defined architecture is designed to indistinctly use a Personal Digital Assistant (PDA) to implement the hand-held analyzer. All the software to analyze data, to implement the user interface and to control the overall operation of the biochip platform is being programmed by using object oriented paradigm and compilers. Only the user interface is dynamically adjustable according to each of the devices is applied. The communication modules are based on off-the-shelf components, following the Universal Serial Bus (USB) standard and the bluetooth wireless technology, by using a globally available frequency band (2.4GHz). This hand-held device acts as a master of the system, allowing the execution of a set of pre-programmed tasks in the biochip acting. It also provides a graphic interface with the user and the software for analysis of data provides the results in a user friendly way. For example, it gives to the geneticist/biologist the relative signal levels that can be directly the ratio of signals V_{mut}/V_{con}, corresponding to the voltage ratio between the chosen mutant target and the chosen control sample.

12.2 SIGNAL AND TEMPERATURE MEASURING TECHNIQUES

The biochip matrix is driven by a constant current or by a constant current with a small a.c. current superimposed. The voltage across each matrix element (TFD + magnetic sensor) is measured. Due to high thermal conductance of the Al_2O_3 biochip passivation layer placed under each biological site over the magnetic sensor and the associated TFD, the temperature of these devices is very close to the biological fluid temperature, T_f. Taking advantage of the very well known electrical properties of the forward constant current biased TFD it is possible to characterize all the fundamental parameters of each matrix element. The biochip matrix allows for the extraction of a differential voltage between each matrix element and reference elements placed in the biochip. However, experimental evaluations shows that, due to the biochip fabrication process, significant mismatches occur between each element and the reference element, leading to voltage offsets that reduce the usable gain range of the reading amplifier. To overcome this problem, each matrix element is driven sequentially with scaled current values of a given reference DC current. The DC measured voltages across each matrix element are used as references for the new measures. The biochip reading involves the following phases: Calibration, Temperature Measuring and Control and Signal Measuring.

12.2.1 Phase Calibration

The ambient temperature is known and no magnetic field is applied to the sensor. Each matrix element is driven with sequentially increased scaled constant current values. This allow to calculate the TFD parameters (n, J_0) and the sensor resistance RS. These values are stored in the microprocessor memory and are used in future biochip readings.

12.2.2 Temperature Measuring and Control Phases

The biochip is fed with the biological fluid and the magnetic particles are carried over each magnetic sensor using the magnetic transportation lines. By using at least two scaled current values, it is possible to obtain the diode temperature T ($\simeq T_f$) and the two element (diode and sensor) voltage temperature sensitivities. Further, the matrix element can be heated by passing high currents through it. As a consequence, the temperature of the biological fluid placed over the sensor also increases. A sequence of heating and temperature measuring phases can be used to control the temperature of the fluid in each site. Preliminary SPICE simulation results using a concentrated thermal parameters model show that this is feasible. However the biochip is a distributed structure and is being modeled using finite elements method. For higher variations in the chip temperature, a heating serpentine is also used.

12.2.3 Signal Measuring Phase

In this phase, the sensor resistance variation due to the magnetic particles captured over each sensor must be measured. A burst of an AC current, superimposed to a DC biasing current, is generated in the MC/DSP and is applied through an analogue multiplexer to each matrix element. The corresponding voltage bursts are originated by the diode incremental resistance, about 125 Ω for 200 μA of DC bias current which leads to about 10 mV variations in a total of 200 mV AC voltages across each matrix element. The use of the voltage in the reference matrix element in a differential amplifier structure reduce the AC common mode voltage to 0 mV, if there is no mismatch, and to about 50 mV if the mismatch resistance (between sensor and reference sensor) is 100 Ω. So the amplifier gain may vary between 15 and 60 in these cases. The sensor signal is digitized and is processed in MC/DSP by implementing digital signal processing algorithms.

12.3 MICROFLUIDIC AND BIOSENSOR CHIP TECHNOLOGY

The i-STAT system is at the cutting edge of diagnostic technology, with the revolutionary integration of biochemical and silicon chip technologies, the system achieved unattainable miniaturization of sophisticated components that provides clinicians a truly portable, nearly instantaneous blood analyzer with no sacrifice in accuracy and reliability. To meet the needs of the diagnostics, a variety of sensor configurations make it possible to perform a number of critical care assays: electrolytes, general chemistries, blood gases and hematology.

The sensors are micro-fabricated with the thin film electrodes. The integration of semiconductor manufacturing process technology with electrochemical principles results in the production of micro-miniaturized sensors with highly reproducible characteristics. Depending on the particular assay, the electrical signals produced are measured by the i-STAT Portable Clinical Analyzer's amperometric, potentiometric or conductometric circuits. Calibrant solution is contained in the cartridge within a foil pouch. The fluid is a pH buffered aqueous solution of the analytes at known concentrations. The test cycle is started by placing the cartridge containing the blood sample into the Analyzer. During the test cycle, the analyzer presses the front of the cartridge, causing a barb to puncture the pouch. The calibrant fluid is released to flow over the sensor array for measurement. When calibration is completed, the analyzer presses the cartridge air bladder which pushes the calibrant solution into the waste reservoir and sends the blood sample over the sensor array for measurement. All blood and calibrant fluid is contained within the cartridge for safe bio-hazard disposal.

The i-STAT sensor arrays insure the integrity of the overall analytical process. The solid-state chips contain biosensors configured to perform specific tests with chemically sensitive membranes and films containing reagent chemicals. Sensors perform other functions such as monitoring the quality of the sample being tested. Silicon-type microfabrication utilizing high quality materials that exhibit exceptional stability allows consistent reproducibility in a high-volume manufacturing environment. This technology ensures that each cartridge offers a high level of accuracy and reliability. When blood samples contact the sensors, they are measured electrochemically as follows:

1. Sodium, Potassium, Chloride, Ionized Calcium, pH and PCO_2 are measured by ion-selective electrode potentiometry. Concentrations are calculated from the measured potential through the Nernst equation.

2. Urea is first hydrolyzed to ammonium ions in a reaction catalyzed by the enzyme urease. The ammonium ions are measured by an ion-selective electrode and concentration is calculated from the measured potential through the Nernst equation.
3. Glucose is measured amperometrically. Oxidation of glucose, catalyzed by the enzyme glucose oxidase, produces hydrogen peroxide. The liberated hydrogen peroxide is oxidized at an electrode to produce an electric current which is proportional to the glucose concentration.
4. The oxygen partial pressure PO_2 is measured amperometrically. The oxygen sensor is similar to a conventional Clark electrode. Oxygen permeates through a gas permeable membrane from the blood sample into an internal electrolyte solution where it is reduced at the cathode. The oxygen reduction current is proportional to the dissolved oxygen concentration.
5. A variety of calculated results are available that include HCO_3, TCO_2, BE, SO_2, Anion Gap and Hemoglobin.

12.4 TYPES

Development in biochip instrumentation, microelectronics and microfluidics makes it possible to design or develop different type of biochips for utilizing in chemical, physical and biological analyte measurements. These biochips have been classified according to their application and which are used to detect or determine analyte or target. The application of many of the phenomena (amperometric, potentiometric and optical fibres) in biochips became possible to develop various configurations. The biochips are used in the modern technology area and set in biochips could be the tool of Big Brother, but they are more likely to become the behavior of choice for the physician of the 21st century. In this section important biochips development, working mechanism and their applications are discussed which are using in different field applications viz., Potentiometric biochips, Amperometric biochips, Optical fibre biochips, DNA biochips, Brain Tumor biochip, Multifunction Microfluidic chip device, RNA biochip, e-NOSE and S4 MS biochips.

12.4.1 Potentiometric Biochips

Electrical signals based on membrane potential (voltage) or membrane transport (current) is not possible to determine directly which ion species is driving the measured electrical event. Therefore, the technique requires multiple ion-replacement control experiments to determine which ion is responsible for the measured electrical signal. The self referencing ion selective electrode technology is used to study the ion channel physiology, developed by Kuhtreiber and Jaffe (1990). This technology involves an ion selective membrane immobilized at the tip of a glass micropipette, which is electrically in contact with a silver or silver chloride electrode through a salt bridge. The potential developed across the ion-selective membrane, because of an ionic concentration, is referenced with another silver or silver chloride electrode. Thus, the potential difference measure which corresponds to the concentration of ions and these sensors are known as potentiometric sensors.

The patch clamp technology is drive towards miniaturizing ion-selective electrode technology and integration on a biochip platform. Hisamoto *et al.* (2006) fabricated a multiple analyte detection biochip by integrating square micro-capillaries into channels formed on a poly-di-methyl-siloxane (PDMS) substrate. Each capillary is filled with a different ion-selective membrane for sensing

Na^+, K^+ and Ca^{2+}. Purvis *et al.* (2003) developed a potentiometric immunosensor biochip 1 mm^2 in size by screen printing gold electrode on a poly ethylene terephthalate (PET) substrate. This chip detects enzyme labeled immuno-complexes formed at the Polypyrrole coating on the screen printed gold electrode. Detection is mediated by a secondary reaction that produces charged products. A shift in potential is measured at the sensor surface, caused by local changes in ionic strength. This magnitude of the difference in potential is related to the concentration of the formed receptor-target complex. This chip is also used for the detection of Hepatitis B surface antigen, Troponin 1 and digoxin.

Integrates H^+ and K^+ ion-selective FET based sensors and a temperature sensors on a needle-like microprobe made from silicon substrate developed by Errachid *et al.*, (2004) Guenat *et al.* (2006) fabricated a two part ion selective biochip that contains silicon nitride micropipettes formed on a silicon substrate with another glass layer bonded below it. This layer contains platinum electrodes and micro-channels and these channels serve as inlets through which Ca^{2+}, K^+ and NH^{4+} selective membranes are filled. The biochip demonstrated Nernstian response and for *in-vitro* physiology applications.

Cell electrophysiology lab-on-a-chip device used for monitoring real time Ca^{2+} currents across the polarized single cells. This device is manufactured using state of the art silicon microfabrication technology. It consists of 16 pores on a silicon substrate each having four silver or silver chloride electrodes at the polar positions. A SU-8 layer (UV cross linked epoxy) is used to form an insulating layer over the electrodes, as well as an encapsulating well around each pore. Finally, the electrodes are coated with a Ca^{2+} selective membrane to impart selectivity for Ca^{2+}. The final configuration creates 16 measurement chambers, each capable of measuring Ca^{2+} ionic activity at four different points around a single cell.

12.4.2 Amperometric Biochips

The amperometric sensors are involved oxidation or reduction of an electro-active species on the surface of a platinum, gold or carbon electrode. The electrode is polarized at a fixed voltage corresponding to the redox peak of analyte. The current generated on the electrode surface as a result of the redox reaction is measured and corresponds to the analyte concentration.

Various interesting applications on a biochip platform have been developed by different scientists for amperometric detection of physiologically relevant analytes. Cui *et al.* (2001) developed a silicon based chip with 25 gold disc electrodes each having a diameter of 30 μm. M9ND and PC12 cells were successfully cultured on chip and potassium evoked dopamine release from the cells was detected amperometrically. Schoning *et al.* (2005) developed a glass substrate amperometric biochip with platinum electrodes. Microfluidic channels for sample containment and capillary electrophoresis based transport were formed in PDMS. This biochip was used for dopamine and catecholamine. Roy *et al.* (2006) utilized the electrical properties of multi-walled carbon nanotubes (MWCNT) on a silicon based biochip. Vertical MWCNTs were grown on silicon on top of which cholesterol oxidase and horseradish peroxidase was immobilized for cholesterol detection.

The most popular type amperometric biochips are the probe type biosensors based on the Michigan probe design, developed by KD Wise *et al.* (2006). The basic technology consists of a silicon micromachined probe. Platinum is usually used as the electrode material. Use of microfabrication technology produces small probes used for studying neuronal physiology. The small size allows them to be minimally invasive so that actual physiological signals rather than injury currents are measured.

12.4.3 Optical Biochips

The progression and application of microfabrication processes to MEMS development, there have been various efforts to incorporate optical components on microchips. Microchips containing optical fibres or spin-on-glass waveguides have been manufactured. Silicon and PDMS have been the most common substrates; recently soda lime glass has been used to fabricate waveguides and couplers. The optical micro-sensor device consists of an optical source, sample chambers, optical detection, signal processing, data acquisition and wireless communications module.

All the modules except the sample chambers will be stationary. The sample chambers designed in an array format to provide high throughput capabilities. The sample chambers can be modified with fluorophores on the bottom surface using an automated microinjection system. The back sides of the sample array include with a microfabricated LED array and emission signals detected using the microfabricated photodiode array. The excitation and emission signals proceed to the signal processing and conditioning module. The data acquisition module acquired the conditioned signals and sends them to the wireless module. The signals are then transmitted wirelessly to a central network, hand-held devices and computers for real time viewing and archiving. All the modules are controlled with a central control software infrastructure with the capability of controlling the progress on each sample chamber in the array. The software will also contain error checking mechanisms.

12.4.4 DNA Biochips

DNA biochip is developed for medical diagnosis. *Eg.* To test blood samples. This chip is used for diagnosis of the AIDS virus, cancer and tuberculosis and also used for some of the environmental applications.The DNA chips possess a DNA covered silicon and the glass or plastic wafers capable of analyzing thousands of genes at a time. *Eg.* Identify the ones that are active in a sample of cells. These DNA micro-arrays emerge balanced to join the war on cancer. These DNA chips recommend a whole new and potentially much earlier, easier and more precise way of detecting cancerous cells. Using the DNA chip based tools to mark the genetic differences among cells; these revealing differences could be used to detect cancerous cells long before symptoms develop and to distinguish one type of cancer from another. The chips will provide a genetic profile of a cancerous cell that can be read like a criminal's rap sheet. The physician will know where the cancerous cell originated, how far it has progressed and which therapies will work best to halt its further growth and spread. A DNA chip based device is used for the genetic changes in the lung cells that are naturally sloughed off into the viscous fluid.

12.4.5 Brain Tumor Biochips

The Brain Tumor Biochip is used to understanding the gene expression in brain tumors, elucidating new criteria for the classification of brain tumors, developing new diagnostic technology and therapeutic principle in the treatment of individuals afflicted with this illness. The Brain Tumor Biochip is the 10K oligonucleotides microarray, which is used to investigate the gene expression in the brain tumor samples and this microarray platform that functions by impedimetric recognition of DNA hybridization. This biochip involves neurosurgery, molecular genetics, engineering and bioinformatics. This chip will help the pathologist in arriving at a molecular diagnostic assessment of excised brain tumors.

DNA biochips are used for diagnostics of brain tumor cancer and prognostics (Fig.12.2). This DNA microarray technology is used for improved diagnosis and treatment of the brain tumor patients. DNA microarrays have enabled to study the expression of cell shape populations in extreme detail, with upwards of 30,000 DNA hybridization samples on a standard microscope slide. Since cancer cells have different expression shape than normal surrounding tissue, this application of DNA microarray is suitable for the identification of a targeted suite of genes associated with brain tumors.

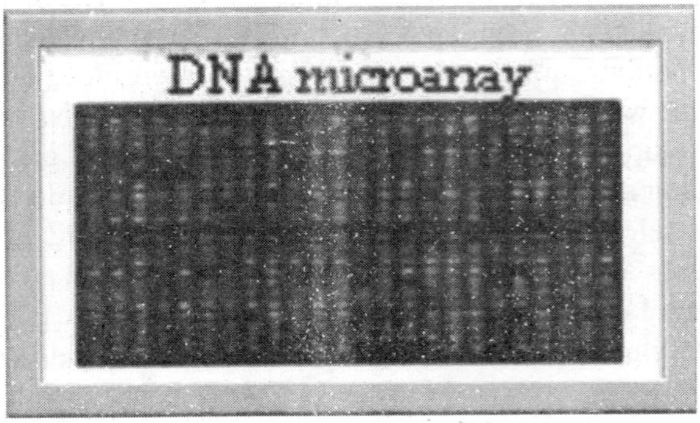

Fig. 12.2 *A Schematic representation of DNA microarray. (Source: Clemson C3B,©1998-2008 Clemson University, Rev 11 Nov 2007)*

The DNA microarrays are used for the identification of a targeted suite of genes associated with *Astrocytomas,* which is used in the Brain Tumor Biochip. The development of a DNA biochip can be described of *Astrocytomas* for enhanced diagnostics and to improved therapeutic effectiveness. Other DNA microarray experiments will be conducted to examine the gene expressions of ovarian and breast cancer tumors. DNA biochips have the potential to revolutionize the way cancer tumors are classified and staged and to pave the way for targeted gene therapy treatments.

12.4.6 Multifunctional Microfluidic Device

The multi-functional microfluidic device is used for the detection of environmental viruses. The device first breaks the membrane of the viruses by using the virus lysis in region 1. Then the sample was moved to region 2, where magnetic particles coated with virus specific DNA primers are mixed. The mixed solution flows through region 3, where a magnet traps the magnetic beads. In this region, the virus RNAs are isolated from the protein rich debris. Then apply a heater to raise the temperature to above 80°C. It breaks the bonding between the RNA viruses and the primers. The purified RNAs are then washed off by clean buffer solution and moved to region 4. In this region, a gold nanoslit microarray is fabricated on the surface. When a light is normally incident into the gold nanostructure, it produces surface plasmon resonance (SPR) in the nano-gap and gold surface. The nano-SPR is very sensitive to surface condition. Different virus specific DNA primers are immobilized on the sensor array. If there are viruses RNAs in buffer solution, they will be bound to the primers. The nano-SPR signals will show the binding events and may indicate how many virus RNAs are detected.

A single chip was integrated with the microfluidic channels, magnets, heater and nano-SPR. The nano SPR signals are measured by a portable reading system, which consists of a LED light source and USB mini-spectrometer. The operation time of the chip for a test round is less than 15 minutes and the sample amount is smaller than 0.5 ml. The quick interaction and low detection volume are the advantages of the chip of microfluidic system and the nano-SPR sensors.

To test the device, the virus sequence is the first target. The fish virus RNAs sequence as a target was demonstrated by Chen's *et al.* (2003). Design the DNA primers and immobilize them on the magnetic beads and the nanoslit array. In the beginning, the interactions between virus RNAs and the designed primers are tested and verified it by using the purified fish viruses. Once the specific bio-affinity between the RNAs and primers is verified, then perform the experiments for the real complicated samples. The reasons for choose fish virus as first target are: (1) it is very important for seafood industry. (2) When the tests are successful, it can be directly applied in the market. (3) If the idea of using the RNA-primer interactions and nano-SPR sensors is feasible in the sensitive virus detection, the same platform can be easily extended to other viruses, especially the human viruses. The human viruses can only be accessed in P3 labs. However, with the experience of the fish virus, can justify the usefulness of the chips and develop the new biosensing chips.

12.4.7 RNA Biochips

RNA chip is a new diagnostic device which promises to put a powerful lab on a dime sized slice of silicon. In 1995, Breaker *et al.,* began to revive this RNA chip in a test tube and successfully engineered RNA based molecular switches. Molecular switch is a molecule that is turned on or off by another compound or molecule. These switches arrange them on a surface and create an array of biosensors that measure or detect compounds. By engineering the RNA switches to detect many different kinds of compounds, the potential of this array could exceed that of a DNA chip, which identifies specific DNA or RNA sequences.

To create the prototype, place the RNA switches on a gold coated silicon surface and arranged them in clusters. Each switch was designed to bind only to a specific molecule and its target and then to release a signal that identifies the target molecule (the prototype RNA switches releases a radioactive signal) and the implications are enticing. The array's ability to simultaneously identify potentially large number of compounds, combined with the precise exclusivity of each switch, adds up to a recipe for a powerful and wide ranging laboratory on a dime sized slice of silicon.

12.4.8 Electronic Nose

An electronic nose is a device which is used to analyze the content of air through the categorization of odors. The electronic nose is far from replacing the human olfactory system, though the possible uses for this technology are endless. Human nose is employed all over the world to test many different products. The odors of food products such as grains, wines, cheeses, and fish are all examined by human nose to determine their quality and freshness. Perfumes and deodorants are also tested to see if they are attractive to the nose.

The sense of smell is even used by doctors to help classify common disorders. Certain problems such as pneumonia or diabetes give the patient's breath or bodily fluids characteristic odors that can be noticed by a trained nose. If these odors could be classified by a machine, then an electronic nose could be employed to do the same job with more possibilities. Another problem electronic nose could solve is the health risk associated with smelling certain chemicals. The toxicity of certain chemicals could prove harmful to someone who was trying to determine an odor using their nose. Even grain samples can give off mold spores which can cause allergic reactions or other disease symptoms. But, electronic noses cannot go, for example, into areas of extreme temperatures, inside the body, inside oil rigs or gasoline tanks, sewer systems, hog farms, or even onto another planet.

In the past, gas chromatography and mass spectrometry have been used as the sensing systems, although these are usually expensive and time consuming. Today, the use of chemical sensors has been established to analyze odors. Essentially, each odor leaves a characteristic pattern or fingerprint of certain compounds. Known odors can be used to build up a database to train a pattern recognition system. One possibility is to have a sensor for every chemical, though this would be costly since there are so many different chemicals.

Of the five human senses, the sense of smell is least understood by the scientists and researchers. Odors can be described as chemicals carried in the air or environment. The scientific challenge is to develop a sensing system capable of detecting trace amounts of chemicals that are associated with a particular class of odor. The electronic Natural Olfactory Sensor Emulator (e-NOSE) is a sensing system along with an artificial neural network to distinguish specific chemicals from certain odors. e-NOSEs consist of two main components, a sensing system and a recognition system. The application of the e-NOSE is to determine the physiological status of shock and trauma patients by monitoring their breath for volatile organic compounds. The e-NOSE to examine improving sensor performance through design and material selection, characterizing the sensing of various compounds and developing a neural network that can identify the presence of specific chemicals by analyzing the electrical signals from the sensor array. In future the e-NOSE platform can have important applications in environmental monitoring and homeland security.

A multi-element array of chemo resistive sensors is being developed to mimic the human nose. Each sensor of the e-NOSE comprises a micro-lithographically fabricated chip that is layered with a chemically responsive polymeric material. The polymer material may be an inherently conductive polymer such as Polypyrrole, Polyaniline or Polythiophane. In all cases, the polymer changes its work electrical impedance characteristics in response to the absorbed or adsorbed vapors. Using Principal Component Analysis (PCA) and Artificial Neural Networks (ANN) the chemo electrical responses of the sensors are processed into odor signatures. Such systems are being developed for monitoring of the expired breaths of patients experiencing shock, various forms of trauma and to differentiate cancerous from non-cancerous tissue. The Artificial Neural Networks (ANNs) are used to detect more chemicals and these were allowed for less selective therefore, less expensive. These are used to distinguish certain odors from certain chemical combinations. Pattern recognition is gained though a given network known odors and classifying them with a signature.

The sensors basically measure the change in voltage due to the presence of certain chemicals. The chemicals in the air change the oxygen content over the sensors, which are measured through electronic circuits. By changing the oxygen content, the resistance across the sensor is changed which can be measured as a voltage drop from the normal or standardized conditions. This analog signal must then be translated into a digital signal by an A/D converter and fed to the computer to understand the information. The number of odor signatures can recognized by the system depends on the number of sensors used and the number of grey levels in the convertor. The maximum signature number is given by ‘gn’, where ‘n’ is the number of sensors and ‘g’ is the number of grey levels. A 10 bit converter has a grey level value of 1024, so an array of three sensors could yield over a billion different signatures. Unfortunately, the actual number is far below this value.

The NOSE is based on a multi-element array of volatile organic compound (VOC) sensors, each based on the Interdigitated Microsensor Electrodes (IMEs). The IME's are covered with a thin, VOC responsive conductive polymer membrane that may be formed from either of Polyaniline, Polypyrrole or Polythiophane. When a VOC or mixture of VOCs is passed over these polymers the vapors are absorbed by the polymer network, causing measurable changes in electrical impedance. These changes in impedance are directly detected across the array of eight sensor elements and recorded as signal data. In addition, temperature and relative humidity are recorded at the same time.

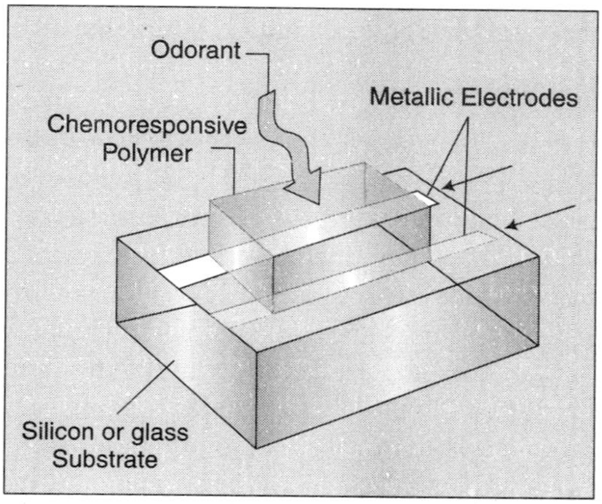

Fig. 12.3 *e-NOSE Operation Process. (Source: Clemson C3B,©1998-2008 Clemson University, Rev 11 Nov 2007)*

Each sensor element is based on a different polymer membrane achieved by the modification of the three basic conductive polymers. Although each sensor is unique, none is specific to any particular vapor. No one sensor element would be able to deliver useful information about the gas. However, the array of sensor elements, taken as a whole, contains all the information. This information is embedded in the group response of the array and is not readily interpreted. Thus, some form of sophisticated data analysis is necessary.

To extract the gas composition and concentration level within a target vapor stream, data from the sensor responses (the signal data) is passed to an artificial neural network (ANN). The ANN has been trained to recognize specific vapors and their concentration levels, over a specified range of relative humidity and temperature, within the air. As long as the vapor has been trained for, the ANN will be able to identify and quantify the concentration of that vapor, or mixture of vapors, in the air down to its ppm detection limit.

The e-NOSE system now being used at the C3B is a test platform designed for the development of new VOC-response-Polymeric membrane materials; investigation of the mechanisms of interaction of vapors with conductive polymers and the development of ANN training sets for specific arrays to targeted vapors. It is far more complex and has many more parts than an actual deployable system. This research platform comprises a gas manifold system for introducing and exhausting gases and vapors. The multi-element sensor array possesses a sensor interrogation and data acquisition subsystem, data analysis and information processing software, and the information output interface.

12.4.8.1 Applications

In future, electronic nose may be applied to environmental problems such as analyzing hazardous waste, fuel mixtures, groundwater, air quality, or factory emissions. The food industry is already using these mechanical sniffers, but the uses could grow to include fish inspection, fermentation processes, container examinations, or verifying if orange juice is natural. The electronic nose could even be used in telemedicine so that a doctor would be able to smell the breath of a patient a thousand miles away by recreating what an electronic nose smells.

12.4.9 Silicon Chip (Thin Nanowires)

Laboratory methods of virus detection involved time consume steps such as taking blood samples, which must be sent to a lab for analysis. To detect the individual virus particles in laboratory by using the optical microscope approaches required purified samples and labeled the viruses with fluorescent markers, which makes them inappropriate for clinical and field diagnostics. They were also not able to detect the viruses in samples with very low virus concentrations or detect multiple types of viruses. In recent years, nanowire and nanotube based transistors (Silicon chip) are developed for the detection of the viruses. To break down the virus particles, the virus DNA can bind to receptors on these devices. It is easy to imagine, adding such sensors to security gates, dangerous viruses can be detected before they spread to other places in the environment.

Quick detection of viral epidemics has been providing the possibility of saving thousands of lives. The pathogens are biological weapons and their detection makes a vital national defense requirement. In 2007, Charles Lieber says that we want to find a single virus before it finds you. Viruses are among the most important causes of human diseases and are of increasing concern as agents for bioterrorism and nanoscale silicon wires (thin nanowires) are used to distinguish between viruses that cause flu, measles and eye infections. Lieber believes future versions will be able to spot HIV, Ebola, SARS, West Nile, hepatitis, bird flu and other dangerous viruses. Nanoscale silicon wires can be configured as detectors that turn on or off in the presence of a single virus particle. Such detectors are formed into arrays capable of sensing thousands of different viruses, ushering in a new era for diagnoses, bio-safety and quick response to viral outbreaks.

One hundred thousand wires, each 20 nanometers long, were fit on the head of a pin. Two-inch-square silicon and metal chip contained an array of nanowires and two pinhead-size entry ports through which blood, saliva or other bodily fluids can enter. Air samples put into a fluid solution can be tested this way. Surfaces of the nanowires, through which a minute current flows, hold specks of protein (antibodies) to which specific viruses bind. These antibodies, produced naturally by the immune system, can be attached to the surface of the nanowires and they bind to viruses and this binding causes a change in current, which signals the presence of viruses.

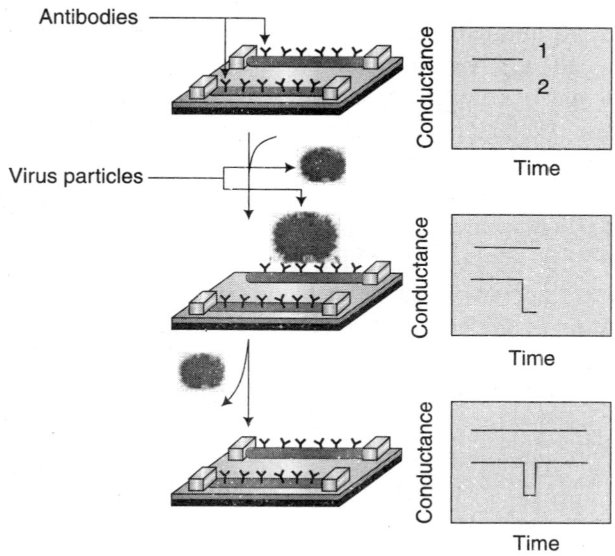

Fig.12.4 *Antibodies bound biochip causing signals in the presence of viruses. (Source: William J-Cromie, early warning for disease and bioterrorism, Harvard News Office, 2007).*

A spiny virus binding to Y-shaped receptors built into ultrathin wires (Silicon chip) carrying a small electric charge (Fig. 12.4). The binding causes a change in electric conductance which immediately identifies the germ. When it becomes unbound the conductance returns to a set value.

Combining selective virus and protein recognition in one nanowire chip, its diagnostic potential can be expanded. Much cancer related proteins are secreted into the blood. *Eg.* Prostate specific antigen (PSA) is used as a marker for the seriousness of prostate cancer. Low PSA levels provide one signal that closely monitors the disease and may be a better choice than surgery or radiation.

Charles lieber *et al.,* validate the selectivity of the sensor chip by using two different viruses. First, they fed the device with two viruses from different families, one that causes influenza A and one responsible for respiratory and eye infections and the detector sense the differences. Then they paired the influenza bug with a closer viral that causes mumps and measles relatively.

Environmental sensors and hand-held devices that quickly and easily detect and identify the individual viruses and it provides the early warning of infections in individuals, the spread of disease in populations, and the biological weapon attacks. The rapid development of nanotechnology in recent years has given researchers tools for building highly sensitive virus detectors. Prototype antibody proteins attached nanowires transistors (detector) are developed for determination of individual virus particles in real time from the unpurified samples. The binding of viruses to the nanowire transistors causes a change in the current. These changes in current produce signals and these signals are used to detect the presence of a virus. The device can also be used to detect individual biomolecules, including DNA and proteins. A prototype device developed by mounting 20-nanometer diameter silicon nanowires, mixing the nanowires with fluid and allowing them into the position across nickel contacts, spaced two microns apart to form nanowire transistors. They coated the nanowires with aldehyde, and then added antibody proteins, which adhered to the aldehyde. They configured a microfluidic channel to flow fluid containing the viruses across the nanowire sensors. When an individual virus binds to a nanowire transistor antibody receptor, the transistor's electrical conductance changes, increasing or decreasing depending on whether the transistor carries positive or negative charge and the virus is positively or negatively charged. This detection method is based on a pure electrical detection of selective binding-unbinding of a single viral particle. The duration of the bind and release cycle depends on the density of the antibody proteins on the nanowires. The cycle averaged just over one second at a low concentration of antibody proteins about 20 seconds at a moderate density and 5 to 10 minutes at a high density. The attached antibody proteins that bind to Influenza A to one portion of the nanowire array and antibody proteins that bind to Adenovirus to another portion and that can detect simultaneous binding of the two different types of viruses. These nanowire detectors are used to facilitate sensing thousands of different viruses simultaneously.

Another technique that has emerged recently is detecting individual virus particles using microscopic cantilevers coated with antibody receptors. The additional mass of an attached virus particle changes the vibration rate of the cantilever, which can be detected electrically. These techniques require high resolution imaging to confirm that only one virus particle has attached to the cantilever.

Biosensors in Clinical, Medical and Healthcare

Chapter 13

13.0 BIOSENSORS IN CLINICAL INDUSTRY

13.0.1 Introduction

There is a great need for the development of efficient and accurate methods related to the detection and identification of chemical and biological agents (analyte) including nucleic acids, proteins, drugs, explosives, toxins, pharmaceuticals, carcinogens, poisons, allergens and infectious agents. Current methods of detecting analytes require extraction of a sample into organic solvents, followed by analysis using stand-alone analytical systems such as gas-liquid chromatography or mass spectroscopy and these methods are time consuming and often expensive.

Three recent advancements in medicine, particularly to expand the potential for the detection of analytes with regard to the diagnosis and treatment of diseases, are nanotechnology, biosensors and biomarkers for specific diseases and/or conditions. Nanotechnology in particular, offers many advantages when used for applications such as the delivery of bioactive agents such as DNA, AIDS drugs, gene therapy, immuno suppressants, chemotherapeutics and drug uptake and degradation (enzyme encapsulation). Nanoparticles have been proposed as providing sites for specific distribution of drugs to, and minimization of loss from, a target site. Nano sized particles have been proposed wherein such particles can be delivered to selected tissues to release their drug load in a controlled and sustained manner. Biosensors are used in naturally occurring and synthetic compounds as highly specific and extraordinarily sensitive detectors of various types of molecules and markers of disease.

Biosensors are to bloom all over the market, merge the computer technology, electrochemistry and enzyme chemistry, and that can perform ultra-microanalysis with unprecedented speed. In clinical industry, the development of a biosensor device that could accurately and efficiently detect the chemical and biological agents therefore will provide a significant cost and time benefit. Healthcare technicians are using the biosensors to obtain immediate and accurate results in many different clinical testings by eliminating number of steps and labor. The best way to grasp the impact of this upcoming technology is through the familiar logistics of clinical laboratory service. Most biosensors consist of two principal components: a molecular recognition element and a transducer or signal-gathering element. The recognition element, which acts as a sensing device, is the biologic component. It may consist of an enzyme, an antibody, a receptor, or an organelle. The second element, the transducer, may be an electrode, a semiconductor, a fibre optic device, or a quartz crystal. Information is transmitted as a signal sent from the sensing region through the transducer to a readout device.

Technology developed towards the synthesis and use of aptamers and other multi-molecular devices as biosensors. Aptamers based biosensors used for the diagnosis and treatment of disease for a wide range of naturally occurring and synthetic compounds. It is desirable to provide low-cost, accurate and timely detecting or measuring the presence of metabolites in a person's bodily fluids in low concentrations via non-invasive methods. Further, in order to effectively apply exhaled breath sensing technology, there is a pressing need for an efficient screening method for determining analytes/biomarkers which are likely to be detectable in exhaled breath. This section will discuss the applications of various types of biosensors in clinical industry viz., detection of nucleic acids, analysis of blood samples, and detection of glucose, lactose and urea.

13.0.2 Detection of Nucleic Acids

The development of biosensors in recent years has provided promising techniques for efficient sequence specific DNA analysis. Biosensors have been used to detect target sequences in PCR products. Most studies have been conducted with the commercially available biosensor BIAcore, which is based on surface plasmon resonance (SPR) technology. In those investigations, either sequence specific oligonucleotide probe or amplified products themselves were immobilized on the sensor chip, limiting their application to one sequence or to one particular sample. Such a strategy suffers several drawbacks: (a) it compromises the automation and high-throughput capability of such instruments when different targets are analyzed; (b) chip-to-chip variation makes it difficult to compare different measurements; and (c) probe immobilization and chip consumption make running costs substantial.

To overcome these drawbacks, a 'one-chip-for-all' strategy has been developed that is capable of, in principle, detecting different target sequences by use of the same sensor chip. Target sequences are amplified by asymmetric PCR using a primer pair in which the low-concentration primer carries a common tag sequence that is identical to that of the oligonucleotide capture probe immobilized on the sensor chip. The PCR product is then injected and captured on the sensor chip, and its sequence identity can be further verified by the use of a target specific probe. This has application by analyzing the following three different genes:

(i) The cystic fibrosis transmembrane conductance regulator gene (CFTR), the mutation of which causes cystic fibrosis

(ii) The hTERT gene, which encodes the catalytic component of the human telomerase complex and

(iii) The human gene for the tumor suppressor p53.

13.0.3 Analysis of Blood Samples

An integrated biosensor cassette has been used for the determination of glucose, lactate, and urea in micro samples of undiluted blood or plasma. The biosensor cassette requires no maintenance and can be used for 1 week or 500 samples after insertion into blood and electrolyte instrument.

The linear analytical ranges of these sensors include the physiological ranges and most of the pathological ranges are shown in the table.

Table 13.1 Linear analytical ranges of sensors.

S.No.	Particulars	Sensor ranges
1	Glucose	up to 40 mmol/L
2	Lactate	up to 20 mmol/L
3	Urea	up to 40 mmol/L

This integrated planar solid state sensor (38 × 20 mm) is produced using thick film technology. The sensor cassette consists of 12 electrode spots (each 1.2 mm in diameter), 9 spots are used for glucose and lactate, and 3 are used for urea. The total fill volume of a cassette is 44 μL, including 6 μl for an external maintenance-free reference electrode for the potentiometric urea sensor system. The integrated biosensor is thermostated in the instrument at 25°C to avoid interference by varying sample temperatures. Results from a whole blood or plasma sample can be obtained within 90 seconds after sample application.

13.0.4 Clinical Chemisty Analysis

Biosensor technology has taken advantage of the high specificity of the Ag-Ab reaction and, a myriad of research work aimed at clinical chemistry analysis was carried out. Thus, when the clinical analyte of interest is an Ag, or it can be chemically modified to behave as an Ag, it is possible to use the complementary Abs as capturing molecules in the bioreceptor system. Abs usually interacts exclusively with their complementary Ags, thus conferring a remarkable specificity to the device. In this case, the biosensor will be called immunosensor, named after the usage of Abs as bioreceptors.

The use of one enzyme-labelled antigen, added together with the sample carrying the analyte, allows performing a competitive assay, in a similar way to competitive, enzyme linked immunosorbent assays (ELISA). In the minimal configuration of amperometric DNA or RNA-based biosensors, a single-stranded (ss-), nucleic acid sequence is anchored to an electrode to capture the analyte, in this case, its complementary DNA or RNA strand. Once the binding partners have met, the event is detected using different approaches, mostly by using an electrochemical reporter of the hybridization.

13.0.5 Aptamer Based Biosensors

Recently, aptamers have also been used as analyte-capturing species. Aptamers are linear nucleotide sequences, usually of 15 to 40 nucleotide length, undergoing many intramolecular interactions, which wrinkle the oligonucleotide chain. Therefore, aptamers present in a three-dimensional shape that confers the molecules and the aptitude to interact either strongly or weakly with partner molecules, depending on their structure. The enormous diversity of aptamer conformational structure, which varies with the nucleotide sequence, allows them to be selected according to the strength of interaction with their partner molecule. Thus, aptamers can be chosen so as to tightly and selectively bind to target molecules, including among them virtually any kind of analyte like proteins, lipids, carbohydrates, DNA, etc. Clinical related aptamer based biosensors have been designed, that are included.

1. Electron transferred based aptasensors to determine small molecules like cocaine in biological fluids.
2. To determine thrombin: Thrombin is the main executioner of the coagulation cascade, and its determination in blood is useful to assess the haemostatic status of patients for the diagnosis of coagulation and thrombotic disorders.
3. Another related aptasensor is the one used to determine simultaneously lysozyme and thrombin; serum lysozyme has been proposed to be a marker of chronic inflammatory diseases, such as sarcoidosis, tuberculosis and liver and bowel inflammatory diseases, as well as of leukemia.

13.0.6 Amperometric Glucose Sensor

The amperometric three electrode glucose sensors, including working, counter, and reference electrodes, is based on the glucose oxidase method using hydrogen peroxide detection with electrochemical recycling of the oxygen, thus making the sensor essentially free from interference from the oxygen content of the sample.

13.0.7 Point of Care (POC) Biosensing Technology

Point of care (POC) systems are viewed as integrated systems that can process clinical samples for a number of different types of biomarkers in a variety of settings, such as clinical laboratories, doctors' offices and eventually, at home. Basically, POC systems make state-of-the-art technology platforms accessible to a large population pool. From a diagnostic or prognostic perspective, POC systems provide the clinician the ability to have access to a wealth of molecular information for providing profiles of cancers using novel technology platforms that in the past have been accessible to only major cancer centers. The development of POC technologies will provide opportunities for better screening of at risk patients, tighter surveillance of disease recurrence and better monitoring of treatment. In addition, POC technologies are by their very nature, low cost in their implementation making large scale screening for disease prevention and more attractive to health care insurers.

POC technologies offer platforms for complex testing by non-specialists and the benefit of increased robustness and reliability due to fewer world to device interfaces. These technologies can contribute to the realization of personalized medicine by creating a link between the diagnosis of disease and the ability to tailor therapeutics to the individual. As biomarkers of disease are discovered and validated through genomics and proteomics research, development of new technology platforms can enable rapid introduction of these discoveries into clinical practice as well as aiding in biomarker discovery efforts.

The basic building blocks of any biosensing platform consist of the biorecognition element, the transducing element and the readout modality. The biosensing platform for cancer diagnosis/prognosis may be a bit more involved due to the complexity of these diseases. Therefore, integration of the biosensor into a functional POC system may be required to provide initial front-end sample preparation. As such, the basic components of the biosensor must be amenable to integration into the POC system and provide multiplexing capabilities if the technology is to be appropriate for cancer diagnostics and prognostics.

13.0.7.1 Molecular recognition elements

The ability to recognize the target or biomarker in a mixed population is viewed as the critical step in any diagnostic assay. The biomarkers are present in intracellular or extracellular cells. Whatever the sampling matrix, the biomarker must be recognized and collected from a heterogeneous population. In addition, the marker can be as complex as a whole cell or as simple as a single molecule (prostate specific antigen). While antibodies can be used for recognizing these biomarkers, more recently, synthetic based recognition elements are being investigated as replacements for antibodies, such as aptamers, peptides, surface-imprinted polymers or molecularly imprinted polymers. Listed below are some examples of recognition materials.

1. The most common approaches for generation of cell specific or molecular ligands are to select a single, well characterized biomarker and generate antibodies or screen libraries for components that bind to it. In the case of generation of cell-based ligands, prior knowledge of the individual cellular components is necessary, which may require production or purification of the target protein. Most efforts in this field have focused on exploiting monoclonal antibodies as specific ligands. Polyclonal antibodies are typically generated against a single antigen and a large body of work is usually required to select the best antigen.
2. For cell-based recognition, high-throughput methods are required to screen libraries of potential ligands for cell binding, even when the nature of the cell surface is undetermined. Using phage display libraries, protocols to identify cell-specific binding peptides has been developed. A remarkable feature of the peptides selected using this procedure is their cell specificity. This high discriminating power of the selected phage suggests that peptides could be identified that selectively bind to different tumor types, even those with similar classification.
3. Several synthetic ligands have been studied for a variety of targets, such as aptamers and peptides. The particular advantages of these synthetic ligands as opposed to antibodies is that they are robust structures that can be placed in diverse settings without losing their specificity, can be made by conventional wet chemistry techniques and are easily modified structurally to support the addition of reporters or immobilization to sensing elements.
   (a) Peptides are smaller than antibodies, can be chemically synthesized in large quantities and are amenable to derivatization. Furthermore, peptides can display high affinities for cell surface components, making them attractive ligands for cell recognition. Phage display has been used to identify peptide ligands that target certain cell types by panning on well-characterized cell surface receptors. In most cases, a purified tumor associated protein has been used as the bait, and peptides are isolated by panning on the purified protein. More recently, peptides with affinity for the tumor antigens HER2/neu receptor, oligonucleotide receptor, ErbB-2, and ICAM- 1 have been isolated by phage display. Peptide bead libraries can also be screened for ligand binding and peptides have been isolated that recognize, and expressed on the prostate cancer line DU145.
   (b) Aptamers can bind to their targets with high affinity and even discriminate between closely related targets. This is due to their adaptive recognition: aptamers, unstructured in solution, fold upon associating with their targets into molecular architectures in which the ligand becomes an intrinsic part of the associated structure. The main advantage over

antibodies is overcoming the use of animal systems for the production of the molecule. In addition, aptamers can be designed in a signaling architecture, in which the structure of the aptamer changes upon association producing a signaling event due to the molecular association. Aptamers are isolated by *in vitro* methods that are independent of animals and can be generated against any target; the immune response to generate antibodies can fail when the target molecule (a protein) has a structure similar to endogenous proteins and when the antigen consists of toxic compounds that can even kill the animal. The production of aptamers is commonly performed by the Systematic Evolution of Ligands by Exponential enrichment (SELEX) process.

13.0.7.2 Transduction process

1. Various optical transducers have been exploited in biosensors including fluorescence spectroscopy, interferometry and spectroscopy of guided modes of optical waveguides and surface plasmons. Fluorescence has been a mainstay in the biosensing arena due to:
   (a) The wide availability of labels that cover a large spectral range and different functional groups.
   (b) Reduction in size and operational complexity of laser induced fluorescence hardware.
   (c) The ability to achieve single molecule detection sensitivity.

   While fluorescence continues to be a major transduction modality in many biosensing platforms and systems using optical readout, alternative technologies are evolving as well. An example is sensors based on spectroscopy of guided modes of optical waveguides (grating coupler, resonant mirror) and surface plasmons, which can be considered label-free measurement technologies. Sensors based on spectroscopy of surface plasmons, sometimes referred to as surface plasmon resonance (SPR) sensors, represent the most advanced type of label-free sensors based on spectroscopy of guided waves.
2. Another technology for transduction in biosensing platforms is piezoelectric, which also possesses the advantage of not requiring the use of labels and can record in real-time the affinity reaction allowing kinetic studies. Piezoelectric affinity biosensors are based on the coupling of the biological recognition element with a piezoelectric element, usually a quartz crystal coated with gold electrodes. Piezoelectric crystals have been used as microbalances and as a micro-viscometer owing to their small size, high sensitivity, simplicity of construction and operation, light weight and the low power requirement. The quartz crystal microbalance (QCM) has traditionally been used in many applications such as thin film deposition control, etching studies, aerosol mass measurements and space system contamination studies.
3. Electrochemical biosensors are extremely useful for delivering biodiagnostic information in a fast, simple, and low cost fashion in connection to POC analyzers. The i-STAT hand-held battery-operated clinical analyzer, combining different amperometric biosensors and potentiometric ISE (on a single disposable cartridge), has been widely used for rapid POC measurements of multiple electrolytes and metabolites in emergency settings. Nanoscale materials offer excellent prospects for designing powerful bioanalytical protocols with remarkable sensitivity and multiplexing/ coding capability. The enormous signal enhancement associated with the use of nanoparticle amplifying labels and with

the formation of nanoparticle-biomolecule assemblies provides the basis for ultrasensitive electrochemical detection. Bioaffinity assays on nanowire transducers offer great promise for label-free simplified detection of biomolecular interactions.

4. Micro-total analysis systems (TAS): An extension of the basic sensor by providing a workable platform for the biosensor is the micro-total analysis system (or lab-on a-chip), which is a device that integrates multiple bioanalytical functions into a small, portable instrument, essentially incorporating functions of a full sized laboratory into a miniaturized device with micro-scale or nanoscale components. The potential benefits of TAS include limited reagent use, faster sample analysis, higher throughput, increased sensitivity, portability, automation of complex bioanalytical processes, ease of use, and closed architectures for minimizing false positives arising from sample environmental contamination.

13.0.7.3 Utility

Biosensors must be used to detect and quantify the presence of biomarkers for cancer to aid clinicians in determining the type of cancer based on a molecular signature or a panel of signatures, and also in determining the stage of the disease and in selecting the most appropriate therapy. Targeted therapies are likely to become more prevalent and they will require patient specific biomarker detection and quantification to ensure selection of the correct therapy.

13.0.7.4 Detection limits

It is likely that the concentration of biomarkers in either serum or urine will be very low. This will require that biosensors for cancer diagnostics will have to possess excellent sensitivity. It also imposes very stringent specificity requirements for the ligands that are used as the biological recognition elements. In the case of intracellular confined biomarkers, high sensitivity will be required as well, since dealing with small clinical samples inevitably means low or even single target cells that must be analyzed.

13.0.8 Brain as a Biosensor

Hypoglycaemia can be defined as an abnormally low blood glucose concentration. In biochemical terms, blood glucose lower than 3.5 mmol/l will often be considered low in diabetes patients treated with insulin or oral hypoglycaemia agents. Both in diabetes patients and in healthy persons, however, spontaneous blood glucose values lower than this threshold may frequently be measured. Blood glucose values down to 2.7 mmol/l or even lower with limited or no symptoms can be measured following long term fasting in healthy humans (Hojlund *et al.,* 2001). Diabetes patients with tight glucose control and recurrent episodes of hypoglycaemia may lack symptoms of hypoglycaemia even at very low glucose levels down to 1 mmol/l.

In clinical terms, hypoglycaemia can be differentiated into mild, moderate or severe events. Mild hypoglycaemia is present when a diabetes patient experiences symptoms of hypoglycaemia such as sweating, shivering or palpitations. The patient is able to react appropriately by eating or drinking and thereby re-establish a normal blood glucose level, avoiding progression into severe hypoglycaemia. Moderate hypoglycaemia is present when the patient may or may not experience hypoglycaemia symptoms but may require help to take action. This could entail simply guiding the patient to eat or drink or a more active approach of giving the patient the food or drink. Severe hypoglycaemia is present when the patient loses consciousness and an active medical approach is needed such as glucose infusion or glucagon injection. The correlation between biochemical and clinical hypoglycaemia is very poor in type 1 diabetes patients (Pramming *et al.,* 1990).

With increased diabetes duration the sympato-adrenal activation may likewise fail resulting in impaired awareness of hypoglycaemia and ultimately, hypoglycaemia unawareness (Howorka *et al.,* 2000). This is defined by a severe cognitive impairment occurring without subjective symptoms of hypoglycaemia. Assuming that the risk and fear of hypoglycaemia is a major hindrance in achieving an optimal glucose control, all possible efforts should be done to reduce them. The first priority must be to optimize the insulin regime. Adjustment of the insulin dose and timing may consequently reduce the risk. Switching from one insulin type to another may ensure a better convergence between insulin concentration and insulin need. The long acting insulin analogues insulin glargine and insulin detemir reduce the risk of hypoglycaemia particularly at night-time (Monami *et al.,* 2009).

Continuous glucose monitoring (CGM) will provide a better protection against severe hypoglycaemia by frequent glucose measurements in the interstitial tissue and alarms based on actual glucose measurements or prediction algorithms. Large clinical studies have shown that the use of CGM enables a more tight glucose control without increased risk of hypoglycaemia, but so far CGM has not been shown to reduce the risk of severe hypoglycaemic events (The Diabetes Control and Complication Trial, 2009; Bergenstal *et al.,* 2010; Tamborlane *et al.,* 2008).

13.0.8.1 Electroencephalogram (EEG)

While hormonal counter regulation and sympatoadrenal symptoms often diminish or disappear with long term diabetes, the devastating effect of low blood glucose on organ function persists. The most important dysfunctions arise from the glycopenic effects on the brain and the heart. Neuroglycopenia results in a gradual loss of cognitive functions. In the early stage, this may only be apparent during systematic cognitive testing. As the glucose concentration falls, the cognitive function continues to decline resulting in slower speed of reaction, blurred speech, loss of consciousness, seizures and ultimately death. The effect of hypoglycaemia on the heart is less well described but comprises prolongation of the QT interval which is a known cause of cardiac arrhythmia. In fact, death among younger patients with insulin treated diabetes is assumed often to be related to malignant cardiac arrhythmia.

The blood glucose threshold at which the organ function is affected varies both between and within diabetes patients. Diabetes patients with tightly controlled blood glucose and frequent hypoglycaemic events may not be severely affected despite a blood glucose level as low as 1.5 mmol/l or even lower. This means, however, that just a slight further reduction in the glucose concentration will result in the serious effects of severe hypoglycaemia.

The concept of a hypoglycaemia alarm based on biosensing involves continuous monitoring of organ function, a real-time signal processing and an alarm device. Preferably, such a biosensor should be able to sense subtle change in brain function (*Eg.* electroencephalography), cardiac function (*Eg.* electrocardiography) or any other organ changes preceding cognitive dysfunction which will preclude the patient from taking action and thereby avoid severe hypoglycaemia.

The electroencephalogram (EEG) is usually measured on the scalp, using surface electrodes that are glued to the scalp with conducting gels. The surface EEG represents the electrical activity taking place inside the brain and originates from the firing neurons, mainly in the superficial part of the brain. When a neuron fires, a very small electrical charge is released, which itself cannot be measured on the scalp. But the macro-pattern that appears when many neurons fire in a synchronized manner, builds up larger electrical signals, which can be measured on the scalp. When measuring the EEG, all the micro changes in the firing pattern disappear due to the averaging effect through the scalp, and only the macro changes remain. The EEG, which is measured outside the scalp, can therefore be used to detect macro changes in the electrical behaviour of the brain. In general, during daytime, the healthy brain is less synchronized than during sleep, and only few daytime phenomena can be characterized and detected. During sleep, the brain is more synchronised and emits many characteristic wave patterns that reflect the different sleep phases (Iber *et al.,* 2007).

Glucose is an essential substrate for brain metabolism. Accordingly, low blood glucose resulting in neuroglycopenia can be assumed to result in EEG changes. Using the brain as a biosensor for hypoglycaemia detection throughout the day requires a stable long-term EEG recording system. The usual 10/20 EEG system (Crespel *et al.,* 2005) with surface electrodes glued to the scalp is not an option, since surface electrodes are highly exposed to movement artefact. Therefore, in our setting, the EEG is measured by electrodes placed in the subcutaneous layer, a few millimetres below the skin, thereby giving the advantage of being more robust to noise and artefact signals. The subcutaneous measurements were tested compared to scalp electrodes and were found to be very similar, showing very high correlation.

EEG patterns have different characteristics depending on the spatial location of the measurement. While some EEG changes are generalized and apparent on the entire surface of the brain, some paradigms are only present in smaller areas, which make detailed measurements in certain locations necessary. The spatial distribution of the hypoglycaemia related EEG changes (HREC) The topographic maximum has been demonstrated to be located in the lateral frontal region during mild hypoglycaemia. This shifts towards the centroparietal and parieto-occipital region in deeper hypoglycaemia (Tribl *et al.,* 1996). Hyllienmark *et al.* (2005) on the other hand studied type 1 diabetes patients with a history of recurrent hypoglycaemia, and the EEG recording was conducted during a period of normal blood glucose. They found similar HREC characteristics as previously described, however predominantly in the frontal region (Hyllienmark *et al.,* 2005). In addition, this could indicate that EEG changes in some cases may become permanent.

In order to detect HREC with a single or a few electrodes, we investigate the spatial distribution of the changes. The hypoglycaemia changes are generally present on most of the scalp area. The spatial distribution of the artefacts particularly derived from muscle activity during facial mimicking, eating, eye movement and sleep related movements, should be taken into account when the optimal electrode placement is to be defined. In contrast to the HREC, these artefacts are more localized, making the location important. Artefact related to electrode movements and the mechanics of the electrode contact are not dependent on the spatial location. The ability to detect the HREC when artefact signals are present is illustrated in Fig. 13.1, where the HREC signal is detected from a single electrode channel on five diabetes patients.

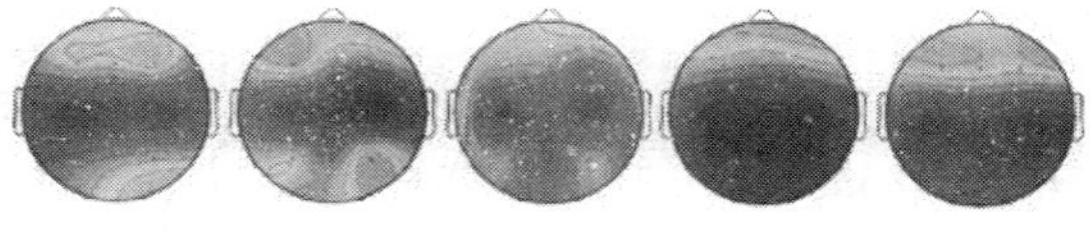

Fig. 13.1 *Illustration of the spatial influence on the ability to detect the HREC paradigm.*

The red areas in the figure indicate that the HREC paradigm detection performance is high, whereas green areas indicate low performance.

Taking into consideration, the spatial influence and the electrode type have to be chosen for the final measurement location as shown in Fig. 13.2.

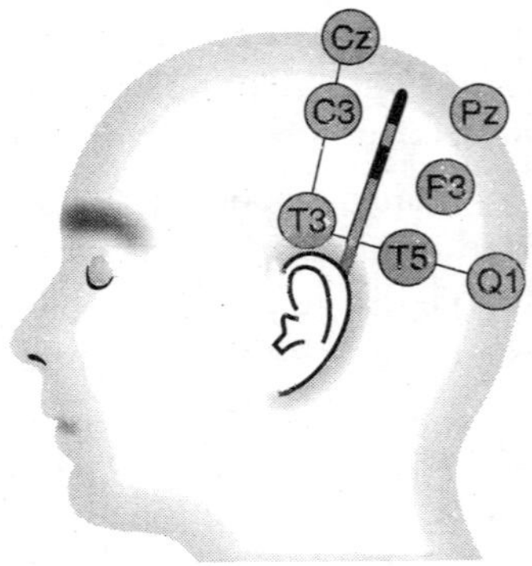

Fig. 13.2 *Location of the subcutaneous EEG electrode. The subcutaneous electrode is inserted in a location behind the ear towards vertex cranii between Cz and Pz. The measurement points are shown in red, giving one differential channel.*

The HREC can be detected by visual inspection by a neurophysiologist, who inspects the waveforms of the EEG. However, if the EEG of the diabetes patients is to be analysed in real-time throughout the day, this must be done automatically using an algorithm. The algorithm structure for hypoglycaemia detection is shown in Fig. 13.3.

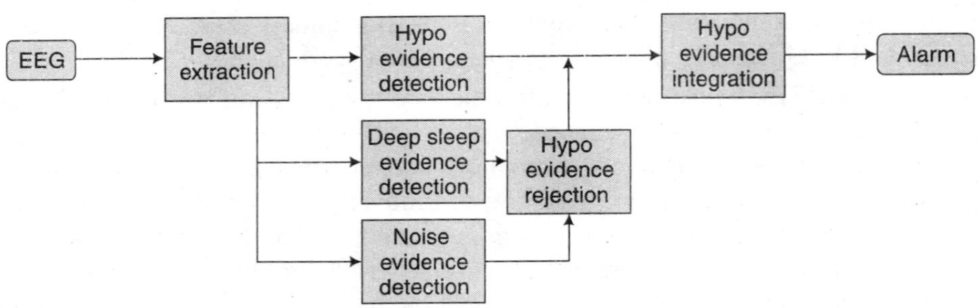

Fig. 13.3 *Structure of the hypoglycaemia detection algorithm.*

The algorithm works in four sequential levels that process the EEG signal and determines whether sufficient evidence of hypoglycaemia is present for an alarm to be triggered. At the first level, the feature extraction process maps the raw EEG into an appropriate feature space in which it is possible to distinguish HREC from normal EEG. The second level consists of three blocks, each of which analyses the features to determine if there is evidence of impending hypoglycaemia, deep sleep patterns, or noise contamination, respectively. At the third level, hypoglycaemia evidence is rejected when deep sleep patterns and/or noise are present.

The EEG based hypoglycaemia alarm system consists of two main parts: An implanted device that captures the EEG signal, and a non-implanted device, which stores and processes the EEG signal. This is illustrated in Fig. 13.4. The inner device is implanted subcutaneously, with the main part behind the ear and the electrode pointing towards the top of the head. The electrode has three measurement points, a length of 8 cm and a diameter of 1.1 mm. Data is transmitted from the inner device to the outer device through a near field communication link. Therefore, the two devices have to be closely aligned for the system to function. The outer device is designed as an ear hanger, illustrated on the right panel of Fig. 13.4. It is therefore easy to wear with a minimum of discomfort for the user. The outer device contains a sound generator and a light indicator to inform the user of critical events, e.g. impending hypoglycaemia. The outer device contains a power source. When the outer device is placed behind the ear, the power source is shared with the inner device through the communication link. When the power source in the outer device is depleted, it must be recharged in a charging station. A full recharge allows for approximately 18 hours of use.

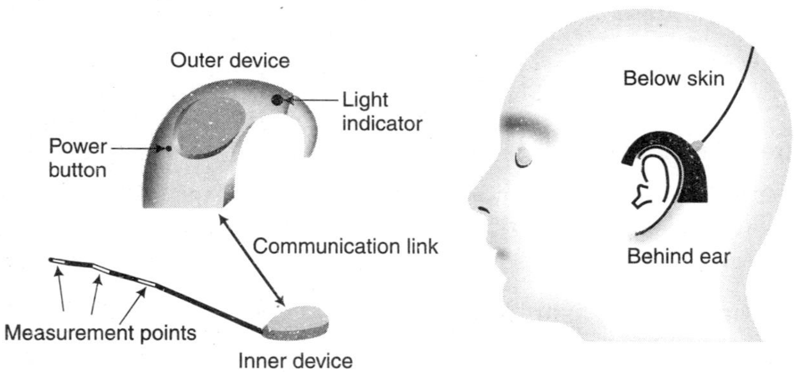

Fig. 13.4 *The EEG based hypoglycaemia alarm system consisting of an inner and an outer device.*

The implantation procedure is simple and takes approximately 15 minutes. The implanted device must be replaced only after two years of use.

The initial clinical studies of continuous EEG recording and real-time data processing during insulin-induced hypoglycaemia in type 1 diabetes patients indicate that it will be possible to predict incidents of severe hypoglycaemia before the patients are severely cognitive impaired both during daytime and sleep. The studies conducted so far, though, have taken place in clinical research units, we are now testing the hypoglycaemia alarm in an out-patient setting.

It is of utmost importance that an alarm device has a high sensitivity and specificity. False alarms may be annoying to the patients yet they are not dangerous. Missed alarms, on the other hand, may render the patient with a false feeling of security. On the other hand, a sensitive and reliable alarm device will allow the patient to achieve a better glucose control with less fear of hypoglycaemia events. The studies conducted so far hold promises that an EEG based device might fulfill these goals.

13.0.9 Biomarkers for Cancer

The utility of any biosensing platform is intimately dependent on the viability of biomarker(s) for producing diagnoses with high confidence. In particular, biomarkers must not only signal the presence of a tumor or cancer, but should also predict the stage of tumor genesis. The explosion in molecular biology and associated technologies has led to a much better understanding of human cancer and malignancy and potential biomarkers that can be used for diagnosis.

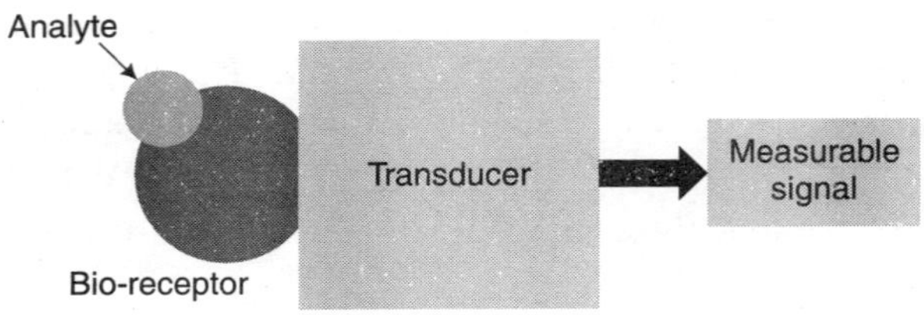

Fig. 13.5 *Schematic representation of a single element biosensor containing the biorecognition element, transducer and the physical output whose magnitude is related to the concentration of the analyte of interest.*

13.0.10 Lactate Sensors

The lactate sensor is constructed identically but with lactate oxidase immobilized rather than glucose oxidase. Acetaminophen is a good model substance for other exogenous and endogenous interfering substances.

13.0.11 Potentiometric Sensors

The potentiometric sensor for urea determinations is based on an ammonium ion selective electrode (ISE) based on nonactin with immobilized urease to convert urea to bicarbonate and ammonium ions.

13.0.12 DNA Amperometric Biosensors

Clinical applications using DNA amperometric based biosensors include devices potentially useful to determine indicators of pathogenic bacteria, virus DNA sequence such as SARS virus, indicators of chronic myelogenous leukemia, and acute promyelocytic leukemia. *Eg:* To diagnose cystic fibrosis, and the presence of pathogenic *Escherischia coli* bacteria.

13.0.13 Magnetic Bead Based Biosensors

Magnetic bead based biosensors based on detection of the analyte subsequent to the separative step. After pairing the target analyte with the biorecognizing element mounted on the beads, the coupled system is magnetically removed from the original sample matrix and transferred to an appropriate interference-free medium for further detection. A number of biosensors have been proposed to detect analytes of clinical interest using this strategy.

Example: To detect BRCA1 breast cancer gene

A number of other biosensors for clinical applications using modified surfaces with Chitosan polymer have been proposed.

Example:

1. *Schistosoma japonicum* antigen
2. Dengue virus oligonucleotide sequences
3. Choline

13.1 BIOSENSORS APPLICATIONS IN MEDICINE AND HEALTHCARE

13.1.1 Introduction

In medicine and healthcare, the monitoring of blood oxygen, blood glucose, blood pressure, heart beat rate, coagulation, pregnancy, infectious disease, stress, pain, immune status, ECG, EMG, EEG, ENG, etc., biosensor based medical instruments can be used for the monitoring of patients during intensive care and surgery. Smart biosensors can be used by patients themselves to improve the management of their chronic diseases. Portable devices integrated with miniature biosensors for consumer use, and those with remote monitoring functions, are considered having very high market potential.

The biosensor industry has come a long way since the mid-1970s, when the Yellow Springs Instrument Co. (Yellow Springs, OH) commercialized the first glucose monitor. The next major breakthrough didn't come until 1987, when MediSense (Cambridge, MA) introduced a pen sized meter for home glucose monitoring.

Healthcare is an important aspect of everyday life hence, it is vital that pervasive computing be extended to healthcare applications. Pervasive computing has the potential to provide low cost, high performance and centric solutions for healthcare and monitoring. The rapid improvements in microprocessor and sensing material technology has led to a development of miniature sensors that can be implanted in the human body. The biosensor based approach to healthcare makes it much more effective by reducing the response time and decreasing the granularity of the application. By this technology, continuous health monitoring of a human body and real time data collection would be possible.

13.1.2 Potentiometric and Amperometric Biosensors

Various types of potentiometric and amperometric biosensors are characterized:

1. Microbial sensors with *Gluconobacter oxydans* cells with potentiometric (pH-sensitive field-effect transistor) and amperometric (Clark-type) electrodes for determining glucose
2. A potentiometric enzymatic electrode with butyrylcholinesterase, which is used in the biosensor designed to detect pesticides
3. Immunosensors with pH sensitive field effect transistors which detect the herbicide 2, 4-D, a biosensor for human immunoglobulin G
4. Biosensors with anaerobic bacteria *Clostridium thermocellum*, chemical and enzymatic sensors containing a photosensitive membrane for determining ammonium ions and urea and

5. Amperometric microbial sensors prepared with *Pseudomonas* cells for determining naphthalene, biphenyl, and polychlorinated benzoates.

13.1.3 Surface Plasmon Resonance Sensors

The SPR technology developed by BIAcore (Piscataway, NJ), a spin-off of the Swedish pharmaceutical firm Pharmacia, uses a single wavelength to induce SPR. It is essentially a laser illumination and what they look for is a change of angle in incident light.

A rough surface permits interaction between the photons and electrons, which must exactly match each other in momentum. Quantech's device uses gold on a grating, very similar to the dispersion grating in a spectrophotometer. The gold can be coated with antibodies to specific target substances. The presence of the antibodies changes the absorption wavelength of the gold. Then, as the antigen binds to the antibody, another wavelength shift occurs, which can then be measured. Quantech's focus is on point-of-care medical diagnostics. According to McKiel, there are currently 20 to 30 diagnostic tests commonly done in hospital labs that would be far more valuable to ER personnel if they could be performed at the patient's bedside.

13.1.4 Biosensors in Medical Diagnostics Industry

When compared to other medical devices or instruments, biosensors are undergoing rapid growth in the industry. The reason behind its proliferation is that it gives researchers and companies exciting challenges to overcome. The principle behind present day industrial sensors, as well as new technologies, is under a process of constant experimentation and corresponding acclimatization in order to develop biosensors that are more effective. As the whole concept of biosensors has basic principles of science such as physics, chemistry and biology as foundations, there is a wide range of factors, which one can manipulate to develop a novel and more efficient sensor or technology platform.

The number of people requiring medical attention is increasing primarily due to population explosion and also because people are succumbing to variegated ailments due to unhealthy day-to-day practices or occupational hazards. Individuals have started attributing more importance to health, especially with diseases such as diabetes, and the need for using biosensors for constantly sensing or monitoring blood glucose levels as well as other physiological factors has increased significantly.

Biosensor in medical diagnostics is part of the technical insights, healthcare subscription and aims at increasing the market and product awareness for key industry participants in this field, aiding research and development of these medical sensors. This study also hopes to increase the awareness among the medical community and patients. By examining the key technologies utilized in this space such as non-invasive sensors, invasive devices, diagnostic test kits, and medical devices, this study provides an in-depth analysis of this industry.

13.1.5 Wireless Body Area Networks

For quality life healthcare is always a big concern for an individual. The growing population of developed countries presents a raising slice of government's budget and presents new challenges to healthcare systems, particularly with elderly people. Generally, health monitoring is performed on a periodic check basis, where the patient must remember its symptoms; the doctor performs some tests and plans a diagnostic, then monitors patient progress along the treatment. Healthcare applications of wireless sensor networks allow in-home assistance, smart nursing homes, clinical trial and research augmentation.

Advances in wireless communication and micro-electro-mechanical systems (MEMS) allow the establishment of a large scale, low power, multi-functional and low cost network. Wireless sensor networks (WSNs) are verdict applications in many areas, such as medical monitoring, emergency response, security, industrial automation, environment, agriculture, infrastructure protection and optimization, automotive and aeronautic applications, and military applications. Wireless sensor networks can be effectively used in healthcare to enhance the quality of life provided for the patients and also the quality of healthcare services. *Eg.* Patients equipped with a wireless body area network (WBAN) need not be physically present at the physician for their diagnostic. A body sensor network proves to be adequate for emergency cases, where it autonomously sends data about patient health so that the physician can prepare for the treatment immediately.

The biosensor based approach to medical care makes it more efficient by decreasing the response time and reducing the heterogeneousness of the application. These sensors are implanted in the human body that forms a wireless network between themselves and some entities which are external to the human body. A wired network requires laying wires within the human body; such a network can be used for a variety of applications, these include both data aggregation and dissemination applications. Biosensors may be used for monitoring the physiological parameters like blood pressure, glucose levels and collecting the data for further analysis. Wearable health monitoring systems allow an individual to closely monitor changes in her or his vital signs and provide feedback to help maintain an optimal health status. If integrated into a telemedical system, these systems can even alert medical personnel when life-threatening changes occur. During the last few years there has been a significant increase in the number of various wearable health monitoring devices, ranging from simple pulse monitors, activity monitors and portable monitors, to sophisticated and expensive implantable sensors.

The biosensor network consists of a group of biosensors implanted inside the human body, external device (control node) placed on the human body and a base station. The biosensors (between themselves) and the control node form a network. The control node is connected to an external base station. A biosensor consists of a processor, memory, transceiver, sensors or simulator and a power unit. These biosensors carry out the tasks like the human body sensing information, processing it, and transmitting it to the control node, receiving external signals to trigger action inside the body. The control node acts as both the data aggregation and dissemination point. The data collected periodically by the control node is sent out to the base station, where it is stored for further processing. As compared to the biosensors, the control node and the base station have significantly higher transmission and processing capabilities. Biosensors have limited amount of power and cannot afford heavy computation and communication. Communication is very expensive than computation, and sensing.

There are three types of wireless communication links in the biosensor healthcare network system. They are the communication links between the biosensors, the biosensor and control node, and control node and the base station (Fig. 13.6). Unlike a wired channel, the wireless links data is accessible to anyone who cares to listen to the channel, therefore all these wireless links are considered to be insecure. As a result, data exchange using any of these communication links has to be secured. The base station and the control node have higher computing and communication capability, hence the link can be secured by means of asymmetric cryptography.

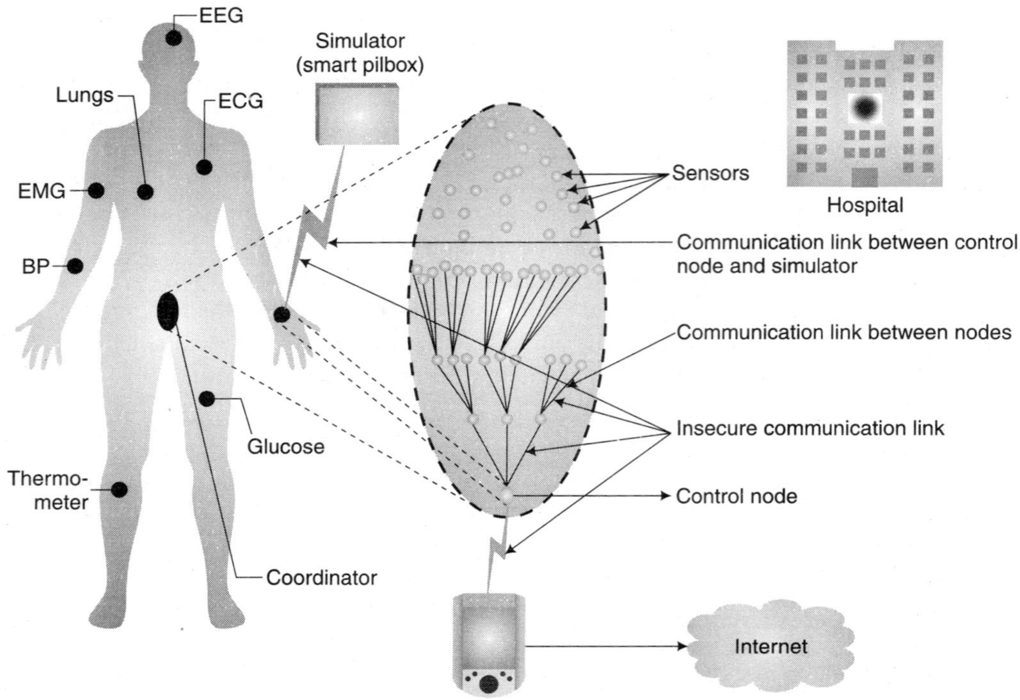

Fig. 13.6 *Wireless communication links in the biosensor healthcare Network System Model. (Source: Sandeep K S Gupta et al., 2003).*

13.1.6 APPLICATIONS

Current healthcare applications of wireless sensor networks target heart problems, asthma, emergency response and stress monitoring. Medical applications benefit from wireless sensor networks in many ways. The recent advances in miniaturization of smart biosensors have opened up new opportunities for continuous monitoring of patients. Non-intrusive, tiny wearable sensors are allowing collection of vast amounts of data automatically and reducing the cost.

13.1.6.1 Cardiovascular diseases

Smart sensor nodes that can be installed on the patient in an unobtrusive way can prevent a large number of deaths caused by cardiovascular diseases. The corresponding medical staff can do treatment preparation in advance as they receive vital information regarding heart rate and irregularities of the heart while monitoring the health status of the patient.

13.1.6.2 Cancer detection

The second leading cause of death in world with rising numbers each year and nowadays one of the biggest threats for human life is cancer. A sensor with the ability to detect nitric oxide (emitted by cancer cells) can be placed in the suspect locations. These sensors have the ability to differentiate cancerous cells between different types of cells.

13.1.6.3 Alzheimer, depression and elderly people monitoring

According to the U.S. Bureau of the Census, the number of adults age 65 to 84 is expected to be doubled from 35 million to nearly 70 million by 2025. This trend is global, so the worldwide population over age 65 is expected to more than double from 357 million in 1990 to 761 million in 2025. In addition, a recent study found that almost one third of U.S. adults, most of whom held full-time jobs, were serving as informal caregivers mostly to an elderly parent. Wireless sensor network can help homebound and elderly people who often feel lonely and depressed by detecting any abnormal situation and alerting neighbors, family or the nearest hospital.

13.1.6.4 Glucose level monitoring

Diabetes can yield other complicated diseases like heart disease, stroke, high blood pressure, blindness, kidney disease, and amputations. A biosensor implanted in the patient could provide a more consistent, accurate and less invasive method by monitoring glucose levels, transmit the results to a fixed terminal and by injecting insulin automatically when a threshold glucose level is reached.

13.1.6.5 Asthma

A wireless sensor network can help those millions of patients suffering from asthma by having sensor nodes that can sense the allergic agents in the air and report the status continuously to the physician and/or to the patient himself.

13.1.6.6 Vital sign monitoring system

Patients at a disaster scene can greatly benefit from technologies that continuously monitor their vital status and track their locations until they are admitted to the hospital. A real-time patient monitoring system that integrates vital signs sensors, location sensors, networking, electronic patient records and web portal technology to allow remote monitoring of patient status.

13.1.6.7 Artificial retina

The eye is one of the most amazing organs in the body. To understand how artificial vision is created, it's important to know about the important role that the retina plays in and how you see. Here is a simple explanation of what happens when you look at an object:

1. Scattered light from the object enters through the cornea.
2. The light is projected onto the retina.
3. The retina sends messages to the brain through the optic nerve.
4. The brain interprets what the object is.

The retina is complex in itself. This thin membrane at the back of the eye is a vital part of your ability to see. Its main function is to receive and transmit images to the brain. These are the three main types of cells in the eye that help perform this function:

1. rods
2. cones
3. ganglion cells

There are about 125 million rods and cones within the retina that act as the eye's photoreceptors. Rods are the more numerous of the two photoreceptors, outnumbering cones 18 to 1. Rods are able to function in low light (they can detect a single photon) and can create black and white images without much light. When enough light is available, the cones give us the ability to see color and detail of objects. Cones are responsible for allowing you to read this article, because they allow us to see at a high resolution. The information received by the rods and cones are then transmitted to the nearly 1 million ganglion cells in the retina. These ganglion cells interpret the messages from the rods and cones and send the information on to the brain by way of the optic nerve.

There are a number of retinal diseases that attack these cells, which can lead to blindness. The most notable of these diseases are retinitis pigmentosa and age-related macular degeneration. Both of these diseases attack the retina, rendering the rods and cones inoperative, causing either loss of peripheral vision or total blindness. However, it's been found that neither of these retinal diseases affect the ganglion cells or the optic nerve. This means that if scientists can develop artificial cones and rods, information could still be sent to the brain for interpretation.

Creating artificial sight

The current path that scientists are taking to create artificial vision received a jolt in 1988, when Dr. Mark Humayun demonstrated that a blind person could be made to see light by stimulating the nerve ganglia behind the retina with an electrical current. This test proved that the nerves behind the retina still functioned even when the retina had degenerated. Based on this information, scientists set out to create a device that could translate images and electrical pulses that could restore vision.

Today, such a device is very close to becoming available to the millions of people who have lost their vision to retinal disease. The artificial silicon retina (ASR), developed by Optobionics, which was in FDA clinical trials as of late 2007, improved vision in 10 subjects over a period of two years. As of late 2007, however, Optobionics was in bankruptcy and awaiting a buyer, which would allow the trials to continue. The ASR is an extremely tiny device, and has a diameter of just 2 mm (0.078 inch) and is thinner than a human hair. In order for an artificial retina to work it has to be small enough so that doctors can transplant it in the eye without damaging the other structures within the eye.

The most significant development in artifical retina research has been the Department of Energy's creation of the Artificial Retina Project, which is led by Mark Humayun. The ARP is a group of public and private companies, universities and research labs that have pooled their efforts to perfect a nano-size device. Since 2002, six blind volunteers have been fitted with the device, which has been successful in helping them perceive light and dark and large objects. The ARP also has two more devices in testing.

Working mechanism of artificial silicon retina

The ASR contains about 3,500 microscopic solar cells that are able to convert light into electrical pulses, mimicking the function of cones and rods. To implant this device into the eye, surgeons make three tiny incisions no larger than the diameter of a needle in the white part of the eye. Through these incisions, the surgeons introduce a miniature cutting and vacuuming device that removes the gel in the middle of the eye and replaces it with saline. Next, a pinpoint opening is made in the retina through which they inject fluid to lift up a portion of the retina from the back of the eye, which creates a small pocket in the subretinal space for the device to fit in. The retina is then resealed over the ASR.

For any microchip to work it needs power and the amazing thing about the ASR is that it receives all of its needed power from the light entering the eye. As you learned before, light that enters the eye is directed at the retina. This means that with the ASR implant in place behind the retina, it receives all of the light entering the eye. This solar energy eliminates the need for any wires, batteries or other secondary devices to supply power.

Another microchip device that would restore partial vision is currently in development by a team of researchers from Johns Hopkins University, North Carolina State University and the University of North Carolina Chapel Hill, called the artificial retina component chip (ARCC), this device is quite similar to the ASR. Both are made of silicon and both are powered by solar energy. The ARCC is also a very small device measuring 2 mm square with a thickness of 0.02 mm (0.00078 inch). There are significant differences between the devices continue.

Unlike the ASR, which is placed between layers of retinal tissue, the ARCC is placed on top of the retina. Because it is so thin, light entering the eye is allowed to pass through the device to strike the photosensors on the back of the chip. However, this light is not the power source for the ARCC. Instead, a secondary device attached to a pair of eyeglasses directs a laser at the chip's solar cells to provide power. The laser would have to be powered by a small battery pack. Wireless Personal Area Networks (WBANs) can also assist blind people. Patients with no vision or limited vision can see at a reasonable level by using retina prosthesis chips implanted within a human eye, as shown in Fig. 13.7

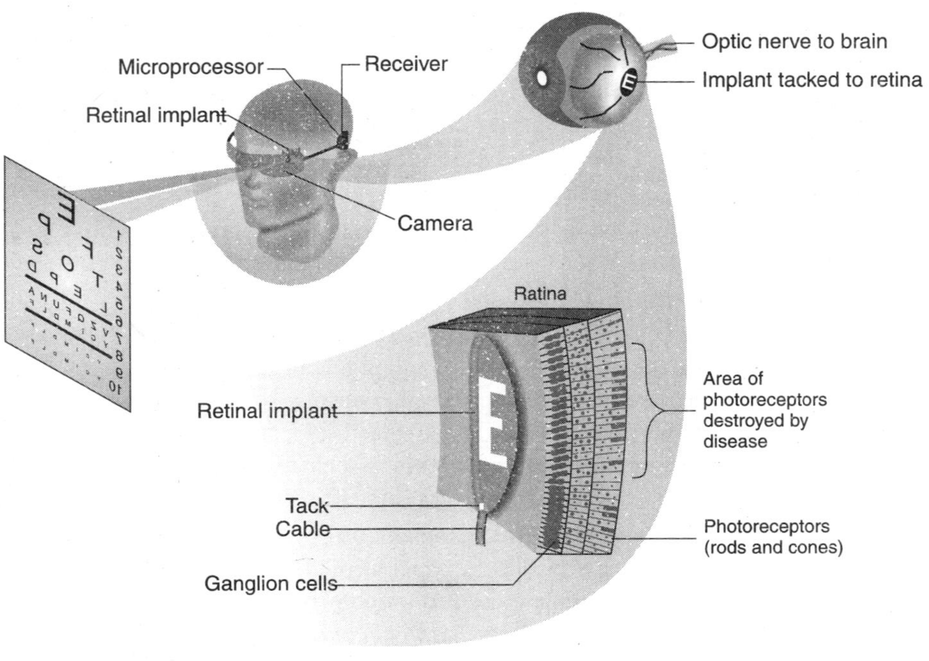

Fig 13.7 *Schematic representation of artificial retina for blind people. (Source: L. Schwiebert et al., 2001)*

13.1.6.8 Life shirt

Life Shirt is a comfortable and completely non-invasive smart garment that gathers data during a patient's daily routine, providing the most complete remote picture of a patient's health status. It enables healthcare professionals and researchers to accurately monitor more than 30 vital life-sign functions in the real-world settings where patients live and work. Life Shirt collects patient data using integrated sensors including respiratory bands (which measure pulmonary function) and an ECG (which record electrical activity of the heart). It also tracks and records posture and physical activity.

13.1.6.9 Polymer microchip

A disposable polymer microchip promises to make medical diagnostics easier and more convenient. Immunoassay is a tool for biomedical diagnostics. In medicine, immunoassay is used in two general classes of applications: (i) For identifying the organism which is responsible for a disease and (ii) For monitoring disease treatment. Even though its success, immunoassay does have the drawbacks which are include long assay time, complex, expensive equipment, need trained technicians and the assay have restricted to centralized laboratories. In this competitive marketplace, medical diagnostic companies are already looking at new technologies that allow fast, quantitative and portable assays with simplified instrumentation.

At present lab-on-a-chip systems are using for the analysis of very low volume of the samples. Microchips should not only greatly simplify the process of immunoassay, but could also allow diagnosis to be carried out in local analytical laboratories, A&E departments, doctors' surgeries or even pharmacies. In this way, they may one day be used by all health professionals at point of care regardless of location and ultimately even by patients in the home.

Electrochemical sensors have to date mainly been devoted to glucose sensing, which is one of the biggest markets for *in vitro* diagnostics. However, lay in developing a portable affinity system using disposable microchips for detecting proteins and these chips to detect the interactions of proteins of interest with the corresponding enzymes or receptors by following small changes in electrical currents.

Polymers (rather than silicon) are good material for the development of chips; proteins are adhering easily to the polymers. These are present considerable flexibility, both in terms of performance and ease of processing and have the advantage of being cheap and readily disposable thereby avoiding any possibility of cross-contaminating samples.

By electrically controlling the flow of solutions along with micro-channels etched into the surface of a polymer chip, it is possible to separate the various solutes electrochemically. Laser capable of drilling micro-holes have polymer sheets are used for photo ablation to produce the micro-channels by using the vaporizing polymer and then sealed these by lamination. The success of electrophoretic separation depended on the generating sufficient charges on the micro-channel.

To develop these microchips as electrochemical sensors, integrated the carbon ink microelectrodes within microchips and use them as transducers capable of detecting electrical current. *Eg.* an oxidation reaction. The same microchips are used to immobilize on the micro-channel walls with a radioactively labelled bacterial toxin, namely *Staphylococcal enterotoxin B* (SEB). Measuring the resulting levels of radioactivity confirmed the efficiency of our immobilization procedure, an essential prerequisite for performing enzyme-linked immunoassays (ELISA). The ability of the chip to perform such ELISA tests, in this case the test involved detecting a particular compound, D-Dimer by following electrochemical changes.

To carry out such an immunoassay, first immobilize an antibody on the walls of the micro-channel that can specifically bind to an antigen of interest. Next, allow a sample of plasma, blood or urine containing this antigen to flow through the system. This antigen binds to the immobilized antibodies to form an antigen-antibody complex and excess sample is washed away. At this stage, introduce a second antibody labelled with an enzyme (*Eg.* alkaline phosphatase, peroxidase or glucose oxidase) into the micro-channel. This secondary antibody further binds to the antigen-antibody complex, and after washing away any excess, add an enzyme substrate (*p*-amino phenyl phosphate in the case of alkaline phosphatase) to the micro-channel. The enzyme transforms this substrate into an electro active product that can be oxidized at the electrodes, releasing electrons that are detected as an electrical current in the process. The magnitude of the current is therefore directly proportional to the concentration of the antigen or other analyte. This is a big advantage of electrochemistry, because the measured current is not limited by the number of analyte molecules present in the system but by its concentration; the response obtained with microchips is therefore identical to that obtained for much larger scale assays.

ELISA type assays with a biotin-alkaline phosphatase (ALP) complex in a buffer solution used to evaluate the detection limit and the dynamic range of the assay. After complex formation of biotin with avidin coated on the micro-channel walls, add an enzyme substrate, *p*-amino phenyl phosphate, to the avidin-biotin-ALP complex and this was transformed by ALP into *p*-aminophenol. This *p*-aminophenol in turn can then be oxidised to quinonimide by applying a small voltage, releasing two electrons for detection. The results showed that it could be possible to detect biotin down to sub-pico-molar ($<10^{-12}$ mol/l) levels - which compares very favorably with other immunoassay detection levels (range: 10^{-13}mol/l to 10^{-6} mol/l).

The main advantages these chips are speed, incubation time less than five minutes and detection limit is within a minute. By reducing the volume of the sample, and hence the distance between molecules, complex formation occurs much faster. The large (microchip) surface to (sample) volume ratio about 100 times larger than with standard well plates also favors the immobilization of one of the binding partners on the chip surface, accelerating antigen-antibody complex formation and increasing the dynamic range of the assay.

13.1.7 Medical Telesensors

The goal is to develop an array of chips to collectively monitor bodily functions. A chip on your fingertip may measure and transmit data on your body temperature. An array of chips attached to your body may provide additional information on blood pressure, oxygen level, and pulse rate. This type of medical telesensor will report measurements of vital functions to remote recorders. The goal is to develop an array of chips to collectively monitor bodily functions. These chips may be attached at various points on a soldier using a non-irritating adhesive like that used in water proof band aids. These medical telesensors would send physiological data by wireless transmission to an intelligent monitor on another soldier's helmet. The monitor could alert medics if the data showed that the soldier's condition fit one of five levels of trauma. The monitor also would receive and transmit global satellite positioning data to help medics locate the wounded soldier.

Wireless monitors attached to the skin could provide valuable information on the physiological condition of intensive-care patients in hospitals, high-risk out patients, babies at risk of suffering sudden infant death syndrome, and police and firefighting personnel in hazardous situations. A 2×2 millimeter (mm) silicon chip attached to the skin can measure body temperature. The chip contains a temperature sensor in an integrated circuit, a lithium thin film battery that supplies the very low level of power required by the circuit and signal processing and transmission electronics, and an antenna that sends the data by radio signals (radio frequency transmission) to a monitor when the chip is queried. This biosensor is called a medical telesensor ASIC (Ferrell) because it uses an application-specific integrated circuit for telemetry automatic measurement and transmission of data from remote sources to receivers for recording and analysis.

A biochip can be developed to measure blood oxygen level. As the blood oxygen level changes, the color of hemoglobin in blood is altered. Such a chip would have a light source and light detector that could measure changes in the color of hemoglobin transmitted when it is illuminated by light. The results are reported by wireless telemetry.

A sensitive detector has been developed for monitoring changes concentrations of calcium ions in the body. It may be useful in diagnosing disease or exposure to chemical warfare agents. This biosensor consists of an optical fibre to which is attached a synthesized hybrid molecule. One half of the hybrid molecule binds calcium ions and the other half fluoresces when calcium ions are bound to the molecule.

Blood pressure and pulse rate may be measured by chips designed to detect pressure changes. The optical fibres made of silicon take advantage of the unique properties, unlike a glass fibre, a silicon fibre is flexible and it can be squeezed or stretched and the amount of compression or expansion can be measured by changes in light transmission through the fibre. Thus, silicon fibres embedded in roads can be used to weigh trucks. If a silicon fibre on a chip can sense pressure at various positions in the body, it may be used for monitoring blood pressure, pulse rate, breathing (chest expansion), knee bending during physical rehabilitation, and foot pressure distribution.

Aging, diseases such as diabetes and Alzheimer's, and chemical warfare agent's cause changes in metal ion concentrations in the body. If these changes could be detected and measured, the information could provide clues about changes in disease states and exposure to toxins. Tuan Vo-Dinh *et al.* (2004) have developed a biosensor using a glass optical fibre and a hybrid molecule. One half of the hybrid molecule binds calcium ions and the other half fluoresces when calcium ions are bound to the molecule. By attaching this molecule to the end of a very small diameter optical fibre, the concentration of calcium ions in a solution can be measured.

13.1.8 Colloid Based Biosensors

Therapeutic drugs can assist some patients. Furthermore, a drug that is normally considered to be safe can have toxic effects in a small proportion of the population, occasionally causing death. Many altered drug responses depend on the genetic constitution of the recipient. The analysis of specific DNA markers within the genome of an individual single nucleotide polymorphism (SNP) could potentially identify subjects at risk in clinical trials and may enable clinicians to individualize medical therapy.

The development of technology for personalized medicine relies on the ability to accurately and rapidly identify SNPs in genetic material. Some of the biotechnology tools for SNP identification are colloid-based biosensors, which are proving to possess several advantages over the traditional microarray format. The current technologies used for SNP genotyping and review colloid-based biosensors which are emerging as a higher throughput, potentially more accurate technology.

13.1.9 Microcantilever Sensors

A micro-cantilever sensor can be adapted to detect physical, chemical, or biological activity. An interesting alternative to the optical fibre is the micro-cantilever, which measures the presence of substances by non-optical methods. It can act as a physical, chemical, or biological sensor by detecting changes in cantilever bending or vibrational frequency.

The micro-cantilever is constructed with silicon plank (100 micrometers (mm) long, 30 mm wide, and 3 to 4 mm thick), when molecules are added to its surface, the extent to which the plank bends can be measured accurately by bouncing a light beam off the surface and measuring the extent to which the light beam is deflected. The vibrational frequency can be induced by piezoelectric transducers and measured with the same laser beam that measured the deflection because it generates an alternating current in the detector. Antibodies were covalently attached to the silicon surface of a cantilever in such a way that the stresses induced in the antibody when it reacted with its antigen were detected.

Detection of biological warfare agents or bacteria and viruses in the hospital laboratory should be expedited with this stressed antibody technique. Additional experiments are under way to demonstrate the usefulness of the micro-cantilever as a biosensor. Because of the small size and versatility of the micro-cantilever, arrays of sensors can be fabricated on a single chip to conceptually mimic the five sensory facilities: sight, hearing, smell, taste, and touch.

13.1.10 Detecting Cancer and Health Abnormalities

Another type of biosensor uses sophisticated technology to detect a specific trait or abnormality in a living organism. Laser fluorescence diagnosis is accurate in over 98% of the cases. A new laser technique is used for non-surgically determining whether tumors in the esophagus are cancerous or benign. This sensor can tell whether a tumor in the esophagus is cancerous or benign. In the past, determining accurately whether a patient has cancer of the esophagus has required surgical biopsy. However, laser-based fluorescence method has eliminated the need for biopsy, reducing pain and recovery time for patients. Laser light of the appropriate wavelength is directed to the inner surface of the esophagus by means of a fibre-optic device that is swallowed by the patient. The epithelial cells and tissue inside the esophagus are fluorescence when excited by the laser light. When the esophagus interior is illuminated with blue light [410 nanometers (nm)], the normal tissue emits light at wavelengths different from those emitted by the cancer cells.

13.1.11 Anthropometry

The biosensors used to measure human body surfaces are called anthropometry. A technique using laser beams and mirrors can determine the shapes of human body parts. Its accuracy could facilitate the creation of clothes that fit. A technique using laser beams and mirrors that measures the topology of a solid surface of a body, using amplitude modulated laser radar which measures arc lengths along with complex inaccessible body contours. Because laser radar measures the phase and amplitude of a reflected, modulated laser beam, only one optical path is required between the sensor and the subject.

Multiple images are combined by integrating information from different virtual viewpoints, any number of which can be created with strategically positioned mirrors. Arc lengths along arbitrary contours, surface areas, and volume estimates all become possible. The accuracy of the measurements is within 1 mm. Because creating clothes that fit may be accomplished using a single camera with several mirrors, a blue jeans manufacturer has shown interest in this methodology.

13.1.12 Drug Discovery

Conventionally, drugs are discovered through a time consuming, tedious process. Biosensors have tremendous potential in speeding up this process.

Biosensors can detect the interaction between a particular target and a possible drug without using markers or the detection of color changes or fluorescence. Biosensors can quickly measure how well a potential drug binds to a target and, by eliminating markers; biosensors eliminate potential causes of interference. Test samples do not need to be purified and can be reused.

Biosensors also improve the accuracy of the results and are far less time consuming. For example Prozac was introduced after 15 years of development, whereas a similar drug took only two trials using biosensing technology. The user simply paints some receptors or targets on the top of the device and runs a test. It has demonstrated applications in refractive index changes, avidin-biotin binding, antibody-antigen dissociation kinetics, thickness of insulators, refractive index of thin dielectrics, specific detection of small molecules, protein binding, concentrations of analytes, attachment of DNA complements and mixture proportions.

13.1.13 Bioluminescent Chips

Critters on a chip the device is deceptively simple, consisting of a coating of bioluminescent bacteria on top of a light-sensitive integrated circuit. The bacteria emit light in direct correlation to the analyte concentration. There are other whole cell biosensors that are similar to the critter chip. Micro physiometers, use a similar light-addressable potentiometric technique; based on micro-machined silicon wafers, they detect the pH change that occurs when organisms secrete organic acid during metabolic processes.

Manufacturing the critter chip is not significantly different from conventional chip making. The possibility that the bacteria could interfere with the electronics, any assay that could be made bioluminescent could potentially benefit from the technology. In conventional Tuberculosis screening, the antibiotic is used to treat the TB requires several days to grow a culture and perform specificity testing. Simpson has been researching a recently developed method for TB screening that uses a chemical compound, firefly Luciferase, to generate a luminescent response. The substance is catalyzed by adenosine triphosphate. Simpson discovered biosensing devices simple enough to use in an emergency vehicle and inexpensive enough to discard after use.

13.1.14 Multi-Functional Biochips (MFB)

The MFB is an integrated multi-array biochip, which is designed by combining integrated circuit elements, an electro optics excitation/detection system, and bioreceptor probes into a self contained and integrated micro device depicts a schematic diagram of the MFB device. The MFB system includes the following elements:

1. An excitation light source
2. Multiple bioprobes having different types of bioreceptors
3. A sampling platform
4. Sensing elements
5. A signal amplification and treatment system.

The development of the multi channel sampling platform involves immobilization of bioprobes on multi array (4 × 4 or 10 × 10 channels for current biochips) substrates, which can be performed on a transducer detection surface to ensure optimal contact and maximum detection. The integrated electro-optic microchip system involved integrated electro optic sensing photo detectors for the biosensor microchips. Such an integrated microchip system with on board integrated circuit (IC) electronics is not currently available commercially. Highly integrated biosensors are made possible partly through the capability of fabricating multiple optical sensing elements and microelectronics on a single integrated circuit.

To develop a biochip system with optimized performance, based on photodiode circuitry, one system having 16 channels (4 × 4 array), and another having 64 channels (8 × 8 array) having four types of electronic circuits on a single platform. The biochips include a large area, n-well integrated amplifier photodiode array that has been designed as a single, custom IC, fabricated for the biochip. This IC device is coupled to the multi-array sampling platform and is designed for monitoring very low light levels.

The individual photodiodes have 900 μm square sizes and are arrayed on a 1-mm spacing grid. The photodiodes and the accompanying electronic circuitry were fabricated using a standard n-well CMOS process. The use of this type of standard process allows the production of photodiodes and phototransistors as well as other numerous types of analog and digital circuitry in a single IC chip. This feature is the main advantage of the CMOS technology in comparison to other detector technologies such as charge coupled devices or charge injection devices. The photodiodes themselves are produced using the n-well structure that is generally used to make resistors or as the body material for transistors. Since the anode of the diode is the p-type substrate material, which is common to every circuit on the IC chip, only the cathode is available for monitoring the photocurrent and the photodiode is constrained to operate with a reverse bias.

Multiplexer allows any number of the elements in the array to be connected to an amplifier. In the final device, each photodiode could be supplied with its own amplifier. The multiplexer is made from 16 cells for the 4 × 4 array device. Each cell has a CMOS switch controlled by the output of the address decoder cell. The switch is open when connecting the addressed diode to an amplifier. This arrangement allows connecting a 4 × 4 (or 10 × 10) array of light sources to the photodiode array and reading out the signal levels sequentially.

With the CMOS technology, highly integrated biosensors are made possible partly through the capability of fabricating multiple optical sensing elements and microelectronics on a single IC. A two-dimensional array of optical detector-amplifiers was integrated on a single IC chip. To evaluate and select an improved IC system for the biochip, which has been designed and fabricated in a chip with an 8 × 8 CMOS sensor array. This microchip contains 4 quadrants, each having a different electronic design, which was evaluated for optimal performance.

In general, biosensors and biochips employ only one type of bioreceptor as probes, i.e., either nucleic acid or antibody probes. Biochips with DNA probes are often called gene chips, and biochips with antibody probes are often called protein chips. An integrated DNA biochip that uses multiple bioreceptors with different functionalities on the same biochip, allowing simultaneous detection of several types of bio-targets on a single platform has been developed. This device is referred to as the multi functional biochip (MFB). The unique feature of the MFB device is the capability to perform different types of bioassay on a single platform using DNA and protein based probes such as antibody probes simultaneously.

Biotargets (*Eg.* gene sequences, bacteria, viral DNA) offers a very high degree of accuracy for identifying DNA sequences complementary to that of the probe. In addition to DNA probes, the MFB uses also another type of bioreceptor, i.e., antibody probes that take advantage of the specificity of the immunological recognition. The results of this study using antibody against *E coli* and DNA probes for *B. anthracis* demonstrate the feasibility of the multi-functional biochip for the detection of multiple biotargets of different functionality (DNA, proteins, etc.,) using a single biochip platform. *Mycobacterium tuberculosis* bacterium or the *p53* cancer proteins have also been detected on the same biochip.

Infectious diseases are moving across borders, becoming a global treat to our lifestyle and well being. Over half of TB cases in some wealthy countries are among foreign born populations. With increasing travel, increasing cases of malaria were reported among travelers. Infectious diseases are becoming a matter of national security for many developing countries. Sustainable development is feasible if countries can tame the infectious diseases that disempowered their people. If these diseases remain unchecked, they could damage the social fabric, diminish agricultural output, affect industrial production, undermine political, social and economic stability, and ultimately contribute to regional and global insecurity.

The MFB, which is a truly integrated biochip system that comprises probes, samplers, detector as well as amplifier, and logic circuitry on board, could enable a rapid and inexpensive test for multiple diseases and for a wide variety of applications. With its multi-functional capability, the MFB technology is a system that allows simultaneous detection of multiple biotargets simultaneously. Such a device could provide information on both gene mutation (with DNA probes) and gene expression (with antibody probes against proteins) simultaneously. It is expected that advances in miniaturization and mass production technology will significantly reduce the cost of fabrication of biochip systems for widespread use worldwide to address the urging need for improved healthcare at a reduced cost.

Biosensors in Agriculture and Food Industry

Chapter 14

14.0 INTRODUCTION

Biotechnology offers efficient and cost effective means to produce an array of novel, value-added products and tools. It has the potential to increase food production, reduce the dependency of agriculture on chemicals, lower the cost of raw materials, and reduce the negative environmental impacts associated with traditional production methods. With improved technology and knowledge about agricultural organisms, processes and ecosystems, opportunities will emerge to produce new and improved agricultural products in an environmentally sound manner. Future gains in agricultural productivity and sustainability depend heavily on the use of biotechnology to improve the health and well-being of agriculturally important plants and animals. Biosensor deals with the application of modern biotechnology in the agriculture sector and it will be highly useful in agriculture and environmental science.

In the food and the agricultural industries, the quality of a product is evaluated through periodic chemical and microbiological analyses which are expensive, slow, need well trained operators and in some cases require steps of extraction or sample pretreatment, increasing the time of analysis. Biosensors can provide rapid, non-destructive and affordable methods for the quality monitoring of a product. Biosensors can reduce assay time and cost or increase the product safety. Biosensors have been adapted to detect or measure analytes in on-line systems. Biosensors have the potential to create an analytical revolution to resolve the problems in the agricultural and the food industries. For quality assessment, grading and sorting of biological products, several types of electronic sensors are being developed for providing rapid and non-destructive determination of internal qualities.

14.1 BIOSENSORS IN AGRICULTURE

Concentrations of herbicides, pesticides and heavy metals in agricultural lands are increasing. Biosensors can be used to measure the levels of pesticides, herbicide and heavy metals in soil and groundwater and can also be used to forecast the possible occurrence of soil disease, which has not been feasible with the existing technology. The biological diagnosis of soil using biosensor means opening the way to reliable prevention and decontamination of soil disease at an early stage.

14.1.1 Soil Diagnosis

The Research Center of Advanced Bionics (RCAB) of the National Institute of Advanced Industrial Science and Technology (AIST) at Tokyo University of Technology has developed world's first biosensor for soil diagnosis. The biosensor is intended not for diagnosing soil infection, but for identifying potential susceptibility of soil. It can forecast the possible occurrence of soil disease, which has not been feasible with the existing technology. The biological diagnosis of soil using biosensor means opening the way to reliable prevention and decontamination of soil disease at an earlier stage. The combination of biosensor technology with conventional physical and chemical measurements will ensure comprehensive diagnosis of soil needed for stable and augmented agricultural production.

Fig. 14.1 *Biosensor for soil diagnosis.*

Soil disease refers to infectious disease of crop plants caused by the infection with soil microorganisms, such as viruses, bacteria, and fungi. However, not all of these microorganisms are responsible to disease risk. Among hundred millions of microbes living in 1 gram of soil, most are not causal agents of diseases in plants (non-pathogenic—good microbes), while some are harmful (pathogenic—bad microbes). The balance between good and bad microbes in soil plays an important role. Healthy soil contains a lot of good microbes. The situation with the intestinal micro-flora applies to the soil environment where vegetables and flowers are cultivated. Farmers have been making efforts to make up healthy soil inducing the proliferation of good microbes by putting in compost or soil conditioner on the basis of their experience. However, soil has often been degraded continuously through repeated cultivation of same crop, unfavorable weather conditions and excessive use of chemical fertilizer. Therefore, the decimating habitat of good microbes leads to soil diseases.

Once soil disease breaks out, crop roots begin to perish, spreading the disease to surrounding areas, and ultimately leading to total wipe out of harvests in the region. *Eg.* A historically famous case is the epidemic of potato blight around 1845, which started from a part of Europe and spread to England and Ireland. The disease propagated further in the next year, causing serious harvest failure. More than a million of people living on potato were starved to death. With this as a turning point, nearly four millions of Irish emigrated to North America and other areas, to break fresh ground, from 1851 to 1905.

The basic principle of soil diagnosis with the biosensor is to estimate the relative activity of good microbes and bad microbes in the soil on the basis of quantitative measurement of differential oxygen consumption in the respiration of two types of soil microorganisms. The measurement proceeds through the following steps: two sensors impregnated with good microbes and bad microbes, respectively, are immersed in a suspension of soil sample in buffer solution. Half an hour later, the oxygen consumption data by two microbes are displayed on a PC screen (Fig. 14.1). By comparing the data, it may be possible to quantitatively decide which microbe favors the soil. It is feasible, therefore, to predict whether or not soil disease is ready to break out in the tested soil beforehand. It is to be emphasized that the biosensor offers an innovative technique of diagnosing soil condition based not on experience but on numerical data.

14.1.2 Amperometric Biosensor for Pesticide Assay

Organophosphorous pesticides are a large group of compounds exerting an important role in different human activities. Their common toxic effect is based on the inhibition of enzymes with cholinesterase activity: acetylcholinesterase (AChE) and butyrylcholinesterase (BChE) that play an important role in human and animal systems. Their mechanism of inhibition is based on phosphorylation of the physiologically active serine hydroxyl group in the active centre of the enzyme. Poisoning by organophosphates leads to malfunction of the cholinergic nervous system.

One of the important tasks in suppressing the harmful effects of methamidophos is timely detection. Widely available techniques are based on mass spectrometry and liquid chromatography. Biosensors could be an alternative to the classical analytical methods. It was described as a useable tool for the analysis of different analytes, including microorganisms and toxic compounds. The suitability of biosensors for organophosphates and carbonates detection was extensively reviewed. Biosensors harboring cholinesterase attached to the electrode surface are the most typical, an assay is based on a decrease in cholinesterase activity due to inhibition by analyte (organophosphate) during the measuring procedure.

Different electrodes and immobilization techniques can be applied. A platinum electrode covered with a composite, including AChE and graphite, cobalt phthalocyanine and acetylcellulose, an adsorbed mixture of cholinesterase, glutaraldehyde and albumin onto a screen printed platinum surface, were found to be convenient. Furthermore, graphite rods and nylon fibres were also considered suitable as a biosensor with cholinesterase construction. AChE was suspended into deionized water: final activity 0.05 nkat/μl. The AChE solution of 1 μl was injected on one out of every four platinum electrodes printed onto ceramic strip and the solution was allowed to dry at the laboratory temperature. In the second step, a whole ceramic strip was placed into the chamber filled with glutaraldehyde vapor and 1 μl of AChE solution per electrode was applied again and allowed to dry for 30 minutes. The created biosensor was gently washed with phosphate buffered saline (PBS).

The biosensor was dipped into a reaction cell (polypropylene, shaped as flat bottom cylinder with maximal volume 2 ml) filled with 990 μl of reaction medium (1 mM ATChCl in PBS). The reaction cell was stirred with a teflon coated magnetic agitator (length 5 mm). A silver wire covered by silver chloride was used as a reference electrode. The applied potential was set on +450 mV for oxidation of the forming thiocholine. The current (i_0) was measured by the amperometric detector MEB. The background current (i_b) was measured in the same manner as i_0; however, the biosensor was displaced by a sensor (strip without immobilized AChE). The current after inhibition was obtained when 10 μl of methamidophos solution in water per one electrode was applied for 10 min. Incubation with methamidophos was realized out of the reaction chamber. After that, the biosensor was dipped into the reaction chamber and measures the concentration of the analytes.

14.1.3 Detection of Diuron, Atrazine and its Main Metabolites

Intensive industrialisation and the use of chemicals in agriculture which are themselves consequences of successful economic and measures to increase prosperity have contributed to the presence of many toxic compounds in air, soil and water, causing environmental degradation and sometimes direct health hazards. Typical molecules most often found during control analysis are s-triazines (atrazine, simazine) and their degradation products, phenylureas herbicides, glyphosate, bromoxynil and paraquat.

Several analytical methods are used to detect these toxic compounds, because the polarity and other chemical properties of the component vary considerably. A variety of extraction techniques have been employed over the years including liquid-liquid extraction, solid phase extraction, solid-phase micro-extraction etc, followed by various chromatographic techniques such as gas and liquid chromatography (GC and LC). Gas chromatography-mass spectroscopy (GC-MS) is the most commonly used technique. All these methods are accurate with a large range of detection. However, they are under constant improvement, time-consuming, expensive, cannot be performed easily outside the laboratory and need very highly skilled technicians. On the other hand, early-warning systems are needed to react in case of accidental pollution.

Enzyme based biosensors for pollutant determination represent potential alternatives to the analysis of pollutants in the environment and also as warning systems, selectivity, relatively low cost of fabrication, good storage and potential for miniaturisation and easy automation, which could lead to the construction of simple portable devices for fast screening purposes and for in-field/on E-site monitoring.

In particular, several biosensors based on tyrosinase were developed for determination of phenols. They include various kinds of electrochemical detection, such as detection of dioxygen consumption, direct reduction of generated *o*-quinone and mediated reduction of *o*-quinone by hexacyanoferrate (II). The tyrosinase has been immobilised on the electrode surface as a thin film or in a membrane on a Clark oxygen electrode, chemically bonded to a solid graphite electrode or controlled pore glass and using electro-polymerisation of an amphilic pyrrole derivative enzyme mixture.

Recent tyrosinase biosensor are based on pH-sensitive field-effect transistors for phenol determination in water solution. The main analytical characteristics were studied under different conditions, as well as the possibility of optimizing these working parameters. Different factors, such as pH of immobilisation, the enzyme loading and time of immobilisation in glutaralaldehyde vapour were investigated with regard to the influence on sensitivity, limit of detection, dynamic range, operation and storage stability. On the other hand, it has also been reported that triazine and phenylurea herbicides could be determined from the inhibition of tyrosinase activity to a substrate using amperometric biosensors.

The enzymatic membrane was prepared on the transducer surface by the cross-linking of enzyme with bovine albumin in saturated glutaraldehyde vapour. A mixture containing 5% (w/w) enzyme, 5% bovine serum albumin, 10% glycerol in 20 mM phosphate buffer (pH 6.0) was deposited on the sensitive surface of the sensor by a drop method, while another mixture of 10% (w/w) bovine serum albumin and 10% (w/w) glycerol in 20 mM buffer (pH 6.0) was deposited on the other electrode, this electrode was then considered to be the reference sensor. The sensor chips were placed in saturated glutaraldehyde vapour for 30 minutes followed by drying in air for 15 minutes at room temperature. The measurements were carried out in daylight, at room temperature in a vessel filled with 5mM phosphate buffer, pH 6.0. The biosensors were then dipped into the magnetic-stirred buffer solution. After soaking in the buffer solution for 20 to 30 minutes, a stabilisation of the base line was reached, and 4-chlorophenol as a substrate was added into the vessel. The differential signal was either registered by a 'home made' conductometric laboratory amplifier for biosensor application or logged by a SR 830 lock-in amplifier, and the steady-state response of the biosensors was plotted against the concentration of 4-chlorophenol. The inhibition rate was determined by comparison of the steady-state response of the conductometric biosensors before and after exposition to a sample solution containing the toxic compounds.

14.1.4 Measurement of Nitrates

Protecting the water quality of rivers and watersheds will require improved methods for detection and analysis of compounds of environmental concern. The new methods will need to be economical and easy to use, so that the frequency of data collection and analysis can be increased. These goals need to be reached without compromising on accuracy, so that the data can be trusted and used. The new methods should also be developed with an emphasis on use of non-hazardous and nontoxic reagents, to keep both the user and the site uncontaminated and safe.

Enzymes are the protein catalysts that speed the chemical reactions and make living systems to run. Enzymatic reactions are fast, efficient, very specific, and require little energy. Nitrate reductase is the enzyme that begins the process by which plants make their proteins. Ironically, it is also one of the physiological reasons that nitrate fertilizers are so effective in increasing crop productivity. These qualities make nitrate reductase a powerful tool for nitrate analysis.

The chemistry of the enzymatic reduction of nitrate involves a flow of electrons from a biological electron donor - such as NADH through the enzyme and to the nitrate, reducing it to nitrite. In a biosensor, the enzyme is immobilized onto a solid support material, and electrons are supplied either directly by an electrode or via an electron-carrying molecule attached to the same support. As nitrate is reduced, there is a tiny flow of electrons, a current, through the enzyme, and this current can be amplified and detected. The amount of current (in nano-or micro-amps) correlates to the concentration of nitrate present in the medium in which the biosensor probe is in contact.

14.1.5 Detection of Nitrogen Fixing Rhizosphere Bacteria

1. A collection of known and potential microorganisms that have been implicated to be able to fix nitrogen in the Rhizosphere of corn or grass species.

2. Engineer multiple bacterial biosensors, each capable of detecting a different concentration range of fixed nitrogen in the form of ammonium nitrate and glutamine, and respond by emitting a specific colour and/or fluorescence.
3. Test biosensor activity on corn/vegetable in gnotobiotc assays where plants are grown in various substrates under limited nitrogen conditions.
4. Evaluate biosensors as a means of screening for nitrogen fixing bacteria under various cultural conditions.
5. Apply the biosensor technology to evaluate bacterial nitrogen-fixation in different maize hybrids and vegetable cultivars, under different field conditions and in the presence of other microbes.

14.1.6 Detection of Avian Influenza Virus

Rapid detection is the key to controlling the spread of avian influenza. Techniques currently used to detect the disease are either time consuming, too expensive or not specific to subtypes of avian influenza viruses. Yanbin Li has developed a portable first biosensor for in field rapid screening of avian influenza virus. This device provides robust and reliable results and introduces the concept of real-time detection to facilitate a coordinated and rapid response. This device specifically and sensitively detects the avian influenza strain H5N1 from poultry tracheal swab samples in less than 30 minutes and could help health officials coordinate a rapid response for the eradication, quarantine and vaccination of animals.

14.1.7 Detection of Chemical Residues

Biosensor may help farmers and regulatory officials to detect herbicides in soil and water samples. The device relies on living organisms or their byproducts to identify traces of chemical residues in a matter of minutes. Heavy applications of herbicides can leave environmentally unsafe residues in soil and water. A liquid sample is passed through the biosensor. If the sample contains an herbicide, the chemical will react with the biosensor's proteins and inhibit oxygen production. The electrode in the biosensor detects oxygen levels and sends the information to a computer that displays the data.

14.2 BIOSENSORS IN FOOD INDUSTRY

Consumers are placing significant importance on the quality of food products that they purchase. This has forced the agricultural and the food industries to place an increased emphasis on quality monitoring of products. Food-borne pathogens pose a risk to food safety and are a threat to the global food supply chain. The detection and identification of pathogens in raw food materials, food products, processing and assembly lines, hospitals and drinking water supplies continue to rely on time consuming conventional culturing techniques. Biosensors have the potential to revolutionize food and water supply monitoring. Biosensors for food analysis take account of the diversity of the food industry, and the specifications which biosensors would have to meet in real applications. It brings food researchers up to date on currently available technology in this growing field, and will be of interest to biochemists, microbiologists and food scientists in research associations and academic institutions.

14.2.1 CARBON BIOSENSOR FOR FOOD INDUSTRY

Carbon biosensor technology consists in a novel amperometric biosensor can be applied in any kind of samples in food industry, during production, packaging and storage stages. Main advantages of the invention are its high selectivity, low cost, high sensitivity, stability and biocompatibility.

The health benefits of phenolic compounds derive from their antioxidant activity. Several methods are available for the analysis of polyphenolic compounds, based on separation techniques such as chromatography or capillary electrophoresis. These methods have disadvantages such as their long operation times or high cost. There is a need for simple techniques for estimating polyphenols for particular purposes, such as the real-time monitoring of polyphenols in food storage, manipulation, or processing. In food quality biosensors are tools for the detection of several types of compounds of interest especially on the determination of phenolic compounds.

Design and development of new electro-analytical sensor has successfully developed a novel biosensor based on the immobilization of Laccase phenoloxidase enzyme. This biosensor employs as the electrochemical transducer a graphite-based sol-gel electrode: Sonogel-Carbon electrode, which is obtained using high-energy ultrasounds. This biosensor can be employed for determining polyphenols in samples and in addition permits the continuous monitoring of such compounds in this matrix during the manufacturing and storage process. These measurements have an outstanding importance provided that these compounds dictate the physical and colloidal stability of some kind of beverages in these stages.

14.2.2 Impedimetric DNA Biosensors for Food Analysis

Impedimetric DNA biosensors for the rapid and reliable detection of DNA in genetically modified plant material and pathogen organisms in the food industry. This analytical method based on micro and nanotechnology incorporates various fields of research such as clean-room technology, surface biochemistry, microelectronics and computing. Based on the design of a microelectrode, that has the property of measuring a variation of impedance in contact with DNA. For this purpose, single stranded DNA sequences are immobilized on the surface of the electrode and tested with a fluorescent-marked complementary DNA strand by fluorescence microscopy (fig. 14.2). Impedance measurements are then performed in different analytical solutions to evaluate the stability and sensitivity of the system.

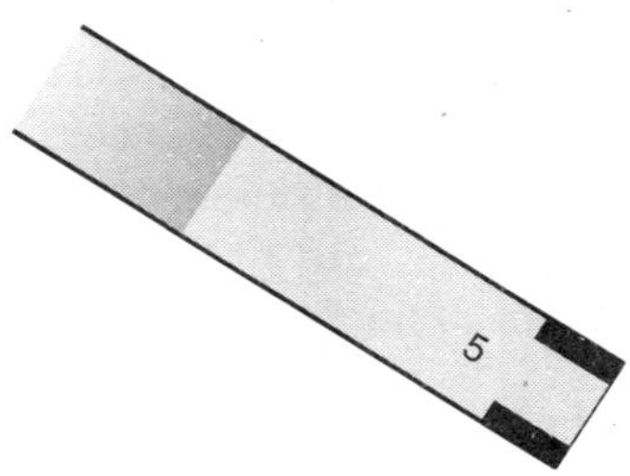

Fig. 14.2 *Design of clean room fabricated 5 μm gap interdigitated microelectrode.*

14.2.3 Mammalian Cell Biosensor for Pathogen Detection

A new biosensor made from mouse white blood cells could provide a quick and easy test for microbes and toxins in food. The sensor can return results in just a few hours. The sensor consists of multi-well plates, each containing mouse lymphocytes suspended in collagen gel. Each well can test one sample, and there are typically 96 wells in a plate, so the technology is easily scaled to test many samples at one time. The sensor works by detecting, when a toxin or bacteria has broken the cell wall of the lymphocytes. This releases an enzyme alkaline phosphatises that can be marked with another chemical to produce a yellow product, which can then be detected using a spectrometer.

The higher the quantity of the toxin or bacteria, the greater the number of lymphocytes killed in this way, and the more intense the yellow colour in the well. The spectrometer can detect the exact intensity of the colour of the well, and through careful calibration it can estimate the quantity of the pathogen in the sample, which can then be used to estimate the extent of the risk. Lymphocytes are a key part of the body's defence against such pathogen, this was not the most important factor when choosing the cells to form the biosensor. One of the limitations is that, mammalian cells need a difficult growth environment. The cell life is limited, meaning it must be used within five or six days. It can be made in the lab, but it must be immediately delivered. The sensor could make use of different types of cells apart from mouse lymphocytes, which would be sensitive to other types of pathogens.

14.2.4 Detection of Toxins in Food

Toxins in food have been detected using an optical fluoro-immunosensor in an array biosensor configuration that is capable of detecting multiple targets (mycotoxin, bacterial toxins). The use of SPR biosensors for detection of food safety related analytes including chemical contaminants, food-borne pathogens and toxins. Optical sensors are used to detect diacetyl vapours that are evolved at the onset of spoilage of meats (Shiers and Honey bourne 1993) and for detection of bacterial contamination and food processing. Gas sensors are used for detection of bacterial volatiles in food analysis and identification of particular microorganisms (Alocilja *et al.,* 2002).

14.2.5 Potentiometric Biosensor in Food Analysis

A potentiometric biosensor for the analysis of isocitrate was developed by using a CO_3^{2-}-selective electrode and enzyme immobilization in flow injection analysis (FIA). The interference effect of major sugars and organic acids on the sensor system was less than 5%. Isocitrate concentrations of some fruit juices analyzed by the isocitrate sensor system were compared with those analyzed by gas chromatography (GC). No significant difference was found between the two analytical methods in any of the fruit juices. This suggests that the isocitrate sensor system is reliable in determining the isocitrate concentrations of foods.

14.2.6 Immunobiosensors in Food Analysis

Immunobiosensors were developed for detection of *Escherichia coli* O157:H7 based on the surface immobilization of monoclone antibodies onto indium tin oxide (ITO) electrodes. The immobilization of antibodies onto ITO chips was carried out by silanization. Silanization is the process of adding organo functional Silane groups to the support and the functional groups of the

Silane can then be used to bind a specific molecule to it. The biosensor could detect the target bacteria with a detection limit of 4 times 10^3 CFU/ml. This biosensor was characterized with high sensitivity, excellent selectivity, short detection time and easy operation. It has a promising application in clinical laboratory diagnoses, environmental detection and food safety.

14.2.7 Amperometric Biosensor in Food Analysis

Histamine could accumulate in sea food when bacteria spoilage commenced and caused histamine poisoning without altering the fish normal appearance and odor. Therefore, a histamine biosensor using immobilized enzyme diamine oxidase (DAO) has been developed for the rapid monitoring of the histamine levels in tiger prawn (*Penaeus monodon*). The histamine biosensor has a response time of less than 1 minute and optimum pH operation was 7.4. The reusable biosensor is simple and can be used for direct histamine determination without further pretreatment, and is suitable for routine analysis of tiger Prawns to monitor spoilage.

14.2.8 Detection of Antibiotics in Food

The presence of banned antibiotics in honey is one of the recent examples why food consumer groups are insistent on better quality assurances and increased testing. Methods for detecting chemical contaminants, *Eg.* streptomycin and chloramphenicol that combine reliability and throughput at the required sensitivity are in demand. Biosensors can analyze the presence of antibiotics reliably, effectively and in a short time.

14.2.9 Vitamin Analysis in Food Products

Biosensor based analysis is becoming more and more important in the food industry and one of the fields of application is in vitamins analysis. The method for vitamin analysis is a label-free inhibition assay. The SPR biosensor monitors interactions of a specific binding protein with the vitamin immobilized on a CM5 sensor chip. The prepared samples are mixed with a fixed concentration of the vitamin binding protein by the auto-sampler and injected over the chip surface. The vitamin present in the sample binds to the protein and subsequently inhibits it from binding to the surface of the sensor chip. The higher concentration of the vitamin is in the sample, the higher level of inhibition and hence the lower the response of the biosensor. A regeneration step prepares the chip surface for the next sample. Quantification is performed by multi-level calibration with the vitamin standards.

14.2.10 Pathogen Detection in Food

As the occurrence of food-borne illness rises to tolerable amounts of contamination in food stuffs. However, conventional methods of pathogen detection require time consuming steps to arrive at a useable measurement. This process, usually taking 7 to 10 days, is much too long for perishable food, which usually enters and leaves the marketplace within 2 to 3 days. Biosensor technology can significantly reduce this time as well as detect even smaller amounts of pathogens with fewer false positives. Using biosensors to detect the cause of food spoilage. Bacteria in the meat packs is releasing gas and causing the packages to explode in transit, biosensors are used for identification of such bacteria. Piezoelectric biosensor are used to detect the bacteria *Listeria monocytogenes*. This bacterium affects wild and domestic animals and humans causing fever, meningitis and encephalitis. The biosensor was applied to the *Listeria* bacteria diluted in milk and is as sensitive as conventional methods yet results can be obtained in less than 15 minutes.

14.2.11 Detection of Multipathogen

Prototype nanotechnology engineered biosensor could help processors to detect multiple pathogens faster and more accurately than current devices, developed by Yang Liu *et al.* This device can also measure the amount of pathogen contamination on a particular food or machine, giving processors more data to determine the extent of a problem. The logic gates are created by patterning antibodies at different spatial locations in an immunosensor assay. Immunosensors are biosensors that use antibodies to recognize the presence of a pathogen.

14.2.12 Detection of Contaminants in Food

Electro-chemical biosensor is used to detect the presence of very small amounts of atrazine in food. One of the most widely used herbicides in agriculture and which also has very long lasting effects on the environment as well as antibiotics in food.

The biosensor is faster, more portable and economic than the expensive laboratory methods which are used to detect contaminants, while having a very similar sensitivity. The system has been tested successfully to detect pesticides in samples of drinking water and commercial orange juice, as well as to detect traces of antibiotics in cow's milk.

14.2.13 Biosensors for Food Safety

Biosensors are being pursued by the military as real-time, field-portable instruments for the detection and identification of pathogenic microorganisms and toxins in foods. Biosensor systems will provide military food inspectors with the ability to rapidly detect pathogens and potential contaminants, including chemical/biological agents. Enhanced screening and surveillance of military food sources will significantly improve food safety, thereby reducing the health risks and medical costs associated with food borne illness.

The Surface Micro Channel Resonator and the electro-spun nano-fibre membrane are two biosensor platform technologies being developed and investigated for military applications. Similar efforts in food sampling strategies are being designed and modified to improve the capture and concentration of a bio-agent from the food matrix. The food safety and biosensor program will serve as a platform for future technology insertions aimed at improving food safety.

14.2.14 Nanoelectronic Biosensors for Food Safety

An electronic biosensor targeted for Agri and Food Bio-safety (Maki *et al.* (2007a,b). The system expertise required for such a device component includes nanotechnology, surface chemistry, organic chemistry, molecular biology and electronics. An electronic nano-transistor based sensor with excellent sensitivity with demonstrated results in the detection of *Staphylococcus aureus.* An integrated electronic molecular sensor used for the detection of *Staphylococcus aureus* which is a versatile and dangerous pathogen in humans. This microorganism also causes severe animal diseases, such as suppurative disease, mastitis, arthritis, and urinary tract infection, which associate with numerous virulence factors, such as the production of extracellular toxins and enzymes. *Staphylococcus aureus* is a major pathogen in bovine intramammary infections of subclinical and chronic nature.

The techniques used for *Staphylococcus aureus* identification include Gram staining, colony morphology, tests for coagulase and urease activities. These conventional methods are time consuming. Therefore, new technology to improve detection of *Staphylococcus aureus* will have a great impact in animal health and food safety. Nano electronic detectors promise high sensitivity, low-cost and portable diagnostics.

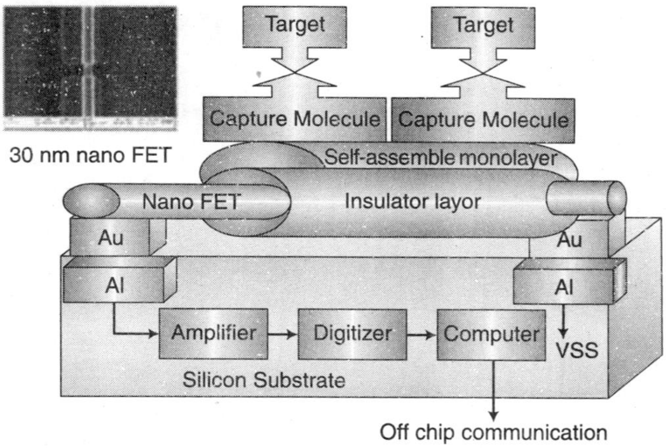

Fig. 14.3 *A schematic representation of electronic nanobiosensor.*

The overview of the final electronic biosensor device system is shown in Fig. 14.3. The system consists of a detector composed of both micro- and nanoelectronic devices in which *Staphylococcus aureus* 16S rRNA and Toxin B are used as target molecules in the nucleic acid and protein detection models.

A unique signal transduction system is used in the target capturing process and generation of signal molecules which are captured and detected on the sensing surface. The presence of signal molecules on the sensing surface will change electronic properties of nano-transistor and generate a detectable electronic signal. The nanoelectronic interface with microelectronics translates low level analog signals to digital signals which are input to an on-chip microprocessor. The microprocessor provides the embedded intelligence for the final user and provides communication ports with numerous devices including cell phones, the Internet, satellites or other computers. Such a system, being portable, has attractive features for animal point of contact healthcare.

Biosensors for Environmental Monitoring

Chapter 15

15.0 INTRODUCTION

One of the approaches for reducing uncertainties in the assessment of human exposure is to better characterize the hazardous wastes that contaminate our environment. Biosensors are excellent analytical tools for monitoring programs working to implement legislation. For monitoring of environmental pollutants, different types of biosensors (commercial devices) are developed. Public concern and legislation are nowadays demanding better environmental control. In addition, the increase in the number of analytes to monitor requires more suitable analytical methods. For conventional off-site analysis, samples need to be sent to a laboratory for testing. Conventional methods are actually reaching, the highest accuracy with low detection limits, expensive, time-consuming, and require the use of highly trained personnel.

The current tendency to carry out field monitoring has driven the development of biosensors as new analytical tools able to provide fast, reliable, and sensitive measurements with lower cost; many of them aimed at on-site analysis. In this sense, biosensors would not compete with official analytical methods, but they can be used both by regulatory authorities and by industry to provide enough information for routine testing and screening of samples.

Biosensors are being developed for environmental control and monitoring, can provide fast and specific data of contaminated sites. They offer other advantages over current analytical methods such as the possibility of portability and working on-site and the ability of measuring pollutants in complex matrices with minimal sample preparation. On the other hand, biosensors offer the possibility of determining not only specific chemicals, but also their biological effects, such as toxicity or endocrine-disrupting effects, which are sample information of great interest. Biosensors are currently being used for detection of environmental pollutants such as phenols, genotoxins, and pesticides such as organophosphates, 2,4-D etc. The majority of reported biosensor research has been directed toward development of devices for clinical markets; however, driven by a need for better methods for environmental surveillance, research into this technology is also expanding to encompass environmental applications.

The unique characteristics of biosensors will allow these devices to complement current field screening and monitoring methods such as immunoassay test kits and chemical sensors. *Eg:* enzyme based biosensors show the potential for continuous monitoring of compounds such as phenolics in process streams, effluents and groundwater. These devices can operate in high concentrations of organics such as methanol and acetonitrile and show promise for *in situ* monitoring of mixed organic wastes. Other potential applications include down hole or perimeter groundwater surveillance as well as process stream monitoring for remediation procedures.

15.1 ENZYME BASED BIOSENSORS FOR PESTICIDES

The extensive use of pesticides for agricultural purposes is the cause of their widespread presence in natural waters, their toxicity and persistence in the environment to set limit on the concentration of pesticides in different environmental waters. The quality of water for human consumption has set a limit for a single pesticide is 0.1 μg/l and for total pesticides is 0.5 μg/l. Conventional techniques (HPLC/MS and GC/MS) gives satisfactory analytical results for pesticide determination, new assays and sensors for cheaper and faster on-site analysis are being developed. The development of biosensors for pesticides is the subject of considerable interest, particularly in the areas of food and environment monitoring (Dzyadevych *et al.,* 2005).

Several enzymes such as cholinesterase enzymes (AChE, BChE) and urease have been used in the design of direct electrochemical biosensors for the detection of pesticides. Analytical devices, based on the inhibition of cholinesterase, have been widely used for the detection of OP compounds (Danet *et al.,* 2003, Crew *et al.,* 2004, Pavlov *et al.,* 2005) and carbamate pesticides (Kok and Hasirci, 2004; Zhang *et al.,* 2005).

Conductimetric AChE biosensor was used for the assessment of the toxicity of methyl parathion and its photodegradation products in water (Dzyadevych *et al.,* 2002). In this study, it was shown for the first time that the inhibition effect of methyl parathion on the AChE activity increases dramatically as soon as the photo-degradation begins. Such a sensor could be employed as part of an "early warning system" for detecting various cholinesterase inhibitors.

Collier *et al.,* (2002) incorporated AChE and BChE into the carbon-based ink used for printing electrodes. The printed electrodes were used for the determination of chlorfenvinphos and diazinon, organophosphate pesticides are known to be used for the control of insect pests in wool. The BChE-based biosensor was shown to be the most sensitive sensor toward a mixture of chlorfenvinphos and diazinon, with a limit of detection equal to 0.5 μg/l. Screen-printed electrodes (SPE) have also been chemically modified with 7,7,8,8-tetracyanoquinodimethane (TCNQ) and used for the mediated electrochemical detection of acetylcholinesterase activity (Del Carlo *et al.,* 2004). In this study, the AChE inhibition has been carried out on samples from different food commodities.

Recently, a novel enzyme capable of selective recognition of OP compounds was employed. In particular, a faster, simpler, and direct biosensing protocol suitable for rapid testing of OP substances can be accomplished utilizing the biocatalytic action of oraganophosphorus-hydrolase (OPH) (Simonian *et al.,* 2004). However, the detection limit obtained using OPH was not very low compared to what is generally obtained by inhibition of cholinesterase enzymes. The organo phosphorous hydrolase (OPH) is able to hydrolyze a number of OP pesticides such as paraoxon and parathion, and chemical warfare agents such as sarin and soman. Hydrolysis of these OP pesticides generates *p*-nitrophenol, which is an electro-active and chromophoric product. Thus, OPH could be combined with an optical transducer to measure the absorbance of *p*-nitrophenol or with an amperometric transducer to monitor the oxidation or reduction current of this product. The capabilities of various pesticides such as cyanide, diethyldithiocarbamates and hydrazines are to inhibit the tyrosinase enzyme. Similarly, diazinon and dichlorvos were detected at limits around 5 μM and 75 μM, respectively, using a Tyrosinase based oxygen sensor. Dithiocarbamate fungicides have been measured by their ability to inhibit the enzyme aldehyde dehydrogenase (AlDH). The detectability of the pesticide level is 1.5μg/l by using bienzymatic system based on the combination of AlDH and diaphorase.

A new strategy for the detection and discrimination of neurotoxins, the anti-acetylcholinesterase activities of mixtures of different neurotoxins, were investigated (Simonian *et al.,* 2001). It was possible to separate the effects of different inhibitors, using a combined recognition/discrimination strategy based on the joint action of acetylcholinesterase and organophosphorus-hydrolase enzymes. They demonstrated the feasibility of eliminating the organophosphate neurotoxins from the different multi-component inhibitor combination via sample pre-treatment with immobilized forms of OPH. Because of such manipulation, the inhibiting influence of non-OP neurotoxins on the AChE can be separated and their true concentration may be determined. In another study, engineered variants of *Drosophila melanogaster* acetylcholinesterase (AChE) were used as biological receptors in AChE-multisensors for simultaneous detection and discrimination of binary mixtures of cholinesterase-inhibiting insecticides (Bachmann *et al.,* 2000). In this method, the system was based on a combination of amperometric multi-electrode biosensors with chemometric data analysis of sensor outputs using artificial neural networks (ANN). The AChE mutants were selected on the basis of displaying an individual sensitivity pattern toward the target analytes (Bachmann *et al.,* 2000). In this work, it was demonstrated that the multisensor was capable of simultaneously detecting and discriminating paraoxon and carbofuran with errors of prediction of 0.4 and 0.5μg/l, respectively. Crew *et al.* (2004) have investigated the determination of OPs using genetically modified *Drosophila melanogaster* AChE. The screen-printed carbon electrodes modified with cobalt phthalocyanine were used as base transducers in the construction of amperometric pesticide biosensors.

Brazil de Oliveira *et al.* (2004) have reported the comparative investigation of biosensors based on different acetylcholinesterases. The enzymes used were genetically modified AChE obtained from *Drosophila melanogaster* and from commercial sources: bovine erythrocytes and human erythrocytes. The screen-printed biosensors designed were used for determination of methamidophos pesticides. The inhibition of acetylcholinesterase is measured by direct or indirect measurement of its activity. In the case of the direct method, the assay is based on the spectrophotometric or electrochemical measurement of thiocholine synthesize from the following reaction:

$$\text{acetylthiocholine} + H_2O \rightarrow \text{acetic acid} + \text{thiocholine} \qquad \textbf{... \textit{eq.}-15.1}$$

However, the measurement of thiocholine production does not accurately reflect the degree of AChE inhibition by pesticides, since the sample may also contain heavy metals. An interference was also observed from copper ions at mg/l levels (Danet *et al.,* 2003) while Ricci *et al.* (2003), have shown that 5μM Cu^{2+}, 1μM Ag^{+}, 0.5μM Hg^{2+} induced a 10% decrease of the amperometric signal of 10 μM thiocholine. Thus, it would seem necessary to use prior separation techniques in the case of multicomponent samples or develop more specific methods for their analysis. A variety of combined systems has been explored without much success to overcome the problems of analyzing complex mixtures. In another multi-enzyme approach, choline oxidase (ChO) was co-immobilized with AChE and coupled to amperometric sensors for the measurement of hydrogen peroxide produced by the choline oxidation (Kok *et al.,* 2002):

$$\text{Acetylcholine} + H_2O \rightarrow \text{Acetic acid} + \text{Choline} \quad \text{... eq.-15.2}$$

$$\text{Choline} + O_2 \rightarrow \text{Betaine} + 2H_2O_2 \quad \text{... eq.-15.3}$$

The performance of an oxygen electrode combined with a bi-enzymatic membrane (AChE and ChO) was tested for the detection of alicarb (AS), carbofuran (CF) and carbaryl (CL) as well as with two mixtures (AS + CF) and (AS + CL) (Kok and Hasirci, 2004). The results indicated that this bi-enzyme system could not differentiate between the different pesticides and the inhibition by total pesticides was not additive. AChE and ChO were also immobilized in a polyurethane polyethylene oxide (PU-PEO) layer which was cast on the top of the probe and covered with a dialysis membrane (Zhang *et al.*, 2001). In this study, the biosensing device was used in either batch or flow analysis mode for paraoxon detection allowing a detection limit down to 10 nM using 0.00025U of AChE. Andreescu *et al.* (2002) have proposed the use of the commercially available phenyl acetate as new substrate for AChE biosensor and tested it in a bienzymatic AChE/tyrosinase system applied to the detection of organophosphorus pesticides. Results were similar to those obtained with the mono-enzymatic AChE system using aminophenyl acetate as substrate. Recent attempts to demonstrate better systems for detection of pesticides have employed different enzymes or enzyme/sensor interfaces. Mazzei *et al.* (2004) have reported the use of a bio-electrochemical system for the determination of pesticides by alkaline phosphatase inhibition. They found a detection limit for malathion (2,4-dochlorophenoxyacetic acid) of 0.5 μg l^{-1}.

A novel optical detection of OP compounds based on reversible inhibition of OPH by copper complexed porphyrins (CuC1TPP) has been developed by Harmon (2004). The absorbance spectrum of the porphyrin enzyme complex is measured via planar waveguide evanescent wave absorbance spectroscopy. Addition of OP compounds displaces the porphyrin from the enzyme resulting in reduced absorbance intensity at 412 nm. Using this method, the OP compounds can be detected at ppt levels (White and Harmon, 2005). Simonian *et al.* (2005) described a novel strategy for the direct detection of OP neurotoxins. Instead of using the pH change associated with enzymatic hydrolysis of the OP substrate as an indicator of the presence of an OP compound, the method described is based on the change in fluorescence of a competitive inhibitor of the OPH enzyme when the inhibitor is displaced by the OP substrate. The change in fluorescence of the inhibitor is produced by the presence of gold nanoparticle attached to the enzyme. Finally, the inhibition of tyrosinase has been also investigated for the determination of carbaryl using amperometric biosensors (Kuusk and Rinken, 2004) with the detection limit for carbaryl concentration obtained with this amperometric biosensor being 0.2 mg l^{-1}. This method also utilised a simple approach of data management which can be automated and used for different biosensors.

Table 15.1 Enzyme inhibition-based biosensors for pesticides.

Inhibitors	*Enzymes*	*Immobilisation matrix*	*Techniques*	*Reference*
Carbaryl, carbofuran	AChE	Immobilization in CPG in presence of GA	Potentiometric	Lee *et al.* (2001)
Atrazine, diuron	Tyrosinase (PPO)	Entrapment in a poly 3,4-ethylene dioxythiphene (PEDT)	Amperometric	V'edrine *et al.* (2003)
Carbofuran	AChE	Lipid films and methylacrylate polymer	Electrochemical	Nikolelis *et al.* (2005)
Trichlorfon	BChE	Adsorption onto a stearylamine monolyer using the electrostatic force	Potentiometric ion sensitive fields effect transistors (ISFET)	Wan *et al.* (2000)
Pirimiphos-methyl-oxon	AChE	Cross-linking with GA	Amperometric	Crew *et al.* (2004)
Carbofuran, carbaryl	AChE	Immobilisation on silica gel by covalent binding	Amperometric	Suwansa-ard *et al.* (2005)
Paraoxon, carbaryl	AChE variants	Cross-linking using GA	Amperometric	Zhang *et al.* (2005)
Coumaphos, chloropyrifos-methyl	BChE + HRP	Cross-linking with GA in polytyramine matrix	Amperometric	Suprun *et al.* (2004)
Chloropyrifos-ethyl	AChE	Immobilization on TCNQ	Amperometric	Del Carlo *et al.* (2002)
Paraoxon	AChE + ChO	Retention with dialysis membrane	Amperometric	Zhang *et al.* (2001)

15.2 HEAVY METAL ION SENSING

Table 15.2 summarizes the characteristics of various biosensors for heavy metal ion sensing, produced by integrating immobilized enzymes with different kinds of transducers. Enzymatic methods are commonly used for metal ion determination, as these can be based on the use of a wide range of enzymes that are specifically inhibited by low concentrations of certain metal ions (Krawczynski vel Krawczyk *et al.,* 2000). For the inhibitive determination of trace mercury, a large number of enzymes has been used: horseradish peroxidase (Han *et al.,* 2001), urease (Kuswandi, 2003; Rodriguez *et al.*, 2004) glucose oxidase (Alexander and Rechnitz, 2000), alcohol oxidase (Pirvutoiu *et al.,* 2002) and glycerol 3-phosphate oxidase (Ciucu *et al.,* 2001), and invertase (Pirvutoiu *et al.,* 2001). Some studies have also focused on the analysis of different organic forms of mercury: phenyl mercury (Doong and Tsai, 2001) using urease, methyl mercury and phenyl mercury using invertase (Mohammadi *et al.,* 2005), and methyl mercury using peroxidase (Han *et al.,* 2001).

Cadmium ion could be monitored by enzymatic sensors since it was found that it induced inhibition of several enzymes such as urease (Lee and Russel, 2003) and butyrylcholinesterase (BChE) (Mourzina *et al.,* 2004). For copper determination, a cholinesterase sensor has been used (Evtugyn *et al.,* 2003). It has been reported that heavy metal ions induce a reversible cholinesterase inhibition.

For heavy metal screening, Rodriguez *et al.* (2004) studied different procedures of urease–glutamic dehydrogenase immobilization. The urease was either cross-linked with glutaraldehyde, fixed by nafion film, entrapped in alginate gel, or immobilized by adsorption. The designed biosensors were then tested for the detection of metal ions (Cu, Hg, Cd and Pb) in soil and water extracts. Immobilization of urease using a nafion film resulted in the best sensitivity for the study of metal inhibition (Rodriguez *et al.*, 2004). Recently, an array-based urease optical biosensor based on the sol–gel immobilization has been used for Hg(II), Cu(II) and Cd(II) determination in tap and river water (Tsai and Doong, 2005). It was reported that mercury had a higher affinity for the cysteine residues of urease and could be detected with a lower limit of 10 nM.

Table 15.2 Survey of enzyme inhibition-based biosensors for heavy metals.

Inhibitors	*Enzymes*	*Immobilisation matrix*	*Techniques*	*Reference*
Hg^{2+}, Cu^{2+}, Cd^{2+}	Urease	Entrapment in sol–gel matrix	Optical	Tsai and Doong (2005)
$Hg(NO3)_2$, $HgCl_2$, $Hg_2(NO_3)_2$, phenyl mercury	Urease	Entrapment in sol–gel film	Potentiometric	Doong and Tsai (2001)
Hg(II), Ag(I), Cu(II), Ni(II), Zn(II), Co(II), Pb(II)	Urease	Immobilization in ultrabindmembrane	Optical fibre biosensor	Kuswandi (2003)
$HgCl_2$, $Hg(NO_3)_2$, $HgCl_2$, methylmercury, phenyl mercury	Invertase	Cross-linkage with GA and deposition on laponit modified electrode	Amperometric	Mohammadi *et al.* (2005)
Chromium(VI)	GOx	Cross-linking with GA and covering with aniline membrane	Amperometric	Guang-Ming *et al.* (2004)
Hg^{2+}	GOx	Immobilized in a polyvinylpyridine (PVP) in presence of 2 -aminoethanethiol mediator	Amperometric	Alexander and Rechnitz (2000)
Cd^{2+}	Urease	Self-assembled monolayer on the gold-coated sensor surface	Optical (SPR)	Lee and Russel (2003)
$HgCl_2$, $Hg(NO_3)_2$, Hg_2Cl_2, Methyl mercury, phenyl mercury	Invertase	Cross-linkage with GA and deposition on laponit modified electrode	Amperometric	Mohammadi *et al.* (2005)

15.3 ENZYME ELECTRODE FOR PHENOLS DETECTION

Electrochemical sensors and biosensors are used for screening of hazardous wastes such as TNT and phenol compounds. Phenolic compounds are hazardous substances, difficult to measure under laboratory conditions and methods for screening these compounds in the field could be of considerable value. Tyrosinase enzyme electrode is sensitive and durable in measuring phenolics in aqueous and organic media. Several technical challenges remain to be solved before this biosensor can be developed as a field method. These challenges involve the development of more robust techniques for immobilization of the enzyme, the calibration of the enzyme electrode response for individual phenolic compounds expected to be found at hazardous waste sites and the characterization of these techniques for the identification of phenols in multianalyte mixtures. Research conducted for focus on characterization of this biosensor with respect to environmentally significant phenolics using relevant matrices (groundwater, soils and sludges). The biosensor system used for measurements of total phenols in groundwater and soil leachate samples.

15.4 OPTICAL DETECTION OF CHEMICALLY INDUCED DNA DAMAGE

Rapid and inexpensive indicator assays that can be used to assess the potential for exposure and that can be related to a biological target (DNA) could be of great benefit to the exposure assessment process. This involves the development and characterization of rapid, sensitive and inexpensive assays for detection of chemically induced damage to surrogate sequences of DNA caused by pollutants. A significant number of short term tests (STTs) for genotoxicity/mutagenicity have been developed to determine the extent of environmental hazards in polluted water and sediments. These bioassays have been used in well constructed batteries of cytotoxicity and genotoxicity tests to complement traditional approaches.

Short term tests (STTs) typically fall into one of several classes depending on their mechanism and end-point. These classes include: bacterial mutagenesis, cultured mammalian cell mutagenesis, chromosomal damage and DNA damage. These assays are expensive to run, require sophisticated technical expertise, and are not well suited to be adapted to screening applications.

The assays detect two types of DNA damage (1) The first type of DNA damage to be analyzed primarily involves single strand breaks in target DNA although this assay may also be sensitive to double strand breaks; adduct formation and base losses. The concept for these assays is as follows: under certain conditions (high temperature or alkaline pH), double stranded DNA will unwind into single strands. The temperature-dependent melting and annealing behavior of both genomic and plasmid DNA are sensitive to various types of ionizing radiation-UV radiation, enzymatic, and chemically induced damage. This melting profile analysis assay is simple, rapid, and sensitive. (2) The second assay approach will measure chemically induced damage caused by the binding or intercalation of genotoxins to the helical structure of DNA. The concept for this assay involves the competitive displacement of an intercalating or groove-binding probe by the genotoxic compounds of interest.

15.5 ORGANOPHOSPHORUS PESTICIDES (OP) DETECTION

The development and characterization of enzyme based bioanalytical and biosensor techniques which can be used to screen for organophosphate pesticides. At present, two enzyme systems include organophosphorus hydrolase and acetylcholinesterase is used for detection of organophosphate pesticides. Each enzyme shows certain advantages and limitations for potential field screening applications. Organophosphorus hydrolase (phosphotriesterase) catalyzes the hydrolysis of a wide spectrum of organophosphate pesticides. Microorganisms that synthesize this enzyme not only protect themselves from the toxic effects of these compounds but have also been used in bioremediation applications.

The development of enzyme techniques capable of rapid screening or *in situ* monitoring of OP pesticides would provide valuable information concerning the effect of field parameters such as pH, temperature, and ionic strength on the microorganism-linked hydrolysis of organophosphates. This assay is based on a substrate dependent change in pH at the local vicinity of the enzyme. The pH change is monitored using fluorescein isothiocyanate (FITC) which is covalently immobilized to the enzyme. This method employs the use of polymethylmethacrylate beads to which the FITC-labeled enzyme is adsorbed. Analytes are then measured using a micro-bead fluorescence analyzer. The dynamic concentration range for the assay extends from 25 μM to 400 μM for paraoxon with a detection limit of 8 μM. Organophosphorus (OP) pesticides measured using this technique includes: ethyl parathion, methyl parathion, dursban, fensulfothion, crotoxyphos, diazinon, mevinphos, dichlorvos, and coumaphos. One potential application for this assay that is being explored involves the process control monitoring of bioremediation.

The Organophosphorus Hydrolase (OPH) technique works by involving the detection of organophosphorus (OP) pesticides by means of measuring the inhibition of the enzyme acetylcholinesterase.

15.6 AMPEROMETRIC SENSORS

Bioassay methods are integrated in biosensors such as the Cell sense is an amperometric sensor that incorporate with the *E. coli* cells for rapid ecotoxicity analysis. It uses ferricyanine, a soluble electron mediator, to divert electrons from the respiratory system of the immobilized bacteria of a suitable carbon electrode. The resulting current is to measure the bacterial respiratory activity and the perturbation by pollutants can be detected as a change in the magnitude of the current. Cell sense has been applied to investigate the toxicity of 3,5-dichlorophenol and other phenols in wastewater, determination of nonionic surfactants and benzene sulfonate compounds, analysis of wastewater treatment works (WWTW) influent and effluent, and testing of toxicity wastewaters and sewage sludge. Moreover, Cell sense has been proposed as one of the newer rapid toxicity assessment methods within the direct toxicity assessment (DTA).

Most environmental biosensors have focused on bacterial systems while eukaryotic biosensors are rare; even rarer is the use of mammalian cells. The mammalian cell, which is more complex than bacteria, can give a more sensitive response when compared to bacteria while responding to the estrogenic effects of chemicals. A recombinant fluorescent Chinese Hamster Ovary cell line, utilizing a fluorescent reporter system, was used to monitor various toxicants, especially endocrine-disrupting compounds (EDCs), in diverse aqueous environments. EDCs have been also analyzed with a multichannel two-stage mini-bioreactor system using genetically engineered bioluminescent bacteria.

15.7 ENDOCRINE EFFECT BIOSENSORS

Many of the environmental contaminants, that produce adverse effects by interfering with endogenous hormone systems, called endocrine-disrupting compounds (EDCs). EDCs constitute a class of substances not defined by chemical nature, but by biological effect and thus, taking advantage of this feature, endocrine effect biosensors have been developed.

Steroid hormones induce different effects in mammalian cells after binding to specific intercellular receptors, which are ligand-dependent transcription factors. Many endocrine disruptors are also believed to bind to the Estrogen Receptor (ER) as agonists or antagonists. Thus, the binding ability of the chemicals toward the ER would be a crucial factor for testing their potential environmental toxicity. Based on ER, several biosensors have been developed which provide significant and useful information about estrogenic potency of the sample. The advantage of receptor assays is that they are quite simple to perform and allow the identification of all endocrine disruptors that act through the corresponding receptor. These biosensors constitute a similar approach as that described above for toxicity biosensors and not a specific substance is monitored. By employing the commonly utilized human estrogen receptor, the SPR biosensor has been applied in the determination of estrogens and xenoestrogens and in binding studies of target compounds. In addition to optical biosensors, electrochemical and piezoelectric biosensors have been developed based on estrogen receptors.

15.8 DETECTION OF HORMONES

Endogenous hormones of human or animal origin have been reaching the environment for thousands of years, even though to an increasing extent they are due to growing population and more intensive farming. Environmental soil and water samples are contaminated as result of the animal excreta and these residues have endocrine disrupting activity in aquatic fauna or even terrestrial. Concentrations as low as ng/l or even pg/l level (ng/l range) of hormones such as estradiol, estrone and ethynil estradiol have been found in water, presence of these hormones to the potential dangerous consequences in the aquatic environment. *Eg.* Estrone along with other organic pollutants such as atrazine and iso-proturon was determined by Rodriguez-Mozaz *et al.,* (2004) with an optical immunosensor in real water samples. The development of new biosensing strategies for the control of hormone residues is an effort to improve food quality controls and to protect public health against the adverse effect of these substances.

Polychlorinated biphenyls (PCBs) are ubiquitous environmental pollutants widely used as industrial chemicals, particularly as dielectric fluids in electrical transformers and capacitors. The high toxicity of some PCB congeners represents a risk for public health as these compounds are still present in the environment. Different biosensor configurations have been designed to determine PCBs in the environment and these include the DNA biosensor with chronopotentiometric detection and various immunosensors with fluorescence, SPR and electrochemical detection principles.

15.9 DETECTION OF DIOXINS

Apart from Polychlorinated biphenyls (PCBs), other polychlorinated compounds of environmental concern are the dioxins which are released as by-products in a number of chemical processes involving chlorine. Thus, processes such as the production of the pesticides, manufacture of PVC plastics, chlorine bleaching of pulp and paper and waste incineration generate dioxins and these are considered carcinogenic and are a potential threat to human health. Conventional dioxins analysis requires laborious multistep clean up procedures that increase the cost of each analysis. Numbers of significant immunoassays are used for measurement of dioxins. The SPR biosensor is developed by Shimomura *et al.* (2001) for the determination of the dioxin. Another biosensor developed for the detection of dioxin-like chemicals such as poly-halogenated dioxins, furans, and biphenyls based on a recombinant mouse hepatoma cell line.

15.10 DETECTION OF PHENOLS

Phenolic compounds that appear in the environment originate from the paper and pulp industry, production of drugs, dyes and antioxidants. Phenolic compounds especially chloro phenols are important environmental pollutants because of their high toxicity and possible accumulation in the environment. They are also considered precursors of the dioxins.

Parellada *et al.* (1998) developed an amperometric biosensor with tyrosinase (Polyphenol oxidase with a relatively wide selectivity for phenolic compounds) immobilized in a higrogel on a graphite electrode, for the determination of phenols in environmental samples. Chlorophenols have been also detected with a flow-injection chemiluminescence fibre optic biosensor, exploiting the ability of certain substituted phenols to enhance the chemiluminescence reaction of luminol, catalyzed by horseradish peroxidases. Lux-based biosensor was used to assess the toxicity of a paper mill sludge being some metals (Cd and Cu) and pentachlorophenol (PCP).

15.11 DETECTION OF SURFACTANTS

Detergent products use surfactants as the basic active component. The anionic surfactants are the most widely used, while the cationic surfactants represent only 5% of the total. An amperometric biosensor for detection of anionic surfactants was constructed with *Pseudomonas rathonis T* (bearing a plasmid for surfactant degradation) as a biological element. Oxygen consumption acted as an indicator of cell metabolism and thus of the surfactant content of the sample. The limit of detection achieved for sodium dodecyl sulfate (SDS) was within 0.25–0.75 mg/l. The sensitivity and selectivity of biosensors based on bacterial strain *Pseudomonas* their ability to degrade the anionic surfactants. The degradation of surfactants by the bacteria caused a decrease in dissolved oxygen and a change in the oxygen electrode current. The microbial biosensor enabled the detection of surfactants with high selectivity, sensitivity, and reproducibility. The lower limit reached for SDS was 0.25 μg/l.

Linear alkylbenzene sulfonates (LASs) are abundantly used in the formulation of household detergents. Despite the efficient removal of LASs in wastewater treatment plants, residues of LASs are still present in surface waters at the low μg/l range. Even though LASs are not severely toxic, they contribute to the permeation of other pollutants into aquatic animals. A combination of two whole cell biosensors was applied to river water samples for the determination of anionic surfactants. The first biosensor was based on the detection of the dissolved oxygen consumed in the degradation of LASs by immobilized LAS-degrading bacteria. The other sensor which used *T. cutaneum* (Yeast) did not respond to LASs. Values for LAS concentrations derived from the combined use of both biosensors correlated well with values determined using conventional methods.

Alkylphenol ethoxylates (APEs) belong to the group of nonionic surfactants, whose detection has gained more importance due to their endocrine disrupting properties. APEs are used on a variety of industrial applications, such as manufacturing of pulp and paper, metals, textiles, paints, resins, adhesives, latex, rubber and plastics. The APEs have been shown to be estrogenic both *in vivo* and *in vitro*. In wastewater treatment processes and in the environment, APEs degrade to alkylphenols (APs), which tend to be more toxic and show greater estrogenic activity. Rose *et al.* (2002) described the capillary based immunoassay (CIA) for APEs and APs, utilizing glucose dehydrogenase (DH) as label.

15.12 AROMATIC COMPOUNDS AND POLYCYCLIC AROMATIC HYDROCARBONS (PAHS)

Contamination of soils and surface and groundwater supplies with petroleum products is a serious environmental problem. Of particular concern for drinking water quality are water soluble aromatic components (*Eg.* benzene, toluene, ethylbenzene and xylenes) of petroleum products. Many of these contaminants are biodegradable they often persist in the environment. A green fluorescent protein based *Pseudomonas fluorescens* strain biosensor was constructed and characterized for its potential to measure benzene, toluene, ethyl-benzene and related compounds in aqueous solutions. Another microbial whole cell biosensor using *E. coli* with the promoter luciferase luxAB gene was developed for the determination of water dissolved linear alkanes by luminescence. The biosensor was used to detect the bioavailable concentration of alkanes in heating oil contaminated groundwater samples.

Polycyclic aromatic hydrocarbons (PAHs) are carcinogenic compounds generally formed during incomplete combustion or pyrolysis of organic matter containing carbon and hydrogen. They are very abundant, ubiquitous and recognized carcinogenic compounds. Amperometric biosensors for naphthalene found in contaminated soils were constructed using *Sphingomonas yanoikuyae* B1. For Benzo(a)Pyrene (BaP) and related adducts a fibre opticfluoro-immunosensor has also been developed in which high sensitivity is achieved by laser excitation and optical detection. The laser radiation reaches the sensor probe and excites the BaP bound to the fibre optic probe.

15.13 DETECTION OF ANTIBIOTICS

Medical substances have been released into the environment and the presence of antibiotics in those substances they promote antibiotic resistance in the environment. The increasing use of antibiotics for therapeutic purposes or as growth promoters in dairy cattle and as feed additives in fish farms or in livestock during the last five decades has caused a genetic selection of more harmful bacteria. The widespread of antibiotics raises significant food safety issues since antibiotic resistance can be transferred to humans on ingestion of affected meat and milk products. Therefore, most of the biosensors developed for the determination of biological or food samples. *Eg.* Commercial biosensor BIACORE 3000 was used to study the cross-reactivity between two sulfonamides (a group of antibiotics): sulfamethazine and furosemide. Sulfonamides cause allergic reactions, whereas their effect in the human inmunosystem is of high interest for their therapeutical application. Sulfamethazine has been also determined with an optical immunosensor by Akkoyun *et al.*, in animal urine. Hansen and Sorensen presented three different reporter gene systems from *V. fischeri*, *E. coli* and *Aequorea victoria* all combined with a tetracycline inducible promoter in the development of three corresponding whole-cell biosensors. They respond to low levels of tetracyclines by producing galactosidase, light or green fluorescent protein, respectively.

15.14 DETECTION OF TOXINS

Toxins are a very heterogeneous group capable of affecting different biochemical processes including membrane function, ion transport, transmitter release, DNA and protein synthesis. In many cases, specific details of the site and mode of action of a toxin at the molecular level are not known. A number of attempts have been made to detect toxins in environmental and clinical samples using receptor sensors. The induction of reporter genes by a promoter, which responds to a wide variety of toxic compounds, can be used to produce biosensors useful as first indicators of the presence of pollutants in the environment. When the cells are exposed to toxic substances, there can be either a reduction (negative signal) or activation (positive) in reporter-protein production. Toxin sensitive cells, those express a reporter gene, which is used as a biosensor for the nonspecific detection of toxics but not for a specific toxin.

Number of specific sensors are used for bacterial toxins and mycotoxins have been developed for food and environmental control. An integrated optical sensor has been developed for the analysis of aflatoxin B in corn. A light addressable potentiometric immunosensor used for the analysis of saxitoxin and ricin. An impedance based immunosensor has been prepared by using an ultrathin platinum film with an immobilized layer of antibodies against the staphylococcal enterotoxin B. Evanescent wave immunosensors have also been developed for detection of botulin. A rapid and sensitive immunosensor developed for the detection of the *Clostridium botulinum* toxin A. This fibre optic based biosensor utilizes the evanescent wave of a tapered optical fibre where antibodies antitoxin. The toxin detected by means of a rhodamine label within a minute at concentrations as low as 5 ng/ml. A portable fibre optic biosensor developed for quantification of the *staphylococcal* enterotoxin B in diverse media.

15.15 DETECTION OF MICROORGANISMS

Bacteria, viruses and other microorganisms are found widely in polluted, untreated and treated water which implies a worldwide public health problem. In the case of untreated water systems, fecal coliforms may enter rivers and streams through agricultural runoff and nonpoint source pollution and via domestic wastewater. Pathogenic compounds may reach humans by various routes such as the use of these waters for recreation or sports, for the irrigation of fruit and vegetables and as drinking water. Therefore, surface waters may play an important role in the transmission of pathogens. Suitable monitoring of the water supply for the presence of pathogens can assist in preventing disease from these sources.

Current methods for detection of pathogenic viruses, bacteria, protozoa and helminthes tend to be inaccurate, time consuming and expensive. As a result, indicator bacteria such as *Salmonella*, total fecal coliforms and total fecal streptococci are commonly used to determine the risk of fecal contamination and the possible presence of pathogens in water and wastewaters. Other emerging pathogenic microorganisms, *Campylobacter*, *Aeromonas* and *Yersinia genera* and also bacteriophages, have been proposed by several groups as indicators for the microbial quality of water since they are more suitable to signal viral contamination of water sources. On the other hand, monitoring of microorganisms is not only of environmental concern; from a military point of view, there are a number of pathogenic bacteria that can be considered possible biological warfare agents.

Conventional analytical methods for microorganisms are based on colony forming unit (CFU) count and require selective culture, biochemical and serological characterization. These methods are sensitive and selective, but time consuming days to weeks are needed to get a result. The increasing public concern over environment safety has led to a search for technologies capable of rapidly identifying contamination problems at source. Biosensors capable of detecting an organism quickly will be important in the environmental monitoring of pathogens in real time. The microbial content of a sample can be determined by monitoring microbial metabolism. The transducer can either detect consumption of oxygen or the appearance/disappearance of an electrochemically active metabolite. However, proliferation of nucleic acid and immuno based detection technologies has provided sensitive and specific detection systems for pathogenic bacteria and viruses. DNA detection may be more specific than immunologically based detection and sensitivity can improve to its combination with polymerase chain reaction (PCR) methods. Gene probes are already finding application in detection of disease causing microorganisms in water supplies, food or in plants, animal or human tissues.

Immunological detection is faster and more robust than DNA detection. Moreover, it has the ability to detect not only contaminating organisms but also their biotoxins. Koubaka *et al.* (2001) reported detection of *Salmonella enteriditis* and *Listeria monocytogenes* in real time using an SPR sensor based on antibodies immobilized on the gold sensor surface. *Salmonella* and *Listeria* were detected by the sensor at concentrations down to 106 cell/ml. *Salmonella typhimurium* detection in liquid samples by an immunosensor based on the acoustic wave principle was reported by Pathirana *et al.* (2000) and their detection limits of detection is 100 cells/ml in less than 2 minutes.

Piezoelectric biosensors are developed for the detection of several microbial contaminants. Piezoelectric biosensors were specific to *Aeromonas hydrophila*, were immobilized onto the surface of a streptavidin coated gold surface of a quartz crystal. This sensor was capable of detecting a PCR product amplified from a specific gene of *A. hydrophila* and of distinguishing between samples that contained the gene in samples. Ercole *et al.* (2002) developed a biosensor for the determination of *E.coli* in water samples by an immunochemical potentiometric alternating biosensor. The monitored change in the redox potential was due to the production of NH_3 by Urease enzyme. *E.coli* antibody conjugate linked with the *E.coli* cells present in the water. Hasebe *et al.* (1997) developed an amperometric Tyrosinase based biosensor for the detection of *E. coli* in wastewater. The detection was based on Tyrosinase catalyzed oxidation of polyphenolic compounds which are produced microbiologically from salicylic acid and the subsequent signal amplification and this sensor detection limit is 103–104 cells/ml. The commercialization of biosensor technology will provide consumers with real time biosensors capable of maintaining sensitivity better than 100 CFU/ml. Advances in antibody production and the recent emergence of phage displayed peptide biosensors offer increased possibilities for the rapid detection of pathogens in water.

15.16 DETECTION OF METALS

The determination of traces of heavy metals such as Cu, Cd, Hg, and Zn in the environment is very important because of their high toxicity, their increasing environmental levels (due to their use in industrial processes), and because metals can bioaccumulate in living organisms, especially in marine organisms. Metals are usually determined after digestion with strong acids.

Common analytical techniques used for determination of heavy metals are ion chromatography, inductively coupled plasma and polarography. Heavy metals can also be determined with ion selective electrodes. However, these methods are not able to distinguish between available (potentially hazardous) and nonavailable (potentially nonhazardous) fractions of metals to biological systems. One advantage of the whole cell sensors is their ability to react only to the available fraction of metal ions. The development of biosensors is relying on intact bacterial cells to monitor toxic metals both nonspecific and specific biosensors that utilize intact cells have been developed for this purpose. Nonspecific microbial biosensors are used to measure general toxicity. Heavy metals are to inhibit the activity of enzymes and application of this phenomenon is used for the determination of these hazardous toxic elements and the inhibition effect is related to its biological toxicity. Durrieu and Tran-Minh described a biosensor for the determination of heavy metals based on inhibition of *Chlorella vulgaris* microalgae. The microalgae cells were immobilized on removable membranes placed in front of the tip of an optical fibre.

The inhibition effect of mercury and other heavy metal ions on urea hydrolysis catalyzed by urease immobilized in a polyvinyl chloride (PVC) layer at the surface of a pH electrode in the form of potentiometric biosensor. Urease is the most frequently applied enzyme for inhibition determination of mercury, as it is relatively cheap and easily available. Because the method was not specific, it was applied for the determination of the total inhibition effect caused by heavy metal ions in water samples.

The application of another urease based optical sensor for monitoring low levels of mercury. Urease immobilized into Nafion film on the surface of an ion-sensitive field effect transistor (ISFET) was also used for the measurement of heavy metal. In this approach, enzyme immobilization into a negatively charged polymer seemed to cause an increase in the inhibition effect of metals due to cation accumulation in the polymeric matrix. A combination of specific additives and selective rewashing techniques to make the urease based biosensor sensitive to mercury ions.

Specific biosensors based on inducible promoters fused to reporter genes *Eg*. Bioluminescence proteins such as luciferase, are more sensitive than both chemical analysis methods and nonspecific toxicity biosensors. The light emitted can be detected by photometers, illuminometers and charge coupled devices. Recombinant luminescent bacterial sensors were used for the determination of the bioavailable fraction of cadmium, zinc, mercury and chromium in soil. Two bacterial recombinant heavy metal sensors were constructed based on two different receptor reporter systems: one was inducible by Zn^{2+}, Cd^{2+} and Hg^{+} and the other by Cr(VI) and Cr(III). The bacterial sensors used were not perfectly specific to one heavy metal, but responded to some non-target metals.

The Merlux gene fusion in *E.coli* was used to estimate bioavailable mercury in soil. The bioavailable fraction was defined here as being part of the water-leachable fraction. The Mer promoter was activated when Hg(II), present in the cytoplasm of the biosensor bacterium, binds to MerR, resulting in transcription of the lux genes and subsequent light emission. The luminescence based bacterial sensor strains *Pseudomonas fluorescens* OS8 (pTPT11) and *Pseudomonas fluorescens* OS8 have also been used for mercury and arsenite detection, respectively, in soil extracts.

Biosensors have been designed based on bioengineered proteins and this sensor monitors conformational changes caused by the binding of the metal ion to the engineered protein. Bontidean *et al.* (2000) used mercuric ion binding regulatory proteins as the biological part of the biosensor, MerR. The conformational change resulting from the binding of the metal ion to the protein caused a change in the capacitance, which was proportional to the concentration of the metal ions determined.

15.17 NITRITE (NO_2) BIOSENSOR

The increasing nitrate levels found in ground and surface waters are of concern because they can harm the water environment. In line with this, urban wastewater treatment regulations aim to reduce pollution, including nitrate pollution from sewage treatment works and industry. A biosensor containing immobilized denitrifying bacteria was used for the determination of NO^{3-} in tap water. Through the reduction of NO^{3-} in a reaction chamber, N_2O was formed and determined by a N_2O microelectrode, which was the sensing element of the biosensor. A micro scale biosensor for nitrate/nitrite determination was used for in-site monitoring in an activated sludge plant. The biosensor was based on the diffusion of nitrate/nitrite through a tip membrane into a dense mass of bacteria converting the ions into nitrous oxide with subsequent electrochemical detection.

Nitrite (NO_2) has a central position in the global nitrogen (N) cycle and is involved in many important biological N transformations. With an intermediate oxidation state, NO_2 acts as an electron donor in nitrification and serves as an electron acceptor in denitrification, dissimilative nitrite reduction to ammonium. In natural ecosystems NO_2 is of interest because of its toxicity for microorganisms and higher organisms. The NO_2 concentrations in natural bulk water bodies are usually very low and in freshwater systems, the average worldwide NO^{2-} concentration has been estimated to be about 1 μg of N liter^{-1}. However, there have been several reports of NO_2 accumulation can be related to imbalances in nitrite production and consumption rates for either aerobic sediment processes (nitrification) or anaerobic sediment processes (denitrification and dissimilatory nitrate reduction to ammonia). In marine systems NO_2 concentrations are normally negligible, but concentrations of 0.3 to 2.5 μM have been reported near the oxic-anoxic boundary of stratified marine water bodies.

In wastewater treatment N removal is traditionally performed by degradation of the organic compounds, followed by oxidation (nitrification) of the liberated NH_4 to NO_3 and subsequent reduction (denitrification) of the NO_3 produced to free nitrogen (N_2). During this type of oxic-anoxic cycling, transient nitrite accumulations have been found and were recognized to be a result of imbalanced nitrification and denitrification processes. Accumulation of nitrite is also presumed to promote release of the greenhouse gas nitrous oxide (N_2O) from wastewaters.

A specific NO_x biosensor for environmental analysis, including wastewater monitoring, has been described; this biosensor is based on bacterial reduction of nitrite and nitrate to N_2O, followed by electrochemical N_2O detection. The strain used in the NO_x biosensor has a truncated denitrifying pathway with termination at N_2O instead of N_2. Nitrite biosensor based on *Alcaligenes faecalis*, which may be used when acetylene is added to block its N_2O reductase.

The Nitrite biosensor consisted of two major compartments: (1) A Clark-type N_2O sensor (Unisense) with a tip diameter of about 0.275 mm was positioned behind a reaction chamber with an inner diameter of 0.3 mm (2) The bacterial chamber was drilled in a 0.35 mm thick polyester plate that was subsequently glued onto a 6 mm inside diameter glass tube. The two compartments were glued together so that the N_2O sensor partly protruded into the hole of the plate, creating a small, 0.15 to 0.3 mm deep reaction chamber in front of the sensor tip.

Bacterial biomass for the sensor was obtained by aerobic growth on a tryptic soy broth (TSB) agar plate. The culture was collected from the plate with a sterile glass rod and subsequently inoculated into the reaction chamber from the front of the sensor through a very thin plastic tube (diameter, 0.1 to 0.2 mm) made by heat drawing a plastic syringe (125 μl). To help retain the culture in the reaction chamber, a 1.3% agarose solution was placed behind the bacterial biomass. To separate the bacterial biomass from the outside environment, a 50 μm thick ion-permeable membrane was attached to the front plate by using double-adhesive tape. A large medium reservoir supplied the culture with all essential growth constituents except the electron acceptor. The standard growth medium inside the sensor and the pH was adjusted to 6.9. Tetracycline (20 mg liter^{-1}) was added to the growth medium when *Stenotrophomonas nitritireducens* was used. Acetylene saturated growth medium was used when the organism was *A. faecalis*. A sterilization procedure in which propylene oxide (sterilizing agent) used to minimize the frequency of contamination with other bacteria, as follows: (i) droplets of propylene oxide were added to the bacterial chamber and allowed to evaporate; (ii) pieces of ion-permeable membrane were bathed in a solution of 10% propylene oxide in water for 24 h before use; and (iii) the glue surface of the sensor tip was exposed to a 10% propylene oxide solution for 4 to 6 h while it was coated with a 5μm thick polyethylene film.

The detection of Nitrite by the biosensor is based on bacterial reduction of NO_2 to N_2O and subsequent detection of this N_2O by the builtin electrochemical N_2O sensor. Nitrites diffuses freely through the ion permeable membrane into the bacterial biomass of the reaction chamber, where it is denitrified to N_2O and subsequently reduced to N_2 at the cathode of the N_2O sensor. There is a linear correlation between the amount of Nitrite reduced and the magnitude of the sensor signal. The N_2O sensor is sensitive to O_2, but bacterial respiration in the outer toxic parts of the chamber prevents O_2 from reaching the cathode. Bacterial biomass in the deeper, anoxic parts of the chamber is responsible for the reduction of Nitrite to N_2O, and the denitrifying capacity of the biomass determines the linear range of the sensor. Three organisms *S. nitritireducens*, *Luteimonas mephitis*, and *Pseudoxanthomonas broegbernensis* are all characterized by denitrifying pathway deficient in both Nitrates and N_2O reductases. These three organisms were tested for use in Nitrites biosensor together with the denitrifying organism *A. faecalis*, which is deficient in a Nitrate reductase but has an active N_2O reductase.

15.18 BIOCHEMICAL OXYGEN DEMAND (BOD)

Biochemical oxygen demand (BOD or BOD5) is defined as the oxygen required to neutralize organic wastes over 5 days at 20°C, and is a parameter widely used to indicate the amount of biodegradable organic material in water. The conventional BOD test has certain benefits such as being a universal method of measuring most wastewater samples and furthermore, no expensive equipment is needed. It has, however, the limitation of being time consuming and it is not suitable for online process monitoring. Thus, it is necessary to develop an alternative method that could circumvent the weakness of the conventional BOD test.

Fast determination of BOD could be achieved by biosensor based methods. A common feature of these sensors is that they consist of microbial films that can bio-oxidize the organic substrate to be quantified, sandwiched between a porous cellulose membrane and a gas permeable membrane as the biological recognition element. The response is usually a change in concentration of dissolved oxygen or other phenomena such as light emission. Most BOD sensors rely on measuring the bacterial respiration rate in close proximity to a transducer, commonly an amperometric sensor for measuring dissolved oxygen developed by Clark in 1956. Most commercially available BOD sensors are flow type systems that can be more easily automated but generally require high maintenance to prevent fouling and clogging. Current BOD biosensor systems still present a series of limitations that restrict their industrial applications:

1. The lack of standardization and legislation in most countries.
2. Complicated maintenance requirement and insufficient resistance to various toxic compounds such as heavy metal ions, CN and phenol in the wastewater.
3. It is possible to eliminate the toxic effects of heavy metal ions by using a chelating agent that complex the ions, *Eg.* Ethylene diamine tetra-acetate (EDTA) and sodium diethyl dithiocarbamate (DIECA). Prevention of contamination by other microbes is also important for a reliable biofilm-type BOD sensor.

Many BOD biosensors have been developed for the determination of high BOD values in industrial wastewater and not adapted to the measurement of low BOD values. An optical fibre biosensor was developed for the evaluation of low BOD values in river waters. The immobilized *Pseudomonas putida* bacterium membrane was placed on the top of an optode, which was linked to a photodiode that detected fluorescence signal and response time was 15 min for chloride upto 1000 mg/l.

15.18.1 Continuous Monitoring of Biochemical Oxygen Demand

Karube *et al.* (1977) first reported on a BOD sensor that used immobilized microorganisms with an oxygen probe. Various BOD sensors developed based on the dissolved oxygen (DO) consumption by immobilized microorganisms, such as Yeast and *Serratia marcescens* using a DO electrode. Such strains are unable to oxidize the entire range of organic contaminants in samples, thus the DO consumption is not always directly proportional to the concentration of biodegradable organics.

The luminous bacterium *Photobacterium phosphoreum* has been found to exhibit a luminescent intensity in proportion to the amount of assimilable organic contaminants in wastewater, thereby enabling the BOD to be determined from the intensity difference measured using a photodiode. A fluorescence technique was developed for the rapid determination of the BOD, which is analyzed by reading the fluorescent intensities of microbial communities growing in wastewater excited by UV at 340 nm.

Microbial fuel cells (MFCs) have also been studied as a BOD sensor. Karube *et al.* (1977) developed a BOD sensor based on an MFC using the hydrogen produced by *Clostridium butyricum* immobilized on the electrode. MFCs with electron mediators have also been studied as BOD sensors, where the mediators are used to facilitate electron transfer from the microbial cells to the electrode. Yet these sensors have a poor long-term stability, as the mediators are generally toxic to microorganisms. MFC can be operated without mediators using an electrochemically active metal reducing bacterium, such as *Shewanella putrefaciens*, where the bacterial cells in the anode of the

MFC consume the substrate as fuel and transfer electrons directly to the electrode. Furthermore, due to its lactate oxidizing activity, *S. putrefaciens* can also be used as a model strain for testing a lactate sensor using an MFC.

15.18.2 Microbial Fuel Cell and its Operation

The fuel cells were constructed using transparent polyacrylic plastic and had anode and cathode compartments separated by a cation exchange membrane (Fig. 15.1). Graphite felt, 20 mm × 120 mm × 5 mm was used for the electrodes. The void volume of each compartment was 20 ml. Platinum wire (0.5 mm diameter) was used to connect the electrodes and the electronic port. The internal resistance between the electrodes and the platinum wire was less than 3 measured by a multi-tester. Throughout the study, the anode and cathode were connected through a voltameter and resistance (*R*load) of 10 to monitor the current under close circuit conditions. The current (*I*) was calculated as $I = V/R$ load, where *V* is the potential drop across *R* load. The wastewater was fed through the injection port of the anode compartment at a rate of 0.13–1.37 ml/min using a peristaltic pump equipped with Marprene II tubing (Watson-Marlow). Air-saturated tap water was fed into the cathode compartment as the oxidant at a feeding rate of 5 ml/min throughout the study, unless stated otherwise, using a peristaltic pump (505S, Watson- Marlow).

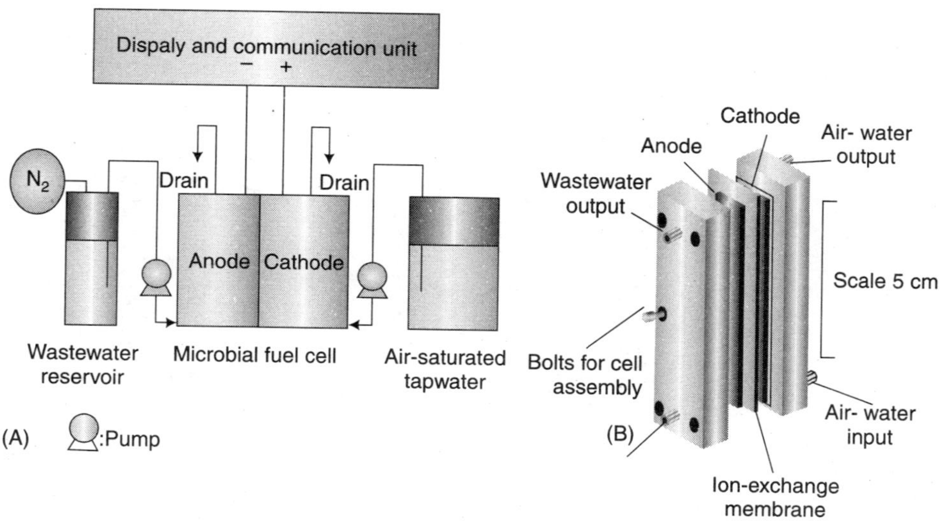

Fig. 15.1 *Schematic diagram of biosensor system (A) and microbial fuel cell (B). (Source: Byung Hong Kim et al., 2004)*

15.18.3 Wastewater

A glucose and glutamic acid check test solution for the BOD was modified and used as the model wastewater throughout the study. This modified artificial wastewater (AW) contained $(NH_4)_2SO_4$, 0.56 g; $MgSO_4{\cdot}7H_2O$, 0.20 g; $CaCl_2$, 15 mg; $FeCl_3{\cdot}6H_2O$, 1 mg; $MnSO_4{\cdot}H_2O$, 20 mg; $NaHCO_3$, 0.42 g; a trace mineral solution, 10 ml; phosphate buffer (1 M, pH 7.0), 50 ml; and distilled water, 940 ml. The wastewater was autoclaved at 110°C for 1 hour before use. The autoclaved wastewater was added along with filter-sterilized glucose and glutamic acid solutions. The BOD of the wastewater was varied from 20 to 200 mg/l. The artificial wastewater was made and maintained under a nitrogen atmosphere by connecting to a nitrogen-containing gas-tight bag with a volume of 15 lit. The MFCs were initially inoculated by activated sludge (10 ml) collected from the sewage treatment plant. The inoculated MFCs were then fed with the AW (10 ml, 100 mg/l BOD) in a batch mode based on activating the feed pump when the current dropped below 0.1 mA at a rate of 1.42 ml/min for 60 minutes. The current was monitored during the enrichment.

15.18.4 Analyses

The chemical oxygen demand (COD) was determined by the standard method using chromate as the oxidant. The dissolved oxygen concentration was measured using a DO meter. The current yield was calculated using the following equation: (observed current/theoretical current) × 100 (%), where the observed current was monitored and the theoretical current calculated based on the BOD consumption rate using following equation:

$$\text{Current (A)} = \text{BOD consumption rate } (\text{mg/s}) \times 12 \ (\text{C/mg BOD})$$

The MFCs were inoculated with activated sludge and enriched for 4 weeks using AW with a BOD of 100 mg/l as fuel. The COD value decreased gradually during the enrichment. After 4 weeks of enrichment, a stable current of over 5 mA was generated.

Table 15.3 BOD removal rate in MFCs fed with AW (BOD of 102.4 mg/l) at various feeding rates.

Feeding speed (ml/min)	*BOD loading rate (mg/h)*	*BOD removal rate (mg/h)*	*Yield*
0.13	0.79	0.74 ± 0.10	97.2 ± 14.0
0.27	1.65	1.17 ± 0.20	87.0 ± 14.5
0.35	2.15	1.28 ± 0.16	95.5 ± 11.6
0.71	4.36	1.81 ± 0.52	88.4 ± 25.4
1.20	7.37	2.27 ± 0.30	77.5 ± 10.6

When the MFCs were operated under various conditions, the current generated was recorded, along with a COD analysis of the effluent. In Stage I, the MFC was operated in a batch mode based on feeding 102.4 mg/l BOD AW at different feeding speeds for 1 h followed by an hour break. The oxidant was supplied continuously at a constant feeding rate of 5 ml/min. The current increased gradually as the flow rate increased from 3.7 mA at a feeding rate of 0.48 ml/min to 5.2 mA at a feeding rate of 1.07 ml/min. However, no significant increase in the current was observed when the feeding rate was raised from 1.07 to 1.37 ml/min. At this fuel feeding rate, the oxidant (air-saturated tap water) flow rate was then increased from 5 to 10, and 15 ml/min to test whether the oxidant was limited in the cathode compartment (Stage II). The maximum current was 5.3, 5.7, and 5.9mA at an oxidant flow rate of 5, 10, and 15 ml/min, respectively, indicating that the cathode reaction was limited by oxidant supply at a high fuel feeding rate, as reported earlier

(Gil *et al.*, 2003). Accordingly, the following experiments were conducted at a fuel feeding speed of less than 1.37 ml/min.

Oxygen supply limitation was also observed when comparing the current generations between batch and continuous runs of the MFCs. In addition, a slightly higher current was repeatedly obtained with a batch operation of the microbial fuel cell compared to a continuous run with a low cathode flow rate. As such, in a batch run, it was assumed that the DO diffused into the porous cathode space during the pause period was used during the feeding period, while in a continuous run; the DO was used before being diffused into the porous space.

15.18.5 Continuous Monitoring of Different Concentrations

A series of AWs with different concentrations ranging from 20 to 200 mg/l as the BOD was prepared and fed into the MFCs to monitor the steady-state response time under a fixed feeding rate of 0.35 ml/min. It took about 60 min to reach a new steady-state current after changing the AW concentration, yet this is too long for real-time monitoring. The long response time was believed to be due to the small initial contact area of the anode, thus altering the configuration of the fuel cell to increase the initial contact area and may reduce the response time.

The feeding sequence did not affect the current generation. The AW generated a similar current after either an up-shift or down-shift in the strength. The current generated from the MFCs fed with an AW with a BOD of 100 mg/l was compared to determine the repeatability, and found to be 3.73 ± 0.36, 3.92 ± 0.25 (MFC1) and 4.09 ± 0.17mA (MFC2). As such, the difference was less than 10%. The MFCs were run using AW with different BOD at a feeding rate of 0.35 ml/min, and then the current was plotted against the BOD concentration to find the reliability. Linearity was observed up to 100 mg/l. Some of the factors that affect current generation may be limited at BOD concentrations higher than 100 mg/l. Several factors affect the performance of an MFC, includes:

1. Fuel oxidation in the anode compartment
2. Electron transfer from the microbial cells to the anode
3. Proton permeability across the membrane
4. Oxygen supply and consumption in the cathode compartment

Consequently, these limitations should be avoided when running an MFC under limited conditions as a BOD sensor. An MFC can be operated at a low anode feeding rate to monitor high BOD values, or the sample can be diluted. Although the MFCs are monitored the BOD concentration as a linear relationship, model fitting (Monod, 1942) has also been applied to predict a BOD concentration higher than 100 mg/l.

15.19 BIOREMEDIATION

Bioremediation is an application of the microbial capacity to transform complex organic molecules into simpler inorganic constituents. Parameters such as nutrient availability, metal ions, pH, dissolved oxygen and temperature, influence the growth of bacteria. Biosensors that can monitor these parameters will help to better control the bioremediation process. Different molecular biosensors are used to monitor these parameters and the biological component in these molecular biosensors is a recombinant plasmid. It has a specific promoter, whose expression is sensitive to a target molecule.

15.20 BIOSENSORS IN INDUSTRIAL MONITORING

Chemical process monitoring has to rely on a conservative selection of well established techniques. It is preferable to have robust, reliable sensor systems especially for on-line control and most work has centered on deriving the required information from relatively indirect measurements, A combination of advances in sensor technology driven primarily by the demands of medicine and the identification of the need for accurate primary information about the concentration of substrates and products. A variety of biosensor configurations have been explored which may find applications ranging from fermentation monitoring and control of separation processes to shelf life prediction and quality assurance. A range of physical and chemical transducers are available to monitor the biological recognition event and can be broadly classified into electrochemical, optical, thermal and piezoelectric devices.

The biosensors have been successfully applied in industrial monitoring and control *Eg.* Number of enzyme electrodes used to product monitoring and these are developed for *in situ* fermentation monitoring. Continuous monitoring of single analytes facilitates novel process control strategies, but simultaneous measurement of a number of parameters offers exciting possibilities in obtaining a fingerprint representative of the more subtle product qualities such as taste and smell.

Blood glucose monitoring is possible to monitor the deterioration of meat by virtue of the glucose profiles that develop. The aerobic metabolism of contaminating bacteria on the surface of the meat set up concentration gradients that may be continuously measured using multiple enzyme electrodes. Mediated electrochemistry may also be used to monitor microbial activity directly. A cocktail of redox mediators is used to divert electron flow from the natural electron acceptor for the organism to an electrode. Pollution control is another important industrial consideration and biosensors are particularly suited to the detection of low concentrations of analytes in aqueous situations. For example, the possibility of detecting herbicide contamination of drinking water using a biosensor constructed using immobilised photosynthetic bacteria. Biosensor work has been carried out exclusively in aqueous solutions for devices suitable for the petrochemical industries also exist. It is possible to detect phenols directly in chloroform using an amperometric biosensor incorporating immobilised polyphenol oxidase. This type of approach has interesting implications both in terms extending the range of samples that may be amenable to analysis and also in improving the stability of biosensors for industrial use.

15.20.1 Industrial Bioprocess (Bioreactor) Control

Real time monitoring of carbon sources, dissolved gases in fermentation processes could lead to optimization of the procedure giving increased yields at decreased materials cost. While real time monitoring with feedback control involving automated systems does exist, currently only a few common variables are measured on-line (*Eg.* pH, temperature, CO_2, O_2) which are often only indirectly related with the process under control. Three different methods of controlling a bioreactor are: (1) Off-line distant - central laboratory coarse control with significant time lapse (2) Off-line local: fine control with short time lapse. (3) On-line: real-time monitoring and control. On-line control is most desirable, which allows the process to follow an ideal pre-programmed fermentation profile to give maximum output. However, many problems exist with on-line measurements including *in situ* sterilization, sensor life-time, sensor fouling, etc. Some of the problems can be overcome if the sensor is situated so that the sample is run to waste, but this causes a volume loss, which can be particularly critical with small volume fermentations.

The following benefits are achievable with process control technology:

1. Improved product quality and reduction in rejection rate.
2. Increased product yield, process tuned in real time to maintain optimum conditions throughout and not just for limited periods.
3. Increased tolerance in quality variation of some raw materials and these variations can be compensated in the process-control management.
4. Reduced reliance on human 'seventh sense' to control process.
5. Improved plant performance-processing rate and line speed automated.
6. Optimized energy efficiency.
7. Facilitated plant automation, cut analysis costs and improve quality control.

Molecular Electronics

Chapter 16

16.0 INTRODUCTION

Molecular electronics, first coined by Mark Ratner in 1974, involves the replacement of a wire, transistor or other basic solid state (silicon) electronic element with one or few molecules. Molecular electronic device exchange information or transfer states with components at the macroscopic level. A simple molecular electronic device consists of organic molecules which are sandwiched between conducting electrodes.

Living cells have capabilities to store and process the information, genes encoded digitally as nucleotide sequences serve as a kind of instruction manual for the chemical processes within the cell and constitute the hereditary information that is passed from parents to their offspring. Information storage and processing within the cell is more efficient by many orders of magnitude than electronic digital computation with respect to both information density and energy consumption. *Eg.* (1) Recombinant DNA technology in which the information processing capabilities of biological molecules are exploited for technological purposes. (2) Combinatorial chemistry in which proteins with a specific activity such as binding to a particular antigen, are synthesized by an iterative process which operates on a large pool of candidate proteins simultaneously *in vitro*. Selecting those proteins that are fittest with respect to the desired activity and then mutating them randomly to obtain a new pool of candidate proteins. Biomolecular computing by using these processes to perform information processing that has no intrinsic connection with the processing of biological molecules and these molecules are used as a medium for general purpose digital computation.

Biomolecular computation was first developed by Len Adleman (1994) and performed through a laboratory experiment in which collection of DNA molecules was constructed representing the possible solutions. Recombinant DNA techniques were used to sift through these molecules to select the correct solution. Study of charge and energy transfer in molecules was developed by Robert Mulliken and Albert Szent-Gyorgi (1940), which is called donor acceptor systems. Theoretical molecular rectifier developed by Mark Ratner and Ari Aviram in 1974. In 1988 Aviram described in detail a theoretical single-molecule field-effect transistor. The individual molecule electronic characteristics direct measurement methods are used for making molecular-scale electrical contacts. The first experiment measuring the conductance of a single molecule was developed by Mark Reed *et al.,* in 1997. In 1974, John Mc Ginness *et al.,* described first experimental demonstration of an operating molecular electronic device and this was a voltage-controlled switch. As its active element, this device used DOPA melanin, an oxidized mixed polymer of polyacetylene, polypyrrole and polyaniline. The ON state of this switch exhibited almost metallic conductivity.

16.1 MOLECULAR ELECTRONICS

Tailoring the graphite sheet to obtain well defined nanometer sized objects to fabricate small graphitic objects that were formed single molecules. After studying the interstellar conditions under which carbon is known to form clusters. Richard *et al.* (2004) developed a method in which graphite was vaporized using laser irradiation. Mass spectrometry revealed that clusters containing specific magic numbers of atoms were stable. In particular, the clusters of 60 atoms are covalently bound with the exact symmetry of a soccer ball. The clusters retained some properties of graphite, such as conductivity and these objects were rapidly envisioned as possible building blocks for molecular electronics. Molecular electronics or moletronics is an interdisciplinary theme that spans physics, chemistry and materials science. The feature of this area is the use of molecular building blocks for the fabrication of electronic components both passive (resistive wires) and active (transistors). The concept of molecular electronics has much excitement both in science fiction and among scientists due to the prospect of size reduction in electronics offered by molecular level control of properties. Molecular electronics provides a means to extend Moore's law beyond the foreseen limits of small scale conventional silicon integrated circuits. Molecular electronics can be split into two sub-disciplines: (i) Molecular materials for electronics utilize the properties of the molecules to affect the bulk properties of a material. (ii) Molecular scale electronics focuses on single molecule applications.

16.1.1 Theory

The theory of single molecule devices is an open quantum system in non-equilibrium driven by voltage. In the low bias voltage regime, the non-equilibrium nature of the molecular junction can be ignored and the current-voltage characteristics of the device can be calculated using the equilibrium electronic structure of the system. However, in stronger bias regimes a more sophisticated treatment is required, as there is no longer a variation principle. In the elastic tunneling case where the passing electron does not exchange energy with the system, the formalism of Rolf Landauer can be used to calculate the transmission through the system as a function of bias voltage. In inelastic tunneling, an elegant formalism based on the non-equilibrium, Meir-Wingreen formulation has been used to great success in the molecular electronics community to examine the more difficult and interesting cases where the transient electron exchanges energy with the molecular system. *Eg.* The electron-photon coupling or electronic excitations systems.

An important issue in molecular electronics is the determination of the resistance of a single molecule both theoretical and experimental. *Eg.* (a) Bumm *et al.,* used STM to analyze a single molecular switch in a self assembled monolayer to determine the condutivity of a molecule. (b) Another problem faced by this field is the difficulty to perform direct characterization since imaging at the molecular scale is often difficult in many experimental devices.

16.1.2 Molecular Memory Array Devices

16.1.2.1 Introduction

Molecular memory devices consist of a layer of molecules sandwiched between metal electrodes. Electric current flows from one electrode to the other through the molecule thick layer. A high voltage jerk flips the resistance of the molecules, making it easier or more difficult for the current to pass through. A low voltage can measure the resistance of the molecules and the two resistance levels can represent the 1s and 0s of computer information. The devices have the potential to store binary information in small numbers of molecules. The smaller electronics would be faster because the currents of electrons that carry signals don't have as far to travel.

A memory cell is provided with an active region including a molecular system and ionic complexes distributed in the molecular system. Pair of write electrodes is arranged for writing information to the memory cell. The active region is responsive to an electric field applied between the pair of write electrodes for switching between an ON state and an OFF state. The active region has a high impedance in the OFF state and low impedance in the ON state. A pair of read electrodes is used to detect whether the active region is in the ON state or in the OFF state to read the information from the memory cell. Read electrodes made of different materials having different work functions are used to reduce leakage current.

Molecular electronics has potential to produce ultra dense and high performance integrated electronic systems. To make the molecular memory, the used beams of electrons to form four parallel gold electrodes, 150 nanometers wide and spaced 200 nanometers apart. A row while 1,500 hydrogen atoms span 150 nanometers, which is about one seventh the diameter of an *E. coli* bacterium. The chemically coaxed single layer of molecules to course made on top of the electrodes and scratched the molecules off on one electrode by using an atomic force microscope. Then they put pre-made palladium nanowires across the electrodes. The 10 to 15 nanometer wide nanowires are ahead of time by coating carbon nanotubes with palladium. Nanotubes are rolled up sheets of carbon atoms and can be smaller than a single nanometer or the span of 10 hydrogen atoms. The nanowires formed a direct contact with the electrode where the monolayer was scratched off and formed junctions with the molecules at the intersections of the other three electrodes.

Forming the top contacts ahead of time makes it possible to gently apply top electrical contacts to a small assembly of molecule coated electrodes. The method eliminates long standing problems like thermal damage and poor yield related to defects in the monolayer. Top contacts are usually added using metal evaporation techniques that subject the device to high temperatures, there are two possibilities (1) Physical twist of the molecules, termed conformation transition and another. (2) Chemical changes a reduction oxidation or redox reaction. This is one of the challenges facing researchers working on nanodevices, electrical and mechanical components whose sizes range down to the atomic scale. Such devices promise to use molecules as super fast computer circuits, store fantastic amounts of information in a minuscule area and sense minute amounts of chemicals and biological materials.

Ultra-dense memory device stores the information by using reconfigurable molecular switches and this is an important toward the creation of molecular computers that are much smaller and could be more powerful than today's silicon based computers. A bit or binary digit is the basic unit of information storage and communication in digital computing. A kilobit is equal to 1,000 bits and is commonly used for measuring the amount of data that is transferred in one second between two telecommunication points. The memory is based on a series of perpendicular, crossing nanowires, similar to a tic-tac-toe board, with 400 bottom wires and another 400 crossing top wires. Sitting at each crossing of the tic-tac-toe structure and serving as the storage elements are approximately 300 bistable-rotaxane molecules. These molecules may be switched between two different states and each junction of a crossbar can be addressed individually by controlling the voltages applied to the appropriate top and bottom crossing wires, forming a bit at each nanowire crossing. The 160 kilobit molecular memory was fabricated at a density of 10^{11} bits per square centimeter.

16.1.2.1.1 Stoddart

Using molecular components for memory or computation or to replace other electronic components holds tremendous promise. This research is the best example of building large molecular memory in a chip at an extremely high density, testing it and working in an architecture that is practical, where it is obvious how information can be written and read. Stoddart's research team is widely considered the world's leader in making molecular switches. Stoddart's group designs and manufactures intricate interlocked molecules in which the relative motions of the interlocked components can be switched in controlled ways. Molecular electronics are using nanoscale molecules as key components in computers and other electronic devices. Molecular self-assembly in sophisticated ways, designing molecules that can be made by templation methods and that orient themselves such that they can be incorporated into solid-state devices in predictable ways. Molecular switches lead to other new technologies beyond molecular electronic computers and they benefit in healthcare, alternative energy and homeland security.

16.1.2.1.2 Zetta core

Zetta Core is the tip of the molecular electronics wedge that evolves into mainstream nano-electronics, displacing highly scaled semiconductors. Zetta core technology is developed by Werner Kuhr *et al.* (2000) used for molecular memory. Zetta Core uses molecules called porphyrins to store the 1s and 0s that comprise digital data. Adding or subtracting electrons changes the molecules' charge, which is used to represent data. The information stored in the less space, such memory could also retain information for minutes or even longer without electricity, which could dramatically reduce the power required by portable devices. Zetta Core's technology is a good technology to nano-circuit manufacturing techniques such as nano-imprinting or self-assembling nanowires for a pure nanomemory device that could scale below the limitations of lithography. It intends to use molecules that can store charge in next-generation memory chips. Thousands or more of the molecules would store electric charge representing information inside each cell of a memory array. The technology uses existing fabrication technologies and processes but accommodates smaller scale features.

16.2 MOLECULAR WIRES

16.2.1 Introduction

Linking single molecules together produces long molecules. Molecular wires tiny electron conductors with huge potential by Laura Mgrdichian. Brookhaven studied how electric charge is distributed in polymer molecule chains several nanometers, or billionths of a meter, in length. Long molecules that can act as molecular wires, of which there are many variations, are one type of nanoscale object with the potential to lead to new technologies, due to their ability to conduct electricity and very small size but unlike conventional metal wires, polymer nanowires need assistance in order to conduct. Using cluster of high energy electrons from an accelerator, can quickly add an extra negative or positive charge to a polymer molecular wire as suggested by Miller. When the end of the wire contains a chemically-attached trap molecule, one where the electrons will be at a lower, more stable energy, the charge moves to it. This allows to see that the wire conducts electrons quickly and over long distances. The Gust and Moore are constructed the dithiol gold circuits for the electronic processes that occur in photosynthesis. Lindsay described charge accumulation on the gold spheres; it is the possibility of both measuring and understanding the electronic properties of a molecule as wired into a metal circuit. Stuart Lindsay *et al.* (2010) have developed the measurement of the resistance of a single strand of a molecular wire soldered to two gold contacts. The molecule was 1,8-octanedithiol, which consists of a chain of eight carbon atoms with a sulfur atom at each end and it was these sulfur that provided the strong connections to the metal through chemical bonding.

Molecular wires or molecular nanowires are molecular scale objects which conduct electrical current. They are the fundamental building blocks for molecular electronic devices. Their typical diameters are less than three nanometers, while their bulk lengths may be macroscopic, extending to centimeters or more. Molecular computer was built with molecular wires and have been connected to its various components includes an insulator, a semiconductor, a conductor and a superconductor.

16.2.2 Tour Wires

James M.Tour's group has been synthesized molecules with aromatic, alkene and alkyne bridges, terminating in thiols at one or both ends. These are known as Tour Wires. A wire is defined as a two-terminal entity that possesses a reasonably linear I(V) curve prior to the breakdown limit (fig. 16.1).

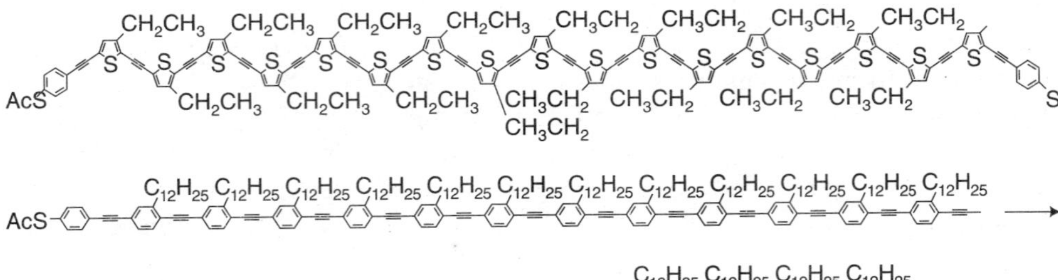

Fig. 16.1 *Oligo(phenylene ethynyfene)s.*

16.3 MOLECULAR DEVICES

Molecular devices are the systems having two or more terminals with current-voltage responses that would be expected to be nonlinear due to intermediate barriers or hetero functionalities in the molecular framework. Two terminal wire with tunnel barrier; wire with a quantum well: RTD; three terminal system: switch; four terminal system: logic gate. P(m, n) refers to the molecular electrostatic potential impedance of a system with *m* 1,4-phenylene moieties and *n* ethynylene moieties (fig. 16.2 & 16.3).

Most of molecular devices are consisted of organic molecules. Higher conductivities originate from highly conjugated systems, while alkane chains are important in understanding basic charge transport and tunneling. Prominent inorganic examples include polymeric materials such as $Li_2Mo_6Se_6$ and $Mo_6S_{9-x}I_x$, and single-molecule extended metal atom chains (EMACs) which comprise strings of late transition metal atoms directly bonded to each other. Molecular wires containing paramagnetic inorganic moieties are interesting, in particular, because they can lead to observations of Kondo peaks.

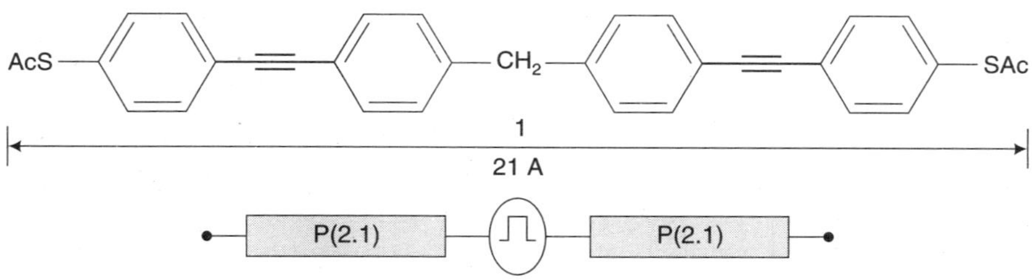

Fig. 16.2 *Molecular wire with a tunnel barrier.*

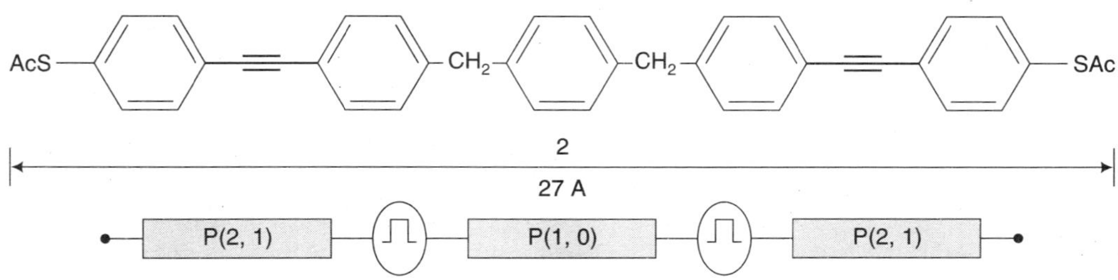

Fig. 16.3 *A wire with two barriers and analogous to a RTD.*

Molecular nanowires are composed of repeating molecular units, which may be organic (DNA) or inorganic (e.g. $Mo_6S_{9-x}I_x$). In the case of DNA, the repeat units are the nucleotides with a backbone made of sugars and phosphate groups joined by ester bonds. Attached to each sugar is one of four types of bases. In case of $Mo_6S_{9-x}I_x$, the repeat units are $Mo_6S_{9-x}I_x$ clusters, which are joined together by sulfur or iodine bridges. EMAC molecular wires consist of distinct molecules that do not aggregate and hence enable control of the exact length of the wire at the atomic scale.

16.4 CONDUCTION OF ELECTRONS

Molecular wires conduct electricity. They typically have non-linear current-voltage characteristics and do not behave as simple ohmic conductors. The conductance follows typical power law behavior as a function of temperature or electric field, whichever is the greater, arising from their strong one-dimensional character. Numerous theoretical methods have been used in an attempt to understand the conductivity of one-dimensional systems, where strong interactions between electrons lead to departures from normal metallic behavior. Effects caused by classical Coulomb repulsion (called Coulomb blockade) have also been found to be important in determining the properties of molecular wires.

16.5 USE OF NANOWIRES IN MOLECULAR ELECTRONICS

1. To be of used for connecting molecules together. The connectors between elements need to be able to self-assemble following well-defined routes and form reliable electrical contacts between them.
2. To obtain reproducible self-assemble complex circuit based on single molecules, it is essential that the connectors which join them have recognitive ability.
3. They should be able to connect to diverse materials, such as gold metal surfaces for connections to outside world, biomolecules for nanosensors, nano-electrodes, molecular switches and most importantly, they must allow branching.
4. The connectors should also be available of pre-determined diameter and length. They should also have covalent bonding to ensure reproducible transport and contact properties.
5. DNA like molecules has specific molecular-scale recognition and can be used very effectively in molecular scaffold fabrication. Very complex shapes have recently been demonstrated but unfortunately metal coated DNA which is electrically conducting is too thick to connect to the individual molecules.
6. Thinner coated DNA lacks electronic connectivity and are not suitable for connecting molecular electronics elements. Some varieties of carbon nanotubes (CNTs) are conducting and connectivity at their ends can be achieved by attachment of connecting groups.
7. For the construction of larger functional circuits using $Mo_6S_{9-x}I_x$ MWs have been demonstrated, either via gold nanoparticles (GNPs) as linkers or by direct connection to thiolated molecules. The two approaches may lead to different possible applications. The use of GNPs (gold nanoparticles) offers the possibility of branching and construction of larger circuits.

16.6 MOLECULAR SWITCH

Molecular switch is a molecule that can be reversibly shifted between two or more stable states. The molecules may be shifted between the states in response to changes in e.g. pH, light, temperature, an electrical current, microenvironment or the presence of a ligand. In some cases, combinations of stimuli are utilized. The oldest forms of synthetic molecular switches are pH indicators, which display distinct colors as a function of pH. Currently synthetic molecular switches are of interest in the field of nanotechnology for application in molecular computers. Molecular switches are also important in biology because many biological functions are based on it, for instance allosteric regulation and they are also one of the simplest examples of molecular machines.

16.6.1 Types

16.6.1.1 *Photochromic molecular switches*

Widely used molecular switches were made with class of photochromic compounds which are able to switch between electronic configurations when irradiated by light of a specific wavelength. Each state has a specific absorption maximum which can then be read out by UV Visible spectrophotometer. Members of this class include azobenzenes, diarylethenes, dithienylethenes, fulgides, stilbenes, spiropyrans and phenoxynaphthacene quinones.

Chiroptical molecular switches are a specific subgroup with photochemcial switching taking place between an enantiometric pairs. In these compounds the readout is by circular dichroism rather than by ordinary spectroscopy. Hindered alkenes such as the one depicted below change their helicity as response to irradiation with right or left handed circularly polarized light. Chiroptical molecular switches that show directional motion are considered synthetic molecular motors.

16.6.1.2 *Host-guest molecular switches*

In host-guest chemistry, the bi-stable states of molecular switches differ in their affinity for guests. Many examples of such systems are based on crown ether chemistry. The first switchable host is described in 1978 by Desvergne & Bouas-Laurent who create crown ether via photochemical anthracene dimerization. In 1980 Yamashita *et al.,* constructed crown ether incorporated the anthracene units (an anthracenophane) and studied ion uptake vs photochemistry.

The anthracene unit was photo-antenna in favor of an azobenzene moiety developed by Shinkai in 1980 and such molecules used for an on-off switch. In this molecule light triggers a trans-cis isomerization of the azo group which results in ring expansion. Thus in the *trans* form the crown binds preferentially to ammonium, lithium and sodium ions while in the *cis* form the preference is for potassium and rubidium both larger ions in same alkali metal group. In the dark the reverse isomerization takes place.

16.6.1.3 *Mechanically interlocked molecular switches*

Some of the most advanced molecular switches are based on mechanically interlocked molecular architectures where the bi-stable states differ in the position of the macrocycle. In 1991 Stoddart devices a molecular shuttle based on a rotaxane on which a molecular bead is able to shuttle between two docking stations situated on a molecular thread. Stoddart predicts that when the stations are dissimilar with each of the stations addressed by a different external stimulus the shuttle becomes a molecular machine. In 1993 Stoddart is scooped by molecular chemistry, delivers a switchable molecule based not on a rotaxane but on a related catenane. This compound is based on two ring systems: one ring holds the photo swichable azobenzene ring and two paraquat docking stations and the other ring is a polyether with to arene rings with binding affinity for the paraquat units. In this system NME Spectroscopy shows that in the azo trans-form the polyether ring is free to rotate around its partner ring but then when a light trigger activates the cis-azo form this rotation mode is stopped.

Stoddart in 1994 modifies his molecular shuttle such a way that an electron poor tetracationic cyclophane bead now has a choice between two docking stations: one biphenol and one benzidine unit. In solution at room temperature NME spectroscopy reveals that the bead shuttles at a rate comparable to the NMR time scale, reducing the temperature to 229K resolves the signals with 84% of the population favoring the benzidine station. However, on addition of trifluoroacetic acid, the benzidine nitrogen atoms are protonated and the bead is fixed permanently on the biphenol station. The same effect is obtained by electrochemical oxidation (forming the benzidine radical ion) and significantly both processes are reversible.

In 2007 molecular shuttles are utilized in an experimental DRAM circuit. The device consists of 400 bottom silicon nanowire electrodes (16 nanometer (nm) wide at 33 nm intervals) crossed by another 400 titanium top-nanowires with similar dimensions sandwiching a monolayer of a bistable rotaxane. Each bit in the device consists of silicon and a titanium crossbar with around 100 rotaxane molecules filling in the space between them at perpendicular angles. The hydrophilic diethylene glycol stopper on the left is specifically designed to anchor to silicon wire (made hydrophilic by phosphorus doping), while the hydrophobic tetraarylmethane stopper on the right does the same to the likewise hydrophobic titanium wire.

In the ground state of the switch, the paraquat ring is located around a tetrathiafulvalene unit but it moves to the dioxynapthyl unit when the fulvalene unit is oxidized by application of a current. When the fulvalene is reduced back a metastable high conductance '1' state is formed which relaxes back to the ground state with a chemical half-life of around one hour. The problem of defects is circumvented by adopting a defect-tolerant architecture. In this way, a circuit is obtained consisting of 160,000 bits on an area the size of a white blood cell translating into 10^{11} bits per square centimeter.

16.6.1.4 Catenane based molecular switch

A solid state bi-stable catenane based molecular switching device was fabricated from a single monolayer of the catenane, anchored with phospholipid counterions and sandwiched between an n-type polycrystalline silicon bottom electrode and a metallic top electrode. The device exhibits hysteretic current/voltage characteristics. The switch is opened at +2 volts, closed at 2 volts and read at ~0.1 volt and may be recycled many times under ambient conditions. A mechano-chemical mechanism for the action of the switch is presented and shown to be consistent with temperature dependent measurements of the device operation.

16.6.1.5 Optimization and chemical control of molecular wires and switches

The design rationale is described with some experimental and theoretical developments concerning rigidly fused porphyrin molecular wires. These materials consist of porphyrin units fused to acene-type bridges and have been synthesized in a range of topologies including linear porphyrin octamers of length ca. 120 Å. For some linear oligoporphyrins, the electronic coupling between the end porphyrin units can be modulated by simple (possibly *in situ*) chemical modulation of the bridging units. The chemical systems considered involve either pH controlled protonation of bridge azines or conversion of bridge quinone or quinone dioxime rings to or from benzenoid or hydroquinone rings. The electronic coupling through oligoporphyrin molecular wires is in simple model in which end to end electronic delocalization is required in order to provide strong long range interactions. By using an appropriate mixture of density functional (B3LYP) and ab initio SCF computational methods the inter orbital coupling can be monitored computanionally. Finally, the bridge modulation of the inter metallic coupling in three homo-valent bis-metallic oligoporphyrin systems, two in which $Ru(CO)_2$ is complexed within terminal porphyrin rings and the other, in which $[Cu(I)Cl_2]^-$ is tethered using phenanthroline end groups.

16.7 MOLECULAR ELECTRONIC DEVICES, SYSTEMS AND TOOLS

Devices and interconnects can be made from a variety of molecules and materials including bistable and switchable organic molecules, carbon nanotubes and single crystal semiconductor nanowires. They can be self assembled into organized structures and attached onto lithographic substrates. Molecular scale devices actually operating include: FETs, junction transistors, diodes and molecular and mechanical switches. Logic gates with voltage gain have been built and many techniques have been demonstrated to assemble nanometer wide wires into large arrays. Programmable and non-volatile devices which hold their state in a few molecules or in square nanometers of material have been demonstrated. Building systems of this size requires a completely new set of approaches to manufacturing and fabrication which will have a significant impact on circuits and architectures. These approaches are characterized by bottom-up molecular-scale fabrication which avoids the needs for lithographic patterning in order to create the nanometer scale dimensions for these devices and assemblies. Cheap giga gate, gigabit, giga Hertz non-volatile FPGAs and RAMs, built with molecular-scale electronic arrays on microelectronic silicon substrates developed for commercial molecular electronics.

16.8 CARBON NANOTUBES

In 1991 Sumio Iijima discovered a tubular variant of the C^{60} bucky ball carbon molecule. A carbon nanotube (NT) is a molecular sheet a single atom thick which has wrapped around into a tube as little as one nanometer wide and up to millimeters long. Since it's a single molecule, nanotubes are extremely strong and flexible. Depending on their lattice geometry, NTs behave as metals or semiconductors, acting as good electrical conductors. Carbon nanotube technology is very new. NTs may become more predictable and manageable in time.

16.9 SINGLE CRYSTAL NANOWIRES

Silicon is the best understood and most manageable of all solids. Single crystal silicon nanowires (NWs) are fabricated in bulk by laser assisted catalytic growth with diameters of 6 to 20 nm and lengths ranging from 1 to 30 microns. Germanium, gold, gallium phosphide, gallium nitride and indium phosphide NWs have also been made. Evaluating nanoscale wires is difficult because the wire/contact/device system is highly intertwined. The wire behaves like a 1-D quantum wire i.e. it transports electrons ballistically and has a length-independent resistance in multiples of $h = e^2$. A perfectly conducting nanotube has a minimal resistance of $h = 4e^2$ or a resistance of approximately 6 kW and the capacitance continues to scale with length.

16.10 TRANSISTORS, DIODES AND NDRS

Nanowires and nanotubes are used as active devices. Metallic wires are grown with molecules which act as two terminal devices embedded in them. Controlled doping of silicon and gallium nitride NWs with phosphorus and boron has been producing p-type and n-type semiconducting nanowires. Carbon nanotubes are also act as diodes or transistors. Two semiconductor NWs, one p-type and one n-type, form a junction diode at their crossing. Three NWs with two crossings form a bipolar junction transistor. Turn-on voltage is 1 V and logic gates are made with crossed nanowire diodes (CNW diodes). CNW diodes are made with 5V turn on by increasing the oxide thickness at the junctions. This oxide can be grown by passing high current through a low turn-on diode in air, oxidizing the junction by joule heating. This device can be used as a one time programmable cross point for ROM or logic arrays or it can be used as an FET.

16.10.1 Resonant Tunneling Diode

Resonant Tunneling Diode (RTD) allows voltage bias to switch on and off the current. Current passes equally well in both directions. Aliphatic groups with high potential energy (PE) establish aromatic ring between them as narrow 'island' of lower PE through which electrons must pass to traverse the entire length of the wire. Smaller the region in which the electrons are confined, farther apart are the allowed quantized energy levels, *Eg.* island and regions to left and right of barrier. The electrons are injected under bias into lowest unoccupied molecular orbital (LUMO) on left hand side (LHS). If the kinetic energy is not enough, no tunneling occurs; switched 'off'. If the bias is high enough, incoming electrons energy resonate with energy levels inside the well and tunneling occurs, etc., switched on.

16.10.2 Negative Differential Resistance (NDR)

A negative differential resistance (NDR) is characterized by a discontinuity in the monotonic increase of current as the voltage is increased. Several of these devices can be combined to give I-V curves with multiple peaks–this behavior has been proposed to lead to multi-state memory and logic devices. Reed and Tour *et al.* (1997) reported the clearest example of molecule-based NDR to date. At 60 K, the assembly was found to display a very strong NDR with a peak-to-valley ratio (PVR) of 1030:1.

In the singly reduced state, the LUMO becomes fully delocalized, allowing enhanced conduction, thus creating the onset of the NDR peak. As the bias voltage is increased the molecule becomes doubly reduced, the LUMO becomes localized across the molecule and decreases the conductivity of the molecule, reducing the current passed through the molecule.

16.11 FABRICATION

The difference between molecular electronics and traditional VLSI is in the methods of fabrication. Molecular electronics is based on bottom up manufacturing as opposed to the traditional top-down approaches used in manufacturing today's chips. Bottom up manufacturing is of necessity a hierarchical process. First the individual devices and wires are manufactured. In this area, there has been great success producing devices in quantity. Then individual devices must be assembled into systems for device and circuit experiments. To create successful molecular electronics systems must be able to assemble, *en masse*, the individual components into larger subunits. These subunits would then be connected together into complete systems.

16.11.1 Techniques

There are many different techniques for assembling and aligning nanoscale components, these include Langmuir-Blodgett films, flow-based alignment, nano-imprinting, electro-magnetic alignment, self-assembled monolayers and catalyzed growth. The common feature of all the assembly techniques, with the possible exception of nano-imprinting, they can only form regular structures. *Eg*. Masses of NTs and NWs have been made to self-assemble into regular arrays by fluidics. Irregular structures have been made using nano-imprinting, however, the process of creating the masters combined with contact printing may limit the smallest pitch to above 100 nm. NWs self assemble into parallel arrays guided by fluid flow. An ethanol fluid carrying NWs in suspension is passed over the substrate guided through a channel in a mold. Average separation can be controlled by varying the flow rate and duration. The NWs adhere well to the silicon substrate, permitting multiple layers to be assembled. Arrays of NWs crossing at right angles have been made this way.

Large regular meshes are not sufficient to create logical circuits. One proposed method of creating a periodic complex structure out of simple meshes is by programmably cutting wires. Crossbar wires can be cut by applying a high voltage which over oxidizes the cross point and breaks the NW, since it's small enough to be consumed by the chemical reaction. This is higher than the voltage used to open or close the cross point.

One potential alternative to the above approaches is to use complementary strands of DNA as smart glue. DNA directed synthesis has been used to connect both nanometer sized components and micron-sized components. In both cases the key is that selective binding of components can be controlled by varying the length and coding of a DNA sequence attached to the components being linked together. DNA based selective binding has also been used to algorithmically guide assembly and define differentiation points within an array at the nanometer scale.

16.11.1.1 Implications

Bottom up manufacturing, and in particular, the use of self assembly as the dominant means of circuit construction imposes the most severe limitation on nano-scale architectures: precise device alignment will be difficult or impossible, to achieve. Chemical self assembly, as a stochastic process, will produce precise alignment of structures only rarely and manipulation of single nano-scale structures to construct large scale circuits is impractical at the best. Furthermore, the defect rates will be high. Thus, bottom up molecular scale fabrication is universally expected to require defect tolerance due to the nature of chemical processes and alignment at that size. Consequently, molecular scale architectures demand a different paradigm for system design than lithographic scale architectures. The lithographic method can be relayed on patterning to specify exactly where each active component and interconnect was placed. In molecular scale, the crystalline structures can be building regularly, with small base feature size, but may not be able to deterministically make each lattice site or wire differ in a controlled manner at manufacturing time.

16.11.2 Structures

The available fabrication primitives imply three classes of architectures: fully deterministic, quasi-regular and completely random. The fully deterministic approaches require precision and control that may one day be available with nano-imprinting or DNA based assembly. Nano-imprint precision is likely to be expensive to build. The quasi-regular approaches assume the ability to assemble some regular structures (*Eg.*, two-dimensional meshes) using self assembly techniques. Finally, the least structured approaches demand only the ability to randomly place molecules in a given area. All of these approaches will require some kind of defect tolerance. The latter two approaches also require that the desired functionality of the circuit be created post fabrication. In other words, the fabricated circuits will be programmable logically *Eg.*, Field-programmable gate array (FPGA).

16.11.2.1 Self assembly and directed self assembly

Self assembly is to adsorb molecules on an electrode. Alligator clips are R-NC, R-S-S-R, R-COOH etc. Alkanethiolates are used as insulating host matrix; electronic properties of embedded molecules can be explored. Directed self assembly: grow rigid substrate molecules normal to the adsorbate by selective insertion into host at defect sites or at step edges (Fig. 16.4).

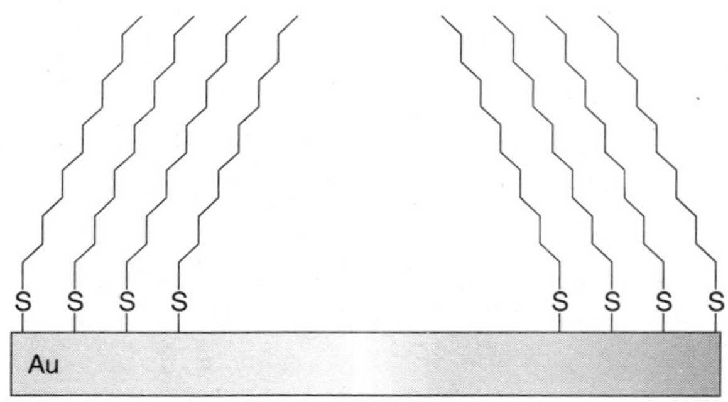

Fig. 16.4 Structure of directed self assembly.

16.11.2.2 Stochastic assembly

Random approaches can yield useful deterministic devices is a demultiplexor created from the random deposition of gold particles. By controlling the chemistry, it is possible to grow a randomly distributed collection of gold particles between the input address wires and the molecular scale nanowires or nanotubes. The set of connections to each nanowire acts as a code to select that nanowire. If the code space and statistics of the random connections are appropriately arranged, all the core nanowires with unique address code for each can be arranged. For example, with 50% of the potential connections randomly connected, a code space with 4log2(N) address bits will allow one to uniquely address 3.

A few lithographed input horizontal metal lines (bottom) connect too many vertical nanowires through randomly distributed gold particles. These nanowires form a programmable crossbar with many output horizontal nanowires. After testing, the crossbar is programmed to select the vertical nanowires which happen to form a binary demultiplexer. Placing parts with molecular scale accuracy is not required. The common feature of these two structures is that, the order necessary to create a useful logical device is discovered after the device is fabricated. Once the functionality is discovered, then the device is programmed to capitalize on what was created.

16.11.2.3 Quasi regular assembley

The quasi-regular classes of architectures are typified by the hierarchical assembly of meshes. Each mesh consists of wires and programmable molecular devices. The meshes are then connected together into a larger mesh of configurable elements. Unlike the random based architectures, the potential functionality of the device does not have to be discovered. However, these architectures do depend on post-fabrication programming in order to create logical circuits. Furthermore, since the fabrication primitives are unlikely to reliably yield perfect meshes, these architectures require re-programmable components in order to provide defect tolerance.

One of the difficulties in creating quasi-regular architectures is in addressing individual wires in the meshes from either the nanoscale or the micronscale. One method is to use the demultiplexer's as described above. Once there is a way to address individual nanoscale wires, these decoders can be assembled with a crossed nanowire memory core, allowing to uniquely addressing each cross point in the memory structure. This same basic structure allows to address and program cross point junctions which serve as wired-OR junctions in crossbars.

16.12 CAD CHALLENGES AND OPPORTUNITIES IN ASSEMBLEY FOR MOLECULAR ELECTRONICS

Molecular electronics, due to its size and assembly methods presents new design automation problems which merit new classes of algorithms and optimizations. While many of these challenges are unique to molecular electronics, many of them are extreme instances of problems found in deep-submicron Very Large Scale Integration (VLSI). Here are briefly presented some of the challenges.

16.12.1 Testing

Traditional way of testing is not possible with molecular electronic devices. Although, traditional fault testing and signature analysis will be important, the defect densities and sheer number of devices on each part will make this a qualitatively different problem. Techniques such as those used in the Teramac form a basis for localizing faults without having to probe each individual device. New techniques, such as described in, suggest that defect rates of up to 10% can be localized in linear time, but these techniques demand new analysis methods and novel test circuits.

Couple the need for fault localization with low bandwidth to the outside world, both through the nanolithographic interface and through the lithographic-I board interface. An external testing alone will become completely prohibitive and some testing will need to be integrated into the lithographic substrate. Furthermore, the test process will need to bootstrap the device, so that it can ultimately test and configure itself.

16.12.2 Novel Devices and Circuits

The sheer numbers of different combinations of even the smallest organic molecules promises to provide a larger variety of active devices have been traditionally available. Determining the best selection of devices and how they are incorporated into circuits is a new challenge. These devices have novel functionality but may lack traditional device functions. For example, many of the devices do not provide gain. In order to use these successfully, they must amalgamated with devices that do provide gain.

16.12.3 Optimizing for Spatial Locality

With chips which are 1000 or more nano-arrays wide running at high clock rates, it will likely take 1000s of cycles to cross between distant nano-arrays. This is the natural scaling of logic speed relative communications which is already forcing to take multiple clock cycles to reach far points on silicon chips. Optimizing the placement of logical functions, so that commonly communicating elements are close will be a first order concern for optimizing performance.

16.12.4 Scalable Place and Route

The sheer size of these systems makes placement and routing times a concern. Add to that the need to accommodate different defect locations for every component and the problem becomes even more acute. The subblock approach may be one piece of the solution hierarchically decomposing the problem into independent intra-subblock placement and routing and inter subblock placement and routing. The small distances and regular structures inside the subblock may make intra sub block placement and routing trivial. *Eg.* Using wire interchange ability within a crossbar. Inter-subblock placement may be eased by exploiting delay independent design for inter-subblock communications; this allows tools to move individual sub blocks. *Eg.* SAMs around without needing to replace and reroute the intra-subblock connections.

16.12.5 Communication Centric Design

As wires ever increasingly dominate design, communication centric design will become more important. This is, again, especially true if subblocks are going to be relocated to support defect and fault tolerance. Delay independent design will allow the computation to deal with varying delays that arise from placement and variable data arrival rates associated with filtering out corrupt data. Timing independence further opens the door to a wide range of communication schemes abstracted from the details of the end point computations and enables dynamic relocation of communicating subblocks to improve spatial locality. Key components for communication centric design appear in Berkeley's SCORE model. SCORE divides the computation into fixed size compute pages which are connected together by persistent data flow links or streams to form a computational graph. SCORE allows compute graphs to have arbitrary size and to evolve during the computation. Data on streams is tagged with presence, providing deterministic, timing independent behavior for the graph.

16.12.6 Design Size

The design time for the billion gate devices molecular electronics enables is also an important concern. If assuming 500 Resistor-Transistor Logic (RTL) gates per designer per day in 2000 and apply the current 21% per year productivity growth rate identified by Sematech, then 2010 productivity is 3300 gates per designer day. A billion gate design will take 1200 designer work years, a completely impractical effort for all but a very few. While a factor of 10 gains in designer productivity will be required for end of the roadmap complementary-symmetry metal–oxide–semiconductor (CMOS), a factor of 100 may be necessary to harness molecular hardware. Design productivity will be the gating resource for developing molecular hardware products.

Today applying standard platform designs are intellectual property (IP) cores and system level design to increase productivity and this may get a factor of ten eventually; another factor of ten will be needed. Possible solutions include parallelism, spatial computations and silicon compilation. Regular parallelism is a well known way to multiply area and design effort into performance. Billion gate designs can easily contain thousands of CPUs (64-bit, million gate CPUs). Applications and algorithms with heavy, high level parallelism will be increasingly favored. Software could be compiled directly into dedicated hardware. An adder for every '+, a mux for every if, etc., for entire major applications.

Photonic Computer

Chapter 17

17.0 INTRODUCTION

The movement of electrons in and out of transistors to do the logic is the concern of today's computers. Optical or photonic computing is anticipated to use photons or light particles produced by using lasers in place of electrons. Compared to the electrons, photons travels faster (30 cm in one nanosecond) and have higher bandwidth. Most of the research work is focused on present computers by replacing their components with optical equivalents and resulting in an optical digital computer system processing binary data. Optical components could be integrated into traditional computers to produce an optical or electronic hybrid. However, optoelectronic devices lose 30% of their energy converting electrons into photons and back. This also slows down transmission of messages. All optical computers eliminate the need of Optical- Electrical-Optical (OEO) conversions.

Photonics is the transmission, control, manipulation and storage of data using photons. Today optical networks are much faster, more efficient and can go further. The light is transmitted through optical fibres using lasers and these optical fibres can carry the equivalent of 300,000,000 simultaneous phone calls. Cities and countries across the world can now be linked together and have the capacity to transmit information to one another at high speed. Many different streams of information can be sent down, the same optical fibre by using different wavelengths of light, much the same as using different radio frequencies to broadcast multiple television and radio stations through the air. Photonic devices have the potential to increase the capacity of existing optical fibre cable networks by 500 times for a relatively small cost.

Photonic technology has many applications in the following areas:

Telephone: Underwater optical fibre cables can be used to link countries together allowing them to communicate via the telephone. These optical fibre cables have the capacity to transmit many thousands of simultaneous phone calls between many different countries.

Internet: Optical fibres can also be used to connect countries together so that they can share and transmit data across the internet. Optical fibres can also be used locally to provide high speed internet connections to homes and business.

Television: Optical fibre networks can be useful for transmitting cable television to homes. They can also be used to provide video and interactive television services, movies and television shows.

Medical Imaging: Images can be transmitted using bundles of a few thousand fibres polished on their ends and arranged very precisely side by side. Each point of the image is projected onto one end of a single fibre and is reproduced on the other end. The entire image can be viewed through the use of a magnifier. This type of image transmission is very useful for seeing inside the bodies of patients.

Local Area Networks (LANs): Optical fibre's can be used to connect many localized computers together for the purpose of sharing data, printers and other computer equipment. Optical network equipment is much more expensive than its copper wire but it has the advantages of higher bandwidth and can be stretched out longer before the signals need to be boosted.

Home Entertainment Equipment: Almost all moderately priced DVD players and audio receivers have optical in/out connectors, some personal audio players and other audio equipment also have these. These connections allow audio data to be transmitted between different audio equipment with virtually zero loss of quality. They are not affected by electrical interference.

Light Computer: One of the distant horizons is using light instead of electricity in computers. This has the potential to enable us to build processers that will blow the current ones out of the water and then some in terms of raw processing speed. Using current technology computer processers can be made a little bit faster before the laws of physics will turn against them. So many transistors can be put onto a computer chip before individual electrons no longer stay in place and just float off. Using light instead of electricity will enable us to keep building faster and faster computers.

17.1 OPTICAL FIBRE

17.1.1 Total Internal Reflection

When light traveling in a transparent material meets the surface of another transparent material two things happen (1) Some of the light is reflected and (2) Some of the light is transmitted into the second transparent material. The light which is transmitted usually changes direction when it enters the second material. This bending of light is called refraction and it depends upon the fact that light travels at one speed in one material and at a different speed in a different material. As a result each material has its own Refractive Index and can be used for the amount of bending which takes place.

Refractive index is defined as: $n = c/V$ ... *eq.*-**17.1**

Where n is the refractive index, c is the speed of light in a vacuum and V is the speed of light in the material.

Two possible cases exist. These are:

1. When light goes from a material with a low refractive index to one with a high refractive index (Fig. 17.1).

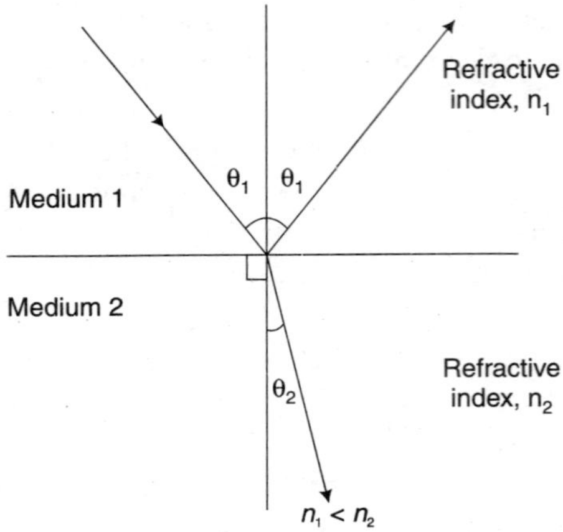

Fig. 17.1

2. When light goes from a material with a high refractive index to one with a low refractive index (Fig. 17.2).

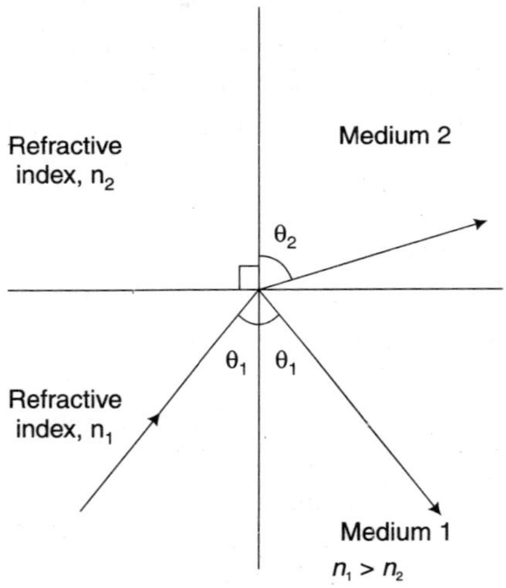

Fig. 17.2

In the second case above, θ_2 is always greater than θ_1 (θ_1 and θ_2 refer to the angles). So, as increase θ_1, eventually θ_2 will reach 90° before θ_1 does. At this point where θ_1 has reached a value called the critical angle (θ_c).

The transmitted ray now tries to travel in both materials simultaneously. This is physically impossible so there is no transmitted ray and all the light energy is reflected. This is true for any value of θ_1, the angle of incidence; equal to or greater than θ_c this phenomenon is called Total Internal Reflection (TIR) (Fig. 17.3).

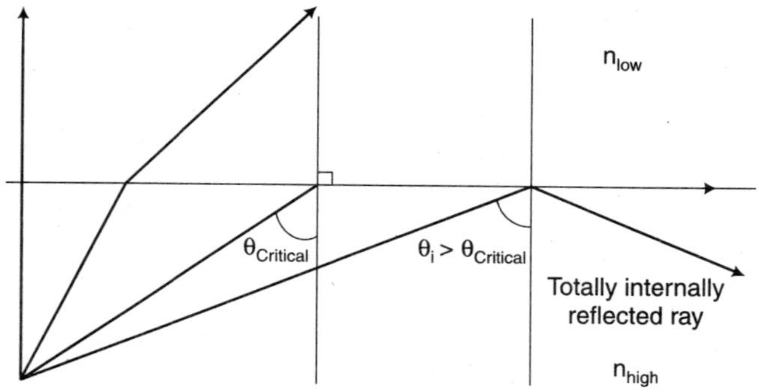

Fig. 17.3 *Total internal reflection.*

Two conditions are necessary for TIR: (1) The refractive index of the first medium is greater than the refractive index of the second one. (2) The angle of incidence, θ_1, is greater than or equal to the critical angle, θ_c. The phenomenon of TIR causes 100% reflection. In no situation in nature, where light is reflected, does 100% reflection occur. So TIR is unique and very useful and is used in optical fibre technology.

17.1.2 Structure of Optical Fibres

An optical fibre is basically made of two types of glass put together in a concentric arrangement so the middle is hollow.

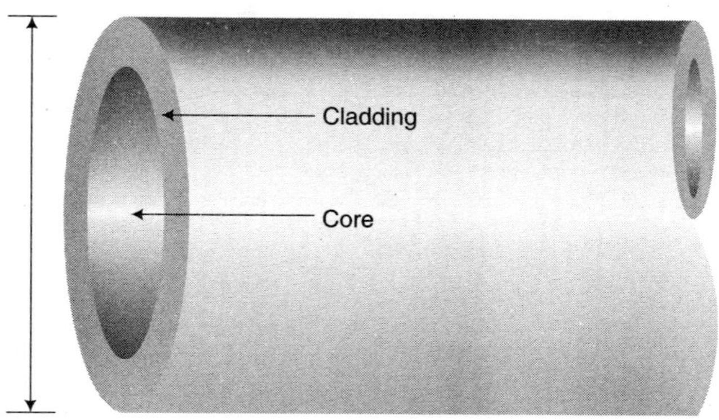

Fig. 17.4 *Structure of an optical fibre.*

The inner circle of glass consists of a glass of higher refractive glass than the outside circle. The outer circle of glass, which is also known as the optical cladding, does not carry light but is essential to maintain the critical angle of the inner glass. The main components of an optical fibre are (Fig. 17.4) (1) Core - Glass centre of the fibre where the light travels. (2) Cladding - Outer layer surrounding the core that reflects the light back into the core. (3) Outer Coating - Plastic cover that protects the fibre from the environment. Many of these optical fibres are put into bundles forming optical cables and these cables are then covered in a plastic jacket.

17.1.3 Propagation of Light in the Fibre

The angle θ in the figure is called the acceptance angle. Any light entering the fibre at less than this angle will meet the cladding at an angle greater than critical angle. If light meets the inner surface of the cladding (the core - cladding interface) at greater than or equal to critical angle then TIR occurs (Fig. 17.5). So all the energy in the ray of light is reflected back into the core and none escapes into the cladding. The ray then crosses to the other side of the core and, because the fibre is more or less straight, the ray will meet the cladding on the other side at an angle which again causes TIR. The ray is then reflected back across the core again and the same thing happens. In this way the light is zig-zags its way along the fibre. This means that the light will be transmitted to the end of the fibre.

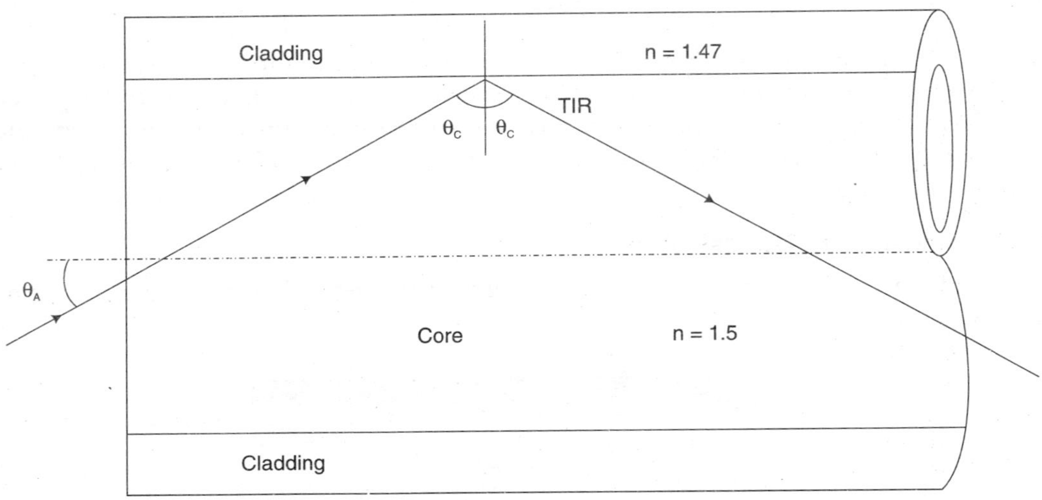

Fig. 17.5 *The propagation of light in optical fibres.*

In reality, the light which enters the fibre is a focused beam, consisting of many millions of rays behaving in a similar way. They are all zig-zag along the core of the fibre, crossing over each other, and filling up the core with light. A pulse of light traveling along the core of the fibre is really a bundle of these rays. Optical fibres act as a guide for light. Once light enters the fibre, it moves in the same direction as the fibre. As the fibre can be bent, light can thus be bent around corners by using optical fibres.

17.1.4 Types of Fibres

There are two main fibre types:

17.1.4.1 Step index fibre

Step index fibre is so called because the refractive index of the fibre "steps" up as we move from the cladding to the core of the fibre. Within the cladding the refractive index is constant, and within the core of the refractive index is constant (Fig. 17.6). This fibres are again classified into two types. They are: (1) Single mode step index fibres and (2) Multimode step index fibres.

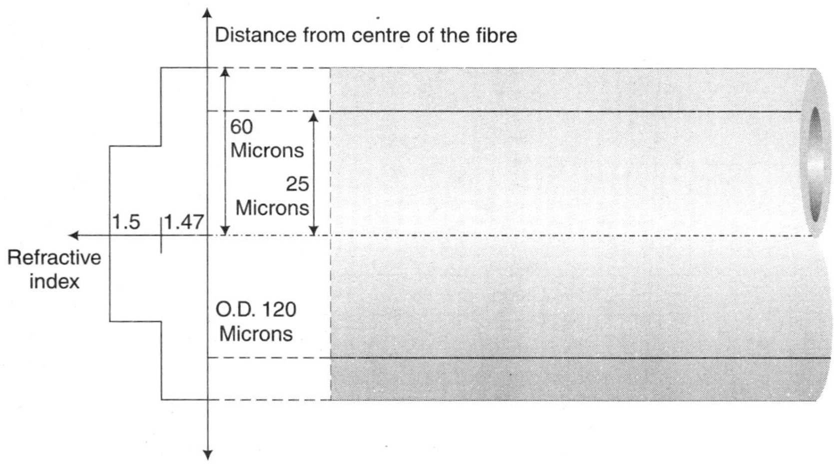

Fig. 17.6 *Step index fibre.*

(i) Single Mode Step Index Fibres

A single mode fibre can support only one mode because its core is narrow and it is also called lowest order mede (Fig. 17.7).

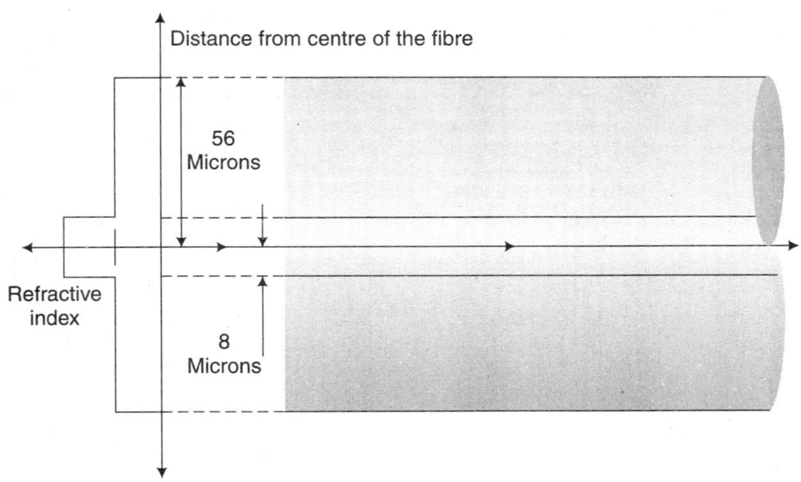

Fig. 17.7 *Single mode step index fibre.*

(ii) Multimode Step Index Fibres

Although it may seem from what we have said about Total Internal Reflection that any ray of light can travel down the fibre, because of the wave nature of light, only certain ray directions can actually travel down the fibre. These are called the Fibre Mode. In a multimode fibre many different modes are supported by the fibre (Fig. 17.8).

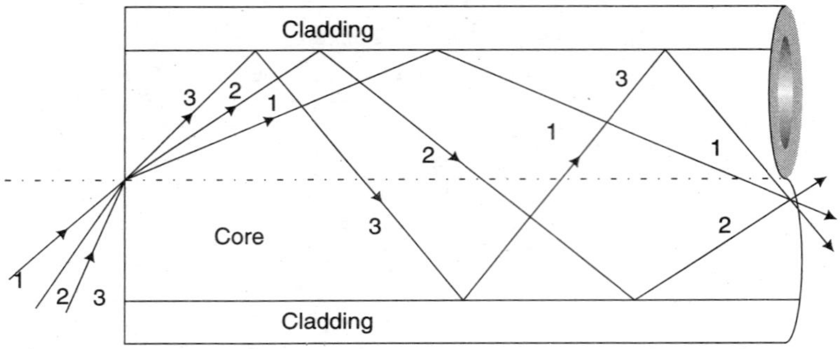

Fig. 17.8 *Multimode step index fibre.*

17.1.4.2 Graded index fibre

Graded Index Fibre has a different core structure from single mode and multimode step index fibres. Whereas in a step-index fibre the refractive index of the core is constant throughout the core, in a graded index fibre the value of the refractive index changes from the centre of the core onwards, infact it has a Quadratic Profile. This means that the refractive index of the core is proportional to the square of the distance from the centre of the fibre.

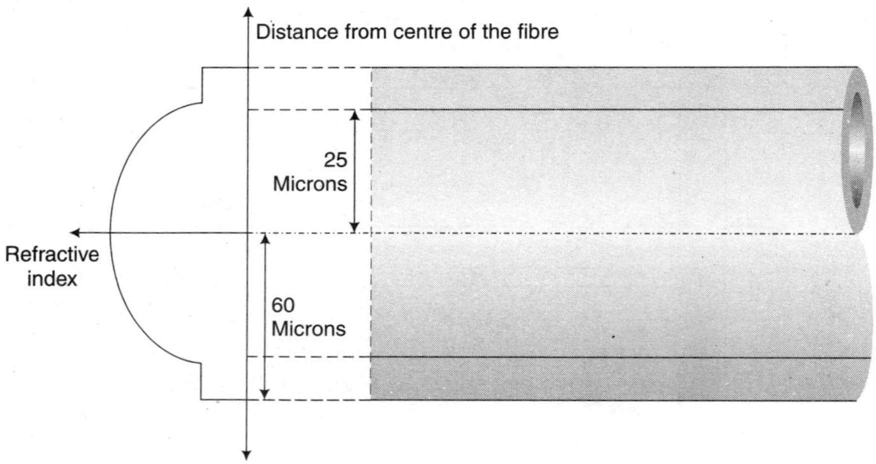

Fig. 17.9 *The graded index fibre.*

In graded index fibre, rays of light follow sinusoidal paths (Fig. 17.9). This means that low order modes, i.e. oblique rays, stay close to the centre of the fibre, high order modes spend more time near the edge of core. Low order modes travel in the high index part of the core and so travel slowly, whereas high order modes spend predominantly more time in the low index part of the core and so travel faster. This way, although the paths are different lengths, all the modes travel the length of the fibre in tandem, i.e., they all reaches the end of the fibre at the same time. This eliminates multimode dispersion and reduces pulse spreading. Graded Index fibre has the advantage that it can carry the same amount of energy as multimode fibre. The disadvantage is that this effect takes place at only one wavelength, so the light source has to be a laser diode which has a narrow line width.

17.1.5 Light Sources

There are two main light sources used in the field of fibre optics (1) Light Emitting Diodes (LED's) (2) Laser Diodes (LD's)

17.1.5.1 Light emitting diodes

An LED is a p-n junction diode (Fig. 17.10) in a transparent capsule usually with a lens to let the light escape and to focus it. LED's can be manufactured to operate at 850 nm, 1300 nm, or 1500 nm. These wavelengths are all in the infrared region. LED's have a typical response time of 8 nanoseconds, a line width of 40 nm, and an output power of tens of microwatts.

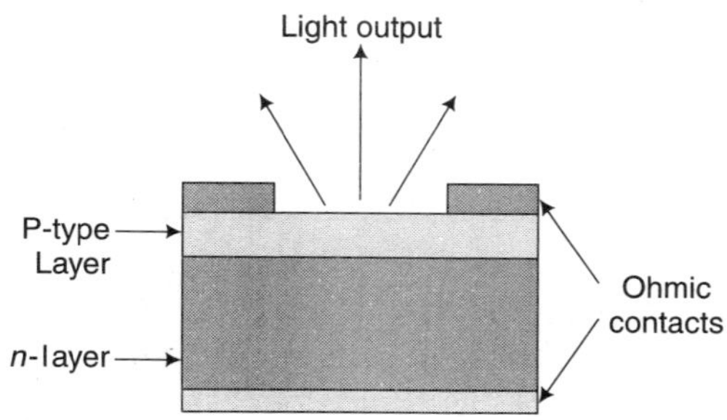

Fig. 17.10 *Planar light emitting diode (LED).*

17.1.5.2 Laser diodes

A laser diode is an LED with two important differences (Fig. 17.11) (1) The operating current is much higher in order to produce OPTICAL GAIN. (2) Two of the ends of the LD are cleaved parallel to each other. These ends act as perfectly aligned mirrors which reflect the light back and forth through the gain medium in order to get as much amplification as possible. The typical response time of a laser diode is 0.5 nS. The line width is around 2 nm with a typical laser power of 10's of milli watts. The wavelength of a laser diode can be 850 nm, 1300 nm, or 1500 nm.

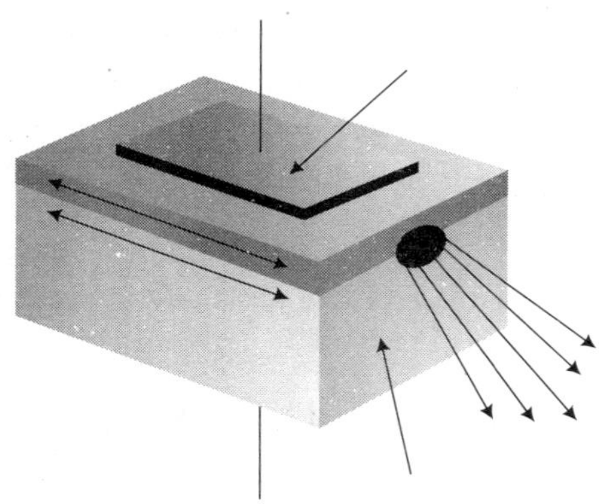

Fig. 17.11 *Laser diode.*

17.1.6 Advantages

1. Cheaper — Optical fibre can be made much cheaper than copper cable.
2. Higher Bandwidth — Because of how thin they are many more optical fibres can be bundled into a cable than copper wires.

3. Less degradation - Light signals travelling through an optical fibre takes longer to degrade than traditional copper cables.
4. Less interference - Unlike copper cables optical fibres do not suffer from electrical interference. Light traveling through one fibre does not affect light traveling in another fibre. This result in higher speed because the same signal only to be sent once and gives a clear TV pictures and phone calls.
5. Zero fire hazard - Since there is no electricity travelling through the fibres hence there is no risk of fire.
6. Low power - Since the signals travelling through the fibre degrade less, low power transmitters can be used instead of high voltage ones in copper networks.

Optical fibres are made of extremely pure silica glass. Optical fibres start life as a perform made by rotating a hollow glass tube on a lathe. The perform is made using a process called modified chemical vapor deposition (MCVD). As the tube is rotated, a flame moves back and forth along the outside while a gas vapor made of oxygen bubbled through solutions of silicon chloride ($SiCl_4$), germanium chloride ($GeCl_4$) and other chemicals are inducted into the tube.

17.2 LIGHT AMPLIFICATION BY STIMULATED EMISSION OF RADIATION (LASER)

Light Amplification by Stimulated Emission of Radiation (LASER) is a concentrated beam of light. The light emitted from a laser is basically have the same wavelength or monochromatic. Unlike light from other sources, such as that from a light bulb, laser light stays as a tight beam over large distances. The waves of light that come from a laser are all lined up perfectly and this is called coherent light, it is vital for many uses of lasers.

Lasers can be classified into two categories: (1) Continuous - this laser is uses gases to generate a continuous beam. (2) Pulsed lasers - this laser is one which uses crystal, glass or a semiconductor as the medium to create a pulsating beam. Depending on their power, lasers are grouped into four classes (1) Class 1 - this group includes lasers that cannot emit levels of radiation above the exposure limits of the eye and considered as safe. (2) Class 2 - emits a visible beam of light, lasers in this group are too bright to stare into for long periods, short viewing is not hazardous. (3) Class 3 - emit any wavelength of light, it is not a fire or skin hazard and this group includes any continuous laser that is not class 1 or 2 and has an output power of less than half a Watt or less. (4) Class 4 - this is any laser that has an output greater than half a Watt, these lasers are fire and skin hazard and have very strict control measures put around them.

17.2.1 Principle of Laser

In order to understand the basic laser operation, absorption and losses, stimulated emission, spontaneous emission and feedback processes are important processes to understand the principle of laser, here are described the basic principle involved in the generation of laser.

17.2.1.1 Absorption, spontaneous emission and stimulated emission

The atoms and molecules can exist only in certain energy states; the states of the lowest energy are called the ground state and the state of the more energy than the ground state and are called excited states. Excited state energy has a fixed amount of energy over and above that of the ground state. Under normal conditions, almost all atoms and molecules are in their ground states. Three types of processes are possible for a two level atomic system.

1. An incoming photon excites the atomic system from a lower energy state into a higher energy state, called absorption or stimulated absorption. In stimulated absorption, the atoms absorb the incident energy at certain frequencies only and it occurs when a photon strikes an atom with exact proper energy to induce an electronic transition between two energy states. *Eg.:* Broad band light is incident on a given two level atomic system, can observe that the complete spectrum is not absorbed but only certain discrete lines are absorbed depending on the difference in their energy levels. This process reduces the lower level population and in the process increases the upper level population. The number of atoms in states E_1 and E_2 at any time N_1 and N_2 respectively. When radiation passes through material, it is absorbed according to:

$$I_x = I_0 e^{-\alpha x} \quad \text{... eq.-17.2}$$

where I_x is the radiance after traveling distance × through the material with absorption coefficient as α and I_0 is the initial intensity of light. The absorption depends on the population difference between N_1 and N_2 and the refractive index of the medium.

Rate of stimulated absorption, R_{12} (abs), from level 1 to 2 is given as:

$$R_{12}\text{ (abs)} = B_{12} N_1 \rho \quad \text{... eq.-17.3}$$

where B_{12} is the Einstein's coefficient for stimulated absorption and has the units as cm^3/s^2J, N_1 is the population in the ground state and ρ is the energy density per unit frequency of the incoming photons.

2. Once the atom has been produced in its excited state, it will emit radiation again and return to their lower energy state, in the process, a photon is emitted. The atoms are spontaneously goes to their lower energy state through the emission of a photon is called spontaneous emission or fluorescence. The rate of spontaneous emission, R_{21} (spon), from level 2 to 1 is given as:

$$R_{21}\text{ (spon)} = A_{21} N_2 \quad \text{... eq.-17.4}$$

where A_{21} is the coefficient of spontaneous emission and has the unit of s^{-1}, N_2 is the number of atoms in level 2.

This spontaneous decay of the upper level takes place in the absence of an electromagnetic field. The rate is proportional to the population of that level and it does not depend on the intensity of the excitation source.

3. Excited atoms can lose their energy not only by spontaneous emission, but also by induced or stimulated emission and therefore the emission output of the system consists of spontaneous and stimulated emissions. The probability of stimulated emission is proportional to the intensity of the energy density of external radiation and the induced emission has a firm

phase relationship with it, unlike spontaneous emission. Since the spontaneous photons have no phase relations with each other, the output is incoherent. But stimulated emission has the same phase, direction, spectral and polarization properties as the stimulating field and both are indistinguishable in all aspects.

Rate of stimulated emission, R_{21} (stim), from level 2 to 1 is given as:

$$R_{21}\,(\text{stim}) = B_{21}\, N_2\, \rho \qquad \text{... eq.-17.5}$$

Where B_{21} is the Einstein's coefficient for stimulated emission and has the dimensions as m^3/s^2J, N_2 is the population in the excited state and ρ is the energy density per unit frequency of the triggering photons.

An ideal material is with only two non-degenerate energy levels, where absorption, spontaneous emission and stimulated emission take place.

Absorption = spontaneous emission + stimulated emission

$$\text{i.e. } B_{12}N_1\, r(n) = A_{21}N_2 + B_{21}N_2\, r \qquad \text{... eq.-17.6}$$

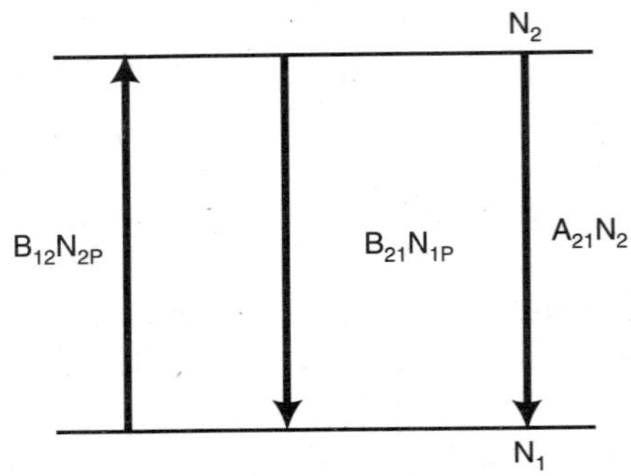

Fig. 17.12 *Spontaneous and stimulated processes in a two level system.*

At any given instance, under normal circumstances, both stimulated and spontaneous emissions may occur (Fig. 17.12), but the probability of stimulated emission is pretty low. One can find out this ratio of spontaneous to stimulated emission using one of the following equations:

$$\left(\frac{R_{spon}}{R_{stim}}\right) = \left(\frac{A_{21}}{\rho B_{21}}\right) = \exp\left(\frac{hv}{kT}\right) - 1 \qquad \text{... eq.-17.6(a)}$$

$$\frac{R_{21}(\text{spon})}{R_{21}(\text{stim})} = \frac{8\Pi\, hv^3}{c^3\rho} \qquad \text{... eq.-17.7}$$

where ρ is the radiation energy density and is equal to Nhn, N being the number of photons of frequency n per unit volume and k is Boltzmann's constant. Considering a case of ordinary bulb having a filament temperature of about 5000K and emitting radiation in the wavelength range of

0.6 micron corresponding to frequency of 5×10^{14} Hz, the probability of stimulated emission is approximately one hundredth of that of the spontaneous emission.

The ratio of the probability of spontaneous to stimulated light emission depends directly on the frequency of emission or inversely to the wavelength. Thus in the microwave region, stimulated emission is more probable than spontaneous, hence the early production of the maser. In the optical region, spontaneous emission is more likely than stimulated emission and this gets worse as we go into the UV and X ray regions of the spectrum.

Under thermal equilibrium, the population N_2 and N_1 of levels E_2 and E_1 respectively governed by the fact that the rate of upward transitions should be equal to rate of downward transitions. The population density of atoms N_1 and N_2 in ground level E_1 and excited state E_2 can be estimated using Boltzmann's relationship as follows:

$$\left(\frac{N_2}{N_1}\right) = \exp - \left(\frac{(E_2 - E_1)}{kT}\right) \qquad \text{... eq.-17.8}$$

$$N_{\text{Total}} = N_1 + N_2 \qquad \text{... eq.-17.9}$$

Since, $(E_2 - E_1)/kT$ is always positive, irrespective of the value of temperature T, N_2 must be less than N_1 if the system is remain at thermal equilibrium. At the most the excited state population N_2 (t) reaches a steady state at $t \rightarrow \infty$, and the highest proportion of atoms that can exist in the excited state $N_2/N_{total}<1/2$. Under these conditions the material always acts as an absorber of incident photons.

In a two level system the number of atoms in the excited state can never exceed the number in the ground state and hence can never work as a laser. If the system is to act as a laser, an incident photon must have a higher probability of causing stimulated emission than of being absorbed i.e. the rate of stimulated emission must exceed that of absorption. The laser action is possible only when $N_2 > N_1$, this non-equilibrium condition is known as population inversion.

17.2.2 Laser Operation

The light is incident on the material, there is competition between absorption, spontaneous emission and stimulated emission processes. Under normal equilibrium conditions, the population of various levels is given by Boltzmann's relationship and thus N_2 always be less than N_1. Stimulated photon emission is much less than the spontaneous photon emission and the absorption. For a system to work as a laser one requires that stimulated emission should exceed photon absorption; it leads us to the following two conditions:

$N_2 > N_1$: i.e. Population Inversion

(a) First condition cannot be achieved under thermal equilibrium conditions. This implies that in order to create population inversion, one must look for non-thermal equilibrium system and thus the need for special laser materials.

(b) The second condition that requires higher value of 'r' necessitates the use of an additional supply of large amount of energy of correct wavelength to excite the desired transition. The process is known as pumping. Various techniques include optical, electrical, chemical, gas dynamic etc.

Population inversion though is the primary condition, but in itself is not sufficient for producing a laser. As there are certain losses of the emitted photons within the material itself in addition to spontaneous emission, one has the geometry that can overcome these losses and there is overall gain. This requires an optical cavity or resonator.

The principle behind the laser, produce a large number of atoms all in excited states. If one of the atoms emitted spontaneously, then the emitted photon would stimulate other atoms to emit. These emitted photons would, in turn, stimulate further emission. The result would be an intense burst of coherent radiation.

A representative laser system is shown in Fig. 17.13. It consists of three basic parts and brief description of each of the components and their basic function are discussed below.

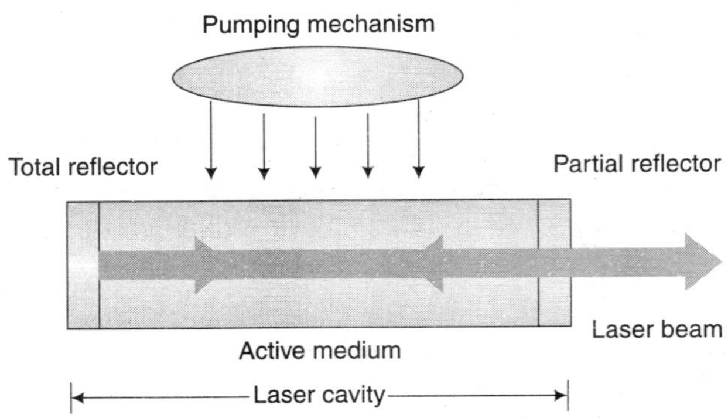

Fig. 17.13 *Basic laser system.*

17.2.3 Active Medium or Gain Medium

An active medium is with a suitable set of energy levels to support laser action. Laser medium is the heart of the laser system and is responsible for producing gain and subsequent generation of laser. It can be a crystal, solid, liquid, semiconductor or gas medium and can be pumped to a higher energy state. The material should be of controlled purity, size and shape and should have the suitable energy levels to support population inversion. it must have a metastable state to support stimulated emission. Most lasers are based on 3 or 4 level energy level systems, which depend on the lasing medium (Figs. 17.14 & 17.15). In case of a three level laser, the material is pumped from level 1 to level 3, which decays rapidly to level 2 through spontaneous emission. Level 2 is a metastable level and promotes stimulated emission from level 2 to level 1.

In a four level laser, the material is pumped to level 4, which is a fast decaying level, and the atoms decay rapidly to level 3, which is a meta stable level. The stimulated emission takes place from level 3 to level 2 from where the atoms decay back to level 1. Four level lasers is an improvement on a system based on three level systems. In this case, the laser transition takes place between the third and second excited states. Since lower laser level 2 is a fast decaying level which ensures that it rapidly gets empty and as such always supports the population inversion condition.

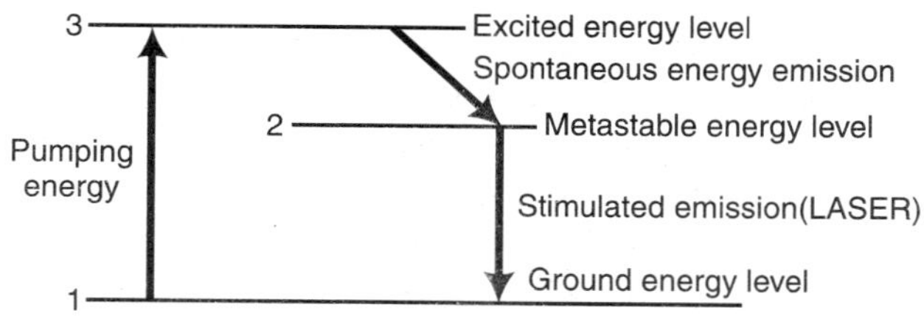

Fig. 17.14 *Energy states of three-levels active medium.*

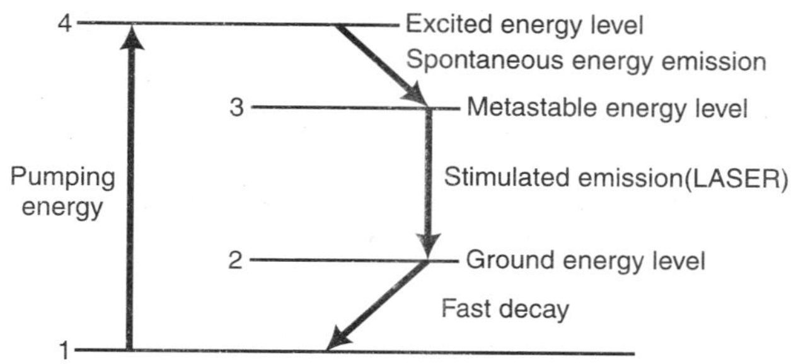

Fig. 17.15 *Energy states of four-levels active medium.*

17.2.4 Excitation or Pumping Mechanism

A source of pumping energy is in order to establish a population inversion. Absorption of the energy by the atoms, electrons, ions or molecules as the case may be, of the active medium is a primary requisite in the generation of laser. In order to excite these elements to higher energy levels, an excitation or pumping mechanism is necessary. It is well known that under the equilibrium state, as per Boltzman's conditions, higher energy levels are much less populated than the lower energy levels. One of the requirements of laser action is population inversion in the levels concerned. i.e. to have larger population in the upper levels than in the lower ones. Otherwise absorption will dominate at the cost of stimulated emission. There are various types of excitation or pumping mechanisms available, the most commonly used ones are optical, electrical, thermal or chemical techniques, which depend on the type of the laser gain medium employed. For example, solid state lasers usually employ optical pumping from high energy xenon flash lamps (*Eg*., ruby) or from a second pump laser or laser diode array. Gas lasers use an AC or DC electrical discharge through the gas medium, or external RF excitation, electron beam bombardment, or a chemical reaction. The DC electrical discharge is most common for small gas lasers (*Eg*., helium-neon, argon ion, etc.). DC most often pumps semiconductor lasers current. Liquid (dye) lasers are usually pumped optically.

17.2.5 Optical Resonator

An optical cavity or resonator is to introduce optical feedback and so maintain the gain of the system overcoming all losses. Optical resonator plays a very important role in the generation of the laser output, in providing high directionality to the laser beam as well as producing gain in the active medium to overcome the losses due to straying away of photons from the laser medium, diffraction losses due to definite sizes of the mirrors, radiation losses inside the active medium due to absorption and scattering etc. In order to sustain laser action, one has to confine the laser medium and the pumping mechanism in a special way that should promote stimulated emission rather than spontaneous emission. Photons are needed to be confined in the system to allow the number of photons created by stimulated emission to exceed all other mechanisms. This is achieved by bounding the laser medium between two mirrors. On one end of the active medium is the high reflectance mirror (100% reflecting) or the rear mirror and on the other end is the partially reflecting or transmissive mirror or the output coupler. The laser emanates from the output coupler, as it is partially transmissive. Stimulated photons can bounce back and forward along the cavity, creating more stimulated emission as they go. In the process, any photons which are either not of the correct frequency or do not travel along the optical axis are lost.

17.2.6 Laser Action

Laser action is preceded by three processes, i.e., absorption, spontaneous emission and stimulated emission - absorption of energy to populate upper levels, spontaneous emission to produce the initial photons for stimulation and finally, stimulated emission for generation of coherent output or laser.

Interaction of electromagnetic radiation with matter produces absorption and spontaneous emission. Absorption and spontaneous emission are natural processes. For the generation of laser, stimulated emission is essential. Stimulated emission has to be induced or stimulated and is generated under special conditions as stated by Einstein in 1917. When the population inversion exists between upper and lower levels among atomic systems, it is possible to realize amplified stimulated emission and the stimulated emission has the same frequency and phase as the incident radiation. Einstein combined Plank law with Boltzmann's statistics in formulating the concept of stimulated emission. In electronic, atomic, molecular or ionic systems the upper energy levels are less populated than the lower energy levels under equilibrium conditions. Pumping mechanism excites, atoms to a higher energy level by absorption (Figs. 17.14 & 17.15).

The atom stays at the higher level for certain duration and decays to the lower stable ground level spontaneously, emitting a photon, with a wavelength decided by the difference between the upper and the lower energy levels. This is referred to as natural or spontaneous emission and the photon is called spontaneous photon. The spontaneous emission or fluorescence has no preferred direction and the photons emitted have no phase relations with each other, thus generating an incoherent light output. But it is not necessary that the atom is always de-excited to ground state. It can go to an intermediate state, called meta stable state with a radiation less transition, where it stays for a much longer period than the upper level and comes down to lower level or to the ground state. Since period of stay of atoms in the meta stable state is large, it is possible to have a much larger number of atoms in meta stable level in comparison to the lower level so that the population of meta stable state and the lower or ground state is reversed, i.e., there are more atoms in the upper meta stable level than the lower level. This condition is referred to as population inversion. Once this is achieved, laser action is initiated in the following fashion. The atom in the meta-stable state comes down to the ground state emitting a photon. This photon can stimulate an atom in the meta stable state to release its photon in phase with it. The photon thus released is called stimulated photon. It moves in the same direction as the initiating photon, has the same wavelength and polarization and is in phase with it, thus producing amplification. Since there are a large number of initiating photons, it forms an initiating electromagnetic radiation field. An avalanche of stimulated photons is generated, as the photons traveling along the length of the active medium stimulates a number of excited atoms in the meta stable state to release their photons. This is referred to as the stimulated emission. These photons are fully reflected by the rear reflector (100% reflective) and the number and consequently the intensity of stimulated photons increases as they traverse through the active medium, thus increasing the intensity of radiation field of stimulated emission. At the output coupler, a part of these photons is reflected and the rest is transmitted as the laser output. This action is repeated and the reflected photons after striking the rear mirror, reach the output coupler in the return path. The intensity of the laser output increases as the pumping continues.

When the input pumping energy reduces, the available initiating and subsequently the stimulated photons decrease considerably and the gain of the system is not able to overcome the losses, thus laser output ceases. Since the stimulation process was started by the initiating photons, the emitted photons can combine coherently, as all of them are in phase with each other, unlike in the case of spontaneous emission and coherent laser light is emitted. Though the laser action will continue as long as the energy is given to the active medium, it may be stated that pulsed laser is obtained if the population inversion is available in a transient fashion and continuous wave (CW) laser is possible if the population inversion is maintained in a steady-state basis. If the input energy is given by say a flash lamp, the output will be a pulsed output and the laser is called a pulsed laser. If equilibrium can be achieved between the number of photons emitted and the number of atoms in the meta stable level by pumping with a continuous arc lamp instead of a flash lamp, then it is possible to achieve a continuous laser output, which is called continuous wave laser.

17.2.7 Uses

Lasers have several uses:

1. They are found in devices such as CD-ROM drives or CD players to read data on CDs.
2. They are also used in checkouts to interpret barcodes in commercial marts.
3. Lasers also have many applications in medicine, such as in the removal of scarring, tattoos and papillomas from a person's skin.
4. In optical networks, lasers are used to transmit information through optical fibres. This is accomplished by flashing or turning on and off the laser thousands of times in a second. The binary system of numbers is normally used to transmit this information. The two states 0 and 1 are represented by either the presence or absence of the laser light.
5. Lasers can be used to measure temperature and speed because a laser does not need delicate probes. They can be extremely accurate and are ideal for use in rugged applications, where a high level of accuracy is needed.

17.2.8 Types

Lasers are normally grouped according to the medium used to create the laser light.

1. Solid state lasers, which are used as a solid laser creating medium. A ruby laser is an example of a solid state laser, using a tube of ruby to create the laser light.

2. Semiconductor lasers or diode lasers, which are very small, use only a low amount of power and very weak. Diode lasers emit a red beam of light between the wavelengths of 630 to 680nm. These lasers are not solid state lasers. They can be found in devices such as laser pointers and in optical networks.
3. Dye lasers, which can be used as organic dyes such as Rhodamine 6G. The dyes are mixed or suspended in a liquid solution and this type of laser can output many different wavelengths of light.
4. Gas lasers, which uses the gases as the laser creating medium. The most common types are helium and helium-neon gas lasers. The output of the gas lasers is red light. Lasers using carbon dioxide output light in the far infra-red wavelength range and are very powerful. They can be used to cut through hard substances such as steel.
5. Excimer lasers or excited dimmers, which are lasers that use reactive gases mixed with non-reactive gases. When stimulated electrically a molecule is produced. This molecule produces ultraviolet light when it is stimulated in a laser. Reactive gases that can be used in an excimer laser are gases such as chlorine and fluorine; non-reactive gases that can be used include argon, krypton and xenon.

17.3 COMPUTER FOR BIOLOGY

Electron computer is the most widespread computer, its primary function is computing the information. However, for the following reasons, its computing has some disadvantages:

1. The computing speed can't be raised without limit because of the electronic quantum effect. To raise the speed the circuits in CPU must be thinner but the quantum effect set a limit for thinning the circuits.
2. Processing unclear information is not realized by electronic computer. Such a limit is clear when compared with brain.

17.4 PHOTONIC COMPUTER

In photonic computer, electrons are replaced with photons and electron conducting material by photon conducting material. It can raise computing speed because of the advantage of photon to electron but its principle does not change.

17.5 BIOCOMPUTER

Three general principles can be used in biocomputers: (1) Based on the current John Von Newman structure, use different biomolecular states to represent 1/0 switch. (2) Use some biomolecular special property to calculate. (3) Use complete different structure, like brain.

The first principles of biocomputer have little advantage to current electron computer, it cannot conduct unclear computing information. Example of the second principle of biocomputer is DNA's parallel computing. The third principle of biocomputer makes a difference, similar with brain, it is expected to have the power to deal with unclear information. However, how brain works is far from clear, so biocomputer with such a principle is far from realized.

17.6 OPTICAL COMPUTER

Electronic computing uses electrons to perform the logic that makes computing work. An optical computer (photonic computer) is a device that uses the photons in visible light or infrared (IR) beams rather than electric current to perform digital computations. An electric current flows at about 10% of the speed of light. This limits the rate at which data can be exchanged over long distances and is one of the factors that led to the evolution of optical fibre. By applying some of the advantages of visible or IR networks at the device and component scale, a computer might be developed that can perform operations 10 or more times faster than a conventional electronic computer.

Visible light and IR beams, unlike electric currents, pass through each other without interacting. Many laser beam scan be show their paths intersect but there is no interference among the beams even when they are confined essentially to two dimensions (Fig. 17.16B). Electric currents must be guided around each other and this makes three dimensional wiring necessary (Fig. 17.16A) . Thus, an optical computer being much faster than an electronic one and might be smaller.

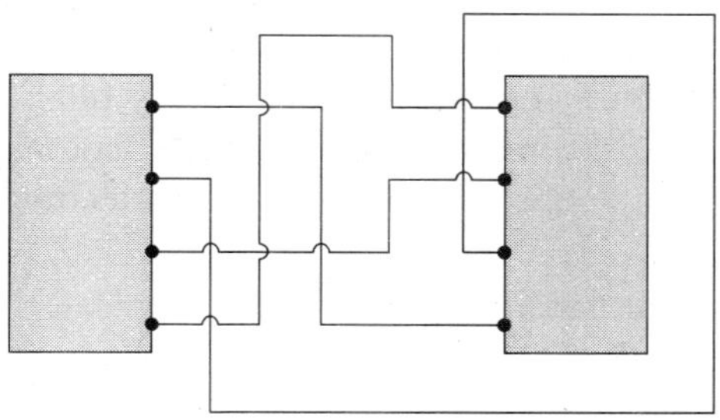

Fig. 17.16A *Electrical current - electrical crossovers requires three dimensions.*

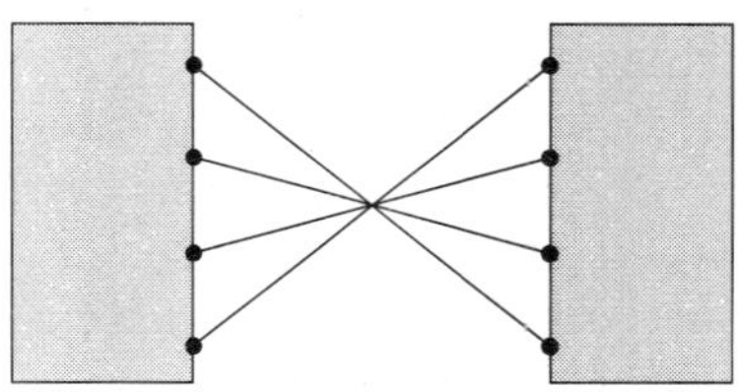

Fig. 17.16B *Optical computer - optical crossovers require only two dimensions because light beams do not interact.*

In optical computing systems, some of the optical integrated circuits are incorporated. Computer has been built using optical circuits. Optical devices can be controlled by electronic currents, even though the impulses carrying the data are visible light or IR. Optical technology has made most significant in digital communications, where fibre optic data transmission has become common place. Photonic network uses visible and IR energy exclusively between each source and destination. Optical technology is employed in CD-ROM drives, laser printers and most photocopiers and scanners. However, none of these devices are fully optical, all rely to some extent on conventional electronic circuits and components.

Photonic computing uses photons of laser light. Electronic transistors are whittled into silicon wafers to make modern computer chips. Today's technology is pushing the electron to its physical limits, as a result the manufacturing processes are becoming inexpensive for producing minor improvements. However, photons are manipulated using inexpensive computer generated holograms made of plastic or glass. Photonic computers will be far more valuable than their slower electronic counterparts and far less expensive to manufacture.

Most of the telephone companies have been investing heavily in the global conversion from copper wire to optical fibre because light does a better job of carrying information than does electricity. This is because photons (the basic unit of light) go faster and have a higher bandwidth than do electrons. Thus, photons are inherently more valuable than electrons. Over 65 major companies have invested heavily in the search for an inexpensive nonlinear crystal able to make one light beam turn another light beam on and off, which is a prerequisite for the production of a completely photonic (optical) computer.

Photonic logic, based on a different physical principle, is proving to be the key to the production of a completely optical computing system. Such a system would completely replace the start and stop surges of electrons with tiny light beams that simply blink on and off, in order to carry information and perform the logic of computing in light speed photonic computers without slowing down the photons in some crystal or enslaving them to some electro optical process.

The Photonic Computer uses only light to collect retrieve and process the data. This makes it operate much faster than electronically based computers because information and processing occurs at the speed of light. Another benefit of the photonic computer is that less energy is used to power the system and it does not have the same interference problems that make it difficult for current chip manufacturers to make smaller transistors. This is because light energy does not interact with other photons. The processor itself is a laser inscribed piece of crystal and the current system uses only a huge amount of RAM to store the information.

The Photonic Computer is also different from a traditional computer in many other ways. First, the hardwired operating system is laser etched into the crystal. This means that the computer itself is not susceptible to computer viruses that affect operating systems. Second, the only way to upgrade the operating system is to get a new processing crystal. The machinery itself to etch the crystals precisely for a modern computer system is also not yet available. This also requires a program that translates the operating system software into the etched crystal that represents it. First Photonic Computer invented by Daniel Nase in 1998.

17.6.1 Photonic Transistor

Photonic Transistor was invented at the Rocky Mountain Research Center, and then tested in the laboratories of the University of Montana, and Montana State in 1989. In 1992, U.S. Patent 5,093,802 was issued to the Rocky Mountain Research Center. Since that time, the entire basis of interference-based photonic computing has been growing.

The advantages of light were recognized for carrying information within the newly emerging computer science. The problem was that, back then, they lacked tools needed to make light compute. As a result, the task fell to electrons, and the electronic computer age was born. Since then, three major events have laid the groundwork for the present effort at producing fully photonic (optical) digital computers.

The first was the invention of the laser. Without the laser it was easy to see why researchers rejected light as a viable computing medium. Ordinary optical signals are a noisy mish-mash of difficult to control electromagnetic energy rather akin to the noisy old spark-gap radio transmitters from the beginning of the 20th century. Lasers, on the other hand, produce much cleaner continuous wave signals that are more like modern radio transmitters that can be used to convey complex information.

The second was the computer generated hologram. Unlike lenses, mirrors and so on, holograms can be calculated into existence using the laws of optical physics. We can input any known optical signal, and calculate the hologram(s) needed to direct and manipulate it in any way that the laws of physics permit. When laser light is directed through the resulting hologram, the light does exactly as we have calculated it to do. In a photonic computer there will be many such optical signals. Each one will have its own characteristics and will need to be redirected into some other configuration in concert with many other such signals. Holograms allow us to interconnect optical components and even create optical components. What's more, they can be mass manufactured of inexpensive glass, plastic and aluminum, just like the holograms on credit cards.

The third background element that has brought forth the photonic transistor at this time rather than 20 some years ago when lasers first became readily available has a more human element. Over the years a considerable investment was made by those 65 companies, and a number of universities, into the optical computing effort. This effort began long before lasers and holograms. Work concentrated on electro-optical and nonlinear optical methods of optical computing and soon realized that have proven insurmountable problems.

First of all, any system that uses electrons in the process cannot be any faster than the electrons themselves. Electronic signals are able to traverse a microscopic electronic transistor and accomplish their assigned task in a little less than a nanosecond. Light on-the-other-hand, travels 30 cm (about 11 3/4 inches) in that same nanosecond (a billionth of a second.)

Nonlinear optics exploits the optical properties of certain (rather expensive) materials that slow light down to two different speeds at the same time. Two problems occur: the first is the light is slowed down. The second is that in order to get light to respond within such crystals so as to perform computing tasks, the lasers have to be so powerful that the components get fried whenever one puts enough of them together to do anything useful. The effort to produce inexpensive nonlinear crystals that switch fast and at reasonable power levels has not been successful. In fact such a substance has been dubbed Unobtainium by those in the field. Their multi-million dollar effort has failed to produce marketable optical computers.

17.6.2 Photonic Transistor Working Mechanism

A photonic computer should use photons. Photons are the basic unit of electromagnetic energy just as electrons are the basic unit of electricity. Unlike the nonlinear optical materials that require a large supply of photons to bias them up to some switching level, Photonic transistors need only signal levels of photons to work. Just as we can see certain things at night that emit only a handful of photons per second, so too, the Photonic Transistor must be able to operate using small amounts of energy.

The next desirable attribute is that they should always operate at the full speed of light. Self Electro-optic Effect Devices (SEEDs) that are popular with, Photonic Transistors do not use electricity in any way shape or form. The fundamental physical control and manipulation processes used do not slow down the light. The only retardation occurs during the very short time that the energy must pass through a dense medium such as a thin hologram. The Photonic Transistor is vacuum compatible, meaning that they can be operated in air or even in a vacuum where there light moves at the universal speed limit.

Optical interference is a process of energy rearrangement that occurs when two laser beams pass through the same point at the same time. The energy pattern forms an interference image that depends upon the wave front pattern, input energy levels, and phase components of each the two input beams, along with the geometry of the encounter. Interference has another very important property. If all of the input parameters of all of the inputs are accurately known, the output interference image formed can be calculated by a process called the linear addition of amplitudes or the vector addition of amplitudes. Since digital computers operate at discrete energy levels, (two levels in the case of binary logic,) each two input photonic logic gate will have 4 possible combinations of its inputs being either high or low on or off. As a result, 4 different images need to be calculated, one for each input combination. During high speed computing, the interference image will switch continually among this set of images, they form a Dynamic Image. At any given location within the dynamic image, the amplitude, and thus the energy level (which is proportional to the square of the amplitude,) will change among 4 different static states as determined by the optical arrangement of the transistor. If an image component separator is placed, such as a mask with a hole in it, at any location within the dynamic image, then any energy that shows up at the hole goes through the mask into the output. Any energy that does not show up at the hole is prevented by the mask from contributing to the output of the transistor.

The output will have a modulated waveform that is dictated by the optical arrangement and controlled by the states of the input beams. The transistor has one other feature of overriding value. The calculated image set produces definite outputs through the hole that define the output energy in all of its states. This precise description of the output signal can then be used to accurately calculate the next transistor, lens, hologram etc. By mathematically stepping from one device to the next, will be calculating into existence entire photonic computers. Usually there are multiple sites in the Dynamic Image that have compatible energy-state series. Holes can be placed at these locations too, and its energy combined with that from the other holes to produce composite transistors with all different kinds of attributes.

17.6.3 Photonic Integrated Circuit (PIC)

Photonic Integrated Circuit (PIC) or Integrated Optics is an optical data transmission device that integrates multiple photonic functions. This technology contains numerous optical components which are bonded together to fulfill complex functions. The chip is made up of indium phosphide, a semiconductor developed for its optical properties. The proposed chip would be light in weight in comparison to electric current. Photonic chips generate, manipulate and detect light rather than electrons. The technology of electronic integrated circuits has reflected numerous achievements such as complex, powerful microprocessors containing million transistors, specialised signal processors or computer memory chips with voluminous data storage capacity. It is expected that all major networking optical function like amplification, mux/demux, switching, transmitting or receiving will be performed by a single chip.

Optical Data Transmission equipments require lasers to send light, multiplexers to split and recombine modulators to encode data and detectors that receive the data, low loss interconnect waveguides and filters. Optical Data Transmission has higher data rates and at the same time eliminates problems resulting from electromagnetic interference. The implementation of silicon based photonic devices will result in smaller and cheaper photonic devices.

Photonic Integrated Circuit (PIC) is considered as a technology for next generation networks. In 2004, David Welch developed a chip with 50 nanoscale optical components patterned into its surface named as Photonic Integrated Circuit. This is the fastest Optical Chip, 2 cm wide strip featuring four patterned gold coloured rectangles, which contain a total of 240 patterned optical components. This developed chip has 1.6 terabit version and is the fastest optical chip. The 1.6 terabit version comprises of 40 lasers, 40 detectors, 40 modulators, 40 channels and every modulator encodes data four times faster. The previously developed chips succeeded in integrating only a few devices on a single chip. The 100 gigabit version was the first device and was efficient to send and receive 100 gigabits of information per second. The 100 gigabit version comprised of 10 lasers, 10 detectors, 10 modulators that encode data by switching light on and off, and 10 waveguides that direct photons into a multiplexer.

17.6.4 Advantages and Disadvantages

PIC has brought forward cost and environmental advantages over circuits composed of discrete components. It allows optical systems to be made compact and perform better than discrete optical components. PIC has a possibility of integration with electronic circuits. Integrated components are reliable, power efficient and thus cut down operating costs. One of the obstacles of PIC is cost, since optical equipment is expensive. Other obstacles are space and time due to speedy developments, life cycles of equipment are shrinking. Optical connections between waveguides and couplers are more critical than electrical connections. They exhibit optical losses and to overcome this problem, an optical amplifier is required.

17.7 PHOTONIC TECHNOLOGY

The primary target of the stakeholder of the computation industry to deliver the data fastly, because of increasingly dependency of all aspects of human life activities based on computers. Many approaches were adopted in order to achieve the targeted goals though it succeeded to a great extent but eventually as due to a certain electron potential barrier there was a limit set over it. Moreover any further enhancement was proving to be exponentially expensive though even if the change in efficiency is very little. All these circumstances began to threaten the future of electronic industry and the quest to get an alternate for the drowning electronic industry initiated.

At present, 300 million transistors approximately can fit on a single chip, it is estimated to be double in period (18-24 months). To enhance the IC performance, the density of transistors on a chip can be increased, which are leading to serious compromises on the reliability of the devices. Though as for the sake of overcoming these factors, invented a variety of methods of increasing chip density ensuring no compromising on the working and reliability of device but all these measure are temporary and would be facing them unless or until find an alternate that will go beyond the barrier of electron.

In the quest for a solution to the mentioned scenario and to achieve our dreamed future, the alternates for electrons are photons. Photon was detected to be an alternate for the sluggish electron a less expensive but performance wise more valuable. While the analysis of photonic instead of electrons, there appear variety of criteria to analyze each other. In photonic circuits, light is shot innovative technological development for the 21st century. As contrast to present electronic based computation, which is to identify presence referred one or absence referred zero one can perform computations at the rate of 2 BPT assuming in to the electron flow limit across system bus. On contrary or simply for ahead, we can identify light (photon) in 16 distinguish states (categorically based on wavelength of light) and if use it, get rate of power of computing to be 16 × BPT. Assuming the general limit of BPT speed to be 32 for electron based computing. It would be 232 or 4,294,967,296 BPT sound huge bits not even an ocean drop if take the same BPT rate for photonic computer it would be 1632 or 340,282,366,920,938,463,463,374,607,431,770,000 ,000 BPT. Fig. 17.17 reflects the optical communication between a photonic based CPU and its peripheral devices. That one could bet if one uses photons instead of electrons, to be straight forward. Photonics shall be replacing majority of the electronic industry and the main reason in contrast to photonic even the latest and most efficient electronic counterpart is far behind.

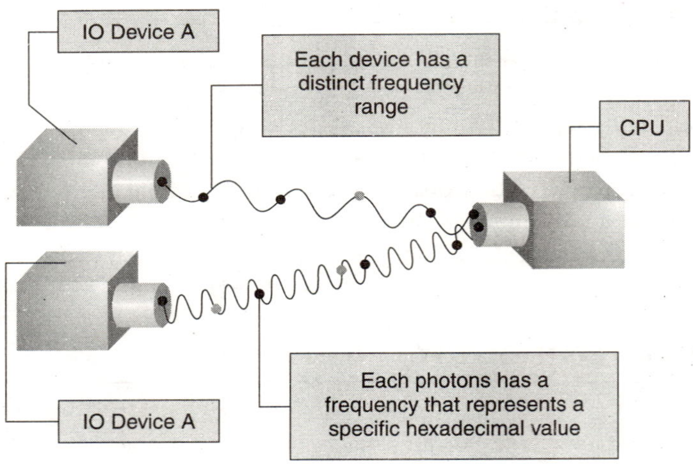

Fig. 17.17 *Reflect the optical communication between a photonic based CPU and its peripheral devices.*

The photonic computer architecture is entirely based on photonics and a light wave with a particular wavelength is given and the signal is then interpreted by the CPU. The photon sent makes the CPU that from where the light signal is from. In case of photon, the light is divided into 16 band gapes based on difference in the wave length and each representing an entirely unique hexadecimal light. Then the signal received by the CPU from the device would be photonic for example 14A3DC. The form of this signal would remain photonic until it reaches the processing chip via the optical medium generally optical bus. The signal would be interpreted and the desired task would be accomplished. All this would be achieved in a speed that one can't even dream of.

17.7.1 Scope

Photons are replacing the electrons as they yield relatively exponentially huge data with far more reliability, speed and cost effectiveness and photonic computing is turning point of technology. The following are the main facts that reflect the photonic technology leap for ahead to an extent one even imagine:

(A) As photons are particles of light so the computation is directly dependent on speed of light which having a clear leap ahead to speed electrons. As the present photonic transistor can switch light equivalent to 1.5 fento seconds (one million of billionth hay a seconds) hence making full use of speed of light via processing the speed of light.

(B) The electron propagation requires field of energy to propagate from one position to the next in a circular path. This is not at all matter of concern while using photons, it would be following a straight path relative to electron circular path which gives photons an edge over electron.

(C) As the technology progressing, the pace of data transfer is required to be quicker. However for electrons, there is a barrier after which the device performance relatively alters though the industry is trying its best and is overcoming it. The only way out is photon which is not at all dependent on properties.

(D) The presence or absence of electrons is the two cases in an electronic computation and hence the data is limited. On contrary light having distinguished color wavelength has independent value. So, it is likely that the number of electrons hold together equivalent to a single beam of photons.

(E) Photonic transistors being component of electromagnetic spectrum. It is estimated that it could have 35 billion non-interacting separate channels referring if using the photon busses they shall be processing 35 billion pathways rather than present max of 64 or 128 KB buses. Fig. 17.18 shows the classification of light on grounds of difference in wave lengths. The portion indicted as communication is the optical domain suited for communication and presently being employed for optical fibre communication.

(F) Photonic transistor is operating on light, hence every device produce shall be processing huge amount with a light speed.

(G) Holographic technique for photonics is far more efficient and liable than equivalent electron chips.

(H) Due to non-interacting property of light it could shrinks far more than that of equivalent electronic chips.

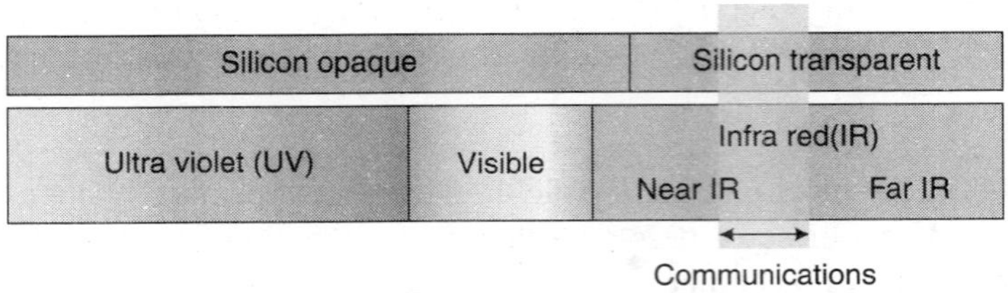

Fig. 17.18 *Fibre optics range studies for computations.*

Physical components for photonic transistor are far cheaper with respect to electronic counterpart. *Eg.* Glass, plastic, aluminum etc. The physical non-interacting properties of light with the transfer medium photon in not address to the issue of power loss, energy remission than that of electronic counterpart. So as there's no space for any limit of barrier set to photon by nature hence there exist any one barrier that limits photons.

17.8 SILICON BASED PHOTONIC COMPUTER

The most basic component of photonics computer is the integrated optic circuit and implementation via different organization, is all what computing is based upon considering the working of optic transistor. Photonic computer is compromised of six building blocks or components, Fig. 17.19 shows operating steps of photonic computer in order to utilize photon for computing.

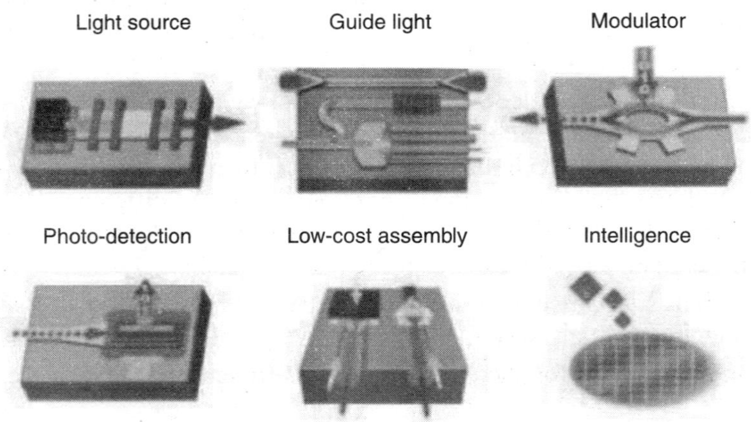

Fig. 17.19 *Modules of photonic computers.*

1. A way in or a trace for light to enter
2. Module that would be guiding the light
3. A module that would be encoding data into optical bits.
4. Interpreter of optic bits.
5. Basic computation operational components.
6. The machine intelligence to interpret the data contained in photons

17.8.1 Wave Guides

Waveguides are considered to be primary concept of photonics based on principal of total internal reflection and the wave guide would be related to the key properties of the matter that shall be chosen. As the wave guide would be dependent on wavelength; gaped silicon which determine the shortest wavelength that would propagate through the silicon and wavelength lie between 1.3 μm – 1.6 μm.

17.8.2 Modulation

Silicon photonic modulation is based on Mach-Zester interconnectors with combination of Kerr effect and Branz effect. Fig. 17.20 shows the modulation of laser light which is initially split, then under goes desired phase shifting and encoded into data.

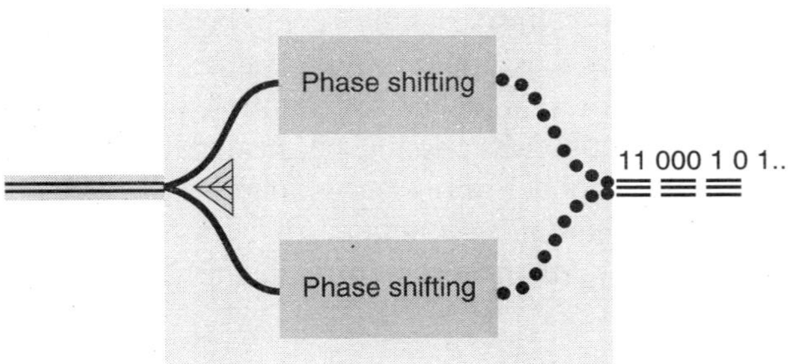

Fig. 17.20 *Modulation of light.*

17.8.3 Silicon Laser

Silicon laser was a most challenging hurdle to overcome for photonics due to indirect band gap, but these issues become resolved by applying principles of Raman effect and eventually a silicon laser was created based on Fabry – Perot resonator.

17.8.4 Photon Detection

The process involves conversion of encoded data of photons to be decoded. For doing so, silicon and germanium were merged. From where once data in form of photons received is forwarded to logical sector.

17.8.5 Intelligence

Silicon has a great edge in order to be chosen under the intelligence module because of compatibility in CMOS structure. The signal received by machine intelligence and then the detector signals is interpreter.

17.8.6 Low Cost Assembly

The components that integrate up to form a photonic chip are inexpensive but the perfection in architecture is key issue, which is ensured by facet preparation technique, precise alignment antireflective thin film location and tapes that ensure both speed and reflection.

17.8.7 Applications

In the modernized world, in every field of life a vital contribution of photonic to make our life more comfort and reliable in our everyday used products. This was made possible by attributes of photons that yield efficiency, reliability and perfection at low cost, high speed and safety.

In our everyday life, we interact with each other by using telecommunication techniques, interact with hospitals their laser surgery endoscopy and a lot for, in military applications infrared sensors, command & control systems, navigation, search & rescue range and lot more in the fields of metrology and spectrometry. All these things are benefits of photonics field.

In short, the photonics could be regarded as booster for information and communication, data storage, Industrial manufacturing and ability, life science and lead it lightening and displays, security, metrology and sense, security and safety, transport, space, aeronautics rather every field of life have been boosted exponentially just due to photonics. Those fields index, the absent photonics would be soon redefining them as well, so that they could become more reliable, provides higher productivity and efficiency, so that they could provide the need of upcoming times. Photonics is no longer an emerging field but a present reality with multiple products readily on the market. The progress in the field of communication was possible because of its optical interconnects other fields including optical data recording, optical fibres xerography based laser printing all such wonders of present time are the practical photonics. Photonics benefits also motivated to replace electric components by photonic counter parts in space craft's due to their greater performance and high reliability.

Carbon Chemistry

Chapter 18

18.0 INORGANIC CARBON

For more than 200 years, chemists have divided compounds into two categories. Those that were isolated from Plants or Animals were called Organic, while those extracted from Ores and Minerals were Inorganic. Organic chemistry is often defined as the chemistry of carbon. But this definition would include calcium carbonate ($CaCO_3$) and graphite, which more closely resemble inorganic compounds. We will therefore define organic chemistry as the study of compounds, such as formic acid (HCO_2H), methane (CH_4), and vitamin C ($C_6H_8O_6$), that contain both carbon and hydrogen.

The chemistry of carbon is dominated by three factors (1) Carbon forms unusually strong C-C single bonds, C=C double bonds and C≡C triple bonds. (2) The electronegativity of carbon (EN = 2.55) is too small to allow carbon to form C^{4-} ions with most metals and too large for carbon to form C^{4+} ions when it reacts with nonmetals. Carbon therefore forms covalent bonds with many other elements. (3) Carbon forms strong double and triple bonds with a number of other nonmetals, including N, O, P and S.

18.1 ELEMENTAL FORMS OF CARBON

Carbon occurs as a variety of allotropes, they are in two forms (a) Crystalline forms- diamond and graphite (b) Non-crystalline forms (amorphous) charcoal, coke and carbon black.

Smithson Tennant (1796) discovered that diamonds consist of carbon. The properties of diamond are the least volatile substances (MP = 3550°C, BP = 4827°C), the hardest substance, it expands less on heating than any other material and logical consequence of its structure. Carbon with four valence electrons forms covalent bonds to four neighboring carbon atoms arranged toward the corners of a tetrahedron (Fig. 18.1). Each of these sp^3 hybridized atoms is then bound to four other carbon atoms, which form bonds to four other carbon atoms and so on. As a result, a perfect diamond can be thought of as a single giant molecule. The strength of the individual C-C bonds and their arrangement in space give rise to the unusual properties of diamond.

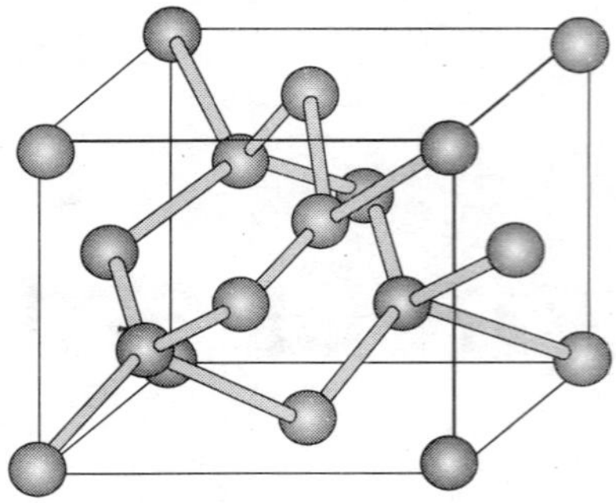

Fig. 18.1 *Carbon with four valence electrons forms covalent bonds to four neighboring carbon atoms arranged toward the corners of a tetrahedron.*

Graphite consists of extended planes of sp^2 hybridized carbon atoms in which each carbon is tightly bound to three other carbon atoms. The strong bonds between carbon atoms within each plane explain the exceptionally high melting point and boiling point of graphite. The distance between these planes of atoms, however, is very much larger than the distance between the atoms within the planes. Because the bonds between planes are weak, it is easy to deform the solid by allowing one plane of atoms to move relative to another. As a result, graphite is soft enough to be used in pencils and as a lubricant in motor oil. The physical properties of graphite can be understood from the structure of the solid (Fig. 18.2).

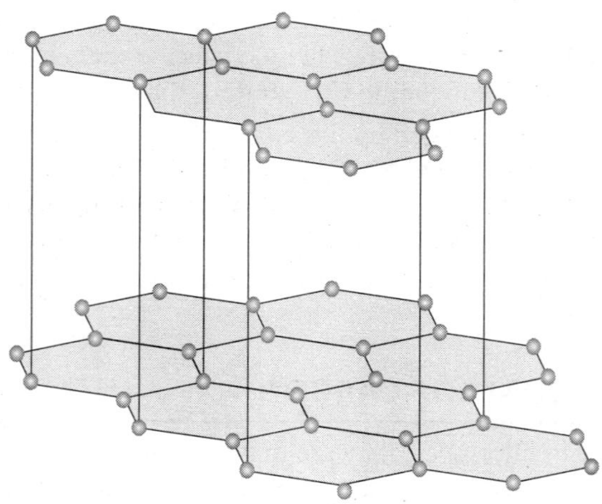

Fig. 18.2 *The physical properties of graphite from the structure of the solid.*

The properties of graphite are like those of diamond. Both compounds are boiling at 4827°C but graphite is also very different from diamond. Diamond (3.514 g/cm^3) is much denser than graphite (2.26 g/cm^3). Whereas diamond is the hardest substance known, graphite is one of the softest. Diamond is an excellent insulator, with little or no tendency to carry an electric current. Graphite is such a good conductor of electricity that graphite electrodes are used in electrical cells.

Lead pencils do not, incidentally, contain lead. This is fortunate because many people chew pencils and lead compounds are toxic. Lead pencils contain graphite or black lead as it was once known, which is mixed with clay (20% to 60% by weight) and then baked to form a ceramic rod. Increasing the percentage of clay makes the pencil harder, so that less graphite is deposited on the paper.

The characteristic properties of graphite and diamond suggest that the diamond is more stable than graphite. The standard enthalpy of formation of diamond (ΔH^{o}_{f} = 2.425 kJ/mol) is slightly larger than the enthalpy of formation of graphite, which is the most stable form of carbon at 25°C and 1 atm pressure. At very high temperatures and pressures, diamond becomes more stable than graphite. In 1955, General Electric developed a process to make industrial-grade diamonds by treating graphite with a metal catalyst at temperatures of 2000 to 3000 K and pressures above 125,000 atm. roughly 40% of industrial quality diamonds are now synthetic.

Both diamond and graphite occur as regularly packed crystals. Other forms of carbon are amorphous they lack a regular structure. Charcoal, carbon black, and coke are all amorphous forms of carbon. Charcoal results from heating wood in the absence of oxygen. To make carbon black, natural gas or other carbon compounds are burned in a limited amount of air to give a thick, black smoke that contains extremely small particles of carbon, which can be collected when the gas is cooled and passed through an electrostatic precipitator. Coke is a more regularly structured material, closer in structure to graphite than either charcoal or carbon black, which is made from coal.

18.1.1 Carbides

Carbon is an inert at room temperature and reacts with less electronegative elements at high temperatures to form compounds known as carbides. When carbon reacts with an element of similar size and electro-negativity, covalent carbide is produced. *Eg.:* Silicon carbide is made by treating silicon dioxide from quartz with an excess of carbon in an electric furnace at 2300 K.

$$SiO_2(s) + 3C(s) \rightarrow SiC(s) + 2CO(g) \quad \text{... eq.-18.1}$$

SiC was first synthesized by Edward Acheson in 1891. Covalent carbides have properties similar to those of diamond. Both SiC and diamond are inert to chemical reactions, except at very high temperatures. Both have very high melting points and both are among the hardest substances known.

Compounds that contain carbon and one of the more active metals are called Ionic carbides.

$$CaO(s) + 3C(s) \rightarrow CaC_2(s) + CO(g) \quad \text{... eq.-18.2}$$

These compounds are contained negatively charged carbon ions: $[Ca^{2+}][C_2^{2-}]$ or $[Al^{3+}]_4[C^{4-}]_3$. The ionic carbides that formally contain the C^{4-} ion react with water to form methane, which is ignited by the heat given off in this reaction.

$$C^{4-} + 2H_2O \rightarrow CH_4 + 4\,OH^- \quad \text{... eq.-18.3}$$

The ionic carbides that formally contain the C_2^{2-} ion react with water to form acetylene, which is ignited by the heat of reaction.

$$C_2^{2-} + 2\,H_2O \rightarrow C_2H_2 + 2\,OH^- \qquad \text{... eq.-18.4}$$

Interstitial carbides such as tungsten carbide (WC), form when carbon combines with a metal that has an intermediate electro-negativity and a relatively large atomic radius. In these compounds, the carbon atoms pack in the holes (interstices) between planes of metal atoms. The interstitial carbides, which include TiC, ZrC, and MoC retain the properties of metals. They act as alloys, rather than as either salts or covalent compounds.

18.1.2 Oxides of Carbon

The different forms of carbon are essentially inert at room temperature, they combine with oxygen at high temperatures to produce a mixture of carbon monoxide and carbon dioxide.

$$2\,C(s) + O_2(g) \rightarrow 2\,CO(g)\ \Delta H^\circ = -110.52\ \text{kJ/mol CO} \qquad \text{... eq.-18.5}$$

$$C(s) + O_2(g) \rightarrow CO_2(g)\ \Delta H^\circ = -393.51\ \text{kJ/mol } CO_2 \qquad \text{... eq.-18.6}$$

CO can also be obtained when red-hot carbon is treated with steam.

$$C(s) + H_2O(g) \rightarrow CO(g) + H_2(g) \qquad \text{... eq.-18.7}$$

These mixtures of gases are formed by the reaction of charcoal or coke with water *(water gas)* and it is also known as town gas because it was once made by towns and cities for use as a fuel. Water gas or town gas was a common fuel for both home and industrial use before natural gas became readily available. The H_2 burns to form water, and the CO is oxidized to CO_2. CO and CO_2 are both colorless gases. CO boils at –191.5°C, and CO_2 sublimes at –78.5°C. Although CO has no odor or taste, CO_2 has a faint, pungent odor and a distinctly acidic taste. Both are dangerous substances but at very different levels of exposure. Air contaminated with as little as 0.002 grams of CO per liter can be fatal because CO binds tightly to the hemoglobin and myoglobin that carry oxygen through the blood. CO_2 is not lethal until the concentration in the air approaches 15%. At that point, it has replaced so much oxygen that a person who attempts to breathe this atmosphere suffocates. The danger of CO_2 poisoning is magnified by the fact that CO_2 is roughly 1.5 times denser than the air in our atmosphere. Thus, CO_2 can accumulate at the bottom of tanks or wells.

18.1.3 CO_2 in Atmosphere

CO_2 influences the temperature of the atmosphere through the greenhouse effect. CO_2 absorbs some of the lower energy, longer wavelength infrared radiation from the sun that would otherwise be reflected back from the surface of the planet. Thus, CO_2 in the atmosphere traps heat. Although there are other factors at work, it is worth noting that Venus, whose atmosphere contains a great deal of CO_2, has a surface temperature of roughly 400°C, whereas Mars, with little or no atmosphere, has a surface temperature of –50°C. There are many sources of CO_2 in the atmosphere. Over geologic time scales, the largest source has been volcanoes. Within the last century, the combustion of petroleum, coal, and natural gas has made a significant contribution to atmospheric levels of CO_2 (Fig. 18.3). Between 1958 and 1978, the average level of CO_2 in the atmosphere increased by 6%, from 315.8 to 334.6 ppm.

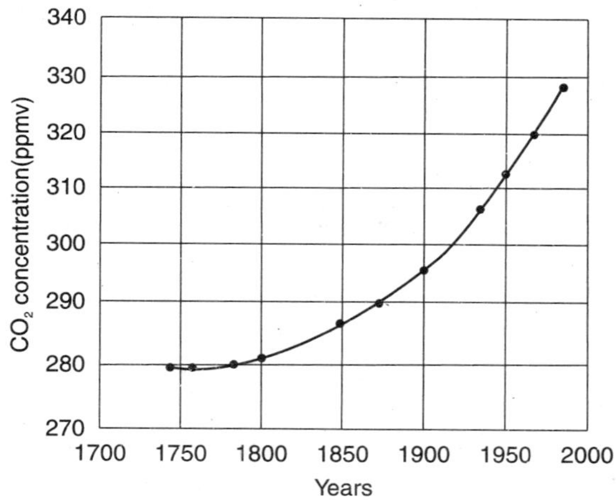

Fig. 18.3 *Atmospheric levels of Carbon oxide (CO_2).*

At one time, the amount of CO_2 released to the atmosphere was not a matter for concern because natural processes that removed CO_2 from the atmosphere could compensate for the CO_2 that entered the atmosphere. The vast majority of the CO_2 was liberated by volcanic action. Eg. CO_2 was captured by calcium oxide or magnesium oxide to form calcium carbonate or magnesium carbonate.

$$CaO(s) + CO_2(g) \rightleftharpoons CaCO_3(s) \quad \text{... eq.-18.8}$$

$$MgO(s) + CO_2(g) \rightleftharpoons MgCO_3(s) \quad \text{... eq.-18.9}$$

$CaCO_3$ is found as limestone or marble or mixed with $MgCO_3$ as dolomite. The amount of CO_2 in deposits of carbonate minerals is at least several thousand times larger than the amount in the atmosphere. CO_2 also dissolves, to some extent, in water.

$$CO_2(g) \overset{H_2O}{\rightleftharpoons} CO_2(aq) \quad \text{... eq.-18.10}$$

It then reacts with water to form carbonic acid, H_2CO_3.

$$CO_2(aq) + H_2O(l) \rightleftharpoons H_2CO_3(aq) \quad \text{... eq.-18.11}$$

As a result of these reactions, the sea contains about 60 times more CO_2 than the atmosphere. A gradual warming of the earth's atmosphere could result from continued increases in CO_2 levels, with adverse effects on the climate and therefore the agriculture of at least the northern hemisphere.

18.2 CHEMISTRY OF CARBONATES (CO_3^{2-} AND HCO_3^-)

Egg shells are almost pure calcium carbonate ($CaCO_3$). $CaCO_3$ can also be found in the shells of many marine organisms and in both limestone and marble. $CaCO_3$ is normally insoluble in water saturated with CO_2, carbonated water or carbonic acid reacts with calcium carbonate to form calcium bicarbonate, which is soluble in water.

$$CaCO_3(s) + H_2CO_3(aq) \rightleftharpoons Ca^{2+}(aq) + 2\ HCO_3^-(aq) \qquad \text{... } \textbf{\textit{eq.}-18.12}$$

If the CO_2 escapes from the water or if some of the water evaporates, solid $CaCO_3$ is redeposited, this happens as water runs across the roof of a cavern, *stalactites*, which hang from the roof of the cave, are formed.

The chemistry of carbon dioxide dissolved in water is the basis of the soft drink industry. The first artificially carbonated beverages were introduced in Europe at the end of the nineteenth century. Carbonated soft drinks today consist of carbonated water, a sweetening agent such as sugar, saccharin or aspartame, an acid to impart a sour or tart taste, flavoring agents, coloring agents and preservatives. As much as 3.5 liters of gaseous CO_2 dissolve in a liter of soft drink. The CO_2 contributes the characteristic bite associated with carbonated beverages.

Carbonate chemistry plays an important role in other parts of the food industry as well. *Baking soda* or bicarbonate of soda is sodium bicarbonate, $NaHCO_3$, a weak base, which is added to recipes to neutralize the acidity of other ingredients. *Baking powder* is a mixture of baking soda and a weak acid, such as tartaric acid or calcium hydrogen phosphate ($CaHPO_4$). When mixed with water, the acid reacts with the HCO_3^- ion to form CO_2 gas, which causes the dough or batter to rise.

$$HCO_3^-(aq) + H^+(aq) \rightleftharpoons H_2CO_3(aq) \rightarrow H_2O(l) + CO_2(g) \qquad \text{... } \textbf{\textit{eq.}-18.13}$$

18.3 FULLERENES

In 1985 Richard Smalley *et al.,* made a stable form of carbon by vaporizing graphite with a laser. The apparatus in which this experiment was performed was designed to create small molecules that were clusters of atoms (Fig. 18.4). In this cluster generator, a pulse of helium gas was swept over the surface of the graphite as it was excited with the laser. The mixture of helium and carbon atoms that vaporized from the graphite surface cooled as the gas expanded and molecules with the formula C_{60} were formed that have a structure that has the symmetry of a soccer ball. Because this structure resembles the geodesic dome invented by R. Buckminster Fuller, C_{60} was named buckminsterfullerene or buckyball. It was formally a new form of pure carbon, C_{60}. C_{60} was synthesized by contrasting this form of pure carbon with diamond and graphite. C_{60} is unique because it exists as distinct molecules, not extended arrays of atoms. C_{60} can be obtained as a pure substance, whereas the surfaces of diamond and graphite are inevitably contaminated by hydrogen atoms that bind to the carbon atoms on the surface.

C_{60} is now known to be a member of a family of compounds known as the fullerenes. Other compounds in this family include C_{32}, C_{44}, C_{50}, C_{58}, and C_{70}. C_{60} is the most important of the fullerenes because it is the most perfectly symmetric molecule possible, spinning in the solid at a rate of more than 100 million times per second. Because of their symmetry, C_{60} molecules pack as regularly as Ping-Pong balls. Initially, it is as soft as graphite, but when compressed by 30%, it becomes harder than diamond. C_{60} therefore has the remarkable property that it bounces back when shot at a metal surface at high speeds. C_{60} also has the ability to form compounds in which it is an insulator, a conductor, a semiconductor or a superconductor. By itself, C_{60} is a semiconductor. When mixed with enough potassium to give a compound with the empirical formula K_3C_{60}, it conducts electricity like a metal. When excess potassium is added, this solid becomes an insulator. When K_3C_{60} is cooled to 18 K, the result is a superconductor. The potential of fullerene chemistry for both practical materials and laboratory curiosities is large enough to explain why this molecule can be described as exocharmic it exudes charm.

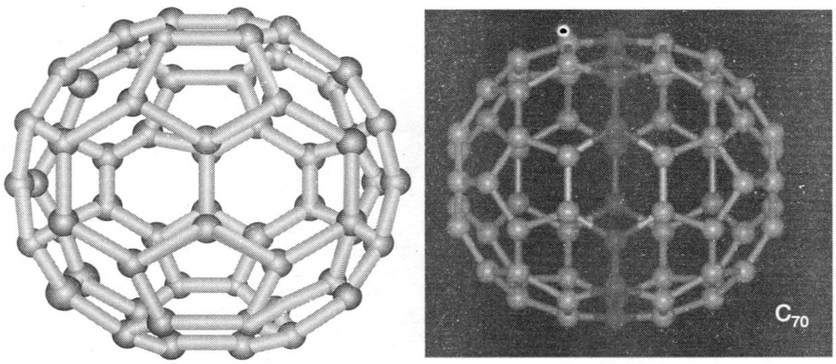

Fig. 18.4 *Fullerenes, buckyballs – 0 Dimension CARBON. (Source: Spherical molecule prepared by Harold Kroto and James Heath, Sean O' Brien, Robert Curl and Richard Smalleyat)*

Table 18.1 C_{60} and C_{70} fullerenes.

C60	C70
Each carbon atom is bonded with 3 others covalently C60 has TWO bond lengths: 1. 6:6 ring bonds between two hexagons (Double Bonds) 2. 6:5 bonds between a Hexagon and a Pentagon	C_{70} has 12 regular pentagonal faces 10 regular hexagonal faces 15 non-regular hexagonal faces 70 vertices 105 edges
Van der Wall diameter : 1 nm Nucleus to nucleus diameter : 0.71 nm Average bond length : 0.14 nm Crystal structure : FCC Density : 1.65 g/cm^3 Lattice constant : 1.415 nm Conductivity : Semiconductor	Shape : Ovel Molecular weight : 840.75 (g/mol) Physical state : Black fine crystalline powder Melting point : >280°C Specific gravity : 1.7 - 1.9

18.4 CARBON NANOTUBES (CNTS)

Carbon Nanotubes (CNTs) are allotropes of carbon with a cylindrical nanostructure. These cylindrical carbon molecules have novel properties, making them potentially useful in many applications in nanotechnology, electronics, optics and other fields of materials science, as well as potential uses in architectural fields. They exhibit extraordinary strength and unique electrical properties and are efficient thermal conductors. Their name is derived from their size, since the diameter of a nanotube is on the order of a few nanometers (approximately 1/50,000th of the width of a human hair), while they can be up to 18 centimeters in length (Fig. 18.5). Nanotubes are members of the fullerene structural family, which also includes the spherical buckyballs. The ends of a nanotube may be capped with a hemisphere of the buckyball structure. Orbital hybridization (Applied quantum chemistry) best describes chemical bonding in nanotubes. The chemical bonding of nanotubes is composed entirely of sp^2 bonds, similar to those of graphite. These bonds, which are stronger than the sp^3 bonds found in alkanes, provide nanotubules with their unique strength. Moreover, nanotubes naturally align themselves into ropes held together by 'van der Waals forces'.

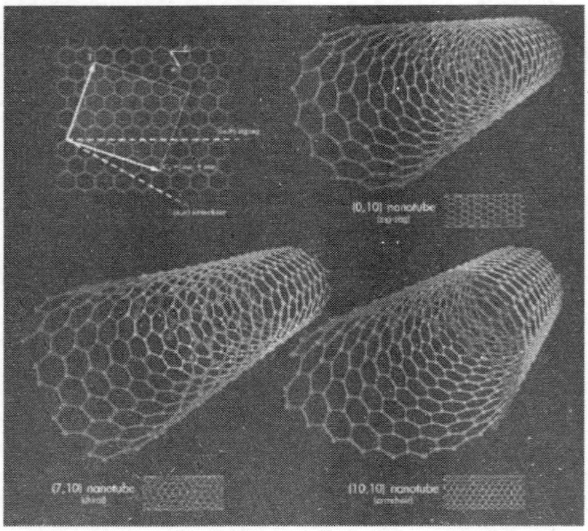

Fig. 18.5 *A schematic representation of nanotube (Nanotube – 1 Dimension carbon).*

Table 18.2 Properties and comparatively of carbon nanotubes.

S.No	*Property*	*Carbon Nanotubes*	*Comparatively*
1	Size	0.6-1.8 nm in diameter	Si wires at least 50nm thick
2	Strength	45 Billion Pascals	Steel alloys have 2 Billion P.
3	Resilience	Bent and straightened without damage	Metals fracture when bent and restraightened
4	Conductivity	Estimated at 10^9 A/cm^2	Cu wires burn at 10^6 A/cm^2
5	Cost	$2500/gram by BuckyUSA in Houston	Gold is $15/gram

18.4.1 Classification

Nanotubes are categorized as single walled nanotubes (SWNTs) and multi-walled nanotubes (MWNTs).

18.4.1.1 Single walled carbon nanotubes

Single walled nanotubes (SWNT) have a diameter of close to 1 nanometer, with a tube length that can be many millions of times longer. The structure of a SWNT can be conceptualized by wrapping a one atom thick layer of graphite called graphene into a seamless cylinder (Fig. 18.7). The way the graphene sheet is wrapped is represented by a pair of indices (n,m) called the chiral vector. The integer's n and m denote the number of unit vectors along two directions in the honeycomb crystal lattice of graphene. If m = 0, the nanotubes are called zigzag. If n = m, the nanotubes are called armchair (Fig. 18.6). Otherwise, they are called chiral. The diameter of a nanotube can be calculated from its (n, m) indices as follows:

$$d = \frac{a}{\Pi}\sqrt{(n^2 + nm + m^2}$$

where $a = 0.246$ nm.

Single walled nanotubes are an important variety of carbon nanotube because they exhibit electric properties that are not shared by the multi walled carbon nanotube (MWNT) variants. In particular, their band gap can vary from zero to about 2 eV and their electrical conductivity can show metallic or semiconducting behavior, whereas MWNTs are zero gap metals. Single walled nanotubes are the most likely candidate for miniaturizing electronics beyond the micro electromechanical scale currently used in electronics. The most basic building block of these systems is the electric wire, and SWNTs can be excellent conductors. One useful application of SWNTs is in the development of the first intra molecular field effect transistors (FET).

Fig. 18.6 *Armchair (n,n Graphene nanoribbon , the translation vector is bent and Zigzag (n,0) while the chiral vector stays straight.*

Fig. 18.7 *Chiral (n,m) n and m can be counted and Graphene nanoribbon at the end of the tube.*

18.4.1.2 Multi walled carbon nanotubes

Multi walled nanotubes consists of multiple rolled layers (concentric tubes) of graphite. There are two models which can be used to describe the structures of multi walled nanotubes (1) The Russian Doll model, sheets of graphite are arranged in concentric cylinders, *Eg.*: A (0,8) single walled nanotube (SWNT) within a larger (0,17) single walled nanotube. (2) The Parchment model, a single sheet of graphite is rolled in around itself, resembling a scroll of parchment or a rolled newspaper. The interlayer distance in multi walled nanotubes is close to the distance between graphene layers in graphite, approximately 3.4 Å.

In the case of SWNT, covalent functionalization will break some C=C double bonds, leaving holes in the structure on the nanotube and thus modifying both its mechanical and electrical properties. In the case of DWNT, only the outer wall is modified. DWNT synthesis on the gram-scale was first proposed in 2003 by the Chemical Vapour Deposition (CVD) technique, from the selective reduction of oxide solutions in methane and hydrogen.

18.4.2 Properties

18.4.2.1 Strength

Carbon nanotubes are the strongest and stiffest materials in terms of tensile strength and elastic modulus respectively. This strength results from the covalent sp^2 bonds formed between the individual carbon atoms. A multi walled carbon nanotube was tested (in 2000) to have a tensile strength of 63 gigapascals (GPa). Since carbon nanotubes have a low density for a solid of 1.3 to 1.4 $g{\cdot}cm^{-3}$, its specific strength of up to 48,000 $kN{\cdot}m{\cdot}kg^{-1}$ is the best of known materials, compared to high carbon steel's 154 $kN{\cdot}m{\cdot}kg^{-1}$.Under excessive tensile strain, the tubes will undergo plastic deformation, which means the deformation is permanent. This deformation begins at strains of approximately 5% and can increase the maximum strain the tubes undergo before fracture by releasing strain energy. CNTs are not nearly as strong under compression. Because of their hollow structure and high aspect ratio, they tend to undergo buckling when placed under compressive, torsional or bending stress.

18.4.2.2 Hardness

Standard single walled carbon nanotubes can withstand a pressure up to 24GPa without deformation. They then undergo a transformation to super hard phase nanotubes. Maximum pressures measured using current experimental techniques are around 55GPa. However, these new super hard phase nanotubes collapse at an even higher, pressure. The bulk modulus of super hard phase nanotubes is 462 to 546 GPa, even higher than that of diamond (420 GPa for single diamond crystal).

18.4.2.3 Kinetic properties

Multi walled nanotubes are multiple concentric nanotubes precisely nested within one another. These exhibit a striking telescoping property whereby an inner nanotube core may slide, almost without friction, within its outer nanotube shell, thus creating an atomically perfect linear or rotational bearing. This is one of the first true examples of molecular nanotechnology, the precise positioning of atoms to create useful machines.

18.4.2.4 Electrical properties

Because of the symmetry and unique electronic structure of graphene, the structure of a nanotube strongly affects its electrical properties. For a given (n,m) nanotube, if n = m, the nanotube is metallic; if n – m is a multiple of 3, then the nanotube is semiconducting with a very small band gap, otherwise the nanotube is a moderate semiconductor. Thus all armchair (n = m) nanotubes are metallic, and nanotubes (6,4), (9,1), etc. are semiconducting. In theory, metallic nanotubes can carry an electric current density of 4×10^9 A/cm^2 which is more than 1,000 times greater than metals such as copper, where for copper interconnects current densities are limited by electro-migration. Multi walled carbon nanotubes with interconnected inner shells show superconductivity with a relatively high transition temperature $T_c = 12$ K. In contrast, the T_c value is an order of magnitude lower for ropes of single-walled carbon nanotubes or for MWNTs with usual, non-interconnected shells.

18.4.2.5 Optical properties

Within materials science, the optical properties of carbon nanotubes refer specifically to the absorption, photoluminescence, and Raman spectroscopy of carbon nanotubes. Spectroscopic methods offer the possibility of quick and non-destructive characterization of relatively large amounts of carbon nanotubes. There is a strong demand for such characterization from the industrial point of view: numerous parameters of the nanotube synthesis can be changed, intentionally or unintentionally, to alter the nanotube quality. The optical absorption, photoluminescence and Raman spectroscopy allow quick and reliable characterization of this nanotube quality in terms of non-tubular carbon content, structure (chirality) of the produced nanotubes, and structural defects. Those features determine nearly any other properties such as optical, mechanical, and electrical properties.

Carbon nanotubes are unique one dimensional system which can be envisioned as rolled single sheets of graphite. This rolling can be done at different angles and curvatures resulting in different nanotube properties. The diameter typically varies in the range 0.4–40 nm (~100 times), but the length can vary ~10,000 times reaching 4 cm. Thus the nanotube aspect ratio, or the length-to-diameter ratio, can be as high as 132,000,000:1. Which is unequalled by any other material. Consequently, all the properties of the carbon nanotubes relative to those of typical semiconductors are extremely anisotropic (directionally dependent) and tunable.

18.4.2.6 Thermal properties

All nanotubes are expected to be very good thermal conductors along the tube, exhibiting a property known as ballistic conduction, but good insulators laterally to the tube axis. Measurements show that a SWNT has a room-temperature thermal conductivity along its axis of about 3500 $W \cdot m^{-1} \cdot K^{-1}$ compare this to copper, a metal well-known for its good thermal conductivity, which transmits 385 $W \cdot m^{-1} \cdot K^{-1}$. A SWNT has a room-temperature thermal conductivity across its axis (in the radial direction) of about 1.52 $W \cdot m^{-1} \cdot K^{-1}$ which is about as thermally conductive as soil. The temperature stability of carbon nanotubes is estimated to be up to 2800°C in vacuum and about 750°C in air.

18.4.2.7 Synthesis: chemical vapor deposition (CVD)

During CVD, a substrate is prepared with a layer of metal catalyst particles, most commonly nickel, cobalt, iron or a combination. The metal nanoparticles can also be produced by other ways, including reduction of oxides or oxides solid solutions. The diameters of the nanotubes that are to be grown are related to the size of the metal particles. This can be controlled by patterned deposition of the metal, annealing, or by plasma etching of a metal layer. The substrate is heated to approximately 700°C. To initiate the growth of nanotubes, two gases are bled into the reactor: a process gas (such as ammonia, nitrogen or hydrogen) and a carbon-containing gas such as acetylene, ethylene, ethanol or methane. Nanotubes grow at the sites of the metal catalyst; the carbon-containing gas is broken apart at the surface of the catalyst particle, and the carbon is transported to the edges of the particle, where it forms the nanotubes. This mechanism is still being studied. The catalyst particles can stay at the tips of the growing nanotube during the growth process, or remain at the nanotube base, depending on the adhesion between the catalyst particle and the substrate. Thermal catalytic decomposition of hydrocarbon has become an active area of research and can be a promising route for the bulk production of CNTs. Fluidized bed reactor is the most widely used reactor for CNT preparation.

CVD is a common method for the commercial production of carbon nanotubes. For this purpose, the metal nanoparticles are mixed with a catalyst support such as MgO or Al_2O_3 to increase the surface area for higher yield of the catalytic reaction of the carbon feedstock with the metal particles. One issue in this synthesis route is the removal of the catalyst support via an acid treatment, which sometimes could destroy the original structure of the carbon nanotubes. However, alternative catalyst supports that are soluble in water have proven effective for nanotube growth.

If plasma is generated by the application of a strong electric field during the growth process (plasma enhanced chemical vapor deposition), then the nanotube growth will follow the direction of the electric field. By adjusting the geometry of the reactor it is possible to synthesize vertically aligned carbon nanotubes (perpendicular to the substrate), a morphology that has been of interest to researchers interested in the electron emission from nanotubes. Without the plasma, the resulting nanotubes are often randomly oriented. Under certain reaction conditions, even in the absence of plasma, closely spaced nanotubes will maintain a vertical growth direction resulting in a dense array of tubes resembling a carpet or forest.

Of the various means for nanotube synthesis, CVD shows the most promise for industrial scale deposition, because of its price/unit ratio, and because CVD is capable of growing nanotubes directly on a desired substrate, whereas the nanotubes must be collected in the other growth techniques. The growth sites are controllable by careful deposition of the catalyst.

18.5 APPLICATIONS

18.5.1 Paper Batteries

A paper battery is a battery engineered to use a paper thin sheet of cellulose infused with aligned carbon nanotubes. The nanotubes act as electrodes; allowing the storage devices to conduct electricity. The battery, which functions as both a lithium-ion battery and a super capacitor, can provide a long, steady power output comparable to a conventional battery, as well as a super capacitor's quick burst of high energy and while a conventional battery contains a number of separate components, the paper battery integrates all of the battery components in a single structure, making it more energy efficient.

18.5.2 Solar Cells

Buckyballs trap electrons, although they can't make electrons flow. Add sunlight to excite the polymers, and the buckyballs will grab the electrons. Nanotubes, behaving like copper wires, will then be able to make the electrons or current flow.

Current applications include:

1. Tips for atomic force microscope probes.
2. In tissue engineering, carbon nanotubes can act as scaffolding for bone growth.
3. In the Kanzius cancer therapy, single-walled carbon nanotubes are inserted around cancerous cells, then excited with radio waves, which causes them to heat up and kill the surrounding cells

18.5.3 Graphene

Graphene is an allotrope of carbon, whose structure is one atom thick planar sheets of sp^2 bonded carbon atoms that are densely packed in a honeycomb crystal lattice (Fig. 18.8). The term graphene was coined as a combination of graphite and the suffixene by HannsPeter Boehm, who described single-layer carbon foils in 1962. The crystalline or flake form of graphite consists of many graphene sheets stacked together.

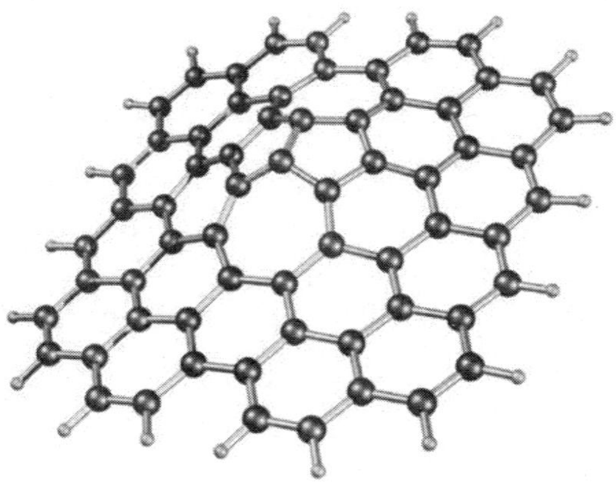

Fig. 18.8 *Graphene – 2 Dimension carbon.*

The carbon-carbon bond length in graphene is about 0.142 nanometers. Graphene sheets stack to form graphite with an interplanar spacing of 0.335 nm, which means that a stack of 3 million sheets would be only one millimeter thick. Graphene is the basic structural element of some carbon allotropes including graphite, charcoal, carbon nanotubes and fullerenes. It can also be considered as an indefinitely large aromatic molecule, the limiting case of the family of flat polycyclic aromatic hydrocarbons.

18.5.3.1 Occurrence and production

In essence, graphene is an isolated atomic plane of graphite. From this perspective, graphene has been known since the invention of X-ray crystallography. Graphene planes become even well separated in intercalated graphite compounds. In 2004 physicists at the University of Manchester and the Institute for Microelectronics Technology, Chernogolovka, Russia, first isolated individual graphene planes by using Scotch tape. They also measured electronic properties of the obtained flakes and showed their unique properties. In 2005 the same Manchester Geim group together with the Philip Kim group from Columbia University demonstrated that quasiparticles in graphene were massless Dirac fermions. These discoveries led to an explosion of interest in graphene.

Since then, hundreds of researchers have entered the area, resulting in an extensive search for relevant earlier papers. Manchester researchers themselves published the first literature review. They cite several papers in which graphene or ultra-thin graphitic layers were epitaxially grown on various substrates. Also, they note a number of pre-2004 reports in which intercalated graphite compounds were studied in a transmission electron microscope. In the latter case, researchers occasionally observed extremely thin graphitic flakes.

This process produces epitaxial graphene with dimensions dependent upon the size of the SiC substrate (wafer). The face of the SiC used for graphene formation, silicon- or carbon-terminated, highly influences the thickness, mobility and carrier density of the graphene. The electronic band structure (so-called Dirac cone structure) has been first visualized in this material. Weak anti-localization is observed in this material and not in exfoliated graphene produced by the pencil trace method. Extremely large, temperature independent mobilities have been observed in SiC epitaxial graphene. They approach those in exfoliated graphene placed on silicon oxide but still much lower than mobilities in suspended graphene produced by the drawing method.

18.5.3.2 Properties

18.5.3.2.1 Atomic structure

The atomic structure of isolated, single-layer graphene was studied by transmission electron microscopy (TEM) on sheets of graphene suspended between bars of a metallic grid. Electron diffraction patterns showed the expected hexagonal lattice of graphene. Suspended graphene also showed "rippling" of the flat sheet, with amplitude of about one nanometer. These ripples may be intrinsic to graphene as a result of the instability of two-dimensional crystals, or may be extrinsic, originating from the ubiquitous dirt seen in all TEM images of graphene. Atomic resolution real-space images of isolated, single-layer graphene on SiO_2 substrates were obtained by scanning tunneling microscopy. Graphene processed using lithographic techniques is covered by photoresist residue, which must be cleaned to obtain atomic-resolution images. Such residue may be the adsorbates observed in TEM images, and may explain the rippling of suspended graphene. Rippling of graphene on the SiO_2 surface was determined by conformation of graphene to the underlying SiO_2, and not an intrinsic effect.

18.5.3.2.2 Electronic properties

Experimental results from transport measurements show that graphene has remarkably high electron mobility at room temperature, with reported values in excess of 15,000 $cm^2V^{-1}s^{-1}$. Additionally, the symmetry of the experimentally measured conductance indicates that the mobilities for holes and electrons should be nearly the same. The mobility is nearly independent of temperature between 10 K and 100 K, which implies that the dominant scattering mechanism is defect scattering. The scattering by acoustic phonons of graphene places intrinsic limits on the room temperature mobility to 200,000 $cm^2V^{-1}s^{-1}$ at a carrier density of 10^{12} cm^{-2}. The corresponding resistivity of the graphene sheet would be 10^{-6} Ω·cm. This is less than the resistivity of silver, the lowest resistivity substance known at room temperature. However, for graphene on SiO_2 substrates, scattering of electrons by optical phonons of the substrate is a larger effect at room temperature than scattering by graphene's own phonons. This limits the mobility to 40,000 cm^2 $V^{-1}s^{-1}$. Graphene exhibits a minimum conductivity on the order of $4e^2/h$. The origin of this minimum conductivity is still unclear. Due to its two-dimensional property, charge fractionalization is thought to occur in graphene. It may therefore be a suitable material for the construction of quantum computers using an ionic circuits.

18.5.3.2.3 Optical properties

Graphene's unique electronic properties produce an unexpectedly high opacity for an atomic monolayer, with a startlingly simple value: it absorbs TTα ≈ 2.3% of white light, where α is the fine-structure constant. This is a consequence of the unusual low-energy electronic structure of monolayer graphene that features electron and hole conical bands meeting each other at the Dirac point, which is qualitatively different from more common quadratic massive bands. Recently it has been demonstrated that the band gap of graphene can be tuned from 0 to 0.25 eV (about 5 micrometre wavelength) by applying voltage to a dual-gate bilayer graphene field-effect transistor (FET) at room temperature. The optical response of graphene nanoribbons has also been shown to be tunable into the terahertz regime by an applied magnetic field. It has been shown that graphene/graphene oxide system exhibits electrochromic behavior, allowing tuning of both linear and ultrafast optical properties

18.5.3.2.4 Thermal properties

The near-room temperature thermal conductivity of graphene was recently measured to be between $(4.84 \pm 0.44) \times 10^3$ to $(5.30 \pm 0.48) \times ss10^3$ $Wm^{-1}K^{-1}$. These measurements, made by a non-contact optical technique, are in excess of those measured for carbon nanotubes or diamond. It can be shown by using the Wiedemann-Franz law, that the thermal conduction is phonon-dominated. However, for a gated graphene strip, an applied gate bias causing a Fermi energy shift much larger than k_BT can cause the electronic contribution to increase and dominate over the phonon contribution at low temperatures. The ballistic thermal conductance of graphene is isotropic.

In graphite, the c-axis (out of plane) thermal conductivity is over a factor of ~100 smaller due to the weak binding forces between basal planes as well as the larger lattice spacing. In addition, the ballistic thermal conductance of a graphene is shown to give the lower limit of the ballistic thermal conductances, per unit circumference, length of carbon nanotubes.

18.5.3.2.5 Mechanical properties

Measurements have shown that graphene has a breaking strength 200 times greater than steel, with a tensile strength of 130GPa (19,000,000 psi). Graphene paper or GP has recently been developed by a research department from the University of Technology Sydney by Guoxiu Wang, which can be processed, reshaped and reformed from its original raw material state. Graphene sheets, held together by van der Waals forces, were suspended over SiO_2 cavities where an AFM tip was probed to test its mechanical properties. Its spring constant was in the range 1–5 N/m and the Young's modulus was 0.5 TPa, which differs from that of the bulk graphite. These high values make graphene very strong and rigid. These intrinsic properties could lead to using graphene for NEMS applications such as pressure sensors and resonators.

As is true of all materials, regions of graphene are subject to thermal and quantum fluctuations in relative displacement. It is believed that a sufficiently large 2D structure, in the absence of applied lateral tension, will bend and crumple to form a fluctuating 3D structure. Researchers have observed ripples in suspended layers of graphene, and it has been proposed that the ripples are caused by thermal fluctuations in the material. As a consequence of these dynamical deformations, it is debatable whether graphene is truly a 2D structure.

18.5.3.3 Potential applications

18.5.3.3.1 Graphene nanoribbons

Graphene nanoribbons (GNRs) are essentially single layers of graphene that are cut in a particular pattern to give it certain electrical properties. Depending on how the un-bonded edges are configured, they can either be in a zigzag or armchair configuration. Calculations based on tight binding predict that zigzag GNRs are always metallic while armchairs can be either metallic or semiconducting, depending on their width. However, recent density functional theory calculations show that armchair nanoribbons are semiconducting with an energy gap scaling with the inverse of the GNR width. Zigzag nanoribbons are also semiconducting and present spin polarized edges. Their 2D structure, high electrical and thermal conductivity, and low noise also make GNRs a possible alternative to copper for integrated circuit interconnects. Some research is also being done to create quantum dots by changing the width of GNRs at select points along the ribbon, creating quantum confinement.

18.5.3.3.2 Graphene transistors

Due to its high electronic quality, graphene has also attracted the interest of technologists who see it as a way of constructing ballistic transistors. Graphene exhibits a pronounced response to perpendicular external electric fields, allowing one to build FETs (field effect transistors). The switching effect is based on a reversible chemical modification of the graphene layer and gives an on-off ratio of greater than six orders of magnitude. These reversible switches could potentially be applied to nonvolatile memories. Although these graphene chips open up a range of new applications, their practical use is limited by a very small voltage gain (typically, the amplitude of the output signal is about 40 times less than that of the input signal). Moreover, none of these circuits was demonstrated to operate at frequencies higher than 25 kHz.

18.5.3.3.3 Integrated circuits

Graphene has the ideal properties to be an excellent component of integrated circuits. Graphene has a high carrier mobility, as well as low noise, allowing it to be used as the channel in a FET. The issue is that single sheets of graphene are hard to produce, and even harder to make on top of an appropriate substrate. Researchers are looking into methods of transferring single graphene sheets from their source of origin (mechanical exfoliation on SiO_2/Si or thermal graphitization of a SiC surface) onto a target substrate of interest. In 2008, the smallest transistor so far, one atom thick, 10 atoms wide was made of grapheme.

18.6 QUANTUM WELLS

A quantum well (Fig. 18.9) is a potential well with only discrete energy values.

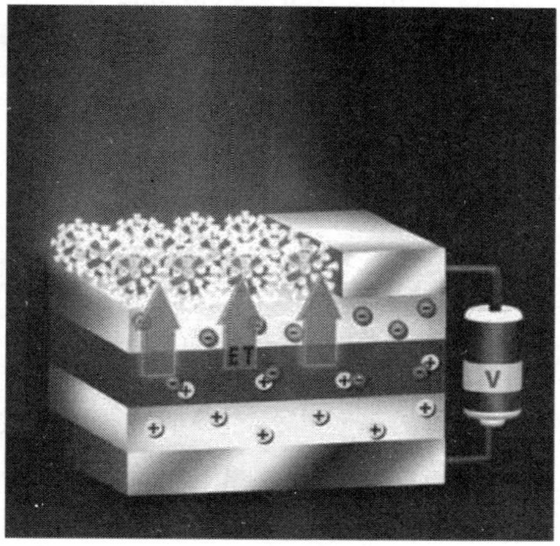

Fig. 18.9 *Quantum well.*

If one makes a hetero structure with sufficiently thin layers, quantum interference effects begin to appear prominently in the motion of the electrons. The simplest structure in which these may be observed is a quantum well, which simply consists of a thin layer of a narrower-gap semiconductor between thicker layers of a wider-gap material. The band profile then shows a `rectangular well as shown below.

Fig. 18.10 shows the energy band profile of a multilayer structure containing three quantum wells showing the confined states in each well. The structure consists of GaAs wells of thickness 11, 8, and 5 nm in $Al_{0.4}Ga_{0.6}$ As barrier layers. The gaps in the lines indicating the confined state energies show the locations of nodes of the corresponding wave functions.

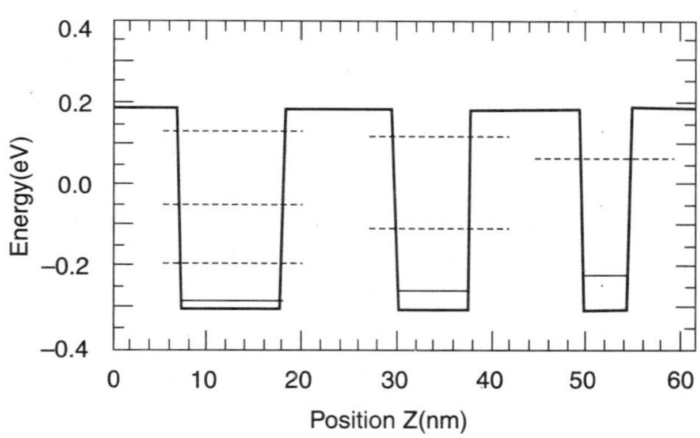

Fig. 18.10 *Energy band profile of a multilayer structure containing three quantum wells.*

The electron wave functions in such a well consist of a series of standing waves, such as a resonant cavity in acoustic, optical or microwave technologies. The energy separation between these stationary states is enhanced by the small effective mass of electrons in the conduction bands of direct-gap semiconductors. With advanced epitaxial techniques, the potential profile of the quantum well need not be rectangular. Because the band-edge energy is usually linear in the composition, $E_{v,c}$ will follow the functional form of the composition. The quantum states in two parabolic wells are illustrated in Fig. 18.11.

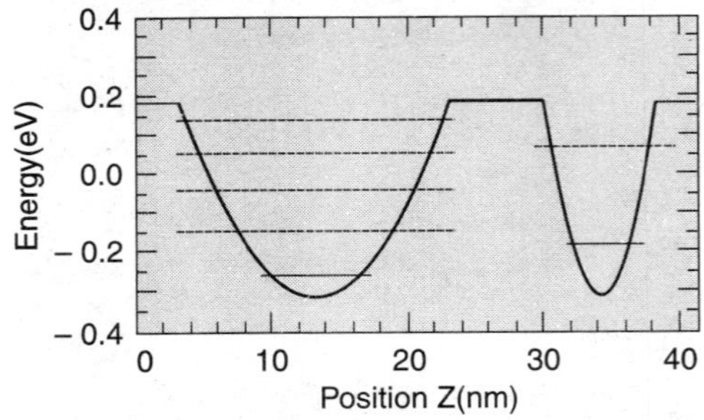

Fig. 18.11 *The quantum states in two parabolic wells.*

Quantum well hetero structures are key components of many optoelectronic devices, because they can increase the strength of electro-optical interactions by confining the carriers to small regions.

18.6.1 Energy Levels

18.6.1.1 The infinite quantum well

The infinite well represents one of the simplest quantum mechanical problems: it consists of a particle in a well which is defined by a zero potential between x=0 and $x=L_x$ and an infinite potential on either side of the well. The potential and the first five energy levels are shown in the Fig. 18.12.

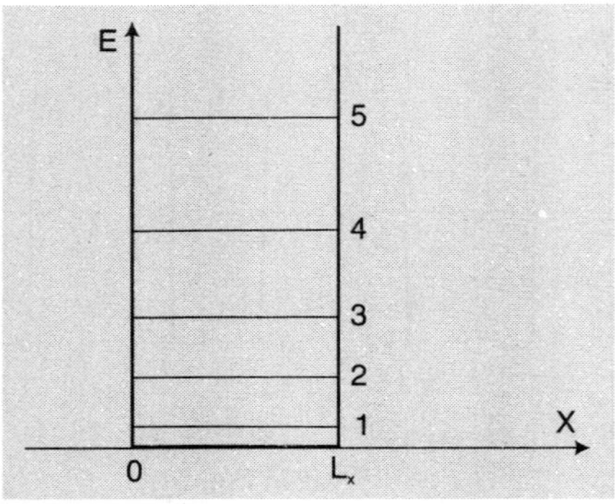

Fig. 18.12 *Potential energy of an infinite well with width L_x. Also indicated are the lowest five energy levels in the well.*

The energy levels in such a infinite well is given by

$$E_n = \frac{h^2}{2m^*}\left(\frac{n}{2L_x}\right)^2, \text{ with } n = 1, 2, \qquad \text{... } \textbf{\textit{eq.}-18.14}$$

where h is Planck's constant and m^* is the effective mass of the particle. n is the quantum number associated with the nth energy level, with energy E_n. Note that the lowest possible energy is not zero even though the potential is zero within the well. Also that, the distance between adjacent energy levels increases as the energy increases. Two electrons with opposite spin can occupy each level as n and s are the only two quantum numbers needed to describe this system (Fig. 18.13).

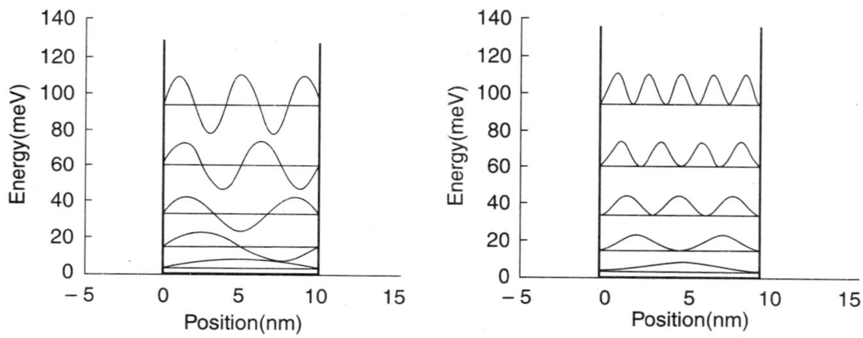

Fig. 18.13 *Energy levels, wave functions (left) and probability density functions (right) in an infinite quantum well. The figure is calculated for a 10 nm wide well containing an electron with mass m_0.*

18.6.1.2 Finite quantum well

The potential and the first three energy levels are shown in the Fig. 18.14 and 18.15:

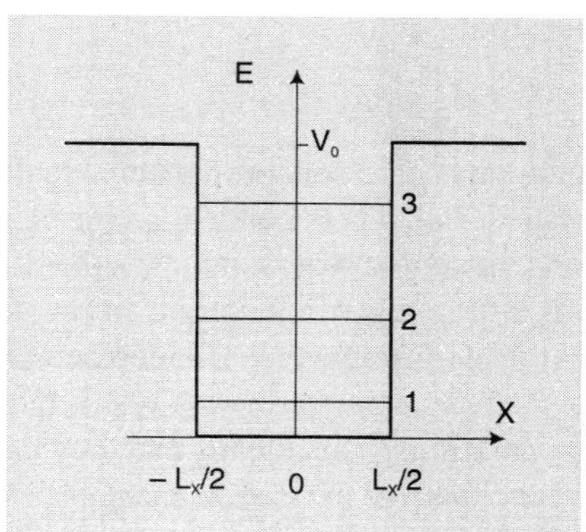

Fig. 18.14 *Potential energy of a finite well with width L_x. Also indicated are the lowest three energy levels in the well.*

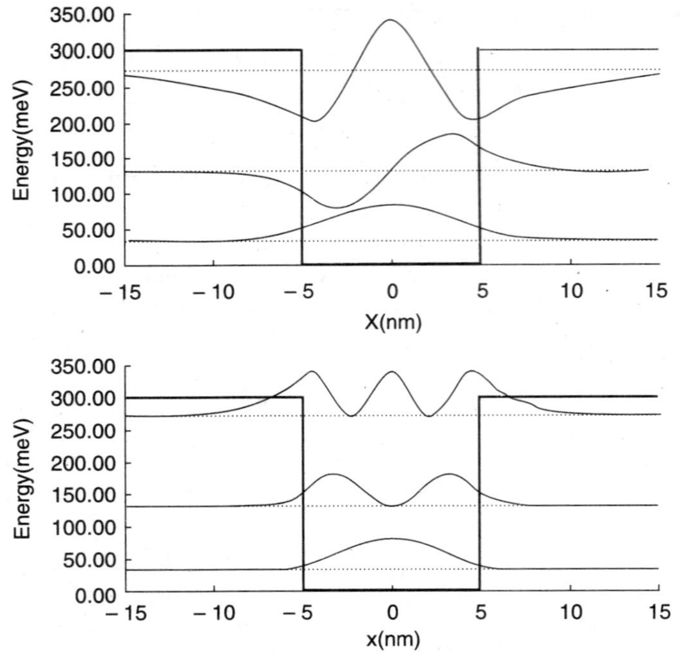

Fig. 18.15 *Energy levels, wave functions (left) and probability density functions (right) in a finite quantum well.*

18.7 QUANTUM WIRE

Quantum wire is an electrically conducting wire, in which quantum effects are affecting transport properties. Due to, the quantum confinement of conduction electrons in the transverse direction of the wire their transverse energy is quantized into a series of discrete values E_0 (ground state energy, with lower value), E_1. One consequence of this quantization is that the classical formula for calculating the electrical resisitivity of a wire:

$$R = \rho\frac{1}{A} \qquad \textbf{...eq.-18.15}$$

The importance of the quantization is inversely proportional to the diameter of the nanowire for a given material. From material to material, it is dependent on the electronic properties, especially on the effective mass of the electrons. In simple words, it means that it will depend on how conduction electrons interact with the atoms within a given material. In practice, semiconductors show clear conductance quantization for large wire transverse dimensions (100 nm) because the electronic modes due to confinement are spatially extended. As a result their Fermi wavelengths are large and thus they have low energy separations. This means that they can only be resolved at cryogenic temperature (few kelvins) where the thermal excitation energy is lower than the inter-mode energy separation.

18.7.1 Optical Properties

The initial motivation was related to the one-dimensional (1D) singularity in the single particle density of states (DOS) that was expected to induce sharp peaks in the optical spectra, thereby leading to structures with improved optical efficiency as compared to their two-dimensional (2D) and three-dimensional (3D) counterparts. Coulomb correlations strongly modify this simple picture and this was first pointed out by Ogawa and Takagahara, using a idealized 1D model. We have implemented a fully 3D approach based on the density-matrix formalism, which results in a set of generalized (multi sub band) semiconductor Bloch equations (SBE), which is able to treat both low-density (excitonic) and high-density (gain) regimes on the same footing, while retaining the full complexity of state-of-the-art samples. For example, it has been used to describe structures obtained by epitaxial growth on non-planar substrates (V-shaped wires) or by cleaved-edge quantum well over growth (T-shaped wires), where the lateral extension of the ground single-particle states is still significant, the excited states gradually approach a 2D-like behavior, and the sub band separation is relatively small, so that the coupling between different sub bands may be important.

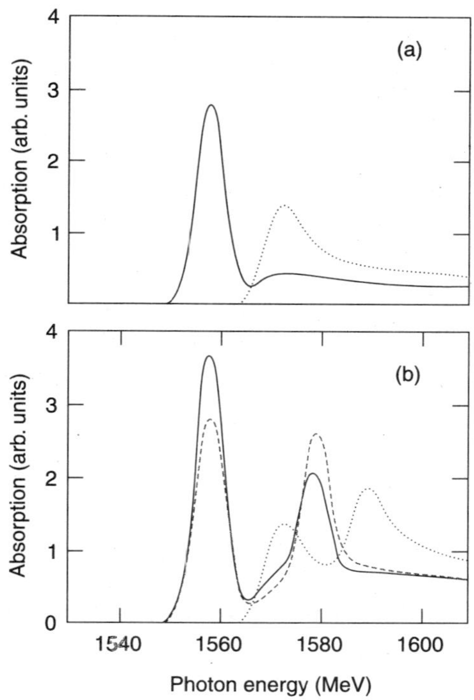

Fig. 18.16 *The variation of optical absorption with photon energy for a V-grooved QWR.*

The figure 18.16 shows the calculated absorption for a V-grooved QWR. Dashed line corresponds to single particle calculations, while full lines show the results including electron-hole correlations. The calculations have shown that:

1. The correlated absorption spectra of realistic wires do show a strong quenching of the 1D single-particle singularity, and this effect does not depend on details of the wire cross section. The Sommerfeld factor, which is greater than unity in the bulk and in quantum

wells (called Coulomb enhancement), is instead smaller than unity in QWRs (Coulomb suppression), thus reducing the influence of dimensionality on the optical spectra.

2. The Coulomb-induced suppression of the 1D singularity is found to hold not only in the linear regime but also at higher carrier densities, and to persist in the gain regime. The above results have had implications for perspective devices, since the initial motivations based on single-particle models are now recognized to be far too simplified.

18.8 QUANTUM DOT

A quantum dot is a portion of matter (*Eg*. semiconductor) whose excitons are confined in all three spatial dimensions. Consequently, such materials have electronic properties intermediate between those of bulk semiconductors and those of discrete molecules. They were discovered at the beginning of the 1980s by Alexei Ekimov in a glass matrix and by Louis E. Brus in colloidal solutions. The term quantum dot was coined by Mark Reed.

Quantum dots are semiconductors whose electronic characteristics are closely related to the size and shape of the individual crystal. Generally, the smaller the size of the crystal, the larger the band gap, the greater the difference in energy between the highest valence band and the lowest conduction band becomes, therefore more energy is needed to excite the dot, and concurrently, more energy is released when the crystal returns to its resting state. For example, in fluorescent dye applications, this equates to higher frequencies of light emitted after excitation of the dot as the crystal size grows smaller, resulting in a color shift from red to blue in the light emitted. In addition to such tuning, a main advantage with quantum dots is that, because of the high level of control possible over the size of the crystals produced, it is possible to have very precise control over the conductive properties of the material. Quantum dots of different sizes can be assembled into a gradient multi-layer nanofilm.

18.8.1 Fabrication

1. Self-assembled quantum dots are typically between 5 and 50 nm in size. Quantum dots defined by lithographically patterned gate electrodes or by etching on two-dimensional electron gases in semiconductor heterostructures can have lateral dimensions exceeding 100 nm.
2. Some quantum dots are small regions of one material buried in another with a larger band gap. These can be so-called core-shell structures, *Eg*., with CdSe in the core and ZnS in the shell or from special forms of silica called ormosil.
3. Quantum dots sometimes occur spontaneously in quantum well structures due to monolayer fluctuations in the well's thickness.
4. Self-assembled quantum dots nucleate spontaneously under certain conditions during molecular beam epitaxy (MBE) and metallorganic vapor phase epitaxy (MOVPE), when a material is grown on a substrate to which it is not lattice matched. The resulting strain produces coherently strained islands on top of a two-dimensional wetting layer. This growth mode is known as Stranski–Krastanov growth. The islands can be subsequently buried to form the quantum dot. This fabrication method has potential for applications in quantum cryptography (i.e. single photon sources) and quantum computation. The main

limitations of this method are the cost of fabrication and the lack of control over positioning of individual dots.

5. Individual quantum dots can be created from two-dimensional electron or hole gases present in remotely doped quantum wells or semiconductor heterostructures called lateral quantum dots. The sample surface is coated with a thin layer of resist. A lateral pattern is then defined in the resist by electron beam lithography. This pattern can then be transferred to the electron or hole gas by etching, or by depositing metal electrodes (lift-off process) that allow the application of external voltages between the electron gas and the electrodes. Such quantum dots are mainly of interest for experiments and applications involving electron or hole transport, i.e., an electrical current.
6. The energy spectrum of a quantum dot can be engineered by controlling the geometrical size, shape, and the strength of the confinement potential. Also, in contrast to atoms, it is relatively easy to connect quantum dots by tunnel barriers to conducting leads, which allows the application of the techniques of tunneling spectroscopy for their investigation.
7. The quantum dot absorption features correspond to transitions between discrete, three dimensional particle in a box and the states of electron and hole, both confined to the same nanometer-size box. These discrete transitions are reminiscent of atomic spectra and have resulted in quantum dots also being called artificial atoms.
8. Confinement in quantum dots can also arise from electrostatic potentials (generated by external electrodes, doping, strain, or impurities).

18.8.2 Optical Properties

An immediate optical feature of colloidal quantum dots is their coloration. While the material which makes up a quantum dot defines its intrinsic energy signature, the nanocrystal's quantum confined size is more significant at energies near the band gap. Thus quantum dots of the same material, but with different sizes, can emit light of different colors. The physical reason is the quantum confinement effect.

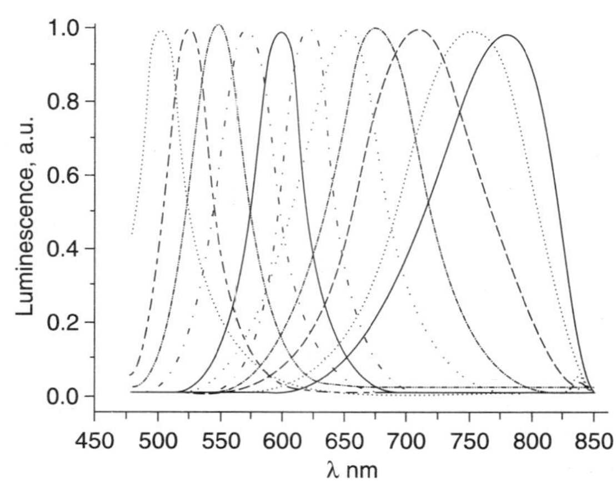

Fig. 18.17 *Fluorescence spectrum of the quantum dot.*

The larger the dot, the higher will be the shift towards red wave length (lower energy) in its fluorescence spectrum. Conversely, smaller dots emit bluer (higher energy) light. The coloration is directly related to the energy levels of the quantum dot (fig. 18.17) Quantitatively speaking, the band gap energy that determines the energy (and hence color) of the fluorescent light is inversely proportional to the size of the quantum dot. Larger quantum dots have more energy levels which are also more closely spaced. This allows the quantum dot to absorb photons containing less energy, i.e., those closer to the red end of the spectrum. Recent articles in nanotechnology and in other journals have begun to suggest that the shape of the quantum dot may be a factor in the coloration as well, but as yet not enough information is available. Furthermore, it was shown that the lifetime of fluorescence is determined by the size of the quantum dot. Larger dots have more closely spaced energy levels in which the electron-hole pair can be trapped. Therefore, electron-hole pairs in larger dots live longer causing larger dots to show a longer lifetime.

As with any crystalline semiconductor, a quantum dot's electronic wave functions extend over the crystal lattice. Similar to a molecule, a quantum dot has both a quantized energy spectrum and a quantized density of electronic states near the edge of the band gap. QD's can be synthesized with larger (thicker) shells (CdSe qdots with CdS shells). The shell thickness has shown direct correlation to the lifetime and emission intensity.

18.8.3 Applications

18.8.3.1 Computation

Quantum dot technology is one of the most promising candidates for use in solid-state quantum computation. By applying small voltages to the leads, the flow of electrons through the quantum dot can be controlled and thereby precise measurements of the spin and other properties therein can be made. With several entangled quantum dots, or qubits, plus a way of performing operations, quantum calculations and the computers that would perform them might be possible.

18.8.3.2 Photovoltaic devices

Quantum dots may be able to increase the efficiency and reduce the cost of today's typical silicon photovoltaic cells. According to an experimental proof from 2006 quantum dots of lead selenide can produce as many as seven excitons from one high energy photon of sunlight. This compares favorably to today's photovoltaic cells which can only manage one exciton per high-energy photon, with high kinetic energy carriers losing their energy as heat. This would not result in a 7-fold increase in final output however, but could boost the maximum theoretical efficiency from 31% to 42%. Quantum dot photovoltaics would theoretically be cheaper to manufacture, as they can be made using simple chemical reactions. The generation of more than one exciton by a single photon is called multiple exciton generation (MEG) or carrier multiplication.

18.8.3.3 Light emitting devices

There are several inquiries into using quantum dots as light emitting diodes to make displays and other light sources, such as QD-LED displays and QD-WLED" (White LED). Quantum dots are valued for displays, because they emit light in very specific gaussian distributions. This can result in a display that more accurately renders the colors that the human eye can perceive. Quantum dots also require very little power since they are not color filtered. Additionally, since the discovery of white light emitting QD, general solid-state lighting applications appear closer than ever. A color liquid crystal display (LCD), for example, is usually powered by a single fluorescent lamp (or occasionally, conventional white LEDs) that is color filtered to produce red, green, and blue pixels. Displays that intrinsically produce monochromatic light can be more efficient, since more of the light produced reaches the eye.

18.8.3.4 Biological applications

In modern biological analysis, various kinds of organic dyes are used. However, with each passing year, more flexibility is being required of these dyes, and the traditional dyes are often unable to meet the expectations. To this end, quantum dots have quickly filled in the role, being found to be superior to traditional organic dyes on several counts, one of the most immediately obvious being brightness as well as their stability. It has been estimated that quantum dots are 20 times brighter and 100 times more stable than traditional fluorescent reporters. For single particle tracking, the irregular blinking of quantum dots is a minor drawback.

The usage of quantum dots for highly sensitive cellular imaging has seen major advances over the past decade. The improved photo stability of quantum dots, for example, allows the acquisition of many consecutive focal-plane images that can be reconstructed into a high-resolution three-dimensional image. Another application that takes advantage of the extraordinary photo stability of quantum dot probes is the real-time tracking of molecules and cells over extended periods of time. Antibodies, streptavidin, peptides, nucleic acid aptamers, or small-molecule ligands can be used to target quantum dots to specific proteins on cells. Researchers were able to observe quantum dots in lymph nodes of mice for more than 4 months.

Semiconductor quantum dots have also been employed for *in vitro* imaging of pre-labeled cells. The ability to image single cell migration in real time is expected to be important to several research areas such as embryogenesis, cancer metastasis, stem-cell therapeutics, and lymphocyte immunology. Scientists have proven that quantum dots are dramatically better than existing methods for delivering a gene silencing tool, known as siRNA, into cells.

First attempts have been made to use quantum dots for tumor targeting under *in vivo* conditions. There exist two basic targeting schemes: active targeting and passive targeting. In the case of active targeting, quantum dots are functionalized with tumor-specific binding sites to selectively bind to tumor cells. Passive targeting utilizes the enhanced permeation and retention of tumor cells for the delivery of quantum dot probes. Fast-growing tumor cells typically have more permeable membranes than healthy cells, allowing the leakage of small nanoparticles into the cell body. Moreover, tumor cells lack an effective lymphatic drainage system, which leads to subsequent nanoparticle-accumulation.

18.9 NANO ROBOTICS

Nanorobotics is the emerging technology field of creating machines or robots whose components are at or close to the microscopic scale of a nanometer (10^{-9} meters). More specifically, nanorobotics refers to the nanotechnology engineering discipline of designing and building nanorobots, with devices ranging in size from 0.1-10 micrometers and constructed of nanoscale or molecular components. The names nanorobots, nanoids, nanites, nanomachines or nanomites have also been used to describe these devices currently under research and development.

Another definition is a robot that allows precision interactions with nanoscale objects, or can manipulate with nanoscale resolution. Such devices are more related to Microscopy or Scanning probe microscopy, instead of the description of nanorobots as molecular machine. Following the microscopy definition even a large apparatus such as an atomic force microscope can be considered a nanorobotic instrument when configured to perform nanomanipulation. For this perspective, macroscale robots or microrobots that can move with nanoscale precision can also be considered nanorobots.

18.9.1 Biomedical Appilications

The enormous potential in the biomedical capabilities of nanorobots and the imprecision and side effects of medical treatments today make nanorobots very desirable. Medical treatment today involves the use of surgery and drug therapy. Surgery is a direct, manual approach to fixing the body. However, no matter how highly trained the specialists may be, surgery can still be dangerous since anesthetics, infections, organ rejection, and missed cancer cells can all cause failure. Surgeons due to lack of fine scale control. From the perspective of a cell, a fine surgical scalpel is as crude as a blunt tool. Invasive surgery wounds peripheral tissue and causes unnecessary harm to the patient.

18.9.2 Properties of Nanomedical Robots

Nanorobots will typically be .5 to 3 microns large with 1-100 nm parts. Three microns is the upper limit of any nanorobot because nanorobots of larger size will block capillary flow. The nanorobot's structure will have two spaces that will consist of an interior and exterior. The exterior of the nanorobot will be subjected to the various chemical liquids in our bodies but the interior of the nanorobot will be a closed, vacuum environment into which liquids from the outside cannot normally enter unless it is needed for chemical analysis. A nanorobot will prevent itself, from being attacked by the immune system by having a passive, diamond exterior. The diamond exterior will have to be smooth and flawless because past experiments have shown that this prevents Ieukocytes activities since the exterior is chemically inert and have low bioactivity. Nanorobots will communicate with the doctor by encoding messages to acoustic signals at carrier wave frequencies of 1-100 MHz. When the doctor gives a command to the nanorobots, the nanorobots can receive the message from the acoustic sensors on the nanorobots and implement the doctor's orders. Replication is a crucial basic capability for molecular manufacturing. However, in the case of nanorobots, we should restrict manufacturing to *in vitro* (in laboratory) replication. Replication in the body (*in vivo*) is dangerous because it might go out of control. If even replicating bacteria can give humans so many diseases, the thought of replicating nanorobots can present unimaginable dangers to the human body. When the nanorobots are finished with their jobs, they will be disposed from the body to prevent them from breaking down and malfunctioning.

18.9.3 Nanorobotic Applications

The availability of advanced nanomedical instrumentalities should not significantly alter the classical medical treatment methodology, although the patient experiences and outcomes will be greatly improved. Treatment in the nanomedical era will become faster and more accurate, efficient and effective.

18.9.3.1 Drug delivery

To be therapeutically effective, drugs need to be protected during their transit to the target action site in the body while maintaining their biological and chemicals properties. Some drugs are highly toxic and can cause harsh side effects and reduced therapeutic effect if they decompose during their delivery. Depending on where the drugs will be absorbed (i.e. colon, small intestine, etc), and whether certain natural defense mechanisms need to be passed through such as the blood-brain barrier, the transit time and delivery challenges can be greatly different. Once a drug arrives at its destination, it needs to be released at an appropriate rate for it to be effective. If the drug is released too rapidly it might not be completely absorbed, or it might cause gastro-intestinal irritation and other side effects. The drug delivery system must positively impact the rate of absorption, distribution, metabolism, and excretion of the drug or other substances in the body. In addition, the drug delivery system must allow the drug to bind to its target receptor and influence that receptor's signaling and action, as well as other drugs, which might also be active in the body.

Drug delivery systems also have severe restrictions on the materials and production processes that can be used. The drug delivery material must be compatible and bind easily with the drug, and be bioresorbable (degrade into fragments after use which are either metabolized or eliminated via normal excretory routes). The production process must respect stringent conditions on processing and chemistry that won't degrade the drug, and still provide a cost effective product.

18.9.3.1.1 Drug encapsulation

One major class of drug delivery systems is materials that encapsulate drugs to protect them during transit in the body. Drug encapsulation materials include liposomes and polymers (Polylactide (PLA) and Lactide-co-Glycolide (PLGA)) which are used as microscale particles. The materials form capsules around the drugs and permit timed drug release to occur as the drug diffuses through the encapsulation material. The drugs can also be released as the encapsulation material degrades or erodes in the body.

Nanoparticle encapsulation is also being investigated for the treatment of neurological disorders to deliver therapeutic molecules directly to the central nervous system beyond the blood-brain barrier, and to the eye beyond the blood-retina barrier. Applications could include Parkinson's, Huntington's, Alzheimer's, ALS and diseases of the eye.

18.9.3.1.2 Functional drug carriers

Another class of drug delivery systems where nanotechnology offers interesting solutions is in the area of nanomaterials that carry drugs to their destination sites and also have functional properties. Certain nanostructures can be controlled to link with a drug, a targeting molecule, and an imaging agent, then attract specific cells and release their payload when required.

18.9.3.2 Drug discovery

Nano and micro technologies are part of the latest advanced solutions and new paradigms for decreasing the discovery and development times for new drugs, and potentially reducing the development costs. Traditional trial-and error methods have contributed to a discovery process lasting 10 years or more for new drugs to reach the market. In recent years, a number of new and complementary technologies have been developed which considerably impact the drug discovery process.

High throughput arrays and ultra-sensitive labeling and detection technologies are being used to increase the speed and accuracy of identifying genes and genetic materials for drug discovery and development. These micro and nano technologies along with information technology solutions such as combinatorial chemistry, computational biology, computer aided drug design, data mining, and data processing tools are addressing the challenges related to eliminating critical bottlenecks in drug discovery replacement. While most types of tissues repair the interaction of stem cells with chemical modulators, there are differences in the ways that various tissues heal.

Hard tissues such as bone and teeth heal by reproducing tissues indistinguishable from the original. However in cases where a dental or artificial bone implant is required, the structural material used in the implant may trigger immune rejection, corrode in the body fluids, or no longer bond to the host bone. This requires additional surgery or result in the loss of the implant's function. In many cases, the failure occurs at the tissue implant interface, which may be due to the implant material weakening its bond with the natural material. To overcome this, implants are often coated with a biocompatible material to increase their adherence properties and produce a greater surface area to volume ratio for the highest possible contact area between the implant and natural tissue. Soft tissues such as skin, muscle, nerves, blood vessels and ligaments repair damaged areas with fibrous tissue. Damaged tissue from various sources such as burns and ulcers can be self-repaired by the body, but can also result in scar formation. Graft material using artificial sheets can replace skin and other tissue with reasonable graft stability and cosmetic outcome.

18.9.3.3 Implantable devics

Nanotechnology offers sensing technologies that provide more accurate and timely medical information for diagnosing disease, and miniature devices that can administer treatment automatically if required. Health assessment can require medical professionals, invasive procedures and extensive laboratory testing to collect data and diagnose disease. This process can take hours, days or weeks for scheduling and obtaining results. Some medical information is extremely time sensitive such as finding out if there is sufficient blood flow to an organ or tissue after transplant or reconstructive surgery, before irreversible damage occurs.

Certain medical tests such as biopsies are subjective and can provide inconclusive or incorrect results. In a false negative result where a needle misses the tumor and then samples a normal tissue, the cancer goes untreated and can impact a patient's chances for long-term survival.

Some tests such as diabetes blood sugar levels require patients to administer the test themselves to avoid the risk of their blood glucose falling to dangerous levels. Certain users such as children and the elderly may not be able to perform the test properly, timely or without considerable pain.

People who are exposed to radiation or hazardous chemicals in their work environment are at a higher risk of illness. Occasional testing is typically done but may not detect a disease in its early stage. Early detection could initiate timely treatment with a higher chance of success, and have a worker removed from the hazardous environment to prevent further damage.

Nanotechnology can new offer new implantable and/or wearable sensing technologies that provide continuous and extremely accurate medical information. Complementary microprocessors and miniature devices can be incorporated with sensors to diagnose disease, transmit information and administer treatment automatically if required. Example applications are as follows.

18.9.3.3.1 Retina implants

Retinal implants are in development to restore vision by electrically stimulating functional neurons in the retina. One approach being developed by various groups including a project at Argonne National Laboratory is an artificial retina implanted in the back of the retina. The artificial retina uses a miniature video camera.

18.9.3.4 *Surgical aids*

18.9.3.4.1 Operating tools

Medical devices that contain nano and micro technologies will allow surgeons to perform familiar tasks with greater precision and safety, monitor physiological and biomechanical parameters more accurately, and perform new tasks that are not currently done.

18.9.3.4.2 Surgical robotics

Robotic surgical systems are being developed to provide surgeons with unprecedented control over precision instruments. This is particularly useful for minimally invasive surgery. Instead of manipulating surgical instruments, surgeons use their thumbs and fingers to move joystick handles on a control console to maneuver two robot arms containing miniature instruments that are inserted into ports in the patient. The surgeon's movements transform large motions on the remote controls into micro-movements on the robot arms to greatly improve mechanical precision and safety.

Glossary

1. ABSOLUTE ERROR The difference between a measured value and the true or accepted value.
2. ABSORBANCE A measure of the amount of light absorbed by a solution. It is equal to the negative logarithm of the ratio of initial and final light intensity.
3. ACCURACY The agreement between a mean measured value and a true or accepted value as quantified by error.
4. ACTINOMETERS An actinometer is a chemical system or physical device which determines the number of photons in a beam integrally or per unit time. Actinometers are instruments used to measure the heating power of radiation and are used in meteorology to measure solar radiation as pyrheliometers.
5. ACTUATOR An actuator is a device that by means of an agent, such as electric field or heat, produces a desired effect on a target or on the surroundings. Actuators are used extensively in diverse industrial applications like controlling flow or position, in photocopying machines, circuit breakers, tightening of textile screens, automobiles etc.
6. AMINO ACID Building block of proteins, around 20 amino acids are present in proteins: alanine, arginine, asparagine, aspartic acid, cysteine, glutamic acid, glutamine, glycine, histidine, isoleucine, leucine, lysine, methionine, phenylalanine, proline, serine, threonine, tryptophan, tyrosine and valine.
7. AMMONIA BIOSENSOR Ammonia oxidizing bacteria (AOB) used in biosensors for determination of NH_4^+ concentrations from the samples. The high sensitivity of bacterial ammonia oxidation to toxic substances may be used for the assay of toxicity. The biosensor consists of a Clark type Oxygen sensor with a layer of immobilized *Nitrosomonas europaea* cells.
8. AMPEROMETRIC A type of biosensor transducer that measures the current produced by a biological reaction *Eg:* an enzyme reaction.
9. AMPEROMETRIC BIOSENSORS Amperometric biosensors function by the production of a current when a potential is applied between two electrodes. These biosensors are generally had response time, dynamic ranges and sensitivities similar to the Potentiometric biosensors. The simplest amperometric biosensors commonly involve the Clark oxygen electrode. This consists of a platinum cathode at which oxygen is reduced and a silver/silver chloride reference electrode. When a potential of -0.6 V, relative to the Ag/AgCl electrode is applied to the platinum cathode, a current proportional to the oxygen concentration is produced.

10. AMPEROMETRIC MICROBIAL BIOSENSOR Amperometric microbial biosensor operates at fixed potential with respect to a reference electrode and involves the detection of the current generated by the oxidation or reduction of species at the surface of the electrode. Amperometric microbial biosensors have been widely developed for the determination of biochemical oxygen demand (BOD) for the measurement of biodegradable organic pollutants in aqueous samples. The microbial strains used as biological sensing element includes *Torulopsis candida, Trichosporon cutaneum, Pseudomonas putida, Klebsiella oxytoca* AS1, *Bacillus subtilis*, *Arxula adeninivorans* LS3, *Serratia marcescens* LSY4, *Pseudomonas* sp., *P. fluorescens*, *P. putida* SG10, *Thermophilic bacteria*, *Hansennula anomala* and Yeast.
11. AMPLIFIER The electronic component of a biosensor that increases the strength of the signal.
12. ANTHROPOMETRY The biosensors used to measure human body surfaces are called anthropometry.
13. ANTIBODY An inducible immunoglobulin protein that is produced by the B lymphocytes of the immune system. It recognizes and binds to a specific antigen molecule of a foreign substance introduced into the organism. Upon binding to corresponding antigens, the antibodies set in motion a process to eliminate the antigens.
14. ANTIGEN Any foreign substance such as virus, bacterium or protein, after introduction into an organism, the antigenic substances elicit an immune response by stimulating the production of specific antibodies.
15. APTAMER Aptamers are linear nucleotide sequences, usually of 15 – 40 nucleotide length, undergoing many intra-molecular interactions, which wrinkle the oligonucleotide chain. *Eg:* Proteins, Lipids, Carbohydrates and DNA.
16. ATOMIC FORCE MICROSCOPE (AFM) A scanning probe microscopy instrument capable of revealing the structure of samples. The AFM uses a sharp metal tip positioned over a conducting or non-conducting substrate and the surface topography is mapped out by measuring the mechanical force exerted on the tip.
17. BIOCHEICAL OXYGEN DEMAND (BOD) BOD gives a measure of the amount of biodegradable material present determined as the mass of oxygen required for degradation per unit volume. The BOD is usually measured in milligrams per liter or in parts per million of oxygen required. An excessive BOD or COD would result in using up the oxygen in rivers and lakes causing fish to die.
18. BIOCHIP Miniaturised substrates onto which a large number of biomolecules are attached with high density and in a defined microarray.
19. BIO-COMPUTER There are three general principles of bio-computer: 1. Based on the current John Von Newman structure, use different bio-molecular states to represent 1/0 switch. 2. Use some bio-molecular special property to calculate. 3. Use complete different structure, somewhat like brain.
20. BIOSYNTHESIS The process by which living organisms produce chemical compounds.
21. BIOLOGICAL ELEMENT A biological element is a material related to biomolecules. The biological material may be a protein such as an enzyme, antibody; a nucleic acid (DNA or RNA), antibody fragment, a whole microbial cell, even a plant or an animal tissue and microbial products.
22. BIOMOLECULAR COMPUTING The field of bio-molecular computing goes further by using these processes to perform information processing that has no intrinsic connection with the processing of biological molecules.
23. BIOSENSOR A biosensor can be defined as *"a device incorporated by a biological sensing element connected to a transducer"*. An analyte of the sensor is a biological component (target analyte) which detects and measures the chemical components (inorganic/organic).

24. BRAIN TUMOR BIOCHIP The Brain Tumor Biochip is used to understanding the gene expression in brain tumors, elucidating new criteria for the classification of brain tumors, developing new diagnostic technology and therapeutic principle in the treatment of individuals afflicted with this illness. The Brain Tumor Biochip is the 10K oligonucleotides microarray, which is used to investigation of the gene expression in the brain tumor samples and this microarray platform that functions by Impedimetric recognition of DNA hybridization. This biochip involves neurosurgery, molecular genetics, engineering and bioinformatics.
25. BULK ACOUSTIC-WAVE RESONATORS (BAW) An array of micro-mechanical resonators made from nano-crystalline diamond. Used to find and hold, a radio frequency.
26. CALIBRATION The process of submitting samples of known value to an instrument, in order to establish the relationship of value to instrumental output.
27. CARBON A nonmetallic element found in all living things. Carbon is part of all organic compounds and in combined form of many inorganic substances. *Eg:* Diamonds, graphite, and fullerenes are pure forms of carbon.
28. CARBON NANOTUBES (CNTs) Carbon Nanotubes (CNTs) are allotropes of carbon with a cylindrical nanostructure. These cylindrical carbon molecules have novel properties, making them potentially useful in many applications in nanotechnology, electronics, optics and other fields of materials science, as well as potential uses in architectural fields.
29. CATALYTIC Relating to, causing or involving catalysis. A catalytic enzyme biosensor involves the enzyme acting on a target analyte to change it to a chemically different product.
30. CELL A small, usually microscopic, membrane bound structure that is the fundamental unit of all living things. Organisms can be made up of one cell (*Eg:* Unicellular- bacteria) or many cells (*Eg.:* multicellular - human beings are made up of an estimated 100,000 billion cells.)
31. CELL RECOGNITION The process by which a cell in a multicellular organism interprets its surroundings.
32. CHARACTERIZATION Analysis of critical features of an object or concept.
33. CHEMICAL CROSS-LINKAGE Biocatalysts can be immobilized through chemical cross-linkage by using homo- and hetero bifunctional agents, among these; glutaraldehyde which interacts with the amino groups through a base reaction has been extensively used. The enzymes or the cells have been normally cross-linked in the presence of an inert protein like gelatin, albumin and collagen.
34. CHEMISORPTION The process by which a liquid or gas is chemically bonded to the surface of a solid.
35. CHEMICAL OXYGEN DEMAND (COD) The amount of oxygen required for total oxidation of organic and oxidizable inorganic compounds in water.
36. CHEMICAL SENSOR Chemical sensor is a device which responds to a particular analyte in a selective way through chemical reactions and can be used for the qualitative or quantitative determination of the analyte. It can transform chemical information ranging from the concentration of a specific sample component to total composition analysis into an analytically useful signal.
37. CHEMILUMINESCENCE Chemiluminescence is the production of light from a chemical reaction in excess of the black body radiation expected from that body. As such, it is often referred to as cold light. The intensity of light produced is proportional to the rate of production of the excited state molecule. Chemiluminescent reactions are relatively rare but often involve strong oxidizers *Eg.:* O_3 or H_2O_2.
38. CHIP An individual integrated circuit built in a tiny, layered square or rectangle on a silicon wafer.

39. COLLOIDS Very fine solid particles that will not settle out of a solution or medium. Smoke is an example of a colloid, being solid particles suspended in a gas. Colloids are the intermediate stage between a truly dissolved particle and a suspended solid, which will settle out of solution.

40. COMPOSITE A material made from two or more components that have properties different from the constituent materials. Composite materials have two phases: matrix (continuous) phase and dispersed phase (particulates, fibers). *Eg:* steel reinforced cement is a composite material. The concrete is the matrix phase and the steel rods are the dispersed phase. The composite material is much stronger than either of the phases separately.

41. CONCENTRATION A measurement of one or more substances dispersed or dissolved in a material or metal working fluid.

42. CONDUCTIVITY The measurement of the ability of a material to conduct an electric current (typically a liquid or solid solution). In metal working fluid, conductivity is dependent on the amount of dissolved ionic material present. The higher conductivity of a material, the more likely the material will conduct an electrical current.

43. CONTINUOUS GLUCOSE MONITORING SYSTEM (CGMS) The Continuous Glucose Monitoring System (CGMS) named as GlucoWatch and other devices are developed such as Glycemic Holter which helps the diabetologist and the ability to provide real-time continuous glucose monitoring. CGMS device has four separate components: the pager sized glucose monitor, sterile disposable subcutaneous glucose sensor, cable connector and communication system, enabling data stored in the monitor to be uploaded to a computer.

44. COPOLYMERIZATION The process of using more than one type of *monomer* in the production of a polymer, resulting in a product with properties different from either monomer.

45. COVALENT BINDING Covalent binding is an extensively used technique for the immobilization of enzymes, though it is not a good technique for the immobilization of cells. The functional groups extensively investigated are the amino, carboxyl, and the phenolic group of tyrosine. Enzymes are covalently linked to the support through the functional groups in the enzymes, which are not essential for the catalytic activity.

46. CROSSLINKING Cross linking is the process of chemically joining molecules by a covalent bond, thereby linking one polymer chain to another.

47. DETECTION LIMIT The smallest concentration or amount of a component of interest that can be detected by a single measurement with a stated level of confidence. Generally regarded as the concentration level with a precision of ±100%.

48. DIP-PEN NANOLITHOGRAPHY (DPN) A method for nanoscale patterning of surfaces by the transfer of a material from the tip of an atomic force microscope onto the surface. Developed by Professor Chad A. Mirkin, the DPN allows researchers to precisely lay down or "write" chemicals, metals, biological macromolecules, and other molecular "inks" with nanometer dimensions and precision on a surface.

49. DEOXYRIBONUCLEIC ACID (DNA) Deoxyribonucleic acid consists of a polysugar-phosphate backbone from which the purines and pyrimidines project. The backbone is formed by bonds between the phosphate molecule and carbon 3 and 5 of the adjacent deoxyribose molecules. The nitrogenous base extends from carbon 1 of each sugar. It forms a double helix hold together by hydrogen bonds between specific pairs of bases (Watson-Crick model) and each strand in the double helix is complementary to its partner with regard to its base sequence.

50. DNA MICROARRAY DNA Microarray also known as a gene chip, DNA chip or biochip. A collection of microscopic DNA spots attached to a solid surface, such as glass, plastic or silicon chip,

forming an array for the purpose of expression profiling, monitoring expression levels for thousands of genes at the same time.

51. DNA RECOGNITION The ability of one DNA molecule to recognize and attach to another molecule that has a complementary shape.
52. DNA REPLICATION The process of making copies of DNA strands prior to cell division using existing DNA as a template for the newly created strands.
53. DOPING In electronics, the addition of impurities to a semiconductor to achieve a desired characteristic, often altering its conductivity dramatically.
54. ELECTROCHEMICAL METHODS Experimental methods used to study the physical and chemical phenomena associated with electron transfer at the interface of an electrode and a solution. Electrochemical methods are used to obtain analytical or fundamental information regarding electroactive species in solution. Four main types of electrochemical methods include potentiometry, voltammetry, coulometry, and conductimetry.
55. ELECTROCHEMICAL PROPERTIES The characteristics of materials that occur when a) an electric current is passed through a material and produces chemical changes and b) when a chemical reaction is used to produce an electric current, as in a battery.
56. ELECTROCHEMICAL SENSORS Electrochemical sensors utilize chemical reactions which induce a current or voltage in the circuit between two electrodes, these electrodes are immersed in an electrolyte and this phenomenon was studied by Faraday in 1833.
57. ELECTRON DIFFRACTION A surface science technique used to examine solids by firing a beam of electrons at a sample and observing the electron deflection from the sample's atomic nuclei.
58. ELECTROLUMINESCENCE (EL) The light produced by some materials (semiconductors) when exposed to an electric field. In this process, the electric field excites electrons in the material, which then emit the excess energy in the form of photons *Eg.:* Light-emitting diodes (LEDs)
59. ELECTRON MICROSCOPY (EM) The visual examination of very small structures with a device that forms greatly magnified images of objects by using electrons rather than light to create an image. An electron microscope focuses a beam of electrons at an object and detects the actions of electrons as they scatter off the surface to form an image.
60. ELECTRON TRANSFER The passage of an electron from one constituent of a system to another, as from one molecule or ion to another. Applications include photography, xerography, and dye-sensitized injection solar cells.
61. ELECTRON TRANSPORT The manipulation of individual electrons. Nanolithography techniques allow single electrons to be transported at very low temperatures in specially designed circuits.
62. ELECTRON TUNNELING The passage of electrons through a barrier that, according to the principles of classical mechanics, cannot be breached. An example of electron tunneling is the passage of an electron through a thin insulating barrier between two superconductors. Electron tunneling is a pure quantum mechanical effect that cannot be explained by a classical theory.
63. ELECTRO-OPTICS The study of the influence of an electric field on the optical properties of matter especially in crystalline form such as transmission, emission, and absorption of light and also known as optoelectronics.
64. ELECTRONIC NOSE An electronic nose is a device which is used to analyze the content of air through the categorization of odors. The electronic nose is being use today from replacing the human olfactory system, the possible uses for this technology are endless.

65. ENDOCRINE DISRUPTING COMPOUNDS Increasing concern regarding many environmental contaminants that produce adverse effects by interfering with endogenous hormone systems, the so called EDCs.

66. ENZYME A protein that speeds up the rate of reaction of a specific substance without being changed itself (a biological catalyst).

67. ENZYME BASED RECOGNITION Catalytic enzyme based sensor recognition elements are very attractive for biosensor applications due to a variety of measurable reaction products arising from the catalytic process, which include protons, electrons, light and heat.

68. EXTRINSIC SENSORS In which the fiber optic is used to couple light usually to and from the region where the light beam is influenced by the measured.

69. FIBER OPTICS Fiber optics have became an important part of sensor technology and their use as a probe or as a sensing element which is increasing in clinical, pharmaceutical, industrial and military applications. Main advantages of Fiber optics biosensors are excellent light delivery, long interaction length, low cost and ability not only to excite the target molecules but also to capture the emitted light from the targets

70. FIELD EFFECT The local change from the normal value produced by an electric field in the charge-carrier concentration of a semiconductor.

71. FIRST GENERATION BIOSENSORS In first generation biosensors, the normal product of the reaction diffuses into the transducer and causes the electrical response. *Eg.:* glucose biosensors.

72. FLUORESCENCE Relaxation route producing radiation from short lived (<10 microseconds) excited states.

73. FLUORESCENT PROBE A stain used for tagging and labeling biological cells to detect structures, molecules, or proteins within the cell. Also single stranded pieces of DNA, with enzymatically incorporated fluorescent tags, affixed in a microscopic array (DNA microarray).

74. FUEL CELL An electrical cell that converts the intrinsic chemical free energy of a fuel into direct-current electrical energy in a continuous catalytic process. Fuel cells extract the chemical energy bound in fuel and, in combination with air as an oxidant, transform it into electricity.

75. FULLERENES Fullerenes are one of only four types of naturally occurring forms of carbon (the other three are diamond, graphite and ceraphite). They are molecules composed entirely of carbon and take the form of a hollow sphere, ellipsoid, or tube. Sperical fullerenes are often called buckyballs (named after Buckminster Fuller, inventor of the geodesic dome), while cylindrical fullerenes are referred to as buckytubes or nanotubes.

76. GAS DETECTOR A gas detector is a device which detects the presence of various gases within an area usually as part of a safety system. This type of equipment is used to detect a gas leak and interface with a control system so a process can be automatically shut down.

77. GAS-PHASE REACTIONS A class of chemical reactions that occur in a single gaseous phase based on the physical state of the substances present. *Eg.:* Combination of common household gas and oxygen to produce a flame.

78. GLASS ELECTRODE OR pH ELECTRODE Electrode with porous non-crystalline glass membrane sensitive to changes in hydrogen ion activity, used with standard reference electrode.

79. GLUCOSE SENSOR The Glucose sensor is fabricated by immobilizing Glucose oxidase (GOD) on the membrane of the oxygen electrode. The Glucose sensor responds as soon as the glucose solution is injected into the buffer solution and reaches a steady state in 5 to 10 minutes.

80. GRAPHENE Graphene is an allotrope of carbon, whose structure is one atom thick planar sheets of sp^2 bonded carbon atoms that are densely packed in a honeycomb crystal lattice.

81. GROUND STATE The lowest energy electronic, vibrational or rotational state. A species may be in the ground electronic state but excited vibrational or rotational states or any other combination of ground and excited states.
82. IMPEDIMETRIC DEVICES The Impedimetric devices follow either impedance (Z) or its components resistance (R) and capacitance (C); inductance has a minimal influence in a typical electrochemical setup. The expression of impedance is the inverse value of resistance called as conductance, such systems named as conductometric. These biosensors are also including two electrodes with applied alternating voltage, from a few to 100 mV and are commonly a functional part of the Wheatstone bridge.
83. INHIBITOR A chemical substance of combination of substances that, when present, reduces corrosion or chemical attack.
84. INTEGRATED CIRCUIT (IC) A chip etched or imprinted with network or electronic components such as transistors, diodes and resistors along with their interconnections.
85. JUNCTIONS In electronics, the interface between two different types of materials within diodes, transistors, and other semiconductor devices.
86. LAB-ON-A-CHIP DEVICES Miniaturized analytical systems that integrate a chemical laboratory on a chip. Lab-on-a-chip technology enables portable devices for point-of-care medical diagnostics and environmental monitoring.
87. LANGMUIR-BLODGETT (LB) Ultrathin films (monolayers and isolated molecular layers) created by nanofabrication. An LB-film can consist of a single layer or many, up to a depth of several visible-light wavelengths. The term Langmuir-Blodgett comes from the names of a research scientist and his assistant, Irving Langmuir and Katherine Blodgett, who discovered unique properties of thin films in the early 1900s. Such films exhibit various electrochemical and photochemical properties. This has led some researchers to pursue LB-films as a possible structure for integrated circuits (ICs). Ultimately, it might be possible to construct an LB-film memory chip in which each data bit is represented by a single molecule. Complex switching networks might be fabricated onto multilayer LB-films chips.
88. LEAF SENSOR A leaf sensor is a phytometric device (measurement of plant physiological processes) that measures water loss or the water deficit stress (WDS) in plants by real time monitoring the moisture level in plant leaves. The first leaf sensor was developed by Leaf Sens, in 2001.
89. LIGHT AMPLIFICATION BY STIMULATED EMISSION OF RADIATION (LASER) Light Amplification by Stimulated Emission of Radiation (LASER) is a concentrated beam of light. The light emitted from a laser is basically all the same wavelength or monochromatic. The waves of light that come from a laser are all lined up perfectly and this is called coherent light, it is vital for many uses of lasers.
90. LIGHT-EMITTING DIODE (LED) A semiconductor device that converts electrical energy into electromagnetic radiation. The LED emits light of a particular frequency depending on the physical characteristics of the semiconductor used.
91. LIPOSOME A type of nanoparticle made from fat molecules surrounding a core of water. Liposomes were the first nanoparticles used to create unique therapeutic agents.
92. LITHOGRAPHY The process of imprinting patterns on materials. Derived from Greek, the term lithography means literally "writing on stone." Nanolithography refers to etching, writing, or printing at the microscopic level, where the dimensions of characters are on the order of nanometers (units of 10-9 meter, or millionths of a millimeter).

93. LUMINESCENCE Cool light emitted by sources as a result of the movement of electrons from more energetic states to less energetic states. There are many types of luminescence. Chemiluminescence is produced by certain chemical reactions. Electroluminescence is produced by electric discharges, which may appear when silk or fur is stroked or when adhesive surfaces are separated. Triboluminescence is produced by rubbing or crushing crystals.

94. MANIFOLD ABSOLUTE PRESSURE SENSORS (MAPS) The Manifold Absolute Pressure sensor (MAP sensor) is one of the sensors used as electronic control system in internal combustion engine's.

95. METAL DECTECOR A metal detector is a device which responds to metal that may not be readily apparent. The simplest form of a metal detector consists of an oscillator producing an alternating current that passes through a coil producing an alternating magnetic field.

96. METAL OXIDE SEMICONDUCTOR SENSOR (MOSS) The H_2S sensor is a metal oxide semiconductor (MOS) sensor which operates by a reversible change in resistance caused by adsorption and desorption of hydrogen sulfide in a film with hydrogen sulfide sensitive material like tin oxide thick films and gold thin films. Current response time is 25 ppb to 10 ppm in less than one minute.

97. MICROCONTACT PRINTING A technique that uses a silicone stamp to deposit molecules on surfaces in patterns with microscale features.

98. MICRO-CANTILEVER SENSOR A micro-cantilever sensor can be adapted to detect physical, chemical, or biological activity. The micro-cantilever is constructed with silicon plank (100 micrometers (mm) long, 30 mm wide, and 3 to 4 mm thick), when molecules are added to its surface, the extent to which the plank bends can be measured accurately by bouncing a light beam off the surface and measuring the extent to which the light beam is deflected.

99. MICROELECTROMECHANICAL SYSTEMS (MEMS) Tiny mechanical devices built onto semiconductor chips, measured in micrometers. They are used to make pressure, temperature, chemical and vibration sensors, light reflectors and switches as well as accelerometers for airbags, vehicle control, pacemakers and games. Sometimes the term MEMS is used interchangeably with nanotechnology since they both deal with microminiaturized objects, but they are vastly different. MEMS deals in devices measured in micrometers, while nanotechnology deals with manipulating atoms at the nanometer level.

100. MICROFLUIDIC DEVICE A device that has one or more channels with at least one dimension less than 1 mm. Common fluids used in microfluidic devices include whole blood samples, bacterial cell suspensions, protein or antibody solutions and various buffers. The small amounts of samples needed and relative inexpensiveness of microfluidic devices make them attractive for biomedical research and creating clinically useful technologies. One of the long term goals in the field of microfluidics is to create integrated, portable clinical diagnostic devices for home and bedside use, thereby eliminating time consuming laboratory analysis procedures.

101. MICROFLUIDICS A multidisciplinary field that studies the behavior of fluids at volumes thousands of times smaller than a drop. Microfluidic components form the basis of "lab-on-a-chip" devices capable of performing several different functions. Microfluidics is critical in the development of gene chip and protein chip technology.

102. MOLECULAR IMPRINTING A process by which functional monomers are allowed to self-assemble around a template molecule and locked into place. The template molecule is then removed, leaving behind a cavity that is complementary in shape and functionality as the template molecule, which will bind molecules identical to the template.

103. MOELCULAR SWITCH A molecular switch is a molecule that can be reversibly shifted between two or more stable states.

104. MOLECULAR ELECTRONICS Molecular electronics (sometimes called moletronics) is an interdisciplinary theme that spans physics, chemistry, and materials science. The unifying feature of this area is the use of molecular building blocks for the fabrication of electronic components, both passive (e.g. resistive wires) and active (*Eg.* transistors). Molecular electronics first coined by Mark Ratner in 1974.

105. MOLECULAR IMPRINTING Molecular imprinting is a method for making selective binding sites in synthetic polymers using molecular templates. Molecular imprinted polymers offer great promise for development of very stable solid-state like artificial biosensing elements.

106. MOLECULAR WIRES Molecular wires tiny electron conductors with huge potential by Laura Mgrdichian. Molecular wires or molecular nanowires are molecular scale objects which conduct electrical current. They are the fundamental building blocks for molecular electronic devices.

107. MULTI FUNCTIONAL BIOCHIP The MFB is an integrated multi-array biochip, which is designed by combining integrated circuit elements, an electro optics excitation/detection system, and bioreceptor probes into a self contained and integrated micro device depicts a schematic diagram of the MFB device.

108. MULTI WALLED NANOTUBES Multi walled nanotubes consists of multiple rolled layers (concentric tubes) of graphite. There are two models which can be used to describe the structures of multi walled nanotubes (1) The Russian Doll model, sheets of graphite are arranged in concentric cylinders, *Eg.*: A (0,8) single walled nanotube (SWNT) within a larger (0,17) single walled nanotube. (2) The Parchment model, a single sheet of graphite is rolled in around itself, resembling a scroll of parchment or a rolled newspaper. The interlayer distance in multi walled nanotubes is close to the distance between graphene layers in graphite, approximately 3.4 Å.

109. NANOCRYSTAL A crystal of nanoscale proportions, whose size, shape, and atomic composition are controlled to yield specific properties such as the absorption of specific wavelengths of light.

110. NANOELECTROMECHANICAL SYSTEMS (NEMS) Tiny mechanical devices measured in nanometers. Sometimes the term MEMS is used interchangeably with NEMS since they both deal with microminiaturized objects, but they are vastly different. MEMS deals in devices measured in micrometers, while nanotechnology deals with manipulating atoms at the nanometer level.

111. NANOMETER A unit of measurement equal to one-billionth of one meter. The head of a pin is about 1 million nanometers across. A human hair is about 60,000 nanometers in diameter, and a DNA molecule is between 2-12 nanometers wide.

112. NANOPARTICLES Particles ranging from 1 to 100 nanometers in diameter. Semiconductor nanoparticles up to 20 nanometers in diameter are often called quantum dots, nanocrystals, or Q-particles.

113. NANOPOROUS MATERIALS Engineered materials with nanoscale holes, used in filters, sensors, and diffraction gratings. In DNA sequencing, nanoporous materials have tiny holes that allow DNA to pass through one strand at a time. In biology, complex protein assemblies that span cell membranes allow ionic transport across the otherwise impermeable lipid bilayer.

114. NANOSENSOR Nanosensor is a device typically comprised of nanoscale semiconductor crystals used in combination with biological probes for a variety of applications.

115. NANOTUBES Long, thin cylinders of carbon, discovered in 1991 by S. Iijima. These large macromolecules are unique for their size, shape, and remarkable physical properties. They can be thought of as a sheet of graphite (a hexagonal lattice of carbon) rolled into a cylinder. The physical properties are still being discovered. Nanotubes have a very broad range of electronic, thermal and structural properties that change depending on the different kinds of nanotube (defined by its diameter, length, and chirality or twist). To make things more interesting, besides having a single cylindrical

wall (Single Walled Nanotubes or SWNTs), nanotubes can have multiple walls (MWNTs) cylinders inside the other cylinders. Applications for carbon nanotubes include high density data storage, nanoscale electronics, and flexible solar cells.

116. OLIGONUCLEOTIDE CHIPS Oligonucleotide chips consist of small glass plates with thousands of short 20-mer oligonucleotide probes attached to their surface. The oligos are synthesized directly onto the surface using a combination of semiconductor-based photolithography and light-directed chemical synthesis. Very large numbers can be probed; at present chips contain over 65,000 probes.

117. OPTICAL BIOSENSORS Optical biosensors are being developed on the reliability of optical methods and number of optical transduction techniques. These biosensors are developed on the basis of linear optical phenomenon which includes absorption, fluorescence, phosphorescence, polarization, diffraction, interference, rotation and also non-linear phenomena.

118. OPTICAL COMPUTER/PHOTONIC COMPUTER Electronic computing uses electrons to perform the logic that makes computing work. An optical computer (photonic computer) is a device that uses the photons in visible light or infrared (IR) beams rather than electric current to perform digital computations. An electric current flows at about 10% of the speed of light.

119. OPTICAL INTERFERENCE: Optical Interference is a process of energy rearrangement that occurs when two laser beams pass through the same point at the same time.

120. OXYGEN SENSOR An oxygen sensor or lambda sensor is an electronic device that measures the proportion of oxygen (O_2) in the gas or liquid being analyzed. It was developed by the Robert Bosch GmbH company during the late 1960s under the supervision of Gunter Bauman.

121. PELLISTOR A Pellistor is a solid state device used to detect gases which are either combustible or which have a significant difference in thermal conductivity to that of air.

122. PEPTIDE NUCLEIC ACIDS (PNA) Peptide nucleic acids (PNA) are synthetic DNA analogues or mimics with a polyamide backbone instead of a sugar phosphate bone. PNAs exhibit superior hybridization characteristics and improved chemical and enzymatic stability compared to nucleic acids.

123. pH The scale that measures the acidity of a solution: pH 1 is very acidic, pH 6 - 7 is neutral and pH 14 is very alkaline.

124. pH SENSOR A pH sensor is a device majorly consists of a glass electrode, which is type of ion selective electrode made of a doped glass membrane that is sensitive to a specific ion. It is an important part of the instrumentation for chemical analysis and physicochemical studies.

125. PHOSPHORESCENCE Relaxation route producing radiation from a long-lived (triplet) state (>10 microseconds to hours).

126. PHOTOLITHOGRAPHY A lithographic technique used to transfer the design of circuit paths onto printed circuit boards, as well as the circuit paths and electronic elements of a chip onto a wafer's surface. A photomask is created with the design for each layer of the board or wafer (chip). The board or wafer is coated with a light-sensitive film (photoresist) that hardens when exposed to light shining through the photomask. The board is then exposed to an acid bath (wet processing), and the unhardened areas are etched away.

127. PHOTON A bundle of electromagnetic energy that exhibits both wave-like and particle-like characteristics.

128. PHOTONIC INTEGRATED CIRCUIT (PIC) Photonic Integrated Circuit (PIC) or Integrated Optics is an optical data transmission device that integrates multiple photonic functions. This technology contains numerous optical components which are bonded together to fulfill complex functions.

129. PHOTONIC NETWORK Optical technology has made its most significant inroads in digital communications, where fiber optic data transmission has become commonplace. The ultimate goal is the so-called photonic network , which uses visible and IR energy exclusively between each source and destination.
130. PHYSCIAL SENSORS Physical sensors for measuring distance, mass, temperature and pressure. The response of a sensor is usually in the form of a physical response.
131. PIEZOELECTRICS Dielectric crystal that produce a voltage when subjected to mechanical stress or can change shape when subjected to a voltage.
132. PIEZOELECTRICITY The ability of certain crystalline materials to produce an electric voltage when subjected to mechanical stress (direct piezoelectric effect). Inversely, the material's ability to change shape when an external voltage is applied.
133. PLASTIC SOLAR One potential application for this finding is in the solar energy industry, particularly in a new field called plastic solar.
134. POTENTIOMETRIC SENSORS Potentiometric sensors, in which the potential of the indicator electrode (ion selective electrode, redox electrode and metaVmeta1 oxide electrode) is measured against a reference electrode.
135. POLYMERASE CHAIN REACTION A technique for copying and amplifying the complementary strands of a target DNA molecule.
136. Ppb Parts per billion: a weight unit of measurement.1% is equivalent to 10,000,000 ppb, 1 ppb is equivalent to 0.001 ppm or 0.0000001%.
137. Ppm Parts per million: a weight unit of measurement.1% is equivalent to 10,000 ppm, 1 ppm is equivalent to 1,000 ppb or 1,000,000 ppt
138. Ppt Parts per trillion: a weight unit of measurement.1% is equivalent to 10,000,000,000 ppt, 1 ppt is equivalent to 0.000001 ppm or 0.0000000001%.
139. PRECISION The reproducibility of a measurement made multiple times in identical ways caused by indeterminate or random errors. Usually quantified by absolute or relative standard deviation.
140. PROTEIN Large organic molecules involved in all aspects of cell structure and function.
141. QUANTUM CONFINEMENT EFFECT Atoms caged inside nanocrystals.
142. QUANTUM DOT A nanoscale crystalline structure made from cadmium selenide that absorbs white light and then re-emits it a couple of nanoseconds later in a specific color. The quantum dot was originally investigated for possible computer applications. Recently, researchers are investigating the use of quantum dots for medical applications, using the molecule-sized crystals as probes to track antibodies, viruses, proteins, or DNA within the human body.
143. RADAN GUN/SPEED GUN A radar speed gun (radar gun and speed gun) is a small doppler radar unit used to measure the speed of moving objects including vehicles, pitched baseballs, runners and other moving objects.
144. RIBONUCLEIC ACID (RNA) A long linear polymer of nucleotides found mainly in the cytoplasm of a cell that transmits genetic information from DNA to the cytoplasm and controls certain chemical processes in the cell.
145. RNA CHIP RNA chip is a new diagnostic device which promises to put a powerful lab on a dime sized slice of silicon. In 1995, Breaker *et al.,* began to revive this RNA chip in a test tube and successfully engineered RNA based molecular switches.
146. SCAFFOLD Three-dimensional biodegradable polymers engineered for cell growth.

147. SCANNING TUNNELING MICROSCOPE (STM) A scanning probe microscopy instrument capable of revealing the structure of samples. The STM uses a sharp metal tip positioned over a conducting substrate with a small potential difference applied between them. The gap between the tip and substrate surface is small enough so that electrons can tunnel between the tip and the surface. The tip is then scanned across the surface and adjusted to keep a contact current flowing. By recording the tip height at each location a map of the sample surface is obtained.
148. SECOND GENERATION BIOSENSORS Second generation biosensors are which involve specific mediators between the reaction and transducer in order to generate improved response.
149. SEISMOMETERS Seismometers are the instruments that measure motions of the ground including those of seismic waves generated by earth quakes, volcanic eruptions and other seismic sources.
150. SEMICONDUCTOR TEMPERATURE SENSOR The semiconductor temperature sensor is an electronic device fabricated in other modern electronic semiconductor components such as microprocessors. The most popular semiconductor temperature sensors are based on the fundamental temperature and current characteristics of the transistor.
151. SENSOR A device able to quantitatively and reversibly detect chemical species and produce an electrical signal. A sensor is a device that detects or measures a physical and chemical property and records, indicates or otherwise responds to it.
152. SILICON (Si) The base material used in chips. Besides oxygen, it is the most abundant element in nature and is found in a natural state in rocks and sand. Its atomic structure and availability make it an ideal semiconductor material. In chip making, it is mined from white quartz rocks and put through a chemical process at high temperatures to purify it. To alter its electrical properties, it is mixed (doped) with other chemicals in a molten state.
153. SINGLE WALLED NANOTUBES (SWNT) Single walled nanotubes (SWNT) have a diameter of close to 1 nanometer, with a tube length that can be many millions of times longer. The structure of a SWNT can be conceptualized by wrapping a one atom thick layer of graphite called graphene into a seamless cylinder
154. STANDARD HYDROGEN ELECTRODE (SHE) Standard reference electrode assigned a value of 0.000 V. Made by dipping an inert Pt wire into a acid solution with hydrogen ion activity of 1.000 M and bubbling H_2 gas around the wire at a partial pressure of 1.000 atm.
155. SUBSTRATE A wafer that is the basis for subsequent processing operations in the fabrication of semiconductor devices.
156. SURFACE PLASMON RESONANCE Surface Plasmon Resonance is physical phenomenon for which the electrons in thin gold film can be excited into collective oscillation by bombarding it with photons (light) of particular frequency. This phenomenon is analogous to resonant frequency in mechanical system and must have specific match between light frequency and momentum, and the characteristics of the gold.
157. SYNTHESIS Any process or reaction for building up a complex compound by the union of simpler compounds or elements.
158. THERMOMETRIC A type of transducer that measures the heat given out by an enzyme reaction.
159. THIN FILM A film one molecule thick; often referred to as a monolayer.
160. THIRD GENERATION BIOSENSORS In third generation biosensors, the reaction itself causes the response and no product or mediator diffusion is directly involved. These biosensors have achieved a direct connection between the enzymatic reaction and the transducer without need for diffusion.

161. TRANSDUCER A device able to quantitatively convert non-electrical domain data into electrical domain data *Eg.:* Photocell. KTransducer is a device that converts energy from one form to another form. In biosensors transducers are converts the biochemical activity into electrical energy. The transducer converts a biorecognition event into a measurable signal that correlates with the quantity or presence of the chemical or biological target.
162. TRANSISTOR An electronic device capable of amplifying electronic signals similar to the vacuum tube but made from a semiconductor material, such as silicon or germanium.
163. ULTRA DENSE Ultra-dense memory device that stores information using reconfigurable molecular switches.
164. VOLATILE ORGNAIC COMPOUNDS (VOCs) Carbon containing chemicals that are liquid at room temperature and pressure but have a low boiling point and thus evaporate readily.
165. ZETTA CORE Zetta Core is the tip of the molecular electronics wedge that evolves into mainstream nano-electronics, displacing highly scaled semiconductors. Zetta core technology is developed by Werner Kuhr *et al.,* (2000) used for molecular memory. Zetta Core uses molecules called porphyrins to store the 1s and 0s that comprise digital data.

References

Abdel Hamid I, Atanasov P and Wilkins E (1994). Development of a Needle Type Glucose Biosensor. Anal. Lett., 27, pp 1453-1473.

Aeraj ul Haque, Mohammad Rameez Chantni, Gang Li and David Marshall Porterfield (2007). Biochips and other microtechnologies for physiomics. Expert, Rev. Proteomics, 4(4), pp 553-563.

Ahmad A, Ramakrishnan A, McLean M A, Li D, Rock M T, Karim and Breau A P (2002). 300, pp 177–184.

Ahn CH, Choi JW, Beaucage G, Nevin, JH Lee, JB Puntambekar A and Lee JY (2004). Disposable Smart lab on a chip for point-of-care clinical diagnostics, Proc, IEEE 92, pp 154–173.

Akkin T, Dave DP, Milner TE and Rylander III HG (2002). Interferometric fibre-based optical biosensor to measure ultra-small changes in refractive index. Proceedings of SPIE, The International Society for Optical Engineering, 4616, pp 9-13.

Akram M, Stuart MC and Wong DKY (2006). Signal Generation at an Electrochemical Immunosensor via the Direct Oxidation of an Electroactive Label Electroanalysis. 18, pp 237-246.

Albery J, Haggett B and Snook D (1986). You know it makes sensors. New Scientist, pp 38-41.

Alcock SJ and Turner APF (1994). IEEE Engineering in Medicine and Biology, 319.

Alexander PW, Rechnitz GA (2000). Enzyme inhibition assays with an amperometric glucose biosensor based on thiolate self-assembled monolayer. Electroanalysis 12, pp 343–350.

Anderson GP and NL Nerurkar (2002). Improved fluoroimmunoassays using the dye Alexa Fluor 647 with the RAPTOR, a fibre optic biosensor. J. Immunol. Methods, 271(1-2), pp 17-24.

Anderson GP, King KD, Cao LK, Jacoby M, Ligler FS and Ezzell J (1998). Quantifying serum anti-plague antibody using a fibre optic biosensor. Clin. Diag. Lab. Imm. 5, pp 609-612.

Anderson GP, King KD, Gaffney KL and Johnson LH (2000). Multianalyte detection using the fibre optic biosensor. Biosensors & Bioelectronics. 14, pp 771-777.

Anderson GP, Shriver-Lake LC, Golden JP and Ligler FS (1992). Fibre-optic-based biosensor: signal enhancement in a production model. Proceedings of SPIE - The International Society for Optical Engineering. 1648, pp 39-43.

Andreescu S, Avramescu A, Bala C, Magearu V, Marty JL (2002a). Detection of organophosphorus insecticides with immobilized acetylcholinesterase-competitive study of two enzyme sensors. Anal. Bioanal., Chem., 374, pp 39–45.

Andreescu S, Fournier D, Marty JL (2003). Development of highly sensitive sensor based on bioengineered acetylcholinesterase immobilized by affinity method. Anal., Lett., 36 (9), pp 1865–1885.

Andreescu S, Magearu V, Lougarre A, Fournier D, Marty JL (2001). Immobilization of enzymes on screen printed sensors via a histidine tail. Application to the detection of pesticides using modified cholinesterase. Anal., Lett., 34 (4), pp 529–540.

Andreescu S, Noguer T, Magearu V, Marty JL (2002b). Screen printed electrode based on AChE for the detection of pesticides in presence of organic solvents. Talanta 57, 169–176.

Anthony Guiseppi-Elie. *In vivo* Biosensors. Dow Chemical Professor Chemical and Biomolecular Engineering, College of Engineering & Science, Clemson University.

Antonik MD, D'Costa NP and Hoh JH (1997). A biosensor based on micromechanical interrogation of living cells, IEEE Engineering in Medicine and Biology, 66.

Aoki H and Umezawa Y (2002). High Sensitive Ion-Channel Sensors for Detection of Oligonucleotides Using PNA Modified Gold Electrodes Electroanalysis.14, pp 1405-1410.

Aravanis AM, DeBusschere BD and Chruscinski AJ *et al.* (2001). A genetically engineeed cell-based biosensor for functional classification of agents, Biosensors and Bioelectronics. 16, pp 571-577.

Arikawa Y, Ikebukuro K and Karube I, Mulchandani A and Rogers KR (Eds.) (1998). Enzyme and Microbial Biosensors: Techniques and Protocols, Humana Press, Totowa, NJ, pp 225.

Ayyagari MS, Pande R, Kamtekar S, Gao H, Marx KA, Kumar J, Tripathy SK, Akkara J A, Kaplan DL (1995). Molecular assembly of proteins and conjugated polymers: Toward development of biosensors. Biotechnology and Bioengineering, 45, pp 116-121.

Azila ABD Aziz (2007). Development of a Biosensor based on Amine Oxidase from Cicer Arietinum for the detection of biogenic amines. Universiti Teknologi Malaysia.

Azila Abdul Aziz (2001). Amperometric Glucose Biosensor: Systematic Material Selection and Quantitative Analysis of Performance. The John Hopkins University: Thesis of Doctor of Philosophy.

Babacan S, Pivarnik P, Letcher S and Rand AG (2000). Evaluation of antibody immobilization methods for piezoelectric biosensor application. Biosens. Bioelectron, 15, pp 615–621.

Babkina SS and NA Ulkhovich (2004) Amperometric biosensor based on denatured DNA for the study of heavy metals complexing with DNA and their determination in biological, water and food samples. Bioelectrochemistry, 63, pp 261-265.

Baeumner AJ (2003). Biosensors for Environmental Pollutants and Food Contaminants. Anal. Bioanal. Chem, 377, pp 434-445.

Bakaltcheva IB, Ligler FS, Patterson CH and Shriver-Lake LC (1999). Multi-analyte explosive detection using a fibre optic biosensor. Anal. Chim. Acta, 399, pp 13-30.

Balslev P, Lynggaard-Jensen A and Nickelsen C (1996). Nutrient sensor based real-time on-line process control of a wastewater treatment plant using recirculation, Water Sci. Tech. 33, pp 183–192.

Bambot SB, Rao G, Romauld M, Carter GM, Sipior J, Terpetchnig E and Lakowicz JR (1995). Sensing oxygen through skin using a red diode laser and fluorescence lifetimes. Biosens. Bioelectron, 10, pp 643–652.

Bang GS, GCho S and Kim BG (2005). A Novel Electrochemical Detection Method for Aptamer Biosensors. Biosens. Bioelectron, 21 pp 863-870.

Barhoumi AH, Maaref A, Rammah M, Martelet C, Jaffrezic N, Mousty C, Vial S and Forano C (2006). Urea biosensor based on Zn3Al-Urease layered double hydroxides nanohybrid coated on insulated silicon structures. Materials Science and Engineering C. 26, 328–333.

Barhoumi AH, Maaref A, Rammah M, Martelet C, Jaffrezic N, Mousty C, Vial S and Forano C (2006). Urea biosensor based on Zn3Al-Urease layered double hydroxides nanohybrid coated on insulated silicon structures. Materials Science and Engineering C, 26, pp 328–333.

Barker SLR *et al.* (1999). Cellular applications of a sensitive and selective fibre optic nitric oxide biosensor based on a dye-labeled heme domain of soluble guanylate cyclase. Anal. Chem, 71(11), pp 2071-2075.

Barker SLR, Tarlov MJ, Canavan H, Hickman JJ and Locascio LE (2000). Plastic microfluidic devices modified with polyelectrolyte multilayers. Anal. Chem., 72, pp 4899–4903.

Baumann WH, Lehmann M, Schwinde A *et al.* (1999). Microelectronic sensor system for microphysiological application on living cells, Sensors and Actuators B, 55, 77.

Baxter GT, Bousse LJ and Dawes TD *et al.* (1994). Microfabrication in silicon microphysiometry, Clin. Chem, 40, 1800.

Bergenstal RM, Tamborlane WV, Ahmann A, Buse JB, Dailey G, Davis SN, Joyce C, Peoples T, Perkins BA, Welsh JB, Willi SM and Wood MA (2010). Effectiveness of sensor-augmented insulin-pump therapy in type 1 diabetes. N Engl J Med, Vol. 363, No. 4, pp 311-320.

Bergmeyer HU (1974). Methods of Enzymatic Analysis (3rd Eds). New York. Verlag Chemie, Academic Press.

Bergveld P (1970). IEEE Trans. Biomed. End, 17, pp 70–71.

Bergveld P (1972). IEEE Trans. Biomed. End, 19, pp 342–351.

Besselink GAJ, Schasfoort RBM and Bergveld P (2003). Modification of ISFETs with a Monolayer of Latex Beads for Specific Detection of Proteins. Biosens. Bioelectron, 18, pp 1109-1114.

Biacore Principles, http://www.biacore.com/principles/detection.html.

Biomolecule Recognition with Electron Transfer: Electrochemical Sensors for DNA Campbell CN, Gal D, Cristler N, Banditrat C and Heller A (2002). Enzyme Amplified Amperometric Sandwich Test for RNA and DNA. Anal. Chem, 74, pp 158-162.

Biran I, Rissin D, Ron E and D Walt (2003). Optical imaging fibre based live bacterial cell array biosensor. Analytical Biochemistry, 315:1, pp 106-113.

Blodgett K B (1934) J. Am. Chem. Soc. 56, 495.

Bluestein BI, Craig M, Slovacek R, Stundtner L, Urciuoli C, Walczak I and Luderer A (1991). Evanescent wave immunosensors for clinical diagnostics. In: Biosensors with Fibreoptics. Humana Press, NJ, pp 181-223.

Blum LJ and Coulet PR (1991) (Eds.). Biosensor Principles and Applications, Marcel Dekker, New York.

Blum LJ, Gautier S M, Berger A, Michel PE and Coulet PR (1995). Multicomponent organized bioactive layers for fibre-optic luminescent sensors. Sensors and Actuators, B: Chemical, B29, pp 1-9.

Boone JH, Wilkins TD, Nash TE, Brandon JE, Macias EA, Jerris RC and Lyerly DM (1999). TechLab and Alexon Giardia enzyme-linked immunosorbent assay kits detect cyst wall protein. J. Clin. Micro. 37, pp 611-614.

Brandenburg A and Gombert A (1993). Grating couplers as chemical sensors: a new optical configuration. Sensors and Actuators B, 17, pp 35-40.

Brandenburg A, Krauter R, Kunzel C, Stefan M and Schulte H (2000). Interferometric sensor for, detection of surface-bound bioreactions. Appl. Optics. 39, pp 6396-6405.

Brandt O and Hoheisel JD (2004). Peptide nucleic acids on microarrays and other biosensors. Trends Biotechnol, 22, pp 617–622.

Brazil de Oliveira P., Nune GS, Santos TCR, Andreescu S, Marty JL (2004). Comparative investigation between acetylcholinesterase obtained from commercial sources and genetically modified *Drosophila melanogaster*. Application in amperometric biosensors for methamidophos pesticide detection. Biosens. Bioelectron, 20, pp 825–832.

Brecht A *et al.* (1995). A Direct Optical Immunosensor for Atrazine Detection. Anal. Chim. Acta, 311(3), pp 289-299.

Brogan K L and Walt D R (2005). Optical fibre-based sensors: Application to chemical biology. Current Opinion in Chemical Biology, 9, pp 494-500.

Brogan KL and Walt DR (2005). Optical fibre-based sensors: Application to chemical biology. Current Opinion in Chemical Biology, 9, pp 494-500.

Brosinger F, Freimuth H, Lacher M, Ehrfeld W, Gedig E, Katerkanp A, Spener F and Cammann K (1997). A label-free affinity sensor with compensation of unspecific protein interaction by a highly sensitive integrated optical Mach-Zehnder interferometer on silicon. Sens, Actuators B, 44, pp 350-355.

Brummel KE, Wright J and Eldefrawi ME (1997). Fibre optic biosensor for cyclodiene insecticides. J Agric, Food Chem, 45(8), pp 3292-3298.

Buckle PE, Davies RJ, Kinning T, Yeung D, Edwards PR and Pollard-Knight D (1993). II Biosensors and Bioelectronics. 8 : 355.

Buerk D G (1993). Biosensor: Theory and Application (First Edition). United State of America. Technomic Publishing Company, Inc.

Bulmus V, Kesenci K and Piskin E (1998). Ply (EGDMA/AAm) Copolymer Beads: A Novel Carrier for Enzyme Immobilization. Reactive Functional Polymers, 38, pp 1-9.

Cai H, Lee TMH and Hsing IM (2006). Label-Free Protein Recognition Using Aptamer-Based Impedance Measurement Assay. Sensor Actuat. B-Chem, 114, pp 433-437.

Cai H, Wang Y He, P Fang Y (2002). Electrochemical Detection of DNA Hybridization Based on Silver-Enhanced Gold Nanoparticle Label. Anal. Chim. Acta, 469, pp 165-172.

Camelia Bala, L Rotariu and V Magearu. Disposable alcohol biosensor based on alcohol dehydrogenase and screen printed electrodes. Analele Universitatii din Bucuresti. Chimie, Anul XII (serie nou), vol. I-II, pp 55–60.

Campbell CN, Gal D, Cristler N, Banditrat C and Heller A (2002). Enzyme-Amplified Amperometric Sandwich Test for RNA and DNA. Anal. Chem. 74, pp 158-162.

Cao L, Van Langen L and Sheldon RA (2003). Immobilised enzymes: carrier-bound of carrier-free. Curr. Opin. Biotechnol, 14, pp 387-394.

Cao LK, Anderson GP, Ligler FS and Ezzell J (1995). Detection of Yersinia pestis F1 antigen by a fibre optic biosensor. J. Clin. Micro. 33, pp 336-341.

Caras S and Janata J (1980). Anal. Chem, 52, pp 1935–1937.

Cardullo F, Diederich F, Echegoyen L, Habicher T, Jayaraman N, Leblanc R.M, Stoddart JF and Wang S (1998). Stable Langmuir and Langmuir- Blodgett films of fullerene-glycodendron conjugates, 14, pp 1955–1959.

Carlson MA, Bargeron CB, RC Benson, Fraser AB, Phillips TE, Velky JT, Groopman JD, Strickland PT and Ko HW (2000). An automated, handheld biosensor for aflatoxin. Biosensors & Bioelectronics. 14, pp 841-848.

Cass AEG, Davis G, Francis GD, O HHA, Aston WJ, Higgins IJ, Plotkin EV, Scott LDL and Turner APF (1984). Ferrocene-Mediated Enzyme Electrode for Amperometric Determination of Glucose. Anal. Chem, 56, pp 667-671.

CehnP Xu B, Tokranova N, Feng X, Castracane J, Gillis KD (2003). Amperometric detection of quantal catecholamine secretion from individual cells on micromachined silicon chips. Anal.Chem, 75, pp 518-524.

Chandrasekaran A and Packirisamy M (2005). MOEMS based integrated microfluidic fibre-optic waveguides for biophotonic applications. Progress in Biomedical Optics and Imaging – Proceedings of SPIE : 5969.

Chara TJ, Rajagopaian R and Heller A (1994). Wired Enzyme Electrodes for Amperometric Determination of Glucose or Lactate in the Presence of Interfering Substances. Anal. Chem.

Chauhan S, Vibhuti Rai and Singh H B (2004). Biosensors. Resonance. pp 33-44.

Chen RJ, Bangsaruntip S, Drouvalakis KA, Kam NWS, Shim M, Li Y Kim and W Utz PJ Dai H (2003). Noncovalent Functionalization of Carbon Nanotubes for Highly Specific Electronic Biosensors. P. Natl. Acad. Sci, USA,100, pp 4984-4989.

Chen Z, Gao HH, Kumar J, Marx K A, Tripathy SK and Kaplan DL (1996). Molecular assembly of multilayer enzyme on fibre optic surface: Toward the development of chemiluminescence based fibre optic biosensors. Proceedings of SPIE - The International Society for Optical Engineering. 2676, pp 156-164.

Chen Z, Kaplan DL, Gao H, Kumar J, Marx KA and Tripathy SK (1996). Molecular assembly of multilayer enzyme: toward the development of a chemiluminescence-based fibre optic biosensor. Materials Science and Engineering C, 4, pp 155-159.

Choi SH and MB Gu (2002). A portable toxicity biosensor using freeze-dried recombinant bioluminescent bacteria. Biosens. Bioelectron, 17(5), pp 433- 440.

Churchouse SJ, Battersby CM, Mullen WH and Vadgama PM (1986). Needle Enzyme Electrodes for Biological Studies. Biosensors, 2, pp 325-342.

Clark L C (1962). Jnr. Ann. NY Acad. Sci, 102, pp 29-45.

Clark LC (1956). Jnr. Trans. Am. Soc. Artif. Intern. Organs, 2, pp 41-48.

Clark LC and Lyons C (1962). Electrode Systems for Continuous Monitoring in Cardiovascular Surgery. Ann. NY Acad. Sci, 102, pp 29-45.

Clark LC Jr (1956). Monitor and control of blood tissue O2 tensions. Transactions of the American Society for Artificial Internal Organs 2, pp 41–84.

Clark LC Jr and Lyons C (1962). Electrode system for continuous monitoring in cardiovascular surgery. Annals of the New York Academy of Sciences. 148, pp 133–153.

Clark LC, Wolf R, Granger D and Taylor Z (1953). Continuous recording of blood oxygen tensions by polarography. J Appl Physiol, 6, pp 189-193.

Clemens AH, Chang PH and Myers RW (1976). Proc. Journes Ann. de Diabtologie de l'Htel-Dieu, Paris.

Clin Chim Acta (2003). D'Orazio, P. Biosensors in Clinical Chemistry, 334, pp 41-69.

Collier WA, Clear MA, Hart AL (2002). Convenient and rapid detection of pesticides of extracts of sheep wool. Biosens. Bioelectron, 17, pp 815–819.

Continuous Glucose Monitoring (2008). National Diabetes Information Clearinghouse. NIH Publication No. 09–4551. http://www.diabetes.niddk.nih.gov/.

Corbisier P *et al.* (1999). Whole cell- and protein-based biosensors for the detection of bioavailable heavy metals in environmental samples. Anal. Chim. Acta, 387(3), pp 235-244.

Corbisiera P, Jib G, Nuytsa G, Mergeaya M and S Silver (1993). LuxAB Gene Fusions with the Arsenic and Cadmium Resistance Operons of Staphylococcus aureus Plasmid pI258.FEMS Microbiology Letters, 110:2, pp 231-238.

Crespel A and Gélisse P (2005). Altas of Electroencephalography. 1st ed. John Libbey Eurotext, Montrouge, France.

Crew A, Hart JP, Wedge R, Marty JL, Fournier D (2004). A screen-printed, amperometric, biosensor array for the detection of organophosphate pesticides based on inhibition of wild type, and mutant acetylcholinesterases, from *Drosophila melanogaster*. Anal. Lett, 37, pp 1601–1610.

Csoregi E, Schmidtke DW and Heller A (1995). Design and Optimization of a Selective Subcutaneously Implantable Glucose Electrode Based on "Wired" Glucose Oxidase. Anal. Chem, 67, pp 1240- 1244.

Cui F, Ye JS, Chen Y, Chong SC and Sheu FS (2006). Microelectrode array biochip: tool for in vitro drug screening based on the detection of a drug effect on dopamine release from PC 12 cells. Anal. Chem, 78, pp 6347-6355.

Cui X, Liu G and Lin Y (2005). Amperometric biosensors based on carbon paste electrodes modified with nanostructured mixed-valence manganese oxides and glucose oxidase. Nanomedicine, 1, pp 30–135.

Cullum BM *et al.* (2000). Intracellular measurements in mammary carcinoma cells using fibre-optic nanosensors. Anal. Biochem, 277(1), pp 25-32.

Cush R *et al.* (1993). The resonant mirror: a novel optical biosensor for direct sensing of biomolecular interations Part I. Biosensors and Bioelectronics, 8, pp347-354.

Cush R, Cronin JM, Stewart WJ, Maule CH, Molloy J and Goddard NJ (1993). II Biosensors and Bioelectronics. 8, pp 347.

D´Agostino G, Alberti G, Biesuz R and Pesavento M (2006). Potentiometric sensor for atrazine based on a molecular imprinted membrane. Biosens. Bioelectron, 22 pp 154–152.

D´Auria S, Gryczynski Z, Gryczynski I, Rossi M and Lakowicz JR (2000). A protein biosensor for lactate. Anal. Biochem, 283, pp 83–88.

Danet AF, Bucur B, Cheregi MC, Badea M, Serban S (2003). Spectrophotometric determination of organophosphoric insecticides in FIA system based on AChE inhibition. Anal. Lett, 36 (1), pp 59–73.

Daniel R Sommers (2003). Design and verification of a Surface Plasmon Resonance Biosensor. Thesis. Georgia Institute of Technology.

Darain F, Park SU and Shim YB (2003). Disposable Amperometric Immunosensor System for Rabbit IgG Using a Conducting Polymer Modified Screen-Printed Electrode. Biosens. Bioelectron, 18, pp 773-780.

Dave BC *et al.* (1994). Sol-Gel Encapsulation Methods for Biosensors. Anal. Chem, 66(22), pp 1120-1127.

Davis F, Hughes MA, Cossins AR and Higson SP (2007). Single gene differentiation by DNA-modified carbon electrodes using an AC impedimetric approach. Anal. Chem, 79, pp 1153–1157.

DeBusschere BD and Kovacs GT (2001). A Portable cell-based biosensor system using integrated CMOS cell-cartridges, Biosensors and Bioelectronics, 16: 543.

Deiss D, Bolinder J, Riveline JP, Battelino T, Bose E, Tubiana-Rufi N, Kerr D and Phillip M (2006). Improved glycemic control in poorly controlled patients with type 1 diabetes using real-time continuous glucose monitoring. Diabetes Care, 29 (12), pp 2730–2732.

Del Carlo M, Mascini M, Pepe A, Diletti G, Compagnone D (2004). Screening of food samples for carbamate and organophosphate pesticides using electrochemical bioassay. Food Chem, 84, pp 651– 656.

Del Carlo M, Pepe A, Mascini M, De Gregorio M, Visconti A, Compagnone D (2005). Determining pirimiphos-methyl in durum wheat samples using an acethylcholinesterase inhibition essay. Anal. Bioanal. Chem, 381, pp 1367–1372.

Dell'Olio F, Passaro VMN (2007). Optical sensing by optimized silicon slot waveguides. Optics Express. 15, 4977-4993.

DeMarco Dr, EW Saaski, DA McCrae and DM Lim (1999). Rapid detection of *Escherichia coli* O157:H7 in ground beef using a fibre-optic biosensor. J. of Food Protection, 62, pp 711-716.

Densmore A, Xu D X, Waldron P, Janz S, Cheben,P, Lapointe J, Delâge A, Lamontagne B, Schmid J H and Post E (2006). Silicon-on-Insulator Photonic Wire Based Evanescent Field Sensor. IEEE Photon. Tech, 18, pp 2520-2522.

Deverkadra R R and McShane M J (2002). Fabrication of a biosensor with embedded waveguides for sensing in microchannels. Annual International Conference of the IEEE Engineering in Medicine and Biology – Proceedings, 2, pp 1720-1721.

Dewa AS and Ko WH (1994). Biosensors, Semiconductor Sensors (Eds) S.M. Sze, New York, Wiley Interscience, pp 415.

Diamond D (1998). Principles of Chemical and Biological Sensors, New York. John Wiley & Sons. vol. 150.

Dinu CZ, Dinca V, Howard J and Chrisey DB (2007). Printing technologies for fabrication of bioactive and regular microarrays of streptavidin. Applied Surface Science. Elsevier BV, 53 pp 8119–8124.

Discher DE, Janmey P and Wang Y (2005). Tissue cells feel and respond to the stiffness of their, substrate. Science, 310, pp 1139-1143.

Divis C (1975). Annals of Microbiology. 126A, pp 175-186.

Doi N and Yanagawa H (1999). Design of generic biosensors based on green fluorescent proteins with allosteric sites by directed evolution. FEBS 453, pp 305– 307.

Dong Wei (2006). Development of a surface plasmon resonance biosensor for The identification of campylobacter jejuni. Auburn, Alabama. MS Auburn University, China.

Doong RA and HC Tsai (2001). Immobilization and characterization of sol-gelencapsulated acetyl-cholinesterase fibre-optic biosensor. Anal. Chim. Acta, 434(2), pp 239-246.

Dostalek J, Ctyroky J, Homola J, Brynda E, Skalsky M, Nekvindova P, Spirkova J, Skvor J and Schrofel J (2001). Surface plasmon resonance biosensor based on integrated optical waveguide. Sens. Actuators B, 76, pp 8-12.

Drapp B, Piehler J, Brecht A, Gauglitz G, Luff BJ, Wilkinson JS and Ingehoff J (1997). Integrated optical Mach-Zehnder interferometers as simazine immunoprobes. Sens. Actuators, B 38, pp 277-282.

Duan C and Meyerhoff ME (1994). Separation-Free Sandwich Enzyme Immunoassays Using Microporous Gold Electrodes and Self-Assembled Monolayer/Immobilized Capture Antibodies. Anal. Chem, 66, pp 1369-1377.

Durrieu C, C Chouteau, J Henry, J Thomazi, JM Chovelon and C Tran-Minh (2004). Bienzymatic whole cell algal biosensors design for environmental monitoring. The 6th workshop on Biosensors and Bioanalytical microtechniques in environmental and clinical analysis, Rome, Italy.

Dzyadevych S, Soldatkin AP, Arkhypova VN, El'skaya AV (2005). Early-warning electrochemical biosensor system for environmental monitoring based on enzyme inhibition. Sens. Actuators B: Chem, 105, pp 81–87.

Dzyadevych SV, Arkhypova VN, Martelet C, Jaffrezic-Renault N, Chovelon JM, El'skaya AV, Soldatkin AP (2004b). Potentiometric biosensors based on ISFETs and immobilized cholinesterases. Electroanalysis, 16, pp 1873–1882.

Dzyadevych SV, Arkhypova VN, Soldatkin AP, El'skaya AV, Martelet C, Jaffrezic-Renault N (2004a). Enzyme biosensor for tomatine detection in tomatoes. Anal. Lett, 37, pp 1611–1624.

Dzyadevych SV, Soldatkin AP, Chovelon JM (2002). Assessement of the toxicity of methyl parathion and its photodegradation products in water samples using conductometric enzyme biosensors. Anal. Chim. Acta, 459, pp 33–41.

Dzyadevych SV, Soldatkin AP, Korpan YI, Arkhypova VN, El'skaya AV, Chovelon JM, Martelet C, Jaffrezic-Renault N (2003). Biosensors based on enzyme field-effect transistors for determination of some substrates and inhibitors. Anal. Bioanal. Chem, 377, pp 496–506.

Earp R, Commercial Biosensors: Applications to Clinical, Bioprocess, and Environmental Samples, John Wiley and Sons. http://www.chem.vt.edu/chem-dept/students/Earp/SPR.html.

Egert U, Schlosshauer B and Fennrich S. *et al.* (1998). A novel organotypic long-term culture of the rat hippocampus on substrate integrated multielectrode arrays, Brain Res. Protocols, 2, pp 229.

Eggins B R (1996). Biosensors: An Introduction. 1st Edition. New york, United States of America: John Wiley & Sons Ltd and BGTeubner.

Egholm M, Buchardt O, Christensen L, Behrens C, Freier SM, Driver DA, Berg RH, Kim SK, Norden B and Nielsen PE (1993). PNA hybridizes to complementary oligonucleotides obeying the Watson- Crick hydrogen-bonding rules. Nature, 365, pp 566–568.

Einstein Lubrin, Elaine Lawrence and Karla Felix Navarro (2006). Architecture for Wearable, Wireless and Smart Biosensors: The MoteCare Prototype Proceedings of the International Conference on Networking, International Conference on Systems and International Conference on Mobile Communications and Learning Technologies (ICNICONSMCL'06). 0-7695-2552-0/06: IEEE.

Elizabeth AH Hall ed. (1991). Biosensors, New York, Prentice Hall.

Ellington AD and Szostak JK (1990). *In vitro* selection of RNA molecules that bind specific ligands. Nature, 346, pp 818–822.

Eltoukhy H, Salama K and Gamal A E (2006). A 018-Mm CMOS bioluminescence detection Lab-on- Chip. IEEE Journal of Solid-State Circuits, 41, pp 651-661.

Eltoukhy H, Salama K and Gamal A E A (2006). CMOS bioluminescence detection Lab-on-Environmental Analysis and Monitoring. Anal. Bioanal. Chem, 386, pp 1025-1041.

Ercole C, Gallo M D, M Pantalone, Santucci S, Mosiello L, Laconi C and Lepidi A (2002). Sens. Actuators, B 83 pp 48–52.

Errachid A, Zine Z, Samitier J and Bausells J (2004). FET-based chemical sensor systems fabricated with standard technologies. Electroanalysis, 16, pp 1843-1851.

Ersoz A, Denizli A, Ozcan A and Say R (2005). Molecularly imprinted ligand-exchange recognition assay of glucose by quartz crystal microbalance. Biosens. Bioelectron, 20, pp 2197–2202.

Esinenco D, Psoma SD, Kusko M, Schneider A and Muller R (2005). SU-8 Micro-biosensor based on Mach-Zehnder interferometer, Rev.Adv.Mater.Sci, 10, pp 295-299.

Esteban O, Gonzalez-Cano A, Diaz-Herrera N and Navarrete M (2006). Absorption as a selective mechanism in surface plasmon resonance fibre optic sensors. Optics Letters, 31, pp 3089-3091, 74, pp 37-46.

Evtugyn GA, Stoikov II, Budnikov HC, Stoikova EE (2003). A cholinesterase sensor based on a graphite electrode modified with 1, 3-disubstituted calixarenes. J. Anal. Chem. 58, pp 1151–1156.

Fan C, Plaxco KW and Heeger AJ (2003). Electrochemical Interrogation of Conformational Changes as a Reagentless Method for the Sequence-Specific Detection of DNA. P. Natl. Acad. Sci. USA, 100, pp 9134-9137.

Fang Y, Ferrie AM and Li G (2005). Probing cytoskeleton modulation by optical biosensors. FEBS Lett. 579, pp 4175-4180.

Fang Y, Ferrie AM, Fontaine NH and Yuen PK (2005). Characteristics of dynamic mass redistribution of EGF receptor signaling in living cells measured with label free optical biosensors. Anal. Chem. 77, pp 5720-5725.

Fang Y, Ferrie AM, Fontaine NH, Mauro J and Balakrishnan J (2006). Resonant waveguide grating, biosensor for living cell sensing. Biophys. J. 91, pp 1925-1940.

Fang Y, Li G and Ferrie AM (2007). Non-invasive optical biosensor for assaying endogenous G protein, coupled, receptors in adherent cells. J. Pharmacol. Toxicol. Methods. 55, pp 314-322.

Fang Y, Li G and Peng J (2005). Optical biosensor provides insights for bradykinin B2 receptor signaling, in A431 cells. FEBS Lett, 579, pp 6365-6374.

Fang Ye (2007). Non-Invasive Optical Biosensor for Probing Cell Signaling. Sensors ISSN 1424-8220, MDPI. www.mdpi.org/sensors.

Fattinger C, Mangold C, Gale M T and Schutz H (1995). Bidiffractive grating coupler: universal transducer for optical interface analytics. Optical Engineering, 34(9), pp 2744-2753.

Ferguson JA *et al.* (1996). A fibre-optic DNA biosensor microarray for the analysis of gene expression. Nat. Biotechnol. 14(13), pp 1681-1684.

Fernandez-Sanchez C, Gallardo-Soto AM, Rawson K, Nilsson O and McNeil CJ (2004). Quantitative Impedimetric Immunosensor for Free and Total Prostate Specific Antigen Based on a Lateral Flow Assay Format. Electrochem. Commun. 6, pp 138-143.

Fernandez-Sanchez C, McNeil CJ, Rawson K and Nilsson O (2004). Disposable Noncompetitive Immunosensor for Free and Total Prostate-Specific Antigen Based on Capacitance Measurement. Anal. Chem. 76, pp 5649-5656.

Ferrer-Miralles N, Feliu J X, Vandevuer S, Muller A, Cabrera-Crespo J, Ortmans I, Hoffmann F, Cazorla D, Rinas U, Prevost M and Villaverde A (2001). Engineering regulable *Escherichia coli* b-galactosidases as biosensors for anti-HIV antibody detection in human sera. J. Biol. Chem. 276, pp 40087–40095.

Ferretti S, Lee SK, MacCraith BD, Oliva AG, Richardson DJ, Russell D, Sapsford KE and Vidal M (2000). Optical biosensing of nitrite ions using cytochrome cd1 nitrite reductase encapsulated in a sol-gel matrix. Analyst. 125, pp 1993-1999.

Fibre Optic Systems, http://blake.sunderland.ac.uk/~ta5ate/fibopt.htm.

Flanagan M T and Pantell RH (1984). Surface plasmon resonance and immunosensors. Electron. Lett, 20, pp 968-970.

Fodor S P, Read J L, Pirrung M C, Stryer L, Lu A T and Solas D (1991). Light-directed, spatially addressable parallel chemical analysis. Science 251, pp 767–773.

Fraaje BA, Appels M, de Boer SH, van Vuurde JWL and van den Bulk RW (1997) Detection of soft rot Erwinia spp. on seed potatoes: conductimetry in comparison with dilution plating, PCR and serological assays. Eur. J. Plant Pathol. 103, pp 183–193.

Frebort I, Skoupa L and Pec P (2000). Amine Oxidase-Based Flow Biosensor for the Assessment of Fish Freshness. Food Control. 11, pp 13-18.

Fuji-Keizai (2004). Biosensor Market, R&D and Commercial Implication.

Gaixia XU, Yicong WU, Rong LI, WANG Ping, YAN Weiming and ZHENG Xiaoxiang (2002). Cell-based biosensors: Towards the development of cellular monitoring. Chinese Science Bulletin Vol. 47 No. 22, pp 1849-1856.

Gao HH, Chen Z, Kumar J, Marx KA, Tripathy SK and Kaplan DL (1996). Multilayer enzyme assembly for the development of a novel fibre optic biosensor. Materials Research Society Symposium Proceedings. 414, pp 125-130.

Gao Z, Agarwal A, Trigg AD, Singh N, Fang C, Tung CH, Fan Y, Buddharaju KD and Kong J (2007). Silicon Nanowire Arrays for Label-Free Detection of DNA. Anal. Chem. 79, pp 3291-3297.

Garg S, Zisser H, Schwartz S, Baile T, Kaplan R, Ellis S and Jovanovic L (2006). Improvement in Glycemic Excursions with a Transcutaneous, Real-Time Continuous Glucose Sensor. Diabetes Care. 29, pp 44-50.

Garjonyte R and Malinaiuskas A (2000). Biosensors & Bioelectronics. 15, pp 445-451.

Gary K Maki, Nirankar Mishra, Shiva Rastogi, Paul Winterrowd, Eric Cameron and Wusi C Maki. Nano Technology Electronic Biosensors for Food Safety. Center for Advanced Microelectronics Biomolecular Research, University of Idaho, Post Falls, ID 83854, USA.

Gautier S M, Blum L J, and Coulet P R (1990). Fibre-optic biosensor based on luminescence and immobilized enzymes: microdetermination of sorbitol, ethanol and oxaloacetate. Journal of bioluminescence and chemiluminescence. 5, pp 57-63.

Gautschi and Gustav H (2002). Piezoelectric Sensorics, Springer.

Gemeiner P, Docolomansky P, Vikartovska A and Stefuca V (1998). Amplification of flow-microcalorimetry signal by means of multiple bioaffinity layering of lectin and glycoenzyme. Biotechnol. Appl. Biochem. 28, pp 155–161.

George P Anderson, Chris A Rowe Taitt and Frances S Ligler (2000). RAPTOR: A portable, automated biosensor. Proceedings of the First Conference on Point Detection for Chemical and Biological Defense.

Gerard M, Chaubey A and Malhotra B D (2002). Biosensors & Bioelectronics 17, pp 345-359.

Ghindilis AL, Atanasov P, Wilkins M and Wilkins E (1998). Immunosensors: electrochemical sensing and other engineering approaches. Biosens. Bioelectron. 13, pp 113–131.

Giaever I and Keese C R (1993). A morphological biosensor for mammalian cells, Nature (London). pp 366: 591.

Gil GC, Chang IS, Kim BH, Kim M, Jang JK, Park HS and Kim HJ (2003). Operational parameters affecting the performance of a mediator-less microbial fuel. Biosens. Bioelectron. 18, pp 327–334.

Gilchrist KH, Barker VN, Fletcher LE *et al.* (2001). General purpose, field-portable cell-based biosensor platform, Biosensors and Bioelectronics. 16, 557.

Golden JP, Anderson G P, Rabbany SY and Ligler (1994). FS An evanescent wave biosensor-Part II: Fluorescence signal acquisition from tapered fibre optic probes. IEEE Transactions on Biomedical Engineering. 41, pp 585-591.

Gooding JJ (2006). Biosensor Technology for Detecting Biological Warfare Agents: Recent Progressand Future Trends. Anal. Chim. Acta. 559, pp 137-151.

Gorecki C, De Labachelerie M and Thiery L (2003). II IEEE Sensors, Journal 3, pp 121.

Gough DA, Lucisano JY and Tse PHS (1985). Two-Dimensional Enzyme Electrode Sensor for Glucose. Anal. Chem. 57, pp 2351-2357.

Graham DL, Ferreira HA and Freitas PP (2004). Magnetoresistive-based biosensors and biochips. Trends Biotechnol. 22, pp 455–462.

Gronewold TMA, Glass S, Quandt E and Famulok M (2005). Monitoring complex formation in the blood coagulation cascade using aptamer-coated SAW sensors. Biosens. Bioelectron. 20, pp 2044–2052.

Gruzina TG, Vember VV, Ul'berg ZR and Starodub NF (2005). Luminescent test based on photobacterium phosphorum b7071 for ionic and colloid gold determination in water. Khimiya i Tekhnologiya Vody. 27, pp 200-208.

Guan JG, Miao YQ and Chen JR (2004). Prussian blue Modified Amperometric FIA Biosensor: One-Step Immunoassay for Fetoprotein. Biosens. Bioelectron. 19, pp 789-794.

Guenat OT, Dufour JF, van der Wal PD, Morf WE, de Rooij NF and Koudelka-Hep M (2005). Microfabrication and characterization of an ion-selective microelectrode array platform. Sens. Actuators B Chem. 105, pp 65-73.

Guenat OT, Generelli S, de Rooij NF, Koudelka-Hep M, Berthiaume F and Yarmush ML (2006). Development of an array of ion-selective microelectrodes aimed for the monitoring of extracellular ionic activities. Anal. Chem. 78, pp 7453-7460.

Guilbault GG and de Olivera Neto G (1985). Immobilised enzyme electrodes. In USES45, Immobilized cells and enzymes a practical approach ed. J.Woodward. Oxford: IRL Press Ltd, pp 55-74.

Guilbault GG and Jordan JM (1988). Analytical Uses of Piezoelectric Crystals: A Review. CRC Crit. Rev. Anal. Chem. 19 (1), pp 1-28.

Guilbault GG and Lubrano GJ (1973). An Enzyme Electrode for the Amperometric Determination of Glucose. Anal. Chim. Acta. 64, pp 439-455.

Guilbault GG and Montalvo J (1969). JACS 91, pp 2164-2569.

Haeussling L, Ringsdorf H, Schmitt F and Knoll W (1991). Biotin-functionalized self-assembled monolayers on gold: surface plasmon optical studies of specific recognition reactions. Langmuir 7.

Hafner F (2000). Cytosensor microphysiometer: technology and recent application, Biosensors and Bioelectronics. 15, pp 149.

Hahm JI and Lieber CM (2004). Direct Ultrasensitive Electrical Detection of DNA and DNA Sequence Variations Using Nanowire Nanosensors. Nano Lett. 4, 51-54. Hybridization. Bioconjugate Chem. 8, pp 906-913.

Hahn EC (1988). Piezoelectric Crystal Detectors and Their Applications. Ph.D, dissertation, University of New Orleans.

Hall DG (1996). Ion-Selective Membrane Electrodes: A General Limiting Treatment of Interference Effects, J. Phys. Chem, 100, pp 7230–7236.

Hall EAH (1986). The developing biosensor arena. Enzyme and Microbial Technology. 8, pp 651-657.

Han S, Zhu M, Yuan Z, Li X (2001). A methylene blue-mediated enzyme electrode for the determination of trace mercury (II), mercury (I), methylmercuryand mercury–glutathione complex. Biosens. Bioelectron. 16, pp 9–16.

Hasebe Y, Yokobori K, Fukasawa K, Kogure T and Uchiyama S (1997). Anal. Chim. Acta 357, pp 51–54.

Hashiguchi Y, Sakakida M, Nishida K, Uemura T, Kajiwara K and Shichiri M (1994). Development of Miniaturized Glucose Monitoring System by Combining a Needle-Type Glucose Sensor with Microdialysis Sampling Method. Diabetes Care.17, pp 387-396.

Hashimoto K, Ito K and Ishimori Y (1994). Sequence-Specific Gene Detection with a Gold Electrode Modified with DNA Probes and an Electrochemically Active Dye. Anal. Chem. 66, pp 3830- 3833.

Heather Strosnider (2003). Whole-Cell Bacterial Biosensors and the Detection of Bioavailable Arsenic. National Network of Environmental Management Studies Fellow. www.clu-in.org.

Heising RA (1946). Quartz Crystal for Electrical Circuits. New York: Van Nostrand, pp 24.

Heller A (1999). Implanted electrochemical glucose sensors for the management of diabetes. Annu. Rev. Biomed. Eng. 1, pp 153–175.

Hianik T *et al.* (2001). Amperometric Detection of DNA Hybridization on a Gold Surface Depends on the Orientation of Oligonucleotide Chains, Bioelectrochemistry. 53, no. 2, pp 199–204.

Hikuma M, Suzuki H, Yasuda T, Karube I and Suzuki S (1979). Amperometric estimation of BOD by using living immobilized yeast. Eur. J. Appl. Microbiol. Biotechnol. 8, pp 289–297.

Hirsch LR, Jackson JB, Lee A, Halas NJ and West JL (2003). A whole blood immunoassay using gold nanoshells. Anal. Chem. 75, pp 2377–2381.

Hirschfeld TB and Block MJ (1984). Fluorescent immunoassay employing optical fibre in capillary tube. USPatent 4, pp 447-546.

Hisamoto H, Yasuoka M and Terabe S (2006). Integration of multiple-ion-sensing on a capillary-assembled microchip. Anal.Chim.Acta 556, pp 164-170.

Hodak J (1997). Layer by layer self assembly of glucose oxidase with a poly (allylamine) ferrocene redox mediator. Langmuir, 13, pp 2708–2716.

Hojlund K, Wildner-Christensen M, Eshoj O, Skjaerbaek C, Holst JJ, Koldkjaer O, Møller Jensen D and Beck-Nielsen H (2001). Reference intervals for glucose, betacell polypeptides, and counterregulatory factors during prolonged fasting. Am J Physiol Endocrinol Metab, Vol. 280, No.1, pp E50-E58.

Hopman W C L, Pottier P, Yudistira D, van Lith J, Lambeck P V, De La Rue R M, Driessen A, Hoekstra HJ W M and de Ridder R M (2005). Quasi one-dimensional photonic crystal as a compact building-block for refractometric optical sensors. IEEE J. Sel. Top. in Quantum Electron.11, pp 11-16.

Hou Y, Jaffrezic-Renault N, Martelet C, Tlili C, Zhang A, Pernollet JC, Briand L, Gomila G, Errachid A and Samitier J *et al.* (2005). Study of Langmuir and Langmuir–Blodgett films of odorant binding protein/ ampphiphile for odorant biosensors. Langmuir. 21, pp 4058–4065.

Hou Y, Tlili C, Jaffrezic-Renault N, Zhang A, Martelet C, Ponsonnet L, Errachid A, Samitier J and Bausells J (2004). Study of mixed Langmuir–Blodgett films of immunoglobulin G/amphiphile and their application for immunosensor engineering. Biosens. Bioelectron. 20, pp 1126–1133.

Howorka K, Pumprla J, Saletu B, Anderer P, Krieger M and Schabmann A (2000). Decrease of vigilance assessed by EEG-mapping in type I diabetic patients with history of recurrent severe hypoglycaemia. Psychoneuroendocrinology, Vol. 25, No.1, pp 85-105.

Hua P, Luff BJ, Quigley GR, Wilkinson JS and Kawguchi K (2002). II Sensors and Actuators B. 87: 250.

Hudder A, Nathanson L and Deutscher MP (2003). Organization of mammalian cytoplasm. Mol. Cell, Biol. 23, pp 9318-9326.

Hughes W S (1922). The potential difference between glass and electrolytes in contact with water. J. Am. Chem. Soc. 44, pp 2860–2866.

Hulanicki Adam, Geab Staniseaw and Ingman Folke (1991). Chemical Sensors definitions and classification. Pure & App/.Chern, Vol. 63, pp 1247-1250.

Hyeong Jae and Oliver Brand (2008). High Q-factor In-Plane-Mode Resonant Microsensor Platform for Gaseous/Liquid Environment. Journal of Microelectromechanical Systems. 17, pp 483-93.

Hyllienmark L, Maltez J, Dandenell A, Ludvigsson J and Brismar T (2005). EEG abnormalities with and without relation to severe hypoglycaemia in adolescents with type 1 diabetes. Diabetologia, Vol. 48, No. 3, pp 412-419.

Hyun CK, Tamiya E, Takeuchi T, Karube I and Inoue N (1993). Novel BOD sensor based on bacterial luminescence. Biotechnol. Bioeng. 41, pp 1107–1111.

Iber C, Ancoli-Isreal S, Chesson A and Quan SF (2007). The AASM Manual for the Scoring of Sleep and Associated Events, Rules, Terminology and Technical Specifications. (1st ed). American Academy of Sleep Medicine, Westchester, Illinois.

Ikebukuro K, Kiyohara C and Sode K (2005). Novel Electrochemical Sensor System for Protein Using the Aptamers in Sandwich Manner. Biosens. Bioelectron. 20, pp 2168-2172.

In Seop Changa, Jae Kyung Jang a, Geun Cheol Gil a, Mia Kima, Hyung Joo Kima, Byung Won Chob and Byung Hong Kima (2004). Continuous determination of biochemical oxygen demand using microbial fuel cell type biosensor. Biosensors and Bioelectronics.19, pp 607–613.

Iris M Wentholt, Marit A Vollebregt, Augustus A Hart, Joost B Hoekstra and J Hans DeVries (2005). Comparison of a Needle-Type and a Microdialysis Continuous Glucose Monitor in Type 1 Diabetic Patients. Diabetes Care. 28, pp 2871–2876.

Ishige Y, Shimoda M and Kamahori M (2006). Jpn. J. Appl. Phys. MDPI, www.mdpi.org/sensors, 45, pp 3776–3783.

Ishigea Yu, Shimodab Maki and Kamahoria Masao (2009). Extended-gate FET-based enzyme sensor with ferrocenyl-alkanethiol modified gold sensing electrode. Biosensors and Bioelectronics. Elsevier BV. 24, pp 1096–1102.

Ito Y (1998) High-spatial resolution LAPS, Sensors and Actuators B. 52, 107.

James P Chambers, Bernard P Arulanandam, Leann L Matta, Alex Weis and James J Valdes. Biosensor Recognition Elements. Horizon Scientific Press. Offprints from www.cimb.org. Curr. Issues Mol. Biol, 10, pp 1–12.

Janata J (1975). An Immunoelectrode. J. Am. Chem. Soc. 97, pp 2914-2916.

Janata J and Moss S (1976). Biomed. Eng. 11, 241–245.

Jayasena S D (1999). Aptamers: An emerging class of molecules that rival antibodies in diagnostics. Clinical Chemistry. 45, pp 1628–1650.

JD Andrade and R Vanwagenen (1983). Process for conducting fluorescence immunoassays without added labels and employing attenuated internal reflection. US Patent 4,368,047.

Jensenius H, Thaysen J, Rasmussen A A, Veje LH, Hansen O and Boisen A (2000). J. Appl. Phys. 76, pp 2615-17.

Ji G and Silver S (1992). Regulation and Expression of the Arsenic Resistance Operon from Staphylococcus aureus Plasmid pI258. J. Bacteriol. 174, pp 3684-3694.

Jiang Y, Fang X and Bai C (2004). Signaling aptamer/ protein binding by a molecular light switch complex. Anal. Chem. 76, pp 5230–5235.

Joachim C (1986). Biochips: dreams...and realities. International Industrial Biotechnology 79:7:12/1, pp 211-219.

Kafi AK, Lee DY, Park SH and Kwon YS (2006). DNA as a support for glucose oxidase immobilization at Prussian blue-modified glassy carbon electrode in biosensor preparation. J. Nanosci. Nanotechnol. 6, pp 3539–3542.

Kajikawa K, Mitsui K (2004). Optical fibre biosensor based on localized surface plasmon resonance in gold nanoparticles. Proceedings of SPIE-The International Society for Optical Engineering. 5593, pp 494-501.

Kanapieniene JJ, Dedinaite AA, Laurinavicius VSA (1992). Miniature Glucose Biosensor with Extended Linearity. Sensor. Actuat. B-Chem. 10, 37-40.

Karube I, Matsunga T, Mitsuda S and Suzuki S (1977). Microbial electrode BOD sensors. Biotechnol. Bioeng. 19, 1535–1547.

Katz E (2002). Surface Plasmon Resonance - a method to analyze interfacial optical properties and to develop biosensors. http://chem.ch.huji.ac.il/~eugeniik/spr.htm.

Kaylee Sill McElroy (2005). Gallium Nitride Nanowire Growth and FET Device and Biosensor Fabrication. NNIN REU Research Accomplishments. NNIN REU Site: Cornell NanoScale Science and Technology Facility, Cornell University. pp 134-135.

Kelley SO, Boon EM, Barton JK, Jackson NM and Hill MG (1999). Single-Base Mismatch Detection Based on Charge Transduction through DNA. Nucleic Acids Res. 27, pp 4830-4837.

Kernevez JP, Konate L and Romette JL (1983). Determination of substrate concentration by a computerized enzyme electrode. Biotechnology and Bioengineering. 25, pp 845-855.

Kholodenko BN (2003). Four-dimensional organization of protein kinase signaling cascades: the roles, of diffusion, endocytosis and molecular motors. J. Exper. Biol. 206, pp 2073-2082.

Kim B, Cha GS and Meyerhoff ME (1990). Homogeneous enzyme-linked binding assay for studying the interaction of lectins with carbohydrates and glycoproteins. Anal. Chem. 62, pp 2663–2668.

Kim BH, Chang IS, Gil GC, Park HS and Kim HJ (2003). Novel BOD (biological oxygen demand) sensor using mediator-less microbial fuel cell. Biotechnol. Lett. 25, pp 541–545.

Kim BH, Kim HJ, Hyun MS and Park DH (1999b). Direct electrode reaction of Fe (III) reducing bacterium, Shewanella putrefaciens. J. Microbiol. Biotechnol. 9, pp 127–131.

Kim DW, Y Zhang, KL Cooper and A Wang (2005). In-fibre reflection mode interferometer based on a long-period grating for external refractive-index measurement. Appl. Opt. 44(26), pp 5368-5373.

Kim HJ, Hyun MS, Chang IS and Kim BH (1999c). Microbial fuel cell type lactate biosensor using a metal-reducing bacterium, Shewanella putrefaciens. J. Microbiol. Biotechnol. 9, pp 365–367.

Kim J H, Cho J H and Cha G S (2000). Biosensors & Bioelectronics. 14, pp 907-915.

Kim, DS, Jeong YT, Park HJ, Shin JK, Choi P, Lee JH and Lim G (2004). Biosens. Bioelectron. 20, pp 69–74.

King KD, Anderson GP, Bullock KE, Regina MJ, Saaski EW and Ligler FS (1999). Detecting staphylococcal enterotoxin B using an automated fibre optic biosensor. Biosensors & Bioelectronics. 14, pp. 163-170.

King KD, Vanniere JM, LeBlanc JL, Bullock KE and Anderson GP (2000). Automated fibre optic biosensor for multiplexed immunoassays. Environ. Sci. & Tech. 34, pp 2845-2850.

Kirby R, Cho EJ, Gehrke B, Bayer T, Park YS, Neikirk DP, McDevitt JT and Ellington AD (2004). Aptamer-based sensor arrays for the detection quantitation of proteins. Anal. Chem. 76, pp 4066–4075.

Kok FN, Bozoglu F, Hasirci V (2002). Construction of an acethylcholinesterase-choline oxidase biosensopr for aldicarb determination. Biosens. Bioelectron. 17, pp 531–539.

Kok FN, Hasirci V (2004). Determination of binary pesticide mixtures by an acetylcholinesterase-choline oxidase biosensor. Biosens. Bioelectron. 19, pp 661–665.

Korpan YI, Dzyadevich SV, Zharova VP and El´skaya AV (1994). Conductometric biosensor for ethanol detection based on whole yeast cells. Ukr. Biokhim. Zh. Pp 66:78–82.

Koubova V, Brynda E, Karasova L, Skvor J, Homola J, Dostalek J, Tobiska P and Rosicky J (2001). Sens. Actuators B 74, 100–105.

KR Rogers (1989). Biosensors for Environmental Applications. Biosensors Bioelectronics. 10, pp 533-541.

Krause M, Ingebrandt S and Richter D *et al.*, (2000) Extended gate electrode arrays for extracellular signal recordings, Sensors and Actuators B. 70, 101.

Krawczynski vel Krawczyk T, Mosszczynska M, Trojanowicz M (2000). Inhibitive determination of mercury and other metal ions by potentiolmetric urea biosensor. Biosens. Bioelectron. 15, pp 681– 69.

Kress-Rogers E (1996). Handbook of Biosensors and Electronic Noses-Medicine, Food and the Environment. CRC Press, Boca Raton, USA.

Kress-Rogers E (1997). Biosensors and Electronic Noses. Chapter 7, CRC Press Inc.

Kretz AP and Styblo D (2003). Toward Continuous Blood Glucose Monitoring, Medical Device & Diagnostic Industry 25, no. 6, pp 78–83.

Kreuzer MP, O'Sullivan CK Pravda M and Guilbault GG (2001). Development of an Immunosensor for the Determination of Allergy Antibody (IgE) in Blood Samples. Anal. Chim. Acta. 442, pp 45-53.

Kricka LJ and Thorpe GHG (1986). Immobilised enzymes in analysis. Trends in Biotechnology. 4, pp 253-258.

Kuhtreiber WM and Jaffe LF (1990). Detection of extracellular calcium gradients with a calcium-specific vibrating electrode. J. Cell.Biol. 110, pp 1565-1573.

Kurihara K, Ohkawa H, Iwasaki Y, Niwa O, Tobita T and Suzuki K (2004). Fibre-optic conical microsensors for surface plasmon resonance using chemically etched single-mode fibre. Analytica Chimica Acta. 523, pp 165-170.

Kuswandi B (2003). Simple optical fibre biosensor based on immobilized enzyme for monitoring of trace heavy metal ions. Anal. Bioanal. Chem. 376, pp 1104–1110.

Kuusk E, Rinken T (2004). Transient phase calibration of tyrosinase based carbaryl biosensor. Enzyme Microb. Technol. 234, pp 657–661.

Kyung-Ah Yoo, Jun-Hyuk Kim, B H Nahm, C J Kang and Yong-Sang Kim (2007). Fabrication and characteristics of microcantilever-based biosensor for detection of the protein-ligand binding. IOP Publishing Journal of Physics. Conference Series, 61, pp 1308–1311.

Lange J and Wittmann C (2001). Enzyme Sensor Array for the Determination of Biogenic Amines in Food Samples. Analytical and Bioanalytical Chemistry.

Langmuir I (1917). J. Am. Chem. Soc. 39, 1848.

Lee M and Walt D R (2000). A fibre-optic microarray biosensor using aptamers as receptors. Anal. Biochem. 282, pp 142–146.

Lee MM, Russel DA (2003). Novel determination of cadmium ions using an enzyme self assembled monolayer with surface Plasmon resonance. Anal. Chim. Acta 500, pp 119–125.

Lee TMH, Li LL and Hsing IM Enhanced Electrochemical Detection of DNA Hybridization Based on Electrode-Surface Modification. Langmuir.19, pp 4338-4343.

Liedberg B, Nylander C and Lundstrm I (1983). Sensors and Actuators. 4, pp 299-304.

Ligler FS, Anderson GP, Davidson PT, Foch RJ, Ives JT, King KD, Page G, Stenger DA and Whelan JP (1998). Remote sensing using an airborne biosensor. Environ. Sci. & Tech. 32, pp 2461-2466.

Ligler FS, Golden JP, Shriver-Lake LC, Ogert RA, Wijesuria D and Anderson GP (1993). Fibre-optic biosensor for the detection of hazardous materials, Immunomethods. 3, pp 122-127.

Lillie JJ, Thomas MA, Jokerst NM, Ralph SE, Dennis KA and Handerson CL (2006). Multimode interferometric sensors on silicon optimized for fully integrated complementarymetal-oxide-semiconductor chemical-biological sensor systems. J. Opt. Soc. Am. B. 23, pp 642-651.

Lin H Y, Tsai W H, Tsao YC and Sheu BC (2007). Side-polished multimode fibre biosensor based on surface plasmon resonance with halogen light. Applied Optics. 46, pp 800-806.

Liss M, Petersen B, Wolf H and Prohaska E (2002). An aptamer-based quartz crystal protein biosensor. Anal. Chem. 74, pp 4488–4495

Liu G and Lin Y (2006). Biosensor based on self-assembling acetylcholinesterase on carbon nanotubes for flow injection/amperometric detection of organophosphate pesticides and nerve agents. Anal. Chem. 78, pp 835–843.

Liu G, Lin YY, Wang J, Wu H, Wai CM and Lin Y (2007). Disposable Electrochemical Immunosensor Diagnosis Device Based on Nanoparticle Probe and Immunochromatographic Strip. Anal. Chem. 79, pp 7644-7653.

Liu J, Bjornsson L and Mattiasson B (2000). Immobilised activated sludge based biosensor for biochemical oxygen demand measurement. Biosen. Bioelectron. 14, pp 883–893.

Loughran MG, Hall JM and Turner APF (1996). Electroanal.7, 1.

Lowe CR (1984). Biosensors. Trends in Biotechnology. 2, pp 59-64.

Lowe CR (1985). An Introduction to the Concepts and Technology of Biosensors. Biosensors, pp 4.

Lu Y, Li X, Zhang L, Yu P, Su L and Mao L (2008). Aptamer-Based Electrochemical Sensors with Aptamer-Complementary DNA Oligonucleotides as Probe. Anal. Chem: 80, pp 1883-1890.

Lubbers DW and Opitz NZ (1975). Naturforsch. C. Biosci. 30c, pp 532-533.

Lucarelli F, L Authier, G Bagni, G Marrazza, T Baussant, E Aas and M Mascini (2003). DNA biosensor investigations in fish bile for use as a biomonitoring tool. Analytical Letters. 36(9), pp 1887-1901.

Lucarelli F, Marrazza G, Turner APF and Mascini M (2004). II Biosensors & Bioelectronics. 19, 515.

Luff BJ, Wilkinson JS, Piehler J, Hollenbach U, Ingenhoff J and Fabricius N (1998). Integrated, optical Mach-Zehnder biosensor, J. Lightwave Technol. 16, pp 583-592.

Lukasz Tymecki Beata Rozum and Robert Koncki (2006). A Very Simple Biosensing System for Educational Purposes. Chem. Anal. (Warsaw), 51, 977.

Luo DH, Levy RA, Hor YF, Federici JF and Pafchek R M (2003). An integrated photonic, sensor for in situ monitoring of hazardous organics. Sens. Actuators B, 92, pp 121-126.

Luo SF and DR Walt (1989). Avidin Biotin Coupling as a General-Method for Preparing Enzyme-Based Fibre-Optic Sensors. Anal. Chem. 61(10), pp 1069-1072.

Luong JHT, Groom CA and Male KB (1991). Biosensors and Bioelectronics 6(7), pp 547–54.

Luzi E, Minunni M, Tombelli S and Mascini M (2003). New trends in affinity sensing – aptamers for ligand binding. Trends in Analytical Chem. 22, pp 810–818.

Magrisso M, Etzion O, Pilch G, Novodvoretz A, Perez-Avraham G, Schlaeffer F and Marks R (2006). Fibre-optic biosensor to assess circulating phagocyte activity by chemiluminescence. Biosensors and Bioelectronics. 21, pp 1210-1218.

Maki G, Mishra N, Rastogi S, Filanoski B, Nelson R, Cameron E, Winterrowd P and Maki W (2007). Electronic non-semiconductor biomolecular pathogen detector. *Nanotechnology Applications in Agriculture and Food Systems*, American Chemical Society.

Maki W, Mishra N, Cameron E, Winterrowd P, Rastogi S and Maki G (2007). A universal signal transduction nano transistor based system for biomolecular detection. Nanoelectronic Devices for Defense & Security Conference.

Marazuela MD and M. Moreno-Bondi (2002). Fibre-optic biosensors - an overview.Anal. Bioanal. Chem.372 (5-6), pp 664-682.

María Espinosa Bosch, Antonio Jesús Ruiz Sánchez, Fuensanta Sánchez Rojas and Catalina Bosch Ojeda (2007). Recent Development in Optical Fibre Biosensors. Sensors. 7, pp 797-859.

Marquette CA and Blum LJ (1999). Luminol electrochemiluminescence-based fibre optic biosensors for flow injection analysis of glucose and lactate in natural samples. Analytica Chimica Acta. 381, pp 1-10.

Marquette CA, Leca B D and Blum LJ (2001). Electrogenerated chemiluminescence of luminol for oxidase-based fibre-optic biosensors. Luminescence. 16, pp 159-165.

Mason WP (1950). Piezoelectric Crystals and Their Application to Ultrasonics. Princeton, NJ: Van Nostrand.

Maxam A M and Gilbert W (1977). A new method for sequencing DNA. Proc, Nat, Acad, Sci, 74, pp 560–664.

Mazzei F, Botr`e F, Montilla S, Pilloton R, Podest`a E, Botr`e C (2004). Alkaline phosphatase inhibition based electrochemical sensors for the detection of pesticides. J. Electroanal. Chem. 574, pp 95–100.

McCauley TG, Hamaguchi N and Stanton M (2003). Aptamer-based biosensor arrays for detection and quantification of biological micromolecules. Anal. Biochem. 319, pp 244–250.

McConnell HM, Owicki JC and Parce J W *et al.* (1992). The cytosensor microphysiometer: Biological application of silicon technology, Science. 257:1906.

Mehrvar and Mehrab (2000). Fibre-Optic Biosensors-Trends and Advances, Analy.Sciences, V16, pp 677-693.

Mehrvar M and Abdi M (2004). Recent developments, characteristics, and potential applications of electrochemical biosensors. Anal. Sci. 20, pp 1113–1126.

Michael Nielsen, Lars Hauer Larsen, Mike S M Jetten and Niels Peter Revsbech (2004). Bacterium-Based NO_2 Biosensor for Environmental Applications. Applied and Environmental microbiology. pp 6551–6558.

Microchem (2002). Nano SU-8, http://www.microchem.com/ products/pdf/SU8_50-100.pdf.

Millan KM and Mikkelsen SR (1993). Sequence-Selective Biosensor for DNA Based on Electroactive Hybridization Indicators. Anal. Chem. 65, pp 2317-2323.

Minunni M, Tombelli S, Gullotto A, Luzi E and Mascini M (2004). Development of biosensors with aptamers as bio-recognition element: the case of HIV-1 Tat protein. Biosens. Bioelectron. 20, pp 1149–1156.

Mirko L, Werner B, Martin B *et al.* (2000). Non-invasive measurement of cell membrane associated proton gradients by ionsensitive field effect transistor arrays for microphysiological and bioelectronical applications, Biosensors and Bioelectronics, 15: 117.

Miroslav Pohanka and Petr Skládal (2008). Electrochemical biosensors – principles and applications. J. Appl. Biomed.6, pp 57–64.

Mishra N, Winterrowd P, Nelson R, Rastogi S, Maki W and Maki G (2007). Fabrication and characterization of nano-field effect transistor for bio-safety, Nano Science and Technology Institute, NanoTech 2007, http://www.nsti.org/ , Vol. 2, pp 453-456.

Misiakos K, Kakabakos SE, Petrou PS and Ruf HH (2004). A monolithic silicon optoelectronic transducer as a real-time affinity biosensor. Analytical Chemistry. 76, pp 1366-1373.

Miyahara Y and Moriizumi T (1985). Sens. Acutuators 7, pp 1–10.

Moatti-Sirat D, Velho G and Reach G (1991). Evaluating in Vitro and in Vivo the Interference of Ascorbate and Acetaminophen on Glucose Detection by a Needle-Type Glucose Sensor. Biosens. Bioelectron. 7, pp 345-352.

Mogensen KB, El-Ali J, Wolff A and Kutter JP (2003). II Applied Optics. 42, 4072.

Mohammadi H, Amine A, Cosnier S, Mousty C (2005). Mercury–enzyme inhibition assays with an amperometric sucrose biosensor based on a trienzymatic–clay matrix. Anal. Chim. Acta. 543, pp 143–149.

Monami M, Marchionni N and Mannucci E (2009). Long-acting insulin analogues vs. NPH human insulin in type 1 diabetes. A meta-analysis. Diabetes Obes Metab, Vol. 11, No. 4, pp 372-378.

Mosbach K and Danielsson B (1974). Biochim. Biophys. Acta. 364, pp 140-145.

Moser and Isabella (2006). BioMEMS for MultiparameterClinical Applications, Proceedings of SPIE. 4982, pp144-155.

Mourzina IG, Yoshinobu T, Ermolenko YE, Vlasov YG, Schoning MJ, Iwasaki H (2004). Immobilization of Urease and cholinesterase on the surface of semiconductor transducer for the development of light-addressable potentiometric sensors. Microchim. Acta, 144, pp 41–50.

Muhammad-Tahir Z and Alocilja EC (2003). Biosensors & Bioelectronics. 18, pp 813-819.

Muhammad-Tahir Z, Alocilja E C and Grooms D L (2005). Sensors Journal. IEEE, 5, pp 757-762.

Mulchandani A, Rogers KR (1998). Enzyme and Micriobial Biosensors: Techniques and Protocols. Humana Press, Totowa, NJ.

Nagase T, Nakata E, Shinkai S and Hamachi I (2003). Construction of artificial signal transducers on a lectin surface by post-photoaffinity-labeling modification for fluorescent saccharide biosensors. Chemistry. 9, pp 3660–3669.

Nakamura H and Karube I (2003). Current research activity in biosensors. Anal. Bioanal. Chem. 377, pp 446– 468.

National Diabetes Information Clearinghouse, NIH. National Diabetes Statistics. http://diabetes.niddk.nih.gov/dm/pubs/statistics/index.htm.

Naughton DP, Grootveld M, Blake BR, Guestrin H R and Narayanaswamy R (1993). An Optical Hydroxyl Radical Sensor. Biosensors and Bioelectronics. 8, pp 325-329.

Newman JD and Turner APF (2005). Home Blood Glucose Biosensors: A Commercial Perspective.Biosens. Bioelectron. 20, pp 2435-2453.

Nicole Jaffrezic Renaul (2006). Tyrosinase Biosensor Based on Interdigitated Electrodes for Herbicides Determination. Int. J. Electrochem. Sci. 1, pp 55-61.

Niculescu M, Nistor C, Ruzgas T, Frebort I, Sebela M, Pec P and Csoregi E (2001). Detection of Histamine and Other Biogenic Amines Using Biosensors Based on Amine Oxidase. Inflamm. Res. 50, pp S146-S148.

Nikolelis D, Krull U, Wang J and Mascini M (1998). Biosensors for Direct Monitoring of Environmental Pollutants in Field (Eds). Kluwer Academic, London.

Nomura Y, Karube I, Williaert R, Baron G and Backer L De (Eds) (1996). Biosensor Using Immobilized Living Cells in Immobilized Living Cell Systems, Wiley, Chichester, UK, pp 345.

North JR (1985). Immunosensors: antibody-based biosensors. Trends in Biotechnology. 3, pp 180-186.

O'Brien MJ, Perez-Luna VH, Brueck SR and Lopez GP (2001). A surface plasmon resonance array biosensor based on spectroscopic imaging. Biosens. Bioelectron. 16, pp 97–108.

O'Connell P and Guilbault G (2001). Future trends in biosensor research. Analytical Letters, 34(7), pp 1063–1078.

Ohnuki H, Saiki T, Kusakari A, Endo H, Ichihara M and Izumi M (2007). Incorporation of glucose oxidase into languir-blodgett films based on Prussian blue applied to amperometric glucose biosensor. Langmuir, 23, pp 4675–468.

Orazio P D (2003). Biosensors in Clinical Chemistry. Clin. Chim. Acta, 334, pp 41-69.

O'Regan T, Pravda M, O'Sullivan CK and Guilbault GG (2003). Development of Biosensor Array for Rapid Detection of Cardiac Markers: Immunosensor for Detection of Free Cardiac Troponin I. Anal. Lett, 36, pp 1903-1920.

Ovchinnikov A N, Ogurtsov V I, Trettnak W, Papkovsky D B (1999). Enzymatic flow-injection analysis of metabolites using new type of oxygen sensor membranes and phosphorescence phase measurements. Analytical Letters, 32, pp 701-716.

Ozsoz M, Erdem A, Kerman K, Ozkan D, Tugrul B, Topcuoglu N, Ekren H and Taylan M (2003). Electrochemical Genosensor Based on Colloidal Gold Nanoparticles for the Detection of Factor V Leiden Mutation Using Disposable Pencil Graphite Electrodes. Anal. Chem. 75, pp 2181-2187.

Palaroan WS, Bergantin Jr J and Sevilla III F (2000). Optical fibre chemiluminescence biosensor for antioxidants based on an immobilized luminol/hematin reagent phase. Analytical Letters. 33, pp 1797-1810.

Parellada J, Narvaez A, Lopez M A, Dominguez E, Fernandez J J, Pavlov V and Katakis I (1998). Anal. Chim. Acta 362, pp 47–57.

Park R (2003). Developing Diagnostics for the War on Bioterrorism, IVD Technology 9, no. 3, pp 29–34.

Park SJ, Taton A and Mirkin CA (2002). Array-Based Electrical Detection of DNA with Nanoparticle Probes. Science. 295, pp 1503-1506.

Parmpi P and Kofinas P (2004). Biomimetic glucose recognition using molecularly imprinted polymer hydrogels, Biomaterials 25, pp 1969–1973.

Pathirana S T, Barbaree J, Chin B A, Hartell M G, Neely W C and Vodyanoy V (2000). Biosens. Bioelectron. 15, pp 135–141.

Patolsky F, Lichtenstein A and Willner I (2000). Electrochemical Transduction of Liposome-Amplified DNA Sensing. Angew. Chem. Int. Edit. 39, pp 940-943.

Paviou and Turner (2000). Sniffing out the Truth: Clinical Diagnosis Using the Electronic Nose. Clin. Chem. Lab. Med. 38(2), pp 99-112.

Pavlov V, Xiao Yi, Willner I (2005). Inhibition of the acetylcholine esterase-stimulated growth of au nanoparticles: nanotechnology-based sensing of nerve gases. Nano Lett, 5, pp 649–653.

Pervez Khan, Md. Asdaque Hussain and Kyung Sup Kwak (2009). Medical Applications of Wireless Body Area Networks. International Journal of Digital Content Technology and its Applications Volume 3.

Piedade M, Sousa L, Germano J, Lemos J, Costa B, Freitas P, Ferreira H, Cardoso F A and Vidal D. Architecture of a Portable System Based on a Biochip for DNA Recognition. INESC-ID/IST. Lisboa, Portugal, pp 1000-029,

Pierson HO (1993). Handbook of Carbon, Graphite, Diamond, and Fullerenes: Properties, Processing, and Applications.

Piliarik M, Homola J, Manikova Z and C tyroky J (2003). Surface plasmon resonance sensor based on a single-mode polarization-maintaining optical fibre. Sensors and Actuators. B: Chemical. 90, pp 236-242.

Pirvutoiu S, Estera D, Bhand S, Ciucu A, Magearu V, Danielsson B (2002). Application of the enzyme thermistor for determination of mercury and other heavy metals using free and immobilised alcohol oxidase. Roum. Biotechnol. Lett., 7, pp 975.

Pirvutoiu S, Surugiu I, Dey ES, Ciucu A, Magearu V, Danielsson B (2001). Flow injection analysis of mercury(II) based on enzyme inhibition and thermometric detection. Analyst, 126, pp 1612–1616.

Pohanka M, Jun D and Kuǒa K (2007c). Amperometric biosensor for evaluation of competitive cholinesterase inhibition by the reactivator HI-6. Anal. Lett, 40, pp 2351–2359.

Pollack SR, Korostoff E, Starkebaum W and Lannicone W (1979). Micro-electrical studies of stress-generated potentials in bone. in Electrical Properties of Bone and Cartilage, (Edit. Brighton CT, Black J & Pollack SR), Grune & Stratton, Inc, New York.

Poscia A, Mascini M, Moscone D, Luzzana M, Caramenti G, Cremonesi P, Valgimigli F, Bongiovanni C and Varalli M (2003). A Microdialysis Technique for Continuous Subcutaneous Glucose Monitoring in Diabetic Patients (Part 1). Biosens. Bioelectron. 18, pp 891-898.

Pradeep Hothur (2008). Glass-Frit Bonding for Selective Functionalization and Packaging of Biosensors. NNIN REU Research Accomplishments. NNIN REU Site: Microelectronics Research Center, Georgia Institute of Technology, Atlanta, GA pp 16-17.

Pramming S, Thorsteinsson B, Bendtson I, and Binder C (1990). The relationship between symptomatic and biochemical hypoglycaemia in insulin-dependent diabetic patients. J Intern Med, Vol.228, No.6, pp 641-646.

Preuschoff F, Spohn U, Janasek D and Weber E (1994). Biosens. Bioelectron. 9, pp 281–295.

Prieto F, Sepulveda B, Calle A, Llobera A, Dominguez C and Lechuga LM (2003). Integrated Mach-Zehnder interferometer based on ARROW structures for biosensor applications. Sens. Actuators B. 92, pp 151-158.

Prieto F, Sepulveda B, Calle A, Llobera A, Dominguez C, Abad A, Montoya A and Lechuga LM (2003). An integrated optical interferometric nanodevice based on silicon technology, for biosensor applications. Nanotechnol Proceedings of SPIE: 5969, 14, pp 907-912.

Psoma S and Turner APF (1994). 3rd World Congress on Biosensors. New Orleans, USA. Elsevier Applied Science, Oxford, UK.

Purvis D, Leonardova O, Farmakovsky D and Cherkasov V (2003). An ultrasensitive and stable potentiometric immnosensor. Biosens.Bioelec. 18, p 1385-1390.

Quian Z adn Tan T C (1998). A model for multicomponent biosensing and its application to a dead cell-based BOD biosensor. Chem. Eng. Sci. 53, 3281-3294.

Quinn JG, O'Neill S, Doyle A, McAtamney C, Diamond D, MacCraith BD and O'Kennedy R (2000). Development and application of surface plasmon resonance-based biosensors for the detection of cell-ligand interactions. Anal. Biochem., 281, pp 135-143.

Radke SM and EC Alocilja (2005). A high density microelectrode array biosensor for detection of E. Coil O157:H7. Biosensors & Bioelectronics. 20, pp1662-1667.

Raitman OA, Katz E, Buckmann AF and Willner I (2002). Integration of Polyaniline/Poly (acrylic acid) Films and Redox Enzymes on Electrode Supports: An in Situ Electrochemical/Surface Plasmon Resonance Study of the Bioelectrocatalyzed Oxidation of Glucose or Lactate in the Integrated Bioelectrocatalytic Systems. J. Am. Chem. Soc. 124, pp 6487-6496.

Rajender Thusu (2010). Biosensors Market, Frost & Sullivan, Leader-Sensors, Measurement & Instrumentation for Frost & Sullivan, San Antonio, TX.

Rajender Thusu Dr (2010). Biosensors Market. Measurement & Instrumentation for Frost & Sullivan, San Antonio, TX.

Ram MK, Bertoncello P, Ding H, Paddeu S and Nicolini C (2001). Cholesterol biosensors prepared by layer-by-layer technique, Biosens. Bioelectron. 16 : pp 849–856.

Ramsay G (Ed.) (1998). Commercial Biosensors: Applications to Clinical, Bioprocess and Environmental Samples, Wiley, Chichester, UK.

Ramsden JJ, Li SY, Prenosil JE and Heinzle E (1994). Kinetics of adhesion and spreading of animal cells. Biotechnol. Bioeng. 43, pp 939-945.

Rao VK Rai GP, Agarwal GS and Suresh S (2005). Amperometric Immunosensor for Detection of Antibodies of Salmonella Typhi in Patient Serum. Anal. Chim. Acta. 531, pp 173-177.

Rasmus Elsborg, Line Sofie Remvig, Henning Beck-Nielsen and Claus Bogh Juhl. Using the Brain as a Biosensor to Detect Hypoglycaemia. Biosensors for Health, Environment and Biosecurity. Hypo-Safe A/S, Odense University Hospital, Sydvestjysk Sygehus Esbjerg Denmark.

Reach G (2003). Glucose monitoring: from the fingertips to continuous monitoring. Symposium: Insulin Delivery Devices and Glucose Sensors. Presented at the 18th International Diabetes Federation Congress. Paris, France.

Renault NJ (2001). New trends in biosensors for organophosphorous pesticides. Sensors. 1, pp 60-74.

Rhodes RK, Shults MC, Updike SJ (1994). Prediction of Pocket-Portable and Implantable Glucose Enzyme Electrode Performance from Combined Species Permeability and Digital Simulation Analysis. Anal. Chem. 66, pp 1520-1529.

Ricci F, Arduini F, Bourais I, Moscone D, Palleschi G, Amine A (2003). Thiocholine mediated oxidation at prussian blue modified electrodes for pesticide and heavy metals detection. In: Proceedings of the First International Workshop on Biosensors for food safety Environmental Monitoring, Marrakech, Morocco (O-4, abstract book).

Riedel K, Mulchandani A and Rogers KR (1998) (Eds). Enzyme and Microbial Biosensors: Techniques and Protocols. Humana Press, Totowa, NJ. pp 199.

Riedel K, Renneberg R, Kühn M and Scheller F (1988). A fast estimation of biochemical oxygen demand using microbial sensors. Appl. Microbiol. Biotechnol. 28, pp 316–318.

Ringeisen BR, Chrisey DB, Pique A, Young HD, Modi R, Bucaro M, Jones-Meehan J, Spargo BJ (2002). Generation of mesoscopic patterns of viable Escherichia coli by ambient laser transfer, Biomaterials, 23, pp 161–166.

Robert O Becker and Andrew A Marino. Electromagnetism & Life, State University of New York Press, Albany, ISBN 0-87395-560-9 Chapter 4, Electrical Properties of Biological Tissue (Piezoelectricity).

Roberto F, J Barnes and D Bruhn (2002). Evaluation of a GFP Reporter Gene Construct for Environmental Arsenic Detection. Talanta. 58:1, pp 181-188.

Roberto Raiteri, Massimo Grattarola, Butt HJ and Skladal P (2001). Sensors and Actuators 79, pp 409-412.

Rodriguez BB, Bolbot JA, Tothill IE (2004a). Development of urease and glutamic dehydrogenase amperometric assay for heavy metals screening in polluted samples. Biosens. Bioelectron. 19, pp 1157–1167.

Rodriguez BB, Bolbot JA, Tothill IE (2004b). Urease–glutamic dehydrogenase biosensor for screening heavy metals in water and soil samples. Anal. Bioanal. Chem. 380, pp 284–292.

Rodriguez M and Alocilja E (2005). IEEE Sensors Journal. 5, pp 733-736.

Rodriguez-Mozaz S, Reder S, Lopez de Alda M, Gauglitz G and Barceló D (2004). Biosens. Bioelectron. 19, pp 633–640.

Rogers *et al.* (1992). Fibre-Optic Biosensors Based on Total Internal-Reflection Fluorescence. American Chemical Society. Ch. 13, pp 165-173.

Rogers K R, Valdes J J and Eldefrawi M E (1991). Effects of receptor concentration, media ph and storage on nicotinic receptor-transmitted signal in a fibre-optic biosensor. Biosensors and Bioelectronics. 6, pp 1-8.

Rosa CC, Cruz H, Vidal M and Oliva AG (2002). Optical biosensor based on nitrite reductase immobilised in controlled pore glass. Biosensors & Bioelectronic, 17, pp 45-52.

Rose A, Nistor C, Emneus J, Pfeiffer D, Wollenberger U (2002). Biosens. Bioelectron. 17, pp 1033–1043.

Roy S, Vedala H and Choi W (2006). Vertically aligned carbon nanotube probes for monitoring blood cholesterol. Nanotechnology. 17, pp S14-S18.

Russell LJ and Rawson KM (1986). The commercialisation of sensor technology in clinical chemistry: An outline of the potential difficulties. Biosensors 2, pp 301-318.

Sara Rodriguez-Mozaz, Maria-Pilar Marco, Maria J Lopez de Alda and Damià Barceló (2004). Biosensors for environmental applications: Future development trends**Pure Appl. Chem.*, Vol. 76, No. 4, pp 723–752.

Sato TK, Overduin M and Emr SD (2001). Location, location, location: membrane targeting directed by, PX domains. Science. 294, pp 1881-1885.

Satoshi Takamatsu, Isao Shitanda, Masayuki Itagaki, Kunihiro Watanabe. Fabrication of Algal Biosensor by Screen Printing. Department of Pure and Applied Chemistry, Faculty of Science, Tokyo University of Science, Noda, Chiba 278-8510, Japan.

Scheller FW, Schubert F and Renneberg R (1985). Biosensors: Trends and commercialization. Biosensors 1, pp 135-160.

Schenck JF (1978). In Theory, Design and Biomedical Applications of Solid State Chemical Sensors; Cheung, PW, Ed., CRC Press. West Palm Beach. pp 165-173.

Schneider BH, Edwards J and Hartman NF (1997). II Clinical Chemistry. 43, pp 1757.

Schöning M J and Poghossian A (2002). Analyst. 127, pp 1137–1151.

Schoning MJ, Jacobs M and Muck A *et al.* (2005). Amperometric PDMS/glass capillary electrophoresis-based biosensor microchip for catechol and dopamine detection. Sens. Actuators B Chem. 108, pp 688-694.

Scognamiglio V, Staiano M, Rossi M and D'Auria S (2004). Protein-based biosensors for diabetic patients. J. Fluores. 14, pp 491–498.

Selkirk JY, Turner APF and Saini S (1996). 4th World Congress on Biosensors. Elsevier Applied Science, Oxford, UK. pp 198.

Shaw GW, Claremont DJ and Pickup JC (1991). *In Vitro* Testing of a Simply Constructed, zighly Stable Glucose Sensor Suitable for Implantation in Diabetic Patients. Biosens. Ioelectron. 6, pp 401-406.

Sheldon RA, Schoevaart R, van Langen LA (2005). Novel method for enzyme immobilization, Biocat. Biotrans, 23 (3/4), pp 141-147.

Shichiri M, Kawamori R, Yamaski R, Hakai Y. and Abe H (1982). Lancet ii, pp 1129-1131.

Shimomura M, Nomura Y, Zhang W, Sakino M, Lee KH, Ikebukuro K and Karube I (2001). Anal. Chim. Acta, 434, pp 223–230.

Shitanda I, Takada K, Sakai Y and Tatsuma T (2005). Analytica Chimica Acta. 530, 191.

Shriver-Lake LC, Patterson CH and Van Bergen SK (2000). New Horizons: Explosive detection in soils with a fibre optic biosensor. Field Anal. Chem. Tech.

Silley P and Forsythe S (1996). Impedance microbiology: a rapid change for microbiologists. J. Appl. Bacteriol. 80, pp 233–243.

Simonian AL, Efremenko EN, Wild JR (2001). Discriminative detection of neurotoxins in multi-component samples. Anal. Chim. Acta, 444, pp 179–186.

Simonian AL, Flounders AW, Wild JR (2004). FET-based biosensors for the direct detection of organophosphate neurotoxins. Electroanalysis. 16, pp 1896–1906.

Simonian AL, Good TA, Wang SS, Wild JR (2005). Nanoparticle based optical biosensors for the direct detection of organophosphate chemical warfare agent and pesticides. Anal. Chim. Acta 534, pp 69–77.

Singh S, Singhal R and Malhotra BD (2007). Immobilization of cholesterol esterase and cholesterol oxidase onto sol-gel films for application to cholesterol biosensor. Anal. Chim. Acta .582, pp335–343.

Skuridin SG, Yevdokimov YM, Efimov VS, Hall JM and Turner APF (1996). Biosensors Bioelectron. 11, 903.

Smith LM, Saunders JZ, Kaiser RJ, Hughes P, Dodd CR, Cornell CR, Heiner C, Kent SBH and Hood LE (1986). Fluorescence Detection in Automated DNA Sequence Analysis, Nature, 321, pp 674-679.

SP Fodor, JL Read, MC Pirrung, L Stryer AT Lu and D Solas (1991). Light-directed, spatially addressable parallel chemical analysis. Science. 251, pp 767–773.

Sriram Cherukuri, Krishna K Venkatasubramanian and Sandeep K S Gupta (2003). BioSec: A Biometric Based Approach for Securing Communication in Wireless Networks of Biosensors Implanted in the Human Body. Proceedings of the 2003 International Conference on Parallel Processing Workshops (ICPPW'03) .IEEE.1530-2016/03.

Stein LY and Arp DJ (1998). Loss of ammonia monooxygenase activity in Nitrosomonas europaea upon exposure to nitrite, Appl. Environ. Microbiol. 64 pp 4098–4102.

Steven A Soper, Kathlynn Brown, Andrew Ellington, Bruno Frazier, Guillermo Garcia-Manero, Vincent Gauf, Steven I Gutmang, Daniel F Haye, Brenda Korte , James L Landersj, Dale Larson , Frances Ligler l, Arun Majumdarm, Marco Mascini , David Nolte, Zeev Rosenzweig , Joseph Wang and David Wilson (2006). Point-of-care biosensor systems for cancer diagnostics/prognostics. Biosensors and Bioelectronics. 21, pp 1932–1942.

Storhoff JJ, Marla SS, Bao P, Hagenow S, Mehta H, Lucas A, Garimella V, Patno T, Buckingham W, Cork W and Muller UR (2004). II Biosensors & Bioelectronics. 19,875.

Subrahmanyam S, Piletsky SA and Turner APF (2002). Application of natural receptors in sensors and assays. Anal. Chem. 74, pp 3942–3952.

Tai DF, Lin CY, Wu TZ and Chen LK (2005). Recognition of Dengue virus protein using epitopemediated molecularly imprinted film. Anal. Chem. 77, pp 5140–5143.

Tamborlane WV, Beck RW, Bode BW, Buckingham B, Chase HP, Clemons R, Fiallo-Scharer R, Fox LA, Gilliam LK, Hirsch IB, Huang ES, Kollman C, Kowalski AJ, Laffel L, Lawrence JM, Lee J, Mauras N, O'Grady M, Ruedy KJ, Tansey M, Tsalikian E, Weinzimer S, Wilson DM, Wolpert H, Wysocki T and Xing Dl (2008). Continuous glucose monitoring and intensive treatment of type 1 diabetes. N Engl J Med, Vol.359, No.14, pp 1464-1476.

Tan HM, Cheong SP, Tan TC (1994). An amperometric benzene sensor using whole cell pseudomonas putida ML2, Biosensors and Bioelectronics. 9:1.

Tan TC, Li F, Neoh K G and Lee Y K (1992). Microbial membrane-modified dissolved oxygen probe for rapid biochemical oxygen demand measurement. Sens. Act. B. 8, pp 167-172.

Tanaka H, Nakamura E, Hoshikawa H and Tanaka Y (1993). Development of the ammonia biosensor monitoring system. Water Sci. Tech. 28 pp 435–445.

Tauriainen S, Virta M and M Karp (2000). Detecting Bioavailable Toxic Metals and Metalloids from Natural Water Samples Using Luminescent Sensor Bacteria. Water Research, 34:10, pp 2661-2666.

Tauriainen S, Virta M, Chang W and M Karp (1999). Measurement of Firefly Luciferase Reporter Gene Activity from Cells and Lysates Using *Escherichia coli* Arsenite and Mercury Sensors. Analytical Biochemistry. 272:2, pp 191-198.

Taylor RF, Marenchic IG and Spencer RH (1991). Antibody- and Receptor-Based Biosensors for Detection and Process Control. Anal. Chim. Acta. 249, pp 67-70.

Tempelman LA, King KD, Anderson GP and Ligler FS (1996). Quantitating staphylococcal enterotoxin B in diverse media using a portable fibre optic biosensor. Anal. Biochem. 223, pp 50-57.

Terry LA, White SF and Tigwell LJ (2005). The Application of Biosensors to Fresh Produce and the Wider Food Industry. J. Agr. Food Chem. 53, pp 1309-1316.

The Rank Brothers Oxygen Electrode (2002). Operating Manual. http://www.rankbrothers.co.uk/.

Thevenot DR *et al.* (2001) Electrochemical biosensors: recommended definitions and classification. Biosens. Bioelectron. 16(1-2), pp 121-131.

Thomas Ming-Hung Lee (2008). Over the Counter Biosensors: Past, Present, and Future, Sensors. DOI: 10.3390/s8095535, ISSN 1424-8220. www.mdpi.org/sensors8, pp 5535-5559.

Tierney MJ, Tamada JA, Potts RO, Jovanovic L, Garg S and Team CR (2001). Clinical Evaluation of the GluoWatch Biographer: A Continual, Non-Invasive Glucose Monitor for Patients with Diabetes. Biosens. Bioelectron. 16, pp 621-629.

Todd A Wells (2007). Construction of a Simple Myoglobin-Based Optical Biosensor. The Chemical Educator, Vol. 12, No. 6, Published on Web 10/7/2007, 10.1333/s00897072086a.

Tombelli S and Mascini M (1998). Electrochemical Biosensors for Biogenic Amines: A Comparison between Different Approaches. Analytica Chimica Acta. 358, pp 277-284.

Tribl G, Howorka K, Heger G, Anderer P, Thoma H and Zeitlhofer J (1996). EEG topography during insulin-induced hypoglycemia in patients with insulin dependent diabetes mellitus. Eur Neurol, Vol.36, No.5, pp 303-309.

Tromberg BJ, Sepaniak MJ, Vo-Dinh T and Griffin GD (1987). Fibre-optic chemical sensors for competitive binding fluoroimmunoassay. Anal.Chem, 59, pp 1225-1230.

Tsai HC, Doong RA (2005). Simultaneous determination of pH, urea, acetylcholine and heavy metals using array-based enzymatic optical biosensor. Biosens. Bioelectron. 20, pp 1796–1804.

Tsiafoulis C G, Prodromidis M I and Karayannis M I (2004). Biosensors & Bioelectronics, 20, pp 620-627.

Tuan Vo-Dinh (2004). Biosensors, nanosensors and biochips: Frontiers in environmental and medical diagnostics. The 1st International Symposium on Micro & Nano Technology, Honolulu, Hawaii, USA.

Tubb AJ C, Payne F P, Millington RB and Lowe CR (1997). Single-mode optical fibre surface plasma wave chemical sensor. Sensors and Actuators, B: Chemical. 41, pp 71-79.

Tuerk C and Gold L (1990). Systematic evolution of ligands by exponential enrichment: RNA ligands to bacteriophage T4 DNA polymerase. Science 249, pp 505–510.

Turner APF (1987). Biosensors: Principles and potential. In Chemical aspects of food enzymes, ed. A.T.Andrews, pp 259-270, London: Royal Society of Chemistry.

Turner APF (1989). Current Trends in Biosensor Research and Development. Sensors Actuators, 17 pp 433-450.

Turner APF (1991, 1992, 1993, 1995.). Advances in Biosensors. I, II, Suppl. I III. JAI Press, London, UK.

Turner APF (2000). Biosensors – sense and sensitivity. Science. 290, pp 1315–1317.

Turner APF, Karube I and Wilson G.S. (1987). Biosensors: Fundamentals and Applications. Oxford University Press, Oxford. 770.

Turner Professor Anthony P F (1996). Biosensors Past, Present and Future. Cranfield University.

Turner RFB, Harrison DJ, Rajotte RV and Baltes HP (1990). A Biocompatible Enzyme Electrode for Continuous *in vivo* Glucose Monitoring in Whole Blood. Sensor. Actuat. B-Chem. 1, pp 561-564.

Umek RM, Lin SW, Vielmetter J, Terbrueggen RH, Irvine B, Yu CJ, Blackburn GF, Farkas DH and Chen YP (2001). Electronic Detection of Nucleic Acids: A Versatile Platform for Molecular Diagnostics. J Mol. Diagn. 3, pp 74-84.

Umer Farooq, M Aqeel Iqbal, Muhammad Ahsan and Mashood Malik (2010). Next Generation High Speed Computing Using Photonic Based Technology.

Vadgama P (1986). Urea pH electrodes: Characterisation and optimisation for plasma measurements. Analyst111, pp 875-878.

Van Bergen SK, Bakaltcheva IB, Lundgren JS and Shriver-Lake LC (2000). On-site detection of Explosives in groundwater with a fibre optic biosensor. Environ. Sci. & Tech. 34, pp 704-708.

Vassanelli S and Fromherz P (1998). Transistor records of excitable neurons from rat brain. Appl. Phys. A. 66: 459.

Verhaegen K, Baert K and Simaels J *et al.* (2000). A high-throughput silicon microphysiometer, Sensors and Actuators B. 82:186.

Vijayendran RA and DE Leckband A (2001). quantitative assessment of heterogeneity for surface-immobilized proteins. Anal. Chem. 73(3), pp 471-480.

Vittorio M N Passaro, Francesco Dell'Olio, Biagio Casamassima and Francesco De Leonardis (2007). Guided-Wave Optical Biosensors. Sensors. 7, p508-536.

Vo-Dinh T (2002). Nanobiosensors: Probing the sanctuary of individual living cells. Journal of Cellular Biochemistry, pp 154-161.

Vo-Dinh T and Allain L (2003). Biosensors for Medical Applications. Biomedical Photonics Handbook, T. Vo-Dinh, Ed. Boca Raton: CRC Press.

Vo-Dinh T, Cullum B M and Stokes DL *et al.,* (2001). Nanosensors and biochips: frontiers in biomolecular diagnostics, Sensors and Actuators B, 74: 2.

Vo-Dinh T, Griffin GD, Alarie JP, Cullum B, Sumpter B and Noid D (2000). Development of nanosensors and bioprobes. Journal of Nanoparticle Research. 2, pp 17-27.

Vo-Dinh T, Stokes D L, Griffin G D, Volkan M, Kim UJ and Simon M I (1999). Surface- Enhanced Raman Scattering (SERS) Method and Instrumentation for Genomics and Biomedical Analysis, J. Raman Spectrosc. 30, pp 785-793.

Vo-Dinh T, Tromberg B J, Griffin G D, Ambrose K R, Sepaniak M J and Gardenhire E M (1987). Antibody-based fibreoptics biosensor for the carcinogen benzo(a)pyrene, Appl. Spectrosc. 41, 735-738.

Volotovsky V, Soldatkin A, Shul'ga AA, Rossokhaty VK, Strikha VI, El'skaya AV (1996). Anal. Chim. 322, pp 77–81.

Vörös J, Graf R, Kenausis GL, Bruinink A, Mayer J, Textor M, Wintermantel E (2000). Spencer II Biosensors & Bioelectronics. 15, pp 423.

Walker PL Jr and Thrower PA ed. Chemistry and Physics of Carbon. Volme: 11, pp 1966–74.

Wang J (1999). Amperometric biosensors for clinical and therapeutic drug monitoring: a review. J. Pharm. Biomed. Anal. 19:47–53.

Wang J (2001) Glucose Biosensors: 40 Years of Advances and Challenges. Electroanalysis: 13, p 983-988.

Wang J and Zhang X (2001). Needle-type dual microsensor for the simultaneous monitoring of glucose and insulin. Anal. Chem. 73, pp 844–847.

Wang J, Ibanez A, Chatrathi MP and Escarpa A (2001). Electrochemical Enzyme Immunoassays on Microchip Platforms. Anal. Chem. 73, pp 5323-5327.

Wang J, Liu G and Merkoci A (2003). Electrochemical Coding Technology for Simultaneous Detection of Multiple DNA Targets. J. Am. Chem. Soc. 125, pp 3214-3215.

Wang J, Rivas G, Cai X, Dontha N, Shiraishi H, Luo D and Valera FS (1997). Sequence-Specific Electrochemical Biosensing of M. Tuberculosis DNA. Anal. Chim. Acta. 337, pp 41-48.

Watanabe M, Kajikawa K (2003). An optical fibre biosensor based on anomalous reflection of gold. *Sensors and Actuators, B: Chemical.* 89, pp 126-130.

Watts Helen J (1994)."Optical Biosensor for Monitoring Microbial Cells" Analytical Chemistry. Vol. 66, No. 15, pp 2465-2470.

Wei-Jen Hoa, Jung-Sheng Chenb, Ming-Dou Ker b, Tung-Kung Wu a, Chung-Yu Wu b, Yuh-Shyong Yang a, Yaw-Kuen Li c and Chiun-Jye Yuan (2007). Fabrication of a miniature CMOS-based optical biosensor. Biosensors and Bioelectronics. Elsevier BV, 22, pp 3008–3013.

Wells M, Gosch M, Rigler R, Harms H, Lasser T, van de Meer JR (2005). Anal. Chem. 77, 2683.

White BJ, Harmon HJ (2005). Optical solid-state detection of organophosphates using organophosphorus hydrolase. Biosens. Bioelectron. 20, pp 1977–1983.

Wilhelm B, Forst S, Weber MM, Larbig M, Pfûtzner A and Forst T (2006). Evaluation of CGMS During Rapid Blood Glucose Changes in Patients with Type 1 Diabetes. Diabet. Technol. Ther. 8, pp 146-155.

Wilkins E and Atanasov P (1996). Glucose Monitoring: State of the Art and Future Possibilities. Med. Eng. Phys. 18, pp 273-288.

Wimmerova M and Macholan L (1999) Sensitive Amperometric Biosensor for the Determination of Biogenic and Synthetic Amines Using Pea Seedlings Amine Oxidase: A Novel Approach for Enzyme Immobilization. Biosensor and Bioelectronics. 14, pp 695-702.

Wise KD (2006). Silicon Microsystems for neuroscience and neural prosthetics. IEEE Eng. Med. Biol. Mag. 24(5), pp 22-29.

Wolfbeis OS (1991). Fibre optic chemical sensors and biosensors. Boca Raton, Fla.: CRC Press.

Wolthuis R, McCrae D, Saaski E, Hartl J and Mitchell G (1992). Development of a medical fibreoptic ph sensor based on optical absorption. IEEE Transactions on Biomedical Engineering. 39, pp 531-537.

Wolthuis RA, McCrae D, Hartl JC, Saaski E, Mitchell GL, Garcin K and Willard R (1992). Development of a medical fibre-optic oxygen sensor based on optical absorption change. IEEE Transactions on Biomedical Engineering. 39, pp 185-193.

Workman RK and Srinivas M (2004). Molecular transfer and transport in noncovalent microcontact printing. Langmuir. 20, pp 805–815.

Wring SA, Hart JP (1992). Chemically modified, carbon-based electrodes and their application as electrochemical sensors for the analysis of biologically important compounds. Analyst, 117, pp 1217–29.

Wu G, Ji H, Hansen K, Thundat T, Datar R, Cote R, Hagan M.F, Chakraborty A K, and Majumdar A (2001). Origin of nanomechanical cantilever motion generated from biomolecular interactions. Proc. Natl Acad. Sci. USA. 98, pp 1560–1564.

Wu Jing and Qu Yang (2006). Mediator-free amperometric determination of glucose based on direct electron transfer between glucose oxidase and an oxidised boron-doped diamond electrode. Anal. Bioanal. Chem. 38 pp 1330-1335.

Wu PK, Ringeisen BR, Callahan J, Brooks M, Bubb DB, Wu HD, Pique A, Spargo B, McGill RA, Chrisey DB (2001). The deposition, structure, pattern deposition, and activity of biomaterial thin-films by matrixassisted pulsed-laser evaporation (MAPLE) and MAPLE direct write, Thin Solid Films. pp 607–614.

Xi C Raskin L and Boppart SA (2005). Evaluation of microfluidic biosensor development using microscopic analysis of molecular beacon hybridization kinetics. Biomed. Microdevices 7, pp 7–12.

Xia Y, Rogers JA, Paul KE, Whitesides GM (1999). Unconventional methods for fabricating and patterning nanostructures, Chem. Rev. 99, pp 1823– 1848.

Xiao Y, Lubin AA, Heeger AJ and Plaxco KW (2005). Label-Free Electronic Detection of Thrombin in Blood Serum by Using an Aptamer-Based Sensor. Angew. Chem. Int. Edit., 44, pp 5456-5459.

Xu J, Grant SA and Pastel RL (2003). Laser-guided direct writing: a novel method to deposit biomolecules for biosensors arrays. IEEE Trans. Biomed. Eng. 50, pp 126–168.

Xu Q, Almeida VR, Panepucci RR and Lipson M (2004). Experimental demonstration of guiding and confining light in nanometer-size low-refractive-index material. Opt. Lett. 29, pp 1626- 1628.

Yalçin A, Popat KC, Aldridge JC, Desai TA, Hryniewicz J, Chbouki N, Little BE, King O, Van V, Chu S, Gill D, Anthes-Washburn M, Selim Unlu M and Goldberg BB (2006). Optical sensing of biomelecules using microring resonators. IEEE J. Sel. Top. Quantum Electron. 12, pp 148-155.

Yamazaki T, Kojima K and Sode K (2000). Extendedrange glucose sensor employing engineered glucose dehydrogenases. Anal. Chem. 72, pp 4689–4693.

Yan X, Xu, X K and Ji HF (2005). Glucose oxidase multilayer modified microcantilevers for glucose measurement. Anal. Chem. 77, pp 6197–6204

Yan Zhang (2005). Dissertation, Miniature fibre-optic multicavity Fabry-Perot interferometric biosensor. Blacksburg, Virginia.

Yeung SSW, Lee TMH and Hsing IM (2008). Electrochemistry-Based Real-Time PCR on a Microchip. Anal. Chem. 80, pp 363-368.

Yildiz HB and Toppare L (2006). Biosensing approach for alcohol determination using immobilized alcohol oxidase. Biosens. Bioelectron. 21, pp 2306–2310.

Yu D, Blankert B, Vire JC and Kauffmann JM (2005). Biosensors in Drug Discovery and Drug Analysis. Anal. Lett. 38, pp 1687-1701.

Yu Lei, Wilfred Chen and Ashok Ulchandani (2006). Microbial biosensors. Analytica Chimica. Acta. 568, pp 200-210.

Zacco E, Adrian J, Galve R, Marco MP, Alegret S and Pividori MI (2007). Electrochemical magneto immunosensing of antibiotic residues in milk. Biosensors and bioelectronics, 22, pp 2184-2191.

Zacco E, Galve R, Marco MP, Alegret S and Pividori MI (2007). Electrochemical biosensing of pesticide residues based on affinity biocomposite platforms. Biosensors and bioelectronics, 22, pp 1707-1715.

Zacco E, Pividori MI, Alegret S, Galve R and Marco MP (2006). Electrochemical magnetoimmunosensing strategy for the detection of pesticides residues. Anal Chem, 78, pp 1780-1788.

Zarini Muhammad Tahir, Evangelyn C Alocilja and Daniel L Grooms (2007). Indium Tin Oxide-Polyaniline Biosensor: Fabrication and Characterization. Sensors. MDPI. 7, pp 1123-1140.

Zarini Muhammad Tahir, Evangelyn C Alocilja and Daniel L Grooms (2007). Indium Tin Oxide-Polyaniline Biosensor: Fabrication and Characterization sensors, MDPI, pp1424-8220.

Zhang J, Ma Y, Stachura S and He H (2005). Assembly of highly aligned DNA strands onto Si chips. Langmuir. 21, pp 4180–4184.

Zhang S, Zhao H, John R (2001). A theoretical model for immobilized enzyme inhibition biosensors. Electroanalysis. 13, pp 1528–1534.

Zhang Wenjun and Li Genxi (2004). Third-Generation Biosensors Based on the Direct Electron Transfer of Proteins. Analytical Sciences. Vol. 20.

Zhang Y, Muench SB, Schulze H, Perz R, Yang B, Schmid RD, Bachmann TT (2005). Disposable biosensor test for organophosphate and carbamate insecticides in milk. Agric. Food Chem.

Zhang Y, Wilson GS (1993). Electrochemical oxidation of hydrogen peroxide on platinum and platinum-iridium electrodes in physiological buffer and its applicability to hydrogen peroxide-based biosensors. J Electroanal Chem. 345, pp 253–71.

Zheng G, Patolsky F, Cui Y, Wang WU and Lieber CM (2005). Multiplexed Electrical Detection of Cancer Markers with Nanowire Sensor Arrays. Nat. Biotechnol. 23, pp 1294-1301.

Zhou CH *et al* (1997). A compact fibre-optic immunosensor for Salmonella based on evanescent wave excitation. Sens. Actuators B Chem. 42(3), pp 169-175.

Zhujun Z *et al.* (1989). Polyvinyl-Alcohol as a Substrate for Indicator Immobilization for Fibre-Optic Chemical Sensors. Anal. Chem. 61(3), pp 202-205.

Zimmermann S, Fienbork D, Stoeber B, Flounders AW and Liepmann DA (2003). Microneedle- Based Glucose Monitor: Fabricated on a Wafer-Level Using in-Deivce Enzyme Immobilization. In Transducers '03, the 12th International Conference on Solid State Sensors, Actuators and Microsystems. pp 99-102.

Index